Camouflage and Behavior
The beautiful camouflage of the flower mantis *Hymenopus coronatus* only works if the mantis is able to find the appropriate flowers on which to wait for prey. The capacity for discriminating habitat selection appears in most cryptically colored animals (Chapter 10). Photograph by E. S. Ross.

An Advantage of Group Living
In a number of amphibian species, vulnerable individuals undergoing metamorphosis to the adult stage clump together. Should a local predator discover this group of tree frogs (*Hyla regilla*), it would probably not be able to eat them all. Frogs that clump together and swamp their predators' feeding capacity may have a better probability of surviving than individuals that are dispersed (see Chapter 10). Photograph by E. S. Ross.

Animal Behavior

ANIMAL

An Evolutionary

BEHAVIOR

Approach • THIRD EDITION

John Alcock
ARIZONA STATE UNIVERSITY

Sinauer Associates, Inc., Publishers
Sunderland, Massachusetts

On the Cover

White-tailed deer raise their tails when they spot a predator. Is this an alarm signal that helps warn other deer? Or, has this action evolved to enable the deer to communicate with the predator, telling the enemy that it has been seen and is unlikely to capture the alert prey? A major goal of research in animal behavior is to test competing hypotheses about the adaptive value of behavioral responses. The cover was designed by Rodelinde Albrecht from a photograph by Jon E. Cates.

Animal Behavior: An Evolutionary Approach, *Third Edition*

Printed in U.S.A.

Library of Congress Cataloging in Publication Data

Alcock, John, 1942–
Animal behavior.

Includes bibliography and indexes.
1. Animal behavior. 2. Evolution. I. Title
[DNLM: 1. Behavior, Animal. QL 751 A354a]
QL751.A58 1983 591.51 83-14420
ISBN 0-87893-021-3

10 9 8 7 6 5 4 3

To my mother and father

Contents

Preface xvii

CHAPTER 1 An Evolutionary Approach to Animal Behavior 1

CURIOSITY ABOUT BEHAVIOR 1

How questions 3
Why Questions 4

NATURAL SELECTION 5

Using Working Hypotheses 8
The Problem with Group Selection 9
Alternative Hypotheses 11

HYPOTHESIS TESTING AND EVOLUTIONARY PREDICTION 14

The Comparative Method 14
Experimental Tests of Evolutionary Predictions 17
Infanticide by Human Males 19

ORGANIZATION OF THE TEST 22

CHAPTER 2 The Genetics of Behavior 25

HEREDITY AND HUMAN BEHAVIOR 25

Are Some Behavior Patterns "Genetically Determined"? 28

THE BEHAVIOR GENETICS OF OTHER ANIMALS 32

Single Gene Effects on Behavior 34
Polygeny 36
Artificial Selection 38
Experiments on Artificial Selection 39

GENES, PHYSIOLOGICAL MECHANISMS, AND BEHAVIORAL DIFFERENCES 42

Genes and Nervous Systems 44
Genes and Geographic Variation in Behavior 45

CHAPTER 3 The Development of Behavior 55

THE DEVELOPMENT OF SEXUAL DIFFERENCES IN BEHAVIOR 55

Hormones, Sensory Stimulation, and the Development of Maternal Behavior 60

DEVELOPMENT AS AN INTERACTIVE PROCESS 62

The Development of Singing Behavior in Birds 62
Experience and the Development of White-Crowned Sparrow Song 63

DEVELOPMENTAL HOMEOSTASIS 70

The Repeatability of the Developmental Process 70
Behavioral Development in Abnormal Conditions 73
Developmental Homeostasis and Human Behavior 78
The Adaptive Value of Developmental Homeostasis 81

CHAPTER 4 Categories of Behavior: A Proximate Classification 85

INSTINCTS AND LEARNING 85

RESTRICTED DEVELOPMENTAL PROGRAMS 89

Innate Releasing Mechanisms 93

SEMIRESTRICTED BEHAVIORAL DEVELOPMENT 96

Imprinting 99
Language Development 101
Language and the Brain 104

PLASTIC DEVELOPMENTAL SYSTEMS 106

Constraints on Conditioning 107

ADAPTATION AND CONSTRAINTS
ON BEHAVIORAL DEVELOPMENT 112

The Development of Predator Avoidance 114
The Development of Courtship Signals: Song Dialects 114
Ecological Correlates of Song Dialects 115
The Adaptive Value of Restricted Song Development 117
Reproductive Success and Developmental Flexibility 118

CHAPTER 5 Nerve Cells and Behavior Patterns 125

HOW DO MOTHS EVADE BATS? 125

Receptor Cell Design and Adaptive Information Gathering 129
Stimulus-Filtering and Behavior 133

STIMULUS-FILTERING AND SELECTIVE PERCEPTION
IN VERTEBRATES 136

The Detection of Ultrasonic Echoes 138
The Identification and Interception of Prey 140

THE NEUROPHYSIOLOGY OF ELECTRIC FISHES 141

How to Avoid Electrical "Jamming" 143
How to Communicate Electrically 144

THE VISUAL WORLD OF THE EUROPEAN TOAD 146

Visual Perception in Other Animals 150

CHAPTER 6 The Organization of Behavior 157

THE PHYSIOLOGY OF BEHAVIORAL CHANGES IN ANOLIS LIZARDS 157

The Control of Receptivity 159
Seasonal Changes in Anole Behavior 160
Climate and Social Interaction 163
The Adaptive Value of the Anole's Regulatory Mechanisms 164

COPING WITH COMPETING STIMULI 165

The Control of Mantis Behavior 166

CYCLICAL CHANGES IN BEHAVIOR 170

Feeding Cycles 171
Circadian Rhythms 174
Long-Term Behavioral Cycles 177

THE ANNUAL CYCLE OF THE WHITE-CROWNED SPARROW 181

The Regulation of Reproduction 183
Environmental Cues and the Timing of Reproduction 186
Social Regulation of Reproduction 189
Social Effects on Mammalian Reproductive Physiology 190

CHAPTER 7 The Ecology of Behavior 195

THE BEHAVIORAL ECOLOGY OF BLACK-HEADED GULLS 195

Ultimate Hypotheses about Black-Headed Gull Behavior 198
Testing Ultimate Hypotheses 199
The Uses of Convergent and Divergent Evolution 202
On the Adaptive Value of Eggshell Removal 207

A FORAGING ADAPTATION: TESTS OF A HYPOTHESIS 208

PARENT-OFFSPRING RECOGNITION: TESTS OF A HYPOTHESIS 210

Learned Egg-Recognition in Birds 212
Divergent Evolution in Fledgling Recognition in Swallows 213
Offspring Recognition in Other Animals 214

THE WEB OF ADAPTATION 217

CHAPTER 8 The Ecology of Finding a Place to Live 223

ACTIVE HABITAT SELECTION 223

Proximate Cues of Habitat Selection 225
Habitat Selection by Honeybees 225
Are Habitat Preferences Adaptive? 227
Habitat Selection and Reproduction in Aphids 229

HOMING 231

Mechanisms of Homing 232
Backup Orientation Mechanisms in Honeybees and Pigeons 234
The Orientation Mechanisms of Ants 236
The Use of Stars versus the Sun 238

MIGRATION 239

The Costs of Migration 241
The Benefits of Migration 244

TERRITORIALITY 246

Comparative Tests of Territorial Function 247
Territories and Reproductive Success in Songbirds 250
A Territorial Mammal 253
Why do Nonterritorial Individuals Acccept Their Status? 254

CHAPTER 9 The Ecology of Feeding Behavior 261

WHAT IS THE ULTIMATE GOAL OF FEEDING BEHAVIOR? 261

Minimizing Foraging Costs 263
Optimal Foraging over a Day 264
Optimal Foraging over Prolonged Periods 267

CONSTRAINTS ON FORAGING EFFICIENCY 269

The Risk of Predation and Foraging Behavior 270
Nutritional Constraints on Foraging 271

HAS COMPETITION SHAPED THE DIETS OF ANIMAL SPECIES? 273

Rodents, Seeds, and Competition 276

HOW TO CAPTURE AND CONSUME DIFFICULT FOODS 278

How to Use Lures to Capture Prey 282
Tool-Using Animals 284
Social Prey Capture 287
Diversity in Social Feeding by Ants 289

CHAPTER 10 The Ecology of Antipredator Behavior 295

MAKING PREY LOCATION MORE DIFFICULT 295

Removing Evidence of One's Presence 298

MAKING CAPTURE MORE DIFFICULT 299

Misdirecting a Predator's Attack 301
How to Make a Predator Hesitate 303

FIGHTING BACK 305

Chemical Repellents of Vertebrates 310
Batesian Mimicry 312
Associating with a Protected Species 314

COOPERATIVE DEFENSE AGAINST PREDATORS 315

Sociality and Alarm Signals 318

Sociality and Improved Vigilance 321
The Selfish Herd Hypothesis 322
The Dilution Effect 324

THE MONARCH BUTTERFLY 325

Monarchs and Their Food Plants 328
Monarchs and Their Mimics 330

CHAPTER 11 The Ecology of Reproductive Behavior: Sexual Selection and Male Competition 335

REPRODUCTION IN A DAMSELFLY 335

THE PUZZLE OF SEXUAL REPRODUCTION 336

Individual Selection and Sexual Reproduction 339

MALE AND FEMALE REPRODUCTIVE STRATEGIES 341

Parental Investment 344
Why Is Parental Care More Often Provided by Females? 346
Sexual Differences and Sexual Selection 347

INTRASEXUAL SELECTION AND MALE REPRODUCTIVE COMPETITION 351

Risk-Taking and Fighting by Males 355
Species Differences in Fighting Intensity 357
How to Cope with Dominant Males 360
Sneak Copulations 363
Alternative Behavioral Traits: Their Evolutionary Maintenance 364
The Coexistence of Two Distinct Strategies 366

THE PROTECTION OF INSEMINATED FEMALES 369

Forced Copulation and Mate Guarding in Birds 371
Mating Plugs and Repellents 373

SEXUAL INTERFERENCE 376

CHAPTER 12 The Ecology of Reproductive Behavior: Female Choice and Mating Systems 383

FEMALE CHOICE 383

Male Genetic Quality and Female Choice 385
How to Pick a Dominant Male 386
Testing Male Dominance 387
The Function of Courtship 389

FEMALE CHOICE AND THE MATERIAL BENEFITS OF COPULATION 393

Access to Monopolized Resources 396
Variation in Male Parental Care and Female Choice 396

MATING SYSTEMS 400

Monogamy 400
Female Defense Polygyny 403
Resource Defense Polygyny 405
Female Reproductive Success and Resource Defense Polygyny 407
Lek Polygyny 409
Scramble Competition Polygyny 413
Polyandry 416
Sex Role Reversal Polyandry 418

CHAPTER 13 The Evolutionary History of Behavior 425

DOES BEHAVIOR EVER FOSSILIZE? 425

The Evolution of Bipedalism in Man 427

THE RECONSTRUCTION OF HISTORICAL PATHWAYS WITHOUT FOSSILS 428

The History of a Fly's Courtship Signals 432
The History of a Bowerbird's Display 435
The History of Honeybee Communication 438
Direction Communication 441
The Evolution of Dance Displays 442
Trends in Bee Communication 446

THE EVOLUTION OF COMMUNICATION SYSTEMS 448

The Origins of Communication Signals 449
Deceitful "Communication" 452
Channels of Communication 453
Predation, Competition, and Communication 460

CHAPTER 14 The Evolution of Societies 467

THE COSTS AND BENEFITS OF SOCIAL LIVING 467

Degrees of Sociality in Prairie Dogs 469
Intraspecific Variation in Group Size 471

THE EVOLUTION OF COOPERATION 473

Altruism and Kin Selection 475
How Might Alarm Calls Evolve? 477
Alarm Calls and Ground Squirrels 478

COOPERATION IN MATE ACQUISITION 481

Cooperation in Mate Acquisition: Individual Selection 483

HELPERS AT THE NEST 486

Helping and Habitat Saturation 487
Reciprocity and Communal Nesting in Green Woodhoopoes 490

THE EVOLUTION OF EUSOCIAL INSECTS 492

The Role of Kin Selection in Eusociality 494
Individual Selection and the Evolution of Eusociality 498

CHAPTER 15 An Evolutionary Approach to Human Behavior 505

THE SOCIOBIOLOGY CONTROVERSY 505

GENES, CULTURE, AND BEHAVIOR 506

The Evolution of Cultures 509
Do Cultures Change Adaptively? 510

THE EVOLUTION OF HUMAN WARFARE 512

Testing the Ecological Hypothesis 516
Relatives, Reciprocity, and the Costs of Warfare 518
Reciprocity among Nonrelatives 521

THE EVOLUTION OF HUMAN REPRODUCTIVE BEHAVIOR 522

The Desire for Sexual Variety 523
Female Choice and the Incest Taboo 524
The Loss of Estrus and the Evolution of Monogamy 526
Female Choice and Economics 527
Sexual Jealousy and Male Parental Care 528
Are There Maladaptive Aspects of Human Reproductive Behavior? 530
Adoption 531
Birth Control 535

HUMAN BEHAVIOR AND EVOLUTIONARY THEORY 538

Bibliography 543

Illustration Credits 579

Film Index 584

Index 589

Preface

This third edition of *Animal Behavior* reflects my continuing belief that evolutionary theory provides the key for understanding animal behavior. Behavioral research is expanding ever more rapidly and is yielding exciting new findings on everything from the behavior genetics of nematodes to the social system of naked mole rats. Given the great variety of topics and the vast literature of animal behavior, how can we make sense of this discipline? The first task is to recognize that there are proximate, or immediate causes of behavior and there are ultimate, or evolutionary, bases for behavior as well. This book employs an organizational scheme that recognizes this fundamental division: there are chapters on behavior genetics, development, and the physiology of behavior, followed by sections on behavioral ecology and the phylogenetic component of behavioral evolution. Throughout the text evolutionary theory is used to integrate these subdisciplines of animal behavior and to show the connections between proximate and ultimate causation.

Furthermore, evolutionary theory provides a fundamental working hypothesis with which to approach the analysis of *any* behavioral trait. This hypothesis is that animals will behave in ways that advance the survival of their genes in the context of the environment in which they live. Recent developments in the field of animal behavior, epitomized by the articles presented in the new journal *Behavioral Ecology and Sociobiology,* show what a wonderfully productive stimulus for scientific research this hypothesis has proved to be. I will be pleased if this book helps readers understand the logic of the evolutionary approach, the ways to test evolutionary hypotheses; and the integrative power of the theory as applied to behavioral research.

In this revision, I have rewritten each chapter completely, updating the examples used and expanding the treatment of some topics such as behavioral development, hormonal control of behavior, the evoluton of reproductive behavior, and the role of cooperation in animal societies. A new feature is the full-color photo essays presented on the endpapers. These were created to highlight important behavioral concepts in a visually dramatic fashion.

As in previous editions, numbers in brackets in the text refer to articles and books cited in full in the bibliography at the end of the book. The original papers develop points more completely than can be presented in the text and give a fuller account of how the research was conducted. The suggested readings at the end of each chapter have been selected in part for readability and in part because the research they describe is unusually satisfying. Each chapter has a concluding summary of the central points in that chapter. The suggested films are both entertaining and instructive in illustrating certain of the major topics discussed in the chapter; the film index provides addresses from which films can be ordered.

Acknowledgments Many individuals have assisted in the preparation of this book, some directly, others indirectly. The gratitude that I have expressed in previous editions still applies fully with respect to the help received from my parents, my wife Sue and sons Joe and Nick; my advisors from my student days, Ernst Mayr and Lincoln Brower; and various of my colleagues including Robert Lockard, Gordon Orians, Shelby Gerking, Ronald Alvarado, Steve Fretwell, and Ron Rutowski.

In writing the third edition, I have benefited from the support of my current departmental chairperson, Kathy Church. Stevan Arnold, David Crews, Randall Lockwood, Del Thiessen and Jeffrey Baylis gave me many useful suggestions based on their reading of the second edition. In the course of revising a draft of the third edition, Robert Gibson, Sarah Lenington, Benjamin Sachs, and Michael Beecher offered constructive criticism. I thank Michael Moore for correcting an early draft of Chapter 6. Numerous persons generously provided photographs and other materials needed for the third edition. Individuals or institutions that granted permissions are credited in a separate section at the back of the book. Once again Becky Payne did a superb job of typing the manuscript. My copyeditor Jodi Simpson worked to improve the clarity of the text. I thank my wife Sue for helping me proofread the galleys of the manuscript. Carlton Brose, as always, skillfully integrated the activities of the many people who work together to produce a book.

John Alcock

Animal Behavior

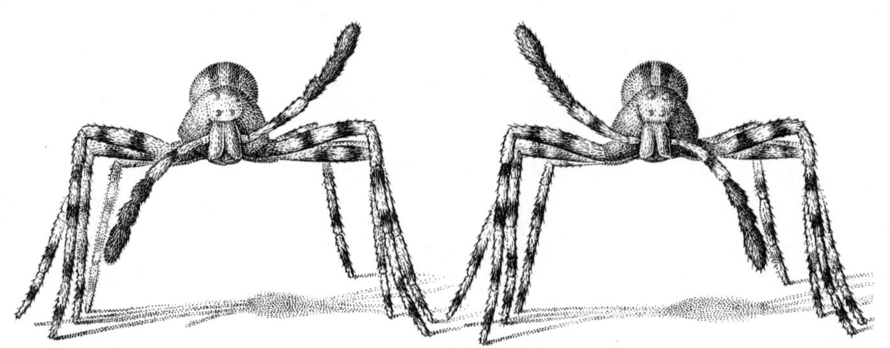

Niko Tinbergen, one of the founders of the study of animal behavior and a great champion of the joy of observing animals under natural conditions, has written, "The curious naturalist often feels sorry for those of his fellow-men who miss such an experience; and miss it so unnecessarily, because it is there, to be seen, all the time. Nor is reading about it anything more than a poor substitute; direct, active observation is the only real thing" [705]. Tinbergen is right. There are countless mysteries of animal behavior in the world around us and watching patiently can solve many of them, much to the intense satisfaction and happiness of the observer. I know this firsthand, having spent a great many hours watching bees and wasps, flies and stinkbugs (often to the bewilderment of other humans observing me). Fortunately, however, Tinbergen's claim that observation was "the only real thing" did not prevent him from writing a number of delightful books. In them he offered an interpretation, as well as a description, of what he had seen in the natural world. His interpretation was based on the belief that animal behavior was adaptive, and this philosophy surely helped his readers see things later that they would have otherwise missed. Partly through reading about evolutionary theory and partly through discussion, I learned some years ago that an "adaptive" behavioral trait might very well *not* help the species survive. This surprised me at first because I had been taught that the preservation of the species was the central focus of evolution. But once I understood the logic of the argument against this position, my interpretation of the living world changed dramatically. Many other biologists have had the same experience. My book is written primarily to explain what "adaptive" means as it applies to the study of animal behavior. I hope that among my readers there will be curious naturalists who will gain more pleasure from their observations after they fully understand an evolutionary approach to animal behavior.

CHAPTER 1

An Evolutionary Approach to Animal Behavior

Curiosity About Behavior

Imagine that you are in southern Arizona wandering about a sandy area in May when you notice a long, thin, reddish wasp carrying a large green caterpillar beneath its body. Unbeknownst to you, the wasp is a female of the species *Ammophila novita*. As you peer at the wasp, she crawls onto a tuft of grass and deposits the caterpillar at the top (Figure 1). She then leaves the immobile grub and flies off low over the ground for some distance before alighting. After walking about for a while, the female begins to gnaw at the soil with her mandibles. She picks up a load of loosened material with her forelegs, presses the load beneath her "chin," and flies up a short distance before releasing the earth, scattering it over the ground. The wasp then returns to her digging site and repeats the process. Soon she has constructed a vertical shaft that can accommodate her body and more. After about 20–30 minutes of excavation, the wasp makes a final flight with a load of sand and then flies back to the spot where she had left the caterpillar. She grasps it with her jaws and drags the prey to the burrow. Depositing the larva near the entrance, she enters the shaft, turns around, grabs the head of her victim and pulls it down into the tunnel. After several minutes, she reemerges and begins to kick sand and pebbles into the burrow (Figure 1), not stopping until the shaft is completely filled. The wasp then flies away.

If you were suitably bewildered by this puzzling series of activities you would probably want to excavate the burrow to find out what had been done to the caterpillar. Having thoughtfully carried a garden trowel, an old spoon, and a wire probe on your walk, you could cautiously cut away the sand, using the probe to push out the new fill in the vertical shaft bit by bit. After following the tunnel down a few inches, you find that it turns to travel horizontally a short way; and very soon the caterpillar appears, resting on its side (Figure 1). On its flank you will find a small white translucent egg. Your work has revealed the answers to some of the questions you had while watching the animal at work. The burrow is an underground nest with a brood chamber that contains a prey placed there for the wasp's offspring. Additional observation and excavations would show that an *Ammophila* larva emerges from the egg and consumes the cater-

1 **Digger wasp nesting behavior.** A female wasp is storing a caterpillar, which she has paralyzed, on a tuft of grass (*top left*). After digging a nest, the wasp has retrieved her prey and is about to drag it into the burrow (*center left*). The wasp has laid an egg on the prey and is now kicking sand into the nest to fill the burrow (*bottom left*). The excavated nest contains the caterpillar coiled in the brood cell (*right*). Photographs by the author.

pillar, after which it metamorphoses into a pupa and finally becomes an adult wasp.

Despite your research, you would probably have many other questions about *Ammophila* behavior. How does the wasp manage to capture a caterpillar? How does it find its way back to its prey after having deposited it on a grassy mound or twig? Why does it scatter the dirt that it excavates from the burrow? Why does this species provision its nests with a single large caterpillar whereas other *Ammophila* provide many small prey to their offspring? Further reflection (it would be best if this were done somewhere in the shade) would reveal that these questions could be placed into two basic groups.

In fact, all questions about behavior are either "how questions" about its PROXIMATE CAUSES or "why questions" about its ULTIMATE CAUSES [549]. How questions ask how an individual manages to carry out an activity; they ask how mechanisms *within* the animal operate to make behavioral responses possible. Why questions ask why the animal has evolved the proximate mechanisms that enable it to do certain things. It is useful and necessary to make a careful distinction between the proximate and ultimate causes of biological phenomena. Otherwise confusion and misunderstanding about possible explanations will result.

For example, imagine an argument between two people on why humans eat so much candy and drink so many soft drinks. One person could claim this happens because these foods taste sweet. Because the sensation of sweetness is rewarding, people learn to consume foods that provide this pleasant experience. The other person might reply that this was entirely wrong. People eat sugary foods because they provide a rich source of calories to fuel the human body's metabolic machinery.

A dispute of this sort is the sterile outcome of a failure to recognize different levels of explanation in biology. The first hypothesis is couched in proximate terms as it deals with the psychological mechanisms within an individual and how they might cause a person to behave in certain ways. The second (ultimate) hypothesis focuses on why humans may have evolved these internal reinforcement mechanisms in the first place. Both ideas could be correct because proximate and ultimate answers complement rather than compete with one another. This point can be amplified through an examination of the proximate and ultimate aspects of digger wasp behavior.

How Questions

How does an *Ammophila* wasp find and catch a caterpillar that is larger than itself? To answer this question, we would have to learn how the wasp's nervous system enables the insect to maneuver in the air and detect the visual or olfactory cues of its prey. There is a physiological foundation for the wasp's behavior, a set of proximate causes based on the operation of its neural networks, muscles, wings, legs, and sense organs. How does the wasp manage to remember where it left a paralyzed prey when it went off to build its nest many meters away? Perhaps, like some other digger wasps [705, 732], it can store information about the visual landmarks along its route, information that it uses to navigate back to a prey storage site.

But how did the wasp get the kind of nervous system, muscles, wings, and legs that enable it to capture its victim and learn about the landmark features of its environment? The *Ammophila* female did not spring from thin air through spontaneous generation. She had a father and a mother and developed from a fertilized egg, which contained genetic instructions donated by each parent. These instructions regulated the way in which development occurred, channeling the proliferation and specialization of cells along pathways that produced a nervous system with special features. Thus, there are genetic-developmental causes, as well as physiological-psychological mechanisms, that account for the distinctive behavior of *Ammophila* wasps.

As is true for the vast majority of animal species, we know little about the proximate causes of the behavior of *Ammophila novita*. Just how the nervous system of a navigating wasp operates is totally mysterious, nor is anything known about the genetic-developmental basis of any wasp's nervous system.

Why Questions

But let's say that you were provided with or that you personally discovered everything there was to know about the proximate causes of *Ammophila novita* behavior. You should not be wholly satisfied. You should still want to know *why* the wasp possesses its special kind of genetic mechanisms, why its brain works the way it does, and why the wasp has some behavioral abilities but not others.

Why questions deal with the evolutionary or *ultimate* reasons why an animal does something. Why does the female wasp hunt for and capture moth larvae? As we have seen, this behavior has a *function*; it provides food for the wasp's offspring, which consume the paralyzed caterpillars and eventually develop into adults. At the appropriate time, they will gnaw their way up to the surface and start a new reproductive cycle, if all goes well.

Each aspect of the wasp's behavior can be examined to determine why behaving that way may help the animal cope with its environment. For example, why does the wasp store the paralyzed caterpillar in a shaded, elevated cache site? Perhaps this prevents the dessication of the prey and hides the larva from various thieves and parasites (such as ants or other hunting wasps), which might more easily locate the caterpillar if it were lying on the ground near the conspicuously digging female. Why does she go to the trouble of flying up with soil from her excavation and scattering the sand widely? Perhaps this prevents the buildup of a mound of earth by the nest entrance and so removes a cue that nest robbers might use to find the burrow and its contents [223].

But why does a female of *A. novita* adopt "functional" responses to its environment? Because of the history of the species to which it belongs. In ancestral populations there were individuals that had slightly different genes and therefore slightly different developmental instructions, physiological systems, and behavioral abilities. Some individuals happened to have genes that helped them develop the kinds of nervous systems and muscles that were the foundation for effective responses to environmental problems. These wasps acquired prey more efficiently, escaped their enemies more regularly, found superior nesting sites, and so on. They tended to have more surviving offspring than those individuals that were less efficient foragers, less capable of evading parasites, and less able to locate productive nesting habitats. Thus, in the population as a whole, the genes of the reproductively successful wasps tended to survive, replacing the genes of less reproductively successful individuals.

Females of each species of *Ammophila* alive today carry in them genetic information that in the past generally conferred a reproductive advantage on the animals that possessed this information. As a result, the developmental options, and therefore the behavioral abilities, of each *A. novita* living today have been defined by differences in gene survival that occurred during the history of the species. The wasp will develop into a creature with a set of capabilities that have proved successful in the past and so are likely to be useful in the present. Living animals possess behavioral abilities because in the past these abilities have helped individuals overcome obstacles to the survival and transmission of their genes.

You should now be able to discriminate between proximate and ultimate causes of behavior and so avoid the confusion that results from failure to make this distinction. All explanations that can be traced to the genetic, developmental, physiological, or psychological mechanisms *within an individual* are proximate explanations. All explanations that deal with the ecological function of behavior, its evolutionary basis, its consequences in terms of reproduction and gene transmission are ultimate explanations.

Natural Selection

The example of *Ammophila* illustrates that the study of behavior involves many diverse phenomena: the action of a gene, the structure of an animal's brain, the relation of the individual to its environment, and evolutionary events occurring over millions of years. Yet an evolutionary approach can hope to integrate these disparate components because the genetic, physiological, and ecological aspects of behavior have all evolved and, therefore, can be treated within an explanatory framework founded on natural selection (Figure 2).

The concept of natural selection is recognized as one of the most important ideas of western culture because of its simplicity, its predictive power, and the vast scope of its application to biological matters. It was Charles Darwin's genius to realize that a few commonplace observations led to a logical conclusion that provided a powerful mechanism for evolutionary change [160, 494]. The Darwinian argument can be summarized as follows:

1. Variation exists in the traits of the members of most species. This is obvious to us in human populations and occurs wherever it has been looked for in other animals (Figure 3).
2. Some of the variation among individuals is heritable. Again, we are all aware that children usually look more like their parents than like other persons. The same observation applies to many other traits in many other species (Chapter 2).

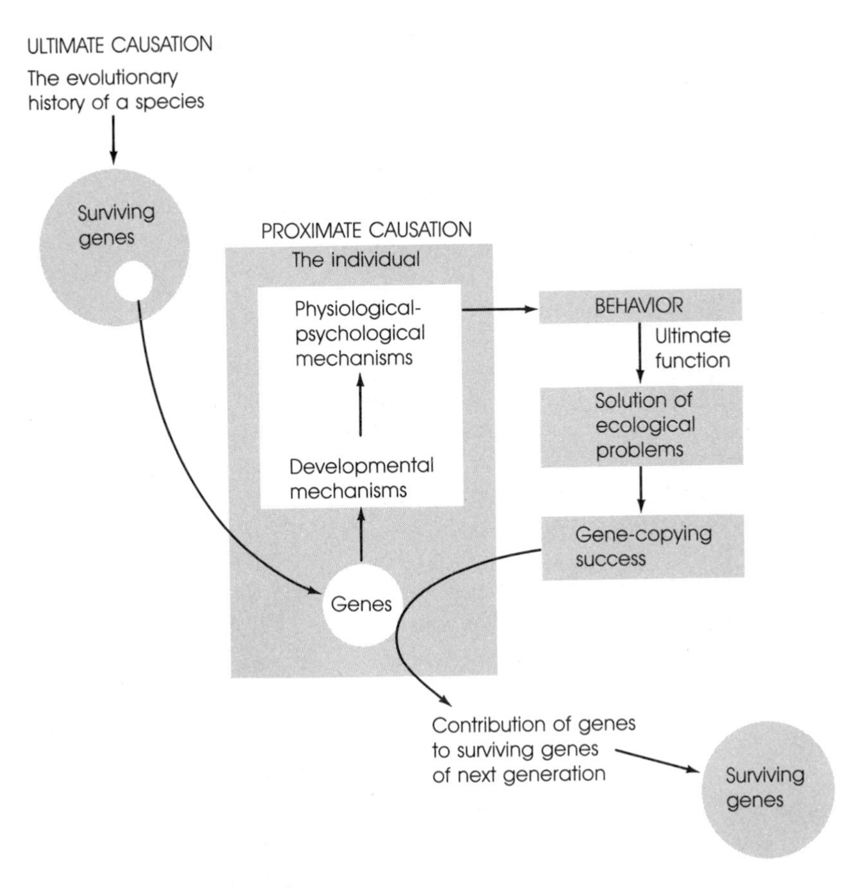

2 **Proximate and ultimate causation of behavior.** Ultimately an animal has within it certain genes because those genes have survived to the present. Genes regulate the development of the proximate mechanisms of behavior. These mechanisms make possible certain behavioral responses to environmental events—responses that will determine the number of copies of the individual's genes that are passed to the next generation. The ultimate function of behavior is to meet ecological pressures in such a way that the individual maximizes its gene-copying success.

3. The reproductive potential of animal species is vast. Darwin pointed out that a single pair of elephants would give rise to 19 million living descendants just 750 years later, provided that every descendant along the way lived to be 100 years old and had six surviving offspring. But elephants and other animal populations rarely increase exponentially. Instead, they remain stable. Therefore, most of the young animals generated by a species die before they can reproduce themselves.
4. Because of their particular heritable attributes, some individuals are likely to cope better with predators or climatic pressures or competition for food or mates. These individuals will tend to survive better

and leave more offspring than others in their species that have different hereditary factors and less successful variant traits.

The differential reproductive success of individuals based on their genetic differences is natural selection. If variation in hereditary makeup affects the production of *surviving, reproducing* offspring, then certain characteristics will be "selected for." The heritable characteristics of those who reproduce more will spread throughout the population (Figure 4). In contrast, the traits associated with reproductive failures will eliminate themselves. The logic of this process enables us to predict that animals will evolve behavioral traits that promote INDIVIDUAL REPRODUCTIVE SUCCESS, as measured by the number of offspring that live to reproduce themselves (this is an individual's PERSONAL FITNESS). Living creatures are the evolutionary end product of unconscious competition among individuals in the past, a competition in which some outreproduced others and so shaped the evolution of the species in their image.

The way in which a reproducing individual influences evolution is really by copying genes within its body and contributing them to its offspring, which may in turn transmit them to their progeny. Darwin developed the theory of natural selection at a time when the concept of genes as the units of heredity was unknown to him. But we can apply Darwinian logic to genes in the same way that Darwin applied it to individuals. This will help

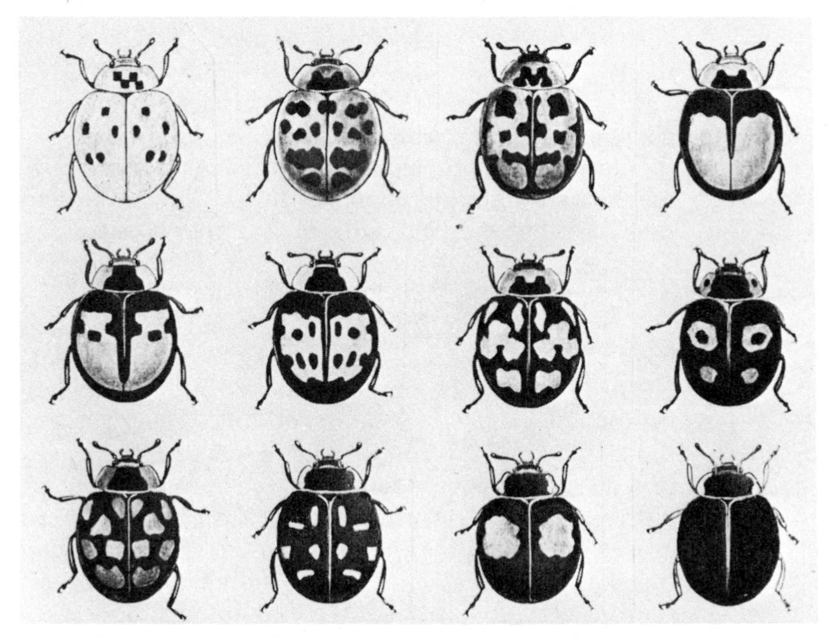

3 **Variation in an animal species.** The ladybird beetle *Harmonia axyridis* exhibits great variation in the color patterns of its wing covers. Much of the variation is due to genetic differences among individuals.

us better understand a modern evolutionary approach to animal behavior [168].

1. Genes are present in all living things; they are nucleic acids that contain coded information about protein synthesis.
2. Many genes occur in two or more alternative forms or ALLELES. This genetic variation results in the production of slightly different forms of the same protein and may have somewhat different effects on the development of individuals.
3. If one allele produces effects on development (through its variant protein) that usually cause its bearers to replicate the allele more often than other individuals with different alleles, then the "successful" allele will become more common in the population. Its competitors could be completely replaced if the relationship between genetic differences and replication success remain constant long enough. (Again we assume that populations remain stable in total numbers so that there are only a finite number of copies of a gene that can exist at any one time.)

The logical conclusion is that selection on individuals should favor alleles that help build bodies that are unusually good at promoting the survival and propagation of the particular alleles they happen to carry. According to this view, individuals are really "survival machines" that promote the survival of their alleles [168], or as E. O. Wilson has put it, a chicken is really a gene's way of making more copies of itself [794].

It is this view that provides most modern behavioral biologists with an approach, a central philosophy, for tackling the practical problem of trying to interpret animal behavior. To develop working hypotheses, they use the assumption that an animal's traits exist in the present because in the past they contributed to individual fitness (that is, success in gene transmission).

Using Working Hypotheses

It is only since 1966, when G. C. Williams published his great book, *Adaptation and Natural Selection* [786], that many biologists have begun to accept the premise that an animal's behavior has the evolved function of propagating the individual's genes rather than contributing to the survival of the *species*. The species-preservation hypothesis requires that species or populations as a whole compete to survive and that traits evolve as a result of this competition that will prevent the extinction of groups. Williams reminded his readers that selection at this level is *not* the same as selection acting on individual differences in genetic success. GROUP SELECTION can occur if populations differ genetically and if these genetic differences are consistently correlated with the long-term survival chance of a group. This will affect the genetic makeup of the species and could even favor individuals that sacrifice their fitness for the long-term survival benefit of the group in which they live [794]. But although genetic self-sacrifice can be favored by group selection, it cannot also be favored by natural (individual) selection, which, by definition, tends to eliminate characteristics that do not help individuals transmit more of their genes than other genetically different individuals. Therefore, group selectionist explanations are often very different from those that are based on an individual selectionist approach.

Group selectionist interpretations of animal behavior are commonplace [808]. A classic example is provided by a familiar hypothesis for the function of dispersal by lemmings. These small arctic rodents leave the area in which they have been living in times of high population density. Because many lemmings die as they travel away from their original homes, the dispersal trait is often said to be suicidal behavior designed to eliminate excess individuals from a population. This will help prevent the destruction of the resource base on which the survival of the species depends (so the group selection argument goes).

A less familiar and more subtle example involves a small scorpionlike creature whose males deposit stalked sperm-carrying structures (spermatophores) on the forest litter. Females find these items and use them to fertilize their eggs. Male pseudoscorpions may also find a spermatophore in their wanderings, and when they do they will eat it and replace it with their own. This observation has been interpreted as an effort by the males to remove older, deteriorated sperm and replace it with fresher gametes, thereby improving the egg fertilization rates of females that find and use the fresher spermatophores. This, in turn, will help maximize the total reproductive output of the females and, therefore, lessen the chance of extinction for the species. Note that in both cases, the group selectionist explanation ignores the disadvantageous effects of the supposedly self-sacrificing action on *individual* reproductive success. This is most obvious in the lemming "suicide" example, because a dead lemming is clearly in no position to reproduce. But the male pseudoscorpion that consumes "old" spermatophores also sacrifices some reproductive chances by spending his time and energy to demolish old sperm packets.

The Problem with Group Selection

Let us imagine a population of lemmings, many of whom truly do commit suicide when population density is high. Now imagine that a mutation occurs in a member of this group that happens to cause its carrier to reproduce whenever possible, even at times of high population density, rather than commit suicide. If the nonsuicidal mutant has higher gene-propagating success than the alternative suicidal genotype, which allele will spread and which will disappear? What behavioral trait will spread and which alternative will disappear (Figure 4), even if this reduces the long-term survival chances of the species as whole?

Similarly one can ask what would happen if, in a population of self-sacrificing male pseudoscorpions that spent their time and energy to improve the quality of sperm available for their females, there arose a mutant

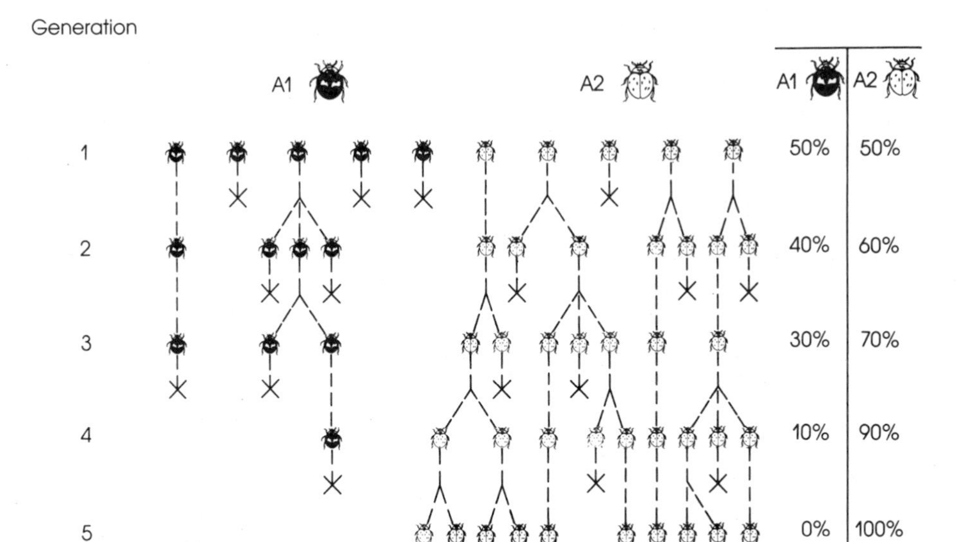

4 **Natural selection.** If individuals with the hereditary trait A1 leave fewer surviving offspring *on average* than carriers of the alternative trait A2, individuals with the A1 trait will eventually disappear from the population.

male that refused to do the work of spermatophore renovation. If the mutant instead invested his time and energy spreading about a greater total number of spermatophores than the competing type and therefore enjoyed greater reproductive success, the mutant allele would spread through the population *even if* the overall fertilization rate of females was somewhat lowered as a result.

The essential point is that if a trait promotes group welfare but reduces individual success in gene propagation, group selection in favor of the self-sacrificing trait will be less strong than individual selection for the "genetically selfish" alternative [786]. Populations consisting of self-sacrificing types are not EVOLUTIONARILY STABLE but are vulnerable to invasion by a selfish mutant that will replace the group-welfare genotype by outreproducing it. For this reason, most biologists consider individual selectionist hypotheses more likely to be correct than explanations founded on the assumptions of group selection.

Individual selectionist hypotheses for the behavior of lemmings and pseudoscorpions are easy to develop. Lemmings may disperse when population levels are high because the chance of reproducing successfully in a location in which most of the food has been or will be eaten is poor. Better to disperse if there is some possibility of finding a patch of less heavily exploited habitat. There are COSTS to dispersing. A traveling lemming is vulnerable to predators, and it may never find suitable habitat. It may even be channeled to deep oceanic fjords by the topography of the mountain valleys of coastal Norway. The rodent may plunge into the ocean as if it were another stream or pond that blocked its path, but it will not come out on the other side as it would when swimming a stream. Nevertheless, the reproductive disadvantages of dispersal may be outweighed if a dispersing lemming has a significantly greater probability of reproducing successfully than one that remains in a crowded area. If so, the trait will spread through the population even though not all dispersing lemmings will survive the search for new habitat.

Similarly, a proponent of individual selection could argue that male pseudoscorpions destroy spermatophores that they find because the time and energy costs are more than compensated by the destruction of sperm of rival males and the nutritional benefits gained by eating competitors' spermatophores. Any benefits to the population as a whole would, according to this viewpoint, be purely incidental side effects of individual selection for males able to outreproduce genetically different opponents.

Alternative Hypotheses

Even if one accepts the argument that individual selection is more powerful than group selection, there is no guarantee that a particular hypothesis based on an individual selectionist approach will be correct. In the first place, it is usually true that one can devise several different explanations for the same phenomenon, all of which are logically consistent with individual selection. Second, and even more fundamentally, some traits may *not* raise individual fitness at all, for a variety of reasons [43]:

1. The trait may have evolved through group selection despite theoretical arguments to the contrary.
2. The trait, although once adaptive under conditions that no longer exist, is currently maladaptive. Its persistence in a present population may stem from the chance failure of a superior mutant allele to appear in the history of the species.
3. The characteristic may be a neutral or maladaptive by-product of the development of another characteristic that is selectively advantageous. As we shall describe in Chapter 2, many genes have more than one effect on the development of an individual and not all of these effects are necessarily positive.
4. The trait, an abnormal or pathological reaction, would never have occurred in the past, but it does in the present because of evolutionarily novel conditions for which the members of the species cannot be adapted.

How does one discriminate among these possibilities? By testing hypotheses. Biologists usually have adopted as a WORKING HYPOTHESIS the position that a particular trait is adaptive in terms of individual genetic success rather than starting with an alternative hypothesis that the trait is maladaptive for one or more of the reasons listed above. They take this position, not because they blindly and naively believe that all traits are actually adaptive, but because the adaptationist working hypothesis has proved the most productive and versatile in the development of TESTABLE PREDICTIONS [392]. I shall illustrate this point repeatedly throughout the text, beginning here with an example of an adaptationist analysis of infanticide by males of an Indian monkey.

Hanumán langurs live in bands, which often consist of one reproductively active male and his harem of adult females and their offspring (Figure 5). Sarah Hrdy, in her studies of this primate, found that other adult males, excluded from troops with females, sometimes challenge a harem master and may oust him. The new male that takes command may harrass and bite the baby langurs in the band, despite the protective efforts of their mothers. Because males are much larger than females, they may succeed in killing some infants [350, 351]. How can we explain this thoroughly unpleasant behavior?

At the proximate level, we might devise hypotheses about the effects of a takeover on the male's testosterone levels and aggressivity or we might consider the possible genetic foundation of the trait. But at the level of ultimate causes, we can develop an evolutionary explanation founded on natural (individual) selection. Sarah Hrdy has provided one idea based on the assumption that behavior should help maximize an individual's reproductive (genetic) success [351]. With the death of each infant sired by the previous harem master, the usurper male directly removes some of his competitor's genes. Moreover, after an infant dies, it is to the mother's own advantage to become sexually receptive promptly. She will be fertilized by the new male, who transmits *his* genes to the resulting offspring. (If the female were to care for the progeny of the displaced harem owner, she would not ovulate again for two to three years.) Thus, Hrdy's hypothesis is that an infanticidal male promotes his personal fitness by eliminating rival genes and by increasing in a recently acquired harem the proportion of females that will be immediately receptive to him.

Hrdy's hypothesis has been subjected to a sophisticated mathematical analysis that indicates that certain conditions are required if an "infanticidal allele" is to spread through a population of langurs and be maintained at 100 percent frequency [134]. Key variables are the interval between pregnancies by females with and without young infants and the rate of inseminations of the females in a harem by its owner. The data gathered on these points by Hrdy are consistent with the hypothesis that infanticide promotes the reproductive success of the infanticidal male.

5 **An extended family of Hanuman langur females and their offspring.** This group is the focus of violent competition among males intent upon acquiring a harem of mates. Photograph by Sarah Hrdy-Blaffer; courtesy of Anthro-Photo. ▶

There are, however, other interpretations for this behavior. For example, a group selectionist argument would be that infanticide occurs at times or in areas with high population density in order to stabilize the growth of the population for the long-term ecological benefit of the species; or that infanticide enables genetically superior males to reproduce more than inferior ones, a result that leads to the genetic improvement of the species as a whole. These hypotheses have the logical difficulties inherent in most group selectionist explanations, but they are nevertheless interpretations that compete with Hrdy's idea.

Still another alternative states that the behavior is *not* adaptive at *either* the individual or the group level. Some primatologists argue that infanticide is a purely pathological response to modern-day conditions that no longer resemble those of the past in which langurs evolved [183, 605]. According to this view, the monkeys have in recent times moved into areas of human habitation, where large quantities of food are available to them. As a result, populations have reached abnormally high densities, and under these atypical conditions the behavior of the monkeys has broken down; in other words, infanticide is a maladaptive reaction to overcrowding and an altered environment.

Hypothesis Testing and Evolutionary Prediction

The existence of so many different competing explanations for male infanticide shows how important it is to try to determine their validity scientifically. Hrdy's hypothesis can be tested, but unfortunately this cannot be done directly by measuring the average number of genes left by two genetically different types of males, one that did and one that did not murder infants fathered by other individuals. Direct measurements of the reproductive success of individuals are notoriously difficult, especially in long-lived species. In the case of the langur, one would have to follow a male over his entire life, record all his copulations, determine the paternity of all the offspring of his mates (if they had several partners prior to a pregnancy), and keep track of all the babies sired by the male to discover how many actually reached reproductive age. Needless to say, this would require an extreme, if not superhuman, degree of dedication.

Fortunately, however, evolutionary ideas about the function of behavioral traits can be tested indirectly in a broad variety of ways. It is not my intention here to present examples of all the major methods, as this will be done later in the book (particularly in Chapter 7). Instead, my goal is simply to demonstrate that there is more than one technique that biologists can use to develop testable predictions based on the working hypothesis that an animal's actions help propagate its genes.

The Comparative Method

The most widely used way to test evolutionary predictions is with THE COMPARATIVE METHOD. To apply this technique to infanticide by male langurs, we must first identify the conditions that appear to make infanticide reproductively profitable for an individual. They are the following: (1) a male can identify who his competitor's offspring are and so can selectively kill just those infants, not his own; (2) the male is larger than the female; therefore, even though she opposes his infanticidal impulse, he is able to impose his will by force on the female at relatively little risk of injury to himself; and (3) females remain under the control of the male that kills their offspring, a situation that leaves them no adaptive option except to resume their sexual cycles and be impregnated by the new harem master.

These factors either tend to raise the benefits of infanticide from an individual male's perspective or lower its costly effects for him. If our reasoning is correct, these conditions should have the same evolutionary effects on other species and therefore we can *predict* that selective murder of infants by adult males will occur in other species when some or all of these three circumstances apply. We can *test* this prediction by searching for other reports of infanticide by adult male animals and comparing the ecology of these other species with that of the langurs.

There are a variety of similar cases of infanticide by males in other nonhuman primates in which infant deaths coincide with a change in leadership of a harem (Figure 6) [21, 685]. In addition, male lions fight for control of prides of females; when a takeover occurs, the new pride owner often kills the young cubs in the group [635] (Figure 25, Chapter 11). The house mouse is still another mammal, unrelated to either primates or lions,

6 **Infanticide in primates.** A male baboon has just violently thrown an infant, which it eventually killed in this manner. Photograph by H. J. Rijksen.

in which infanticide by males occurs in the context of competition for control of female harems [742]. A nonmammalian analog to langurs is provided by some poison-arrow frogs (so named because their highly toxic skin was used as a source of poison for hunting arrows by certain tropical American tribes). Males of *Dendrobates pumilio* defend territories and call to attract receptive females to them. The female lays a clutch of five to ten eggs on a horizontal leaf moistened by the seminal fluid of the territory owner, who is present at the start of egg laying. A male does not always remain by any one of his egg clutches but visits the leaves in his territory that carry his fertilized eggs at least once a day in order to moisten them with additional fluid. In captivity (and probably in nature), males can discriminate between eggs they have fertilized and those fertilized by rivals. They care for the former and eat the latter (Figure 7) [775]. In nature this probably occurs most often when an intruder male ousts a territory owner from his property.

Thus, in species as different as langurs, lions, mice, and poison-arrow frogs, males selectively kill young of their species as part of a process of reproductive competition with rival males. Although infanticide occurs in many other contexts and may have as many as eight distinct evolved functions [516, 656], there is enough evidence to suggest that infanticide by adult male langurs fits a particular adaptive pattern. The fact that this "takeover pattern" is not isolated but appears in several ecologically similar

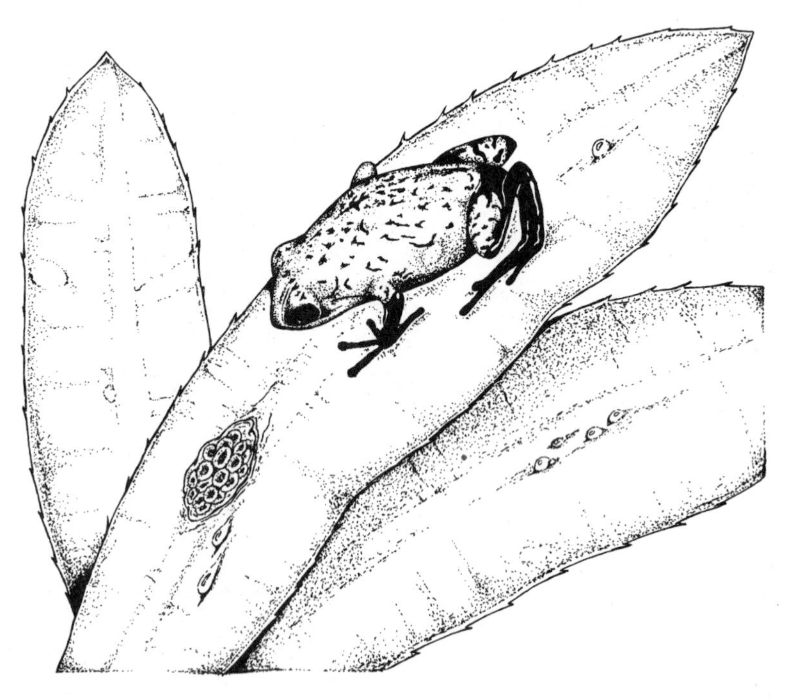

7 **Infanticide in frogs.** A male *Dendrobates* frog is about to consume a clutch of eggs, the offspring of a male whose territory he has invaded.

species supports the prediction that infanticide by males is an adaptive response to certain socioecological pressures.

The comparative method is a kind of "natural experiment" in which the observer, by choosing a sufficiently large sample of *unrelated* species, tests whether there is a lawful relationship between a particular trait and specified environmental conditions. In our example, we find that the occurrence of infanticide by males is linked with (1) opportunities for some males to take a harem or a female-attracting territory from other males, (2) low risks of physical injury while dispatching an infant, and (3) little chance that the male would kill his own progeny.

Experimental Tests of Evolutionary Predictions

We can also test the prediction that these factors promote the evolution of infant-killing males by conducting "controlled experiments" in which the observer *manipulates* key conditions to test their effect on the behavior of some animals. Work of this sort is often done in the laboratory, where one can precisely regulate all the features of an environment and thereby remove the chance that something outside the observer's control causes the animals to behave as they do.

For example, we have hypothesized that if a male can successfully identify who his young are, he can avoid killing them and instead can practice infanticide on the progeny of rival males. This should make infanticide more "profitable" for a male and therefore more likely to evolve. This hypothesis leads to the prediction that infanticidal males will possess mechanisms that help them discriminate between their own progeny and another male's offspring. This prediction has been tested experimentally using house mice [460, 742] and another rodent, the collared lemming [460]. In the experiment with collared lemmings, pregnant females were separated from their mates. Then, the day after giving birth to a litter of little lemmings, either (1) the biological father of the litter was reintroduced into the cage with his mate or (2) a strange male was placed in the cage with the female. All other conditions were kept the same in a laboratory setting. The survival rate of the young lemmings was dramatically different in the two cases (Table 1). The strange males killed many of the baby lemmings (but did not eat them), whereas the stud males did not kill any of their progeny. The prediction that males would practice infanticide selectively (to their genetic benefit) was supported by the experiment.

Essentially the same result was secured in experiments with house mice, although the procedure here was different. In one experiment, Jay Labov permitted 30 males to mate and then separated the pairs. Eight to seventeen days later he placed each male in a cage with a different pregnant female (not the *original* mate). Thirty-two other males that were not given an opportunity to mate before the experiment were also paired with females at the same stages of pregnancy as in the first group. Of the young produced by females in group 1 with sexually experienced males, 78 percent survived; but in the second group, only 70 percent of the progeny lived to the fifth

Table 1
Effect of Copulation on the Tendency of Male Lemmings to Practice Infanticide

Condition	Number of Offspring of Females	Number That Survive Contact with Male
Males have copulated with female (n = 16)	43	41 (95%)
Males have not copulated with female (n = 32)	85	47 (55%)

Source: Mallory and Brooks [460].

day. The difference, although not huge, is statistically significant. Labov found more mouse pups bitten and killed, but not eaten, in the cages with the males that had not copulated prior to pairing [400].

Apparently male house mice in some way store information about how recently they have mated, and this influences whether or not they will practice infanticide. In nature, dominant males control territories in which one or more females live. These females mate only with the territory owner. If a male has copulated with a female in his territory and she has a litter three weeks later (the gestation period is 21–22 days), the odds are excellent that the baby pups are his offspring and, therefore, that he would gain nothing by killing them. In the laboratory, the tendency of males to use mating experience as a regulator of their behavior can be exploited by the experimenter to fool a male house mouse into treating another individual's progeny as if they were his own. (Labov had also painted each male's copulatory partner with the urine, and therefore the distinctive odor, of the male's future cagemate. This may have been another factor that "encouraged" some males to treat their cagemate as if she had actually been the female he had recently inseminated.)

More recently, Frederick vom Saal and Lynn Howard [742] have experimentally confirmed the importance of recent mating experience on the treatment of infants by male house mice. They first identified a sample of dominant males by pairing individuals against rivals and seeing which mouse could subdue the other in fights. They divided the dominant males into two groups: a sexually experienced set that was permitted to mate with two different females each and a sexually "naive" set that was not permitted to copulate before the experiment. Three weeks later, each mouse was placed in a cage with two newborn mice (not fathered by him). The results of the experiment appear in Table 2. Most sexually naive males attacked one or both pups, whereas the sexually experienced males tended to behave in a parental fashion, remaining close to the pups as if guarding them.

This experiment shows again that mating prepares male house mice for parental behavior at the appropriate time. As a result, a dominant male living in the wild would be unlikely to kill his own pups. But if he had not mated and found pups in his domain, they would surely have been fathered by some other individual. Under these conditions infanticide might well be an adaptive response. Typically in nature this would happen when a newly dominant male had acquired another male's territory. Not only does an infanticidal newcomer eliminate rival genes, but he also shortens the time to receptivity in the females living in the usurped territory. Vom Saal and Howard paired 40 sexually naive males with females that had given birth within 24 hours. Some males killed the entire litter; others did not. Those females whose pups were completely eliminated delivered a new litter sired by the infanticidal male in about 22 days; in cases in which the male did not kill the female's progeny, she did not bear a new litter until 30 days had passed. Thus, infanticide in house mice as well as langurs speeds reproduction by the killer male and should increase his lifetime production of surviving progeny.

Infanticide by Human Males

The comparative and experimental methods offer two different ways to test evolutionary predictions. There is yet another method in which one first makes a prediction about what an animal *should* do if its behavior has evolved to maximize individual fitness. Then one can collect the observational data that are needed to test whether the prediction is correct. As long as one does not know the outcome in advance, this is a perfectly legitimate way to test the ultimate value of a trait. Let me give an example of a hypothesis about human behavior derived from the logic of our analysis of infanticide in langurs and other animals. We have argued that adult males in these animals may maltreat the progeny of females with whom they associated (1) if the abused young are likely to have been fathered by a different male and (2) if the infants will divert the reproductive energies

Table 2
Treatment of Two Newborn Mice Held in a Cage for 30 Minutes with a Dominant Male House Mouse

	Sexually Naive Males[a]		Sexually Experienced Males[b]	
Action	Number	Percentage	Number	Percentage
Infanticidal response	28	82	5	15
Parental response	4	12	25	76
Pups ignored	2	6	3	9

Source: vom Saal and Howard [742].
$\chi^2 = 31.4$; $P < .001$.
[a] Males that have not copulated for a period of at least three weeks.
[b] Males that mated three weeks before the experimental test.

of the female from potential offspring of the male. We know that severe child abuse leading to murder occurs sometimes in human societies. This has led several evolutionary biologists to predict that the assault and murder of children by their "caretakers" will occur significantly more often when the caretaker is a stepfather than when the caretaker is the child's biological father. A stepparent has no direct genetic investment in the children he acquires by marriage. Under some circumstances he may, therefore, not care for them as well as possible and in extreme cases may attempt to dispose of the acquired offspring.

This is a *prediction* based on the logic of natural selection, presented in an effort to test an evolutionary hypothesis. We are dealing with a possible ultimate explanation of a behavior. To try to explain something is not to try to justify or excuse it. We are simply talking about the possible genetic consequences of an act. In addition, there is no implication that a murderous stepfather is any more aware of the genetic consequences of his or her behavior than a male langur is. Just what motivates a human being to do something is a question about the proximate causes of behavior and requires an investigation of the developmental-physiological-psychological foundations for the trait. This would be interesting research, but we do not have to do it in order to test a prediction about an ultimate cause of the behavioral action [13].

To test the prediction we might use statistics available on the relative risk of child abuse in families in which (a) both biological parents live together with their children and (b) a biological parent lives with a stepparent. About 70 percent of all children living in the United States live with both biological parents; they contribute less than 50 percent of all cases of child abuse. In contrast, only about 10 percent of all children live with a stepparent, and yet this family situation is associated with about 20–25 percent of all validated cases of child abuse in a number of studies (Figure 8). Thus, a stepchild is at substantially greater risk of being attacked than a child living with both biological parents [156]. This is consistent with our evolutionary prediction, as is the finding that a stepparent is more likely to be the perpetrator than a biological parent in cases of child abuse [413]. Additional data are needed to determine unequivocally that stepfathers are more likely than biological fathers to engage in child abuse (but see [421]). Many difficulties complicate the collection, analysis, and interpretation of these data, but the point is that the prediction can be tested. If our hypothesis is wrong, we have ways to discover this and discard the explanation as false. If our hypothesis is right, our predictions should be supported by our observations, comparisons, and experiments.

We can achieve even greater confidence in a hypothesis by devising a set of alternative ideas and then performing tests that discriminate among them. For a given behavioral characteristic, there may be several plausible evolutionary hypotheses that may generate some of the same predictions. In these cases it is therefore critically important to devise tests that will enable one to discard all but one of the possible hypotheses. For examples

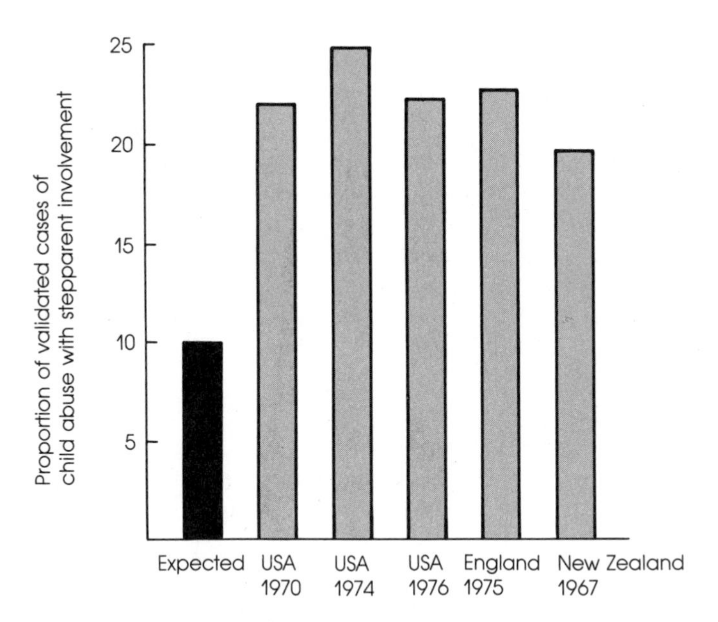

8 **Test of an evolutionary prediction about human behavior.** Although only about 10 percent of all children 17 years or younger live with one or two stepparents in the United States, England, and New Zealand, stepparents were involved in about 20 percent of all cases of child abuse in samples collected in five different years.

of this problem and its solution, see the discussion of flocking in gulls (Chapter 7) or the use of alarm calls by ground squirrels (Chapter 14).

Complete certainty, however, is *never* achieved in science. What appears to be an ironclad experiment or an unassailable hypothesis one day may prove the next to have critical flaws that require its rejection. Scientists are continually squabbling about each other's ideas; and as long as these arguments are done with good humor (which is not always the case, unfortunately), the squabbling can be positive, stimulating, and conducive to better science. Sometimes even when the arguments get nasty, science advances. The uncertainty about the truth that scientists accept (at least in theory) often makes nonscientists nervous, in part because scientific results are often presented to the public as if they were absolute truths written in stone for eternity. That this is incorrect will be obvious to anyone who has conducted even the most cursory and superficial review of the history of a scientific field. New ideas emerge, old ones are replaced or modified, and science evolves and sometimes is revolutionized. It is even possible that Darwinian evolutionary theory will some day be overthrown just as Newtonian theory in physics has been.

The constantly changing content of science is not a defect but a reflection of its ability to incorporate new insights and discoveries. For the moment, Darwinian theory offers the most comprehensive and practical way in which to approach the totality of biology. There is plenty of room for disagreement, exploration, and invention within its framework. If natural selection is ultimately shown not to be a major force behind the evolution of the characteristics of all living things, it will probably be because of discoveries made while using the working hypothesis of the evolutionary approach

[440]. Until a replacement theory is available, I believe that an understanding of the operation of natural selection should be part of the education of every person—the better to enjoy the world around us.

Organization of the Text

The chapters that follow present a variety of testable ideas of evolutionary biologists about the proximate and ultimate causes of behavior. Chapters Two, Three, and Four examine the underlying genetic and developmental mechanisms of behavior from an evolutionary perspective, and Chapters Five and Six are devoted to the physiological bases of behavior. The emphasis then shifts to the ultimate causation of animal behavior in Chapters Seven through Fourteen. Chapter Seven discusses in more detail the use of evolutionary theory to develop hypotheses about the adaptive significance of a behavioral trait and ways to test these hypotheses. There follow separate chapters on the ecology of habitat selection, feeding, antipredator behavior, and male and female reproductive strategies. Chapter Thirteen deals with techniques for tracing the possible evolutionary history of a behavior pattern, and Chapter Fourteen analyzes the evolution of animal societies. The final chapter of the book outlines an evolutionary approach to some elements of human behavior. The organization of the text recapitulates our sketch of the analysis of *Ammophila* wasp behavior with its genetic, physiological, ecological, and historical components. I hope that the basic structure of the book will be easy to keep in mind and will prove to be a useful system for organizing the many diverse topics that together make up the study of animal behavior. As another organizational aid, let me suggest that the reader scan a chapter's summary as a guide to its central issues before reading the chapter itself.

Summary

1 Basic questions about animal behavior fall into two categories. HOW QUESTIONS require answers about how the proximate mechanisms within an individual—its genetic-developmental and physiological-psychological systems—enable the individual to carry out its behavior patterns. WHY QUESTIONS require answers about why ultimate factors—long-term ecological and evolutionary events—have produced the proximate mechanisms that cause individuals to behave in certain ways.

2 There are several possible candidates for the selective forces that have influenced the evolution of behavior. First, natural (individual) selection occurs if genetically different *individuals* have different numbers of surviving offspring. Second, group selection will occur if genetically different *groups* differ significantly in their long-term survival chances because of genetic differences among them.

3 A trait favored by group selection may lead an individual to pass on fewer of its genes than other members of its species with alternative

genes and alternative traits. If so, the group-benefiting allele should usually be replaced over time by the alternative allele that promotes individual genetic success. Therefore, the ultimate function of a behavioral trait should usually be to promote gene propagation by the individual rather than the welfare and survival of the group to which the individual belongs.

4 Because genes have been selected for their ability to survive, the *working hypothesis* of an individual selectionist is that all the evolved characteristics of an animal should tend to promote the survival of the genes that underlie these characters. Specific hypotheses about the possible contribution of a trait to individual fitness (gene propagating success) often can be used to develop concrete predictions that can be tested by making additional observations, or by applying the comparative method, or by conducting controlled experiments. The appeal of an evolutionary approach is that it is based on a logical concept, that it produces plausible hypotheses about an enormous array of biological phenomena, and that many of these hypotheses can be tested.

Suggested Reading

For more information on the behavior of digger wasps, see *The Wasps* [226] by Howard Evans and Mary Jane West Eberhard.

Other books that capture the sense of curiosity and excitement that students of animal behavior have about the animals they observe include Archie Carr's *So Excellent a Fish* [129], Vincent Dethier's *To Know a Fly* [172], Howard Evans's *Life on a Little Known Planet* [224], Konrad Lorenz's *King Solomon's Ring* [448], George Schaller's *The Year of the Gorilla* [634], and two books by Niko Tinbergen, *Curious Naturalists* and *The Herring Gull's World* [705, 707].

The best book on the logic of natural selection and the evolutionary approach to biology is G. C. Williams's *Adaptation and Natural Selection* [786] (especially Chapters 1 and 2). Richard Dawkins's *The Selfish Gene* [168] is an extremely readable account of the application of gene thinking to an understanding of animal behavior. For a critique of gene thinking, see Sewall Wright's [807] article on the subject.

Suggested Film

Evolution. Color, about 20 minutes. An amusing animated film largely about the historical sequence of events in the evolution of life on earth, but with segments that illustrate the action of natural selection.

One of the central problems in an evolutionary approach to animal behavior is to establish clearly the relation between genes and particular behavior patterns. As we saw in Chapter 1, selection will not lead to evolutionary change unless there are genetic differences that in some way induce among individuals physiological, structural, or behavioral differences that affect their fitness. To claim that a behavioral trait is adaptive and has evolved in response to an environmental pressure is to imply that the trait has a genetic foundation that differs from that of some alternative characteristic. Is this claim likely to be true?

This chapter examines the proximate basis of behavior at the genetic level in order to determine how genetic factors "cause" behavioral attributes. Our first goal is to demonstrate that it is possible both in theory and in practice to show that there is a correspondence between genes and behavior and that individuals that differ with respect to even a single allele may behave in significantly different ways. But behavior genetics has moved beyond the stage in which its primary objective was to prove that there was a genetic foundation for behavioral acts. It has begun to link genetic differences between behaviorally different individuals to variation in their nervous or hormonal systems, identifying the proximate ties among specific genes, physiological traits, and behavioral attributes. A few workers have even shown how ecological differences among populations might be responsible for the spread of different genes, different physiological mechanisms, and different behavioral traits in geographically separated populations of the same species. The chapter concludes with a description of these innovative achievements by behavior geneticists.

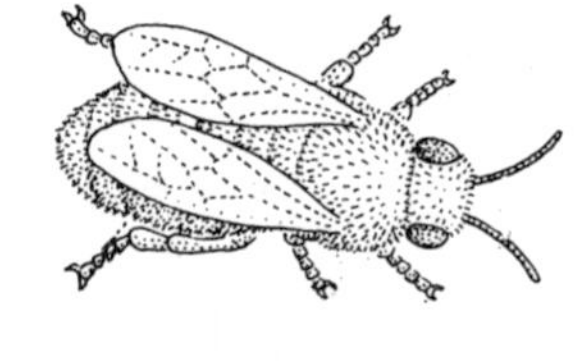

CHAPTER 2

The Genetics of Behavior

Heredity and Human Behavior

Oskar Stohr was raised as a Catholic in Nazi Germany by his grandmother. Jack Yufe grew up on various Caribbean islands with his Jewish father. One can hardly imagine more diverse environments for the development of two human beings, and yet these men are remarkably similar in appearance (Figure 1) *and* behavior. They "like sweet liqueurs, . . . store rubber bands on their wrists, read magazines from back to front, dip buttered toast in their coffee and have highly similar personalities" [327]. They are identical twins who were separated after they were born and did not meet again until they were 47 years old.

The separation of identical twins early in life provides a very rare natural experiment with which to test a hypothesis about the proximate basis of behavior. The hypothesis is that some of the *differences* in the behavior of humans are due to the genetic differences among them. A prediction that follows is that identical (monozygotic) twins should be more similar behaviorally than nonidentical or fraternal (dizygotic) twins. This is because identical twins are derived from the very same fertilized egg (which divides and gives rise to two separate, but genetically identical, embryos), whereas fraternal twins are produced from two different fertilized eggs. The probability that fraternal twins will possess the same allele is only ½.

In humans and other animals, each gene may be represented by one or many slightly different forms (alleles). Each gene codes for a distinctive protein such as a hormone or enzyme (e.g., amylase, which facilitates the chemical reaction that breaks down one sugar, glycogen, to smaller molecules). Each allele of the amylase gene carries information for the construction of a slightly different form of the amylase molecule, each form having its own chemical properties and abilities. Human beings are diploid organisms. Each of us has two copies of every gene. The two alleles of any particular gene possessed by a person may be the same or they may be different. Imagine a population in which there were four alleles of the amylase gene (*amy 1, amy 2, amy 3,* and *amy 4*). In this population there might be a woman with copies of *amy 1* and *amy 2,* and she might have a husband with copies of *amy 3* and *amy 4.* The eggs produced by the female

1 **Identical twins reared apart.** Jack Yufe (left) and Oskar Stohr (right): JY was raised as a Jew in Trinidad, OS was raised in Nazi Germany during World War II. Photograph by Bob Burroughs.

are haploid (i.e., have only a single copy of each gene) and so their genetic makeup or GENOTYPE may be either *amy 1* or *amy 2*. These eggs will be fertilized by sperm, which are also haploid and therefore have one copy of the amylase gene, either *amy 3* or *amy 4* in this case. The sons and daughters produced by these individuals are equally likely to have one of the following four genotypes: *amy 1,amy 3* or *amy 2,amy 3* or *amy 1,amy 4* or *amy 2,amy 4* (Figure 2).

Pick one of these genotypes (e.g., *amy 1,amy 3*) and compare it with each of the four equally probable genotypes of its siblings (in the sequence listed above). The proportion of shared alleles between them is 1.0, 0.5, 0.5, and 0.0. The average shared proportion of alleles is $(1.0 + 0.5 + 0.5 + 0.0) \div 4 = 0.5$. Although this is just one example, the probability that two siblings (other than identical twins) will inherit the same allele from a parent when two different alleles are represented in the parent's gametes is always ½. As a result, if behavioral development is influenced by the presence of a particular allele, fraternal twins should often differ (because they will often have different forms of the gene), whereas identical twins should be the

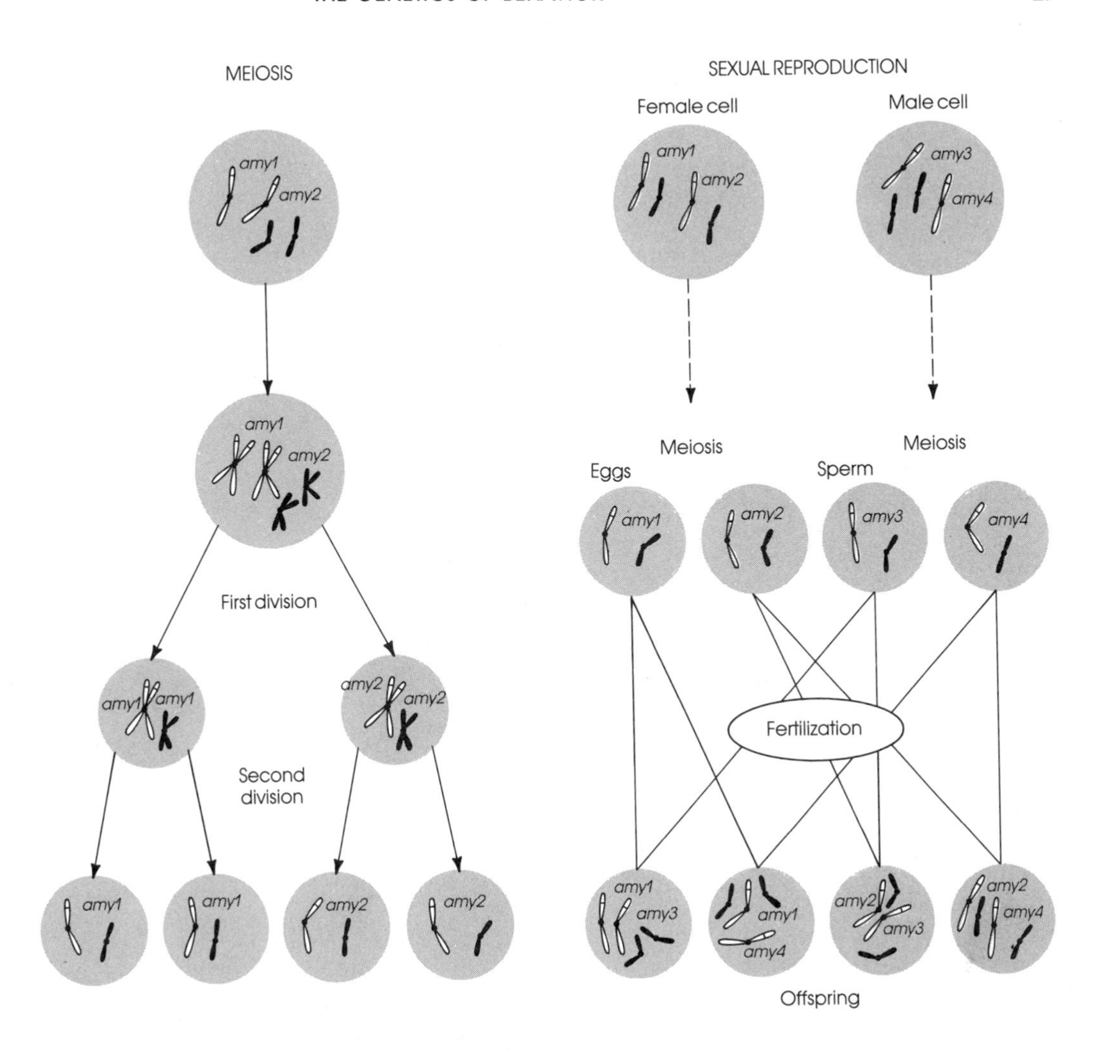

2 **Meiosis and sexual reproduction.** The cells that give rise to eggs or sperm undergo a series of meiotic divisions that produce cells with one copy of every gene in the parental genotype. When an egg fuses with a sperm in sexual reproduction, the resulting offspring has two copies of each gene.

same because they share all the same alleles. In fact, identical twins are more similar than fraternal twins with respect to a host of traits including appearance, personality measures, IQ scores, and other behavioral abilities.

But behavioral development is influenced, not only by the genes the individual inherits, but also by his or her environment and experiences. Some people have suggested that because identical twins look alike, parents treat them more similarly than fraternal twins (which are no more similar in appearance than any other pair of ordinary siblings). This environmental influence, so the argument goes, could be the reason why identical twins are so similar behaviorally. The evidence that identical twins are actually subjected to more similar environmental influences than fraternal twins is not strong [581], but to be on the safe side it would be interesting to observe the behavior of identical twins that were reared apart. If they developed similar behavior, it could not be because they were encouraged to do so by their parents and a shared home environment. This is why Jack Yufe and Oskar Stohr and others like them provide such valuable information.

Although the results of the most recent study of monozygotic twins reared separated are still preliminary, they strongly suggest that these individuals are remarkably alike (even to choice of jewelry, toothpaste, clothing, and the names given to their children!). If this study confirms that identical twins are more similar than fraternal twins on a broad range of behavioral characteristics, it will support the hypothesis that certain behavioral differences between fraternal twins can be traced in part to their different alleles.

Are Some Behavior Patterns "Genetically Determined"?

As indicated in the preceding section, there is already evidence that identical twins are generally more similar than fraternal twins in their scores on IQ tests [80]. When identical twins are reared together, the CORRELATION between their scores is usually between 0.8 and 0.9. If the IQ scores of pairs of identical twins were always exactly the same, the correlation would be 1.0. Correlations in the 0.8 to 0.9 range mean that if one twin has an IQ of 115, the odds are excellent that the other will be close to this score (say, between 110 and 120), and there would be only a very small chance that the other twin would score 95 or 135. If the correlation were zero, the score of one twin would be utterly useless as a predictor of the score of the other.

Identical twins, even if they have been reared apart, have an IQ correlation that is higher on average than that of dizygotic twins reared together (Table 1). Further evidence that there is a genetic contribution to the differences in IQ scores comes from a comparison of the IQs of parents and their biological children, which are substantially more similar than those of foster parents and their adopted children (Table 1). Each parent donates half its genotype to each child so that the proportion of shared genes between a mother and her son (for example) is 0.5. But individuals that have adopted children obviously have not endowed their adoptees with any of their genes and so have a genetic relatedness of zero.

The fact that there is connection between the proportion of genes shared by descent and the degree of similarity in IQ scores is sometimes interpreted to mean that intelligence is "genetically determined." Leaving aside the serious problem of establishing the relation between IQ and intelligence (about which much ink has been spilt), this interpretation is incorrect. The results of the familial comparisons within populations outlined in Table 1 show just one thing: that genetic differences among people contribute to

Table 1
Familial Correlations for IQ Scores: Predicted versus Actual Correlations

Category	Predicted Correlation[a]	Number of Studies	Actual Median Correlation
Monozygotic twins reared together	1.00	34	0.85
Monozygotic twins reared apart	1.00	3	0.67
Dizygotic twins reared together	0.50	41	0.58
Siblings reared apart	0.50	69	0.45
Nonbiological sibling pairs	0.00	5	0.29
Parent–biological child	0.50	32	0.39
Parent–adopted child	0.00	6	0.18

Source: Bouchard and McGue [80].
[a] Predicted correlations if the differences between individuals were solely due to the genetic differences between them.

the differences in the scores they achieve on IQ tests. This is an interesting finding, but before we explain its utility it is critical to realize what these results do *not* mean.

First, IQ is *not* "genetic," or "genetically determined," or "inherited" in the sense of not requiring environmental influences for its development. It should be obvious that taking IQ tests requires a brain that, in turn, is the product of a breathtakingly complex interaction between the genetic information in a fertilized egg and the "environment," which includes the nongenetic materials in the egg and the egg's surroundings. These materials are utilized by the growing embryo and are absolutely essential for the construction of brain tissue. Moreover, as the brain develops and begins to generate electrical activity and receive messages from outlying sensory receptors linked to it, it is creating and receiving "experiences" that can influence the way in which its structural and functional properties continue to develop. Without these inputs the brain would fail to mature, and no IQ score would be possible. (Chapter 3 explores the issue of behavioral development in more detail.)

Second, it is *not* true that these intrafamily comparisons show that the differences between individuals are the *sole* result of genetic differences among family members. If the environment played no role in contributing to the differences between people, then the correlation between the IQs of identical twins should be 1.0; but it is less than this. Identical twins are not behaviorally identical because their development is shaped by somewhat different environments. Each twin experiences slightly different surroundings in its mother's uterus, each consumes somewhat different foods, and each has its own unique set of social interactions (this is especially true of twins separated early in life). Likewise, if genes shared by descent were the sole basis for the similarities between people, then the correlation between the IQ of a parent and its adopted child should be zero. But it is greater than this because these two individuals share similar environments, which influence to some extent the development of the abilities that underlie IQ scores.

What then is the significance of familial comparisons as a behavior genetics technique? One can point first to the discovery that there are many differences in *specific* behavioral abilities of humans that have a genetic component. The mental development scores of very young monozygotic twins are not only similar, they follow a remarkably similar, age-related pattern (Figure 3) [797]. Later in life, the sensitivity of identical twins to the odor of acetic acid has been shown to be much more highly correlated than that of fraternal twins [355]. If one identical twin cannot detect low concentrations of acetic acid, the odds are pretty good that the other twin will also be insensitive. The correlation is much less when one compares fraternal twins. Likewise, left-handedness is much more likely to occur in children if one or more of their biological parents is left-handed. But if a child whose biological parents are right-handed is adopted into a family with a left-handed foster parent, the child's environment does not increase the probability that it will exhibit left-handedness [130].

The occurrence of certain mental illnesses and other afflictions has also been related to genetic differences among people. The children of a schizophrenic, manic-depressive, or alcoholic parent have half the parent's genotype *and* a much greater risk of developing the condition than the children of parents who do not have these diseases [323, 507, 641]. One might, however, argue (not unreasonably) that living with a person suffering from schizophrenia or alcoholism creates an environment that induces the trait in others. Once again, one can avoid this complication by the right comparisons, in this case by following the life histories of children separated early in life from their biological parents (and a possibly disturbing home environment) and reared in foster homes. Those children with a biological parent with the illness show an incidence of schizophrenia four or five times greater than adopted individuals whose parents were not schizophrenic. If the foster child had an alcoholic parent, it too is at special risk of developing the condition, even if it is unaware of the history of its parent.

I emphasize again that schizophrenia, manic-depression, and alcoholism are not "genetically determined." A child of a schizophrenic (or alcoholic or manic-depressive) may or may not have inherited the critical allele or alleles from the affected parent (see Figure 2). If the child is not a carrier, it has no genetic factors that increase its risk of developing the disease. Even if the child has received the key allele(s), he or she will not automatically be afflicted no matter what the environment. Most carriers are behaviorally normal; but, as a population, they are at somewhat greater risk

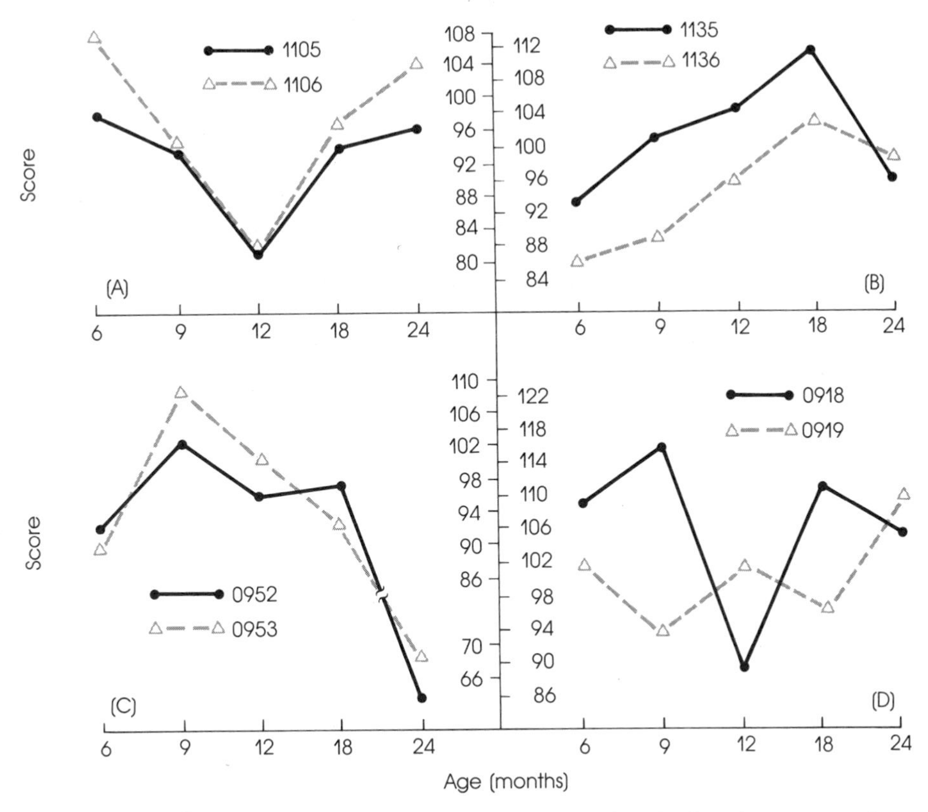

3 **Mental development scores for identical twins at ages 6 to 24 months.** Most identical twins show highly similar patterns of development (e.g., A–C); very rarely, a pair will exhibit distinct patterns (e.g., D).

of developing the condition presumably because of greater susceptibility to the environmental factors that trigger the syndrome [323].

Studies of human behavior genetics indicate that there is considerable genetic variation within modern populations, a genetic variation that contributes to the behavioral variation among humans. This makes it more plausible to argue that there has been behavioral variation in the past in human populations and that this behavioral variation could have resulted in the evolution of particular behavioral traits. But studies of the behavior of relatives leaves unanswered many other intriguing questions: How many alleles are responsible for specific behavioral differences among people? Why do alleles that promote the development of schizophrenia persist in human populations over time (as they would seem to reduce individual fitness)? How does possession of any one allele exert its impact on the development of a behavioral trait? This last problem is particularly relevant for persons interested in the proximate basis of behavior, and yet very little progress has been made on this point in studies of human behavior genetics.

At best there are only tantalizing suggestions.

For example, postmortem analyses of 20 schizophrenic persons showed that in certain regions of their brains the nerve cells were 50–75 percent more likely to bind with the chemical substance dopamine than the brain cells of normal individuals [409]. There is indirect evidence from another study that some schizophrenic patients have unusually low levels of an enzyme that breaks down dopamine molecules [684]; therefore, they may have high levels of the substance circulating through their nervous systems. Dopamine is a NEUROTRANSMITTER that relays messages from one cell to another in certain networks that have been implicated in the regulation of our emotions. It is possible (but highly speculative) that in schizophrenic brains these networks become "overstimulated" because of their attraction for dopamine or because of an excess of this neurotransmitter. The hyperactivity of these neurons could help produce the emotional disturbances and delusions that characterize schizophrenic behavior. In theory, the presence of even one distinctive allele could promote the development of schizophrenia if, for example, this allele coded for a defective enzyme that failed to deactivate dopamine molecules in the biologically appropriate fashion.

A more definite relationship between the enzymatic product of a gene and psychiatric disorders has been established in the case of monoamine oxidase (MAO) [112]. The activity of this enzyme varies over a broad range among individuals. Very high and very low activity levels characterize particular families and are presumably related to the presence of specific alleles. In families with low MAO levels, the incidence of suicide and attempted suicide is eight times that of families whose members have high MAO levels. Research findings of this sort may be an initial step toward an understanding of the way in which human genes influence the development of human behavior, but much more work lies ahead.

The Behavior Genetics of Other Animals

Research on the behavior genetics of human beings is exciting but is often marred by controversy and criticism, in part because there is no ethical way to conduct truly controlled breeding experiments with humans. Twins are extremely rarely separated at birth and are never placed randomly in different environments. Likewise, foster children studies are complicated by the tendency to place children with foster families whose socioeconomic status is similar to that of the child's biological parents. Therefore, one cannot rule out the possibility that what looks like the effect of genetic similarity may actually be the product of subtle environmental similarities.

The great advantage in studying behavior genetics of mice and fruit flies is the freedom to carry out carefully designed experiments that control for environmental influences. For example, if one wants to test the hypothesis that the differences between two species are in part the products of genetic differences, it may be possible to create hybrids and to compare their behavior with that of the parental species. William Dilger performed such an experiment with two closely related parrots [181]. One of these, *Aga-*

4 **Lovebird, *Agapornis roseicollis.*** The parrot holds a strip of paper it has just cut (*left*). The bird is tucking the nest material into its flank feathers prior to flying back to its nest (*right*). Photographs courtesy of William C. Dilger.

pornis personata, builds its nest by gathering bits of bark in its beak and carrying them back to a nest site. The other species, *A. roseicollis,* has the bizarre habit of transporting nest material by tucking it into the feathers on its flank (Figure 4). By keeping females of one lovebird in cages with males of the other, Dilger produced some hybrids that exhibited a remarkable intermediate pattern. These birds would pick up material in their beaks, place it in their flank feathers, remove it again, and repeat the pattern over and over again before finally flying back to the nest with the paper either held in their beaks or in their feathers. Because the hybrids were reared in the same laboratory setting as the parental generation, the difference in the way they transported nest material can be safely attributed to genetic differences between hybrid and parental individuals.

This result in itself does not tell us a great deal about the nature of the relationship between genes and nest-building behavior in lovebirds. For example, the hybrids refuse to mate with either parental line and so one cannot test whether one or many genes are involved in this behavioral difference between species. Some hybrid studies with other species, however, have shown that certain behavioral differences are due to multiple genetic differences [61]. But research on hybrid animals is not the only tool of behavior geneticists. Various other experimental techniques have led to the following results:

1. Confirmation that single allelic differences can cause behavioral differences between individuals.
2. Experimental demonstration that (artificial) selection for certain behavioral traits can be highly effective in altering the behavior of a population over time.
3. Discovery of some physiological effects of genetic differences between individuals that are responsible for their distinctive behavioral characteristics.
4. Finding that the differences in the genetic and physiological characteristics of populations of the same species may be related to variation in the ecological pressures operating in different areas.

Let us examine some examples of each of these important contributions of behavior genetics, contributions that were made by studying nonhuman animals.

Single Gene Effects on Behavior

In honeybees there is variation among hives in the ability of the worker bees to open capped cells that contain diseased pupae and to remove them from the hive. The members of some colonies will perform this task; the residents of others will not. W. C. Rothenbuhler crossed queens from one type with drones from the other kind of colony and then performed some additional crosses with the "hybrid" generation. These experiments revealed that each behavior pattern was under the control of separate genes called *U* and *R*. Workers endowed with two copies of the recessive allele of each gene (*uu rr*) would both uncap cells and remove the pupae from the hive. But workers with one or two copies of the dominant form of each gene (e.g., *Ur Rr*) would do neither. Through his mating experiments, Rothenbuhler produced some bees with the genotype (*Uu rr*). These bees failed to uncap cells with dead pupae, but they would dispose of the deceased offspring if Rothenbuhler lifted off the wax cell caps for them [617].

This is one of the most famous experiments in behavior genetics, in part because it deals with behavior patterns of clear adaptive value and in part because it shows that the development of complex behavioral traits may be influenced by the presence of a single allele in an animal's genotype. There have been other demonstrations of single gene effects, often using laboratory mice and the geneticist's favorite insect, the fruit fly *Drosophila* [202]. For example, if one looks at sufficiently large numbers of fruit flies, particularly if their parents have been exposed to substances that cause mutations, one can sometimes find behavioral mutants. By performing the appropriate breeding experiments, some behavioral abnormalities have been traced to the alteration of a single gene [63]. There are specific mutations that cause a fly to move away from light instead of toward light as normal flies do, or to hold its wings above its body instead of flat over the abdomen. Other types of mutant flies have earned the titles "stuck" (males fail to dismount after the normal 20 minutes of copulation), "coitus interruptus" (the mutant males disengage after just 10 minutes of copulating), and "tko" (when the mutant fly experiences a mechanical jolt, it falls over, twitches, and then goes into a coma for a few minutes before recovering). Still another mutation causes a condition that has been labeled "amnesiac" because flies with the mutant allele can learn to avoid a specific odor that is associated with an electrical shock and 45 minutes later they have forgotten what they learned. In contrast, individuals that lack this allele and that have experienced shocks in conjunction with a distinctive scent will not move into a tube with this odor for hours (Figure 5) [594].

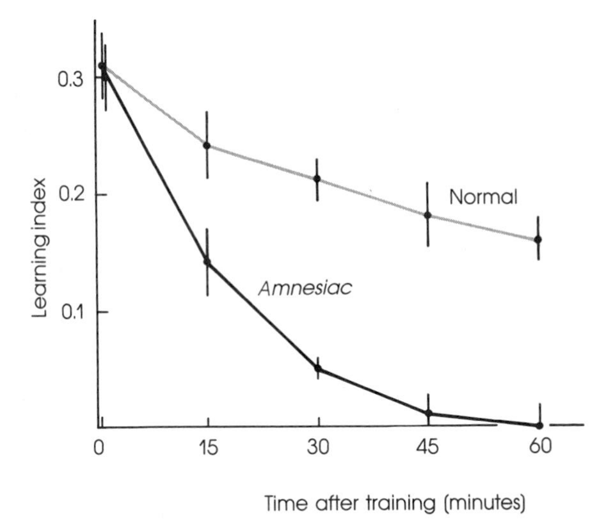

5 **Differences in memory retention between populations of normal (gray line) and *Amnesiac* flies.** Both groups of flies, which differ by a single allele, are able to learn to avoid tubes with certain odors associated with electric shock. But the *Amnesiac* flies fail to retain what they have learned as well as the normal flies.

One of the earliest behavioral mutants to be studied is called "yellow" because the mutant alleles (*y*-*y*) produce a fly with an abnormally yellow body. In addition (*y*-*y*) males behave in unusual ways not obviously related to this change in body color. When they are placed in a cage containing males and females with the same genotype except for this allele, they have poor mating success relative to the nonmutant males [47]. Because visual cues are not that important in fruit fly mating (the effect holds if the flies are tested in complete darkness), the yellow body pigment cannot be responsible for lowered breeding success by the (*y*-*y*) males. Careful observation of these males revealed that they vibrated their wings more slowly in the courtship sequence than the normal males. This made them less attractive to females and substantially lowered their reproductive success. This example illustrates an important point. An allele may and usually does contribute to the development of more than one characteristic of an individual (this is called PLEIOTROPY; Figure 6). A gene does *not* make a trait; typically it codes for an enzyme that catalyzes a particular reaction.

In the case at hand, individuals with (*y*) enzymes develop yellow bodies (the enzyme evidently contributes to the formation or deposition of a pigment in the exoskeleton of the fly) and slower wingbeat in courtship (the enzyme may play a role in the construction of abnormal neural circuitry responsible for wingbeat control or the enzyme may disrupt normal muscle development—two of many possible causes of the behavioral effect of the mutation).

Pleiotropic effects are the rule, not the exception, in the expression of a gene. To take another example from fruit flies, individuals with the allele *Hk* ("hyperkinetic") (1) are more active than flies without the allele, (2) exhibit rapid leg-movement when under anesthesia, (3) have a shorter life span, (4) engage in aberrant mating behavior, and (5) have a violent jumping reaction to shadows passed over them. Moreover, special genetic techniques have shown that the (*Hk*) allele has effects both in the region of the thoracic ganglia and also on the brain of the fly [297]. Therefore, when one considers the fitness effects of an allele, one must take into account the possibility that the allele will have an entire battery of developmental consequences; one or a few effects may be major, many others minor. The existence of pleiotropic effects, which may vary depending on the environment of the individual and the other genes in its genotype, greatly complicates making predictions about the genetic consequences of selection. An allele that tends to have one major positive effect on personal fitness may have a number of other negative effects that are of variable magnitude and that are difficult to identify but that may have considerable influence on the propagation of that allele over time.

Polygeny

Another point that requires special emphasis is that dramatic, single-gene effects are rare. This is because most characteristics, especially behavioral ones, stem from the integrated action of a large number of gene products. The technical term for the involvement of many genes in the development of a single character is POLYGENY (Figure 6). When a single gene effect does occur, it does not mean that there is one gene for the courtship song of a fruit fly, another for uncapping a cell by a honeybee, and still another for the removal of a dead pupa from the hive. Almost certainly single gene effects on behavior occur because one gene's product can play a key role in a complex developmental process that requires the regulated interaction of dozens or thousands of genes. A mutant allele's effects may be analogous to giving a worker on a long assembly line an improper tool. Although many persons participate in building a car engine, if just one worker omits or damages even a single component of the engine, the performance of the machine may be seriously impaired. (Note that most single-gene effects are deleterious and that the responsible mutation would normally be quickly eliminated by selection in nature.)

Some genes are known to play a major role in regulating development by coordinating the activity of batteries of other (structural) genes whose

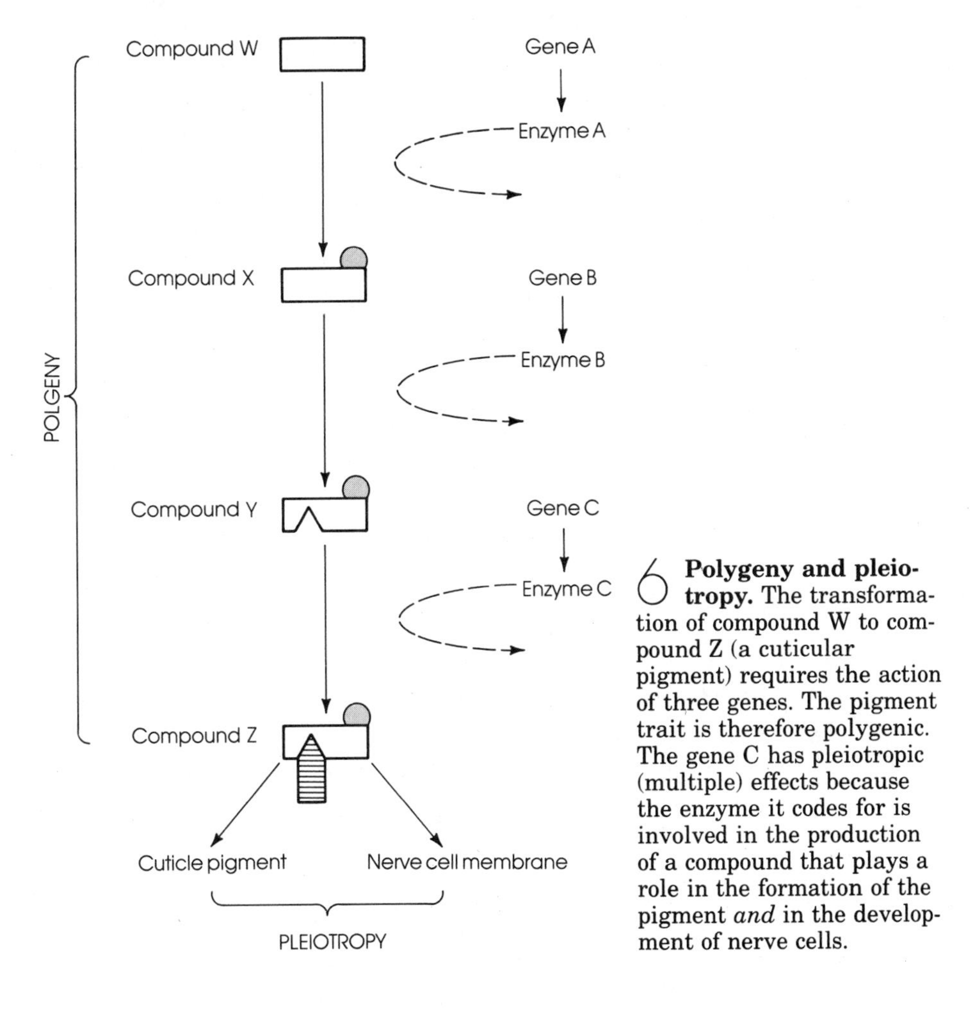

6 **Polygeny and pleiotropy.** The transformation of compound W to compound Z (a cuticular pigment) requires the action of three genes. The pigment trait is therefore polygenic. The gene C has pleiotropic (multiple) effects because the enzyme it codes for is involved in the production of a compound that plays a role in the formation of the pigment *and* in the development of nerve cells.

job it is to produce enzymes when this is appropriate. For example, humans and chimpanzees have very similar chromosomes (Figure 7) and produce many, perhaps even most, of the same kinds of enzymatic proteins. Of a sample of 44 proteins tested, all 44 were found in the cells of both species and most had the identical or a very similar sequence of amino acid building blocks [380]. If a research scientist from the University of Mars had only these data, he might conclude that chimps and humans should be classified as the same species or, at the very least, as exceptionally closely related species. Most of us would dispute this claim, pointing indignantly to the differences in appearance and behavior that distinguish us from chimpanzees. Although our cells have many of the same enzymes at their disposal, these enzymatic tools are used in very different ways to construct the distinctive nervous and structural systems for each species that underlie our separate behavioral abilities. Chimps and humans almost surely have

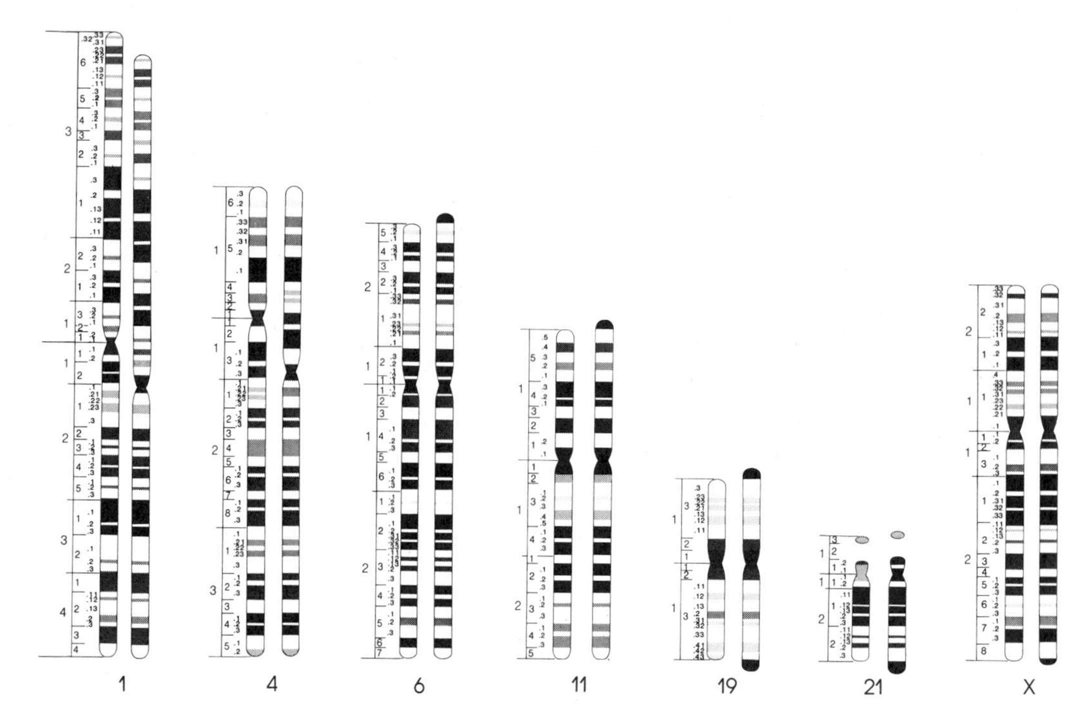

7 **Human and chimpanzee chromosomes.** The diagram shows the banding patterns of a selected sample of chromosomes present in humans (the right hand member of each pair) and our closest relative, the chimpanzee. Courtesy of Jorge Yunis.

different regulatory genes that control when and where in our developing cells certain enzymes will be produced [380]. Mutations in regulatory genes should almost always have significant effects on development by affecting the timing of enzyme production and the role played by groups of enzymes in cell metabolism. Single-gene effects involving a regulatory gene may be analogous to changing the foreman of an assembly line, thus altering the rules governing the activity of all the workers on the line.

In any case, the fact that a change in just one gene can produce a specific behavioral change suggests that some, if not all, behavioral traits may be accessible to natural selection. If a mutant gene produces a variation in behavioral ability that confers a reproductive advantage on its possessor (rare though this may be), the genetic basis for the change could quickly spread throughout the population.

Artificial Selection

That behavior can evolve has been repeatedly demonstrated in the artificial selection experiments of behavior geneticists [202]. If individuals in a population exhibit variation in a behavioral character, then it is possible to test whether the variation has a genetic component. This is done experimentally by permitting individuals with certain attributes to interbreed (raising their fitness) and preventing others from breeding (reducing their fitness to zero). By rearing each generation in identical environmental conditions, one can show that a trend toward the exaggeration of a particular ability must stem from a concentration of genetic factors contributing to the trait and not because of changing environmental factors. In the process of domestication, artificial selection for behavioral responses has produced homing pigeons, house dogs that are especially gentle with children, watchdogs that bark and bite readily, and so on. The often spectacular changes induced over a relatively short period by humans intent on breeding a safer pet or superior race horse is powerful evidence that natural selection can also lead to evolutionary change, a point that Darwin relied on heavily in the development of his theory [160].

Artificial selection by humans can occur unintentionally, as the U.S. Department of Agriculture demonstrated in its attempts to develop a biological control program against the screwworm fly [116]. (The flies lay their eggs in the wounds of cattle and the larvae feed on the flesh of the animals.) The USDA raised literally billions of sterilized flies and then released them to flood natural populations with the sterile individuals. The eggs of a wild female inseminated by a sterile male will not hatch. In order to rear huge numbers of flies in a small space, many sheets of paper were hung in the rearing rooms to provide more surface area on which the flies could rest. This created a maze of obstacles to flight and evidently acted as a selective force favoring flies that walked to wherever they wanted to go. In addition, the temperature in the rearing rooms were kept high to speed development of the larvae. There were correlated genetic changes in the population, especially with respect to one gene controlling the production of an enzyme important in the metabolism of the flight muscles of the fly. The allele of the gene present in wild flies was rapidly replaced by another form among the reared flies. Flies with this new allele produced a variant muscle enzyme that worked well at high temperatures but less well at field temperatures. The reared flies were incapable of prolonged rapid flight under natural conditions and were probably at a competitive disadvantage among wild flies, whose males must fly many miles to find widely scattered receptive females. Thus, accidentally, the USDA selected for a strain of flies that reduced the effectiveness of their control program.

Experiments on Artificial Selection

Geneticists have performed many carefully controlled experiments in which they have deliberately attempted to select for or against a particular behavioral attribute. They usually succeed. For example, a pioneer in behavior genetics, R. C. Tryon, conducted a long series of selection experiments on laboratory rat behavior. His most famous exercise consisted of an attempt to select for maze-running ability in the animals [721]. First, he tested a diverse population of rats and divided them on the basis of their performance in a maze into three groups: (1) "maze-bright" rats, that is, those that learned quickly the characteristics of the maze during a number of initial trials and therefore ran through the maze with few detours into dead-end alleys; (2) "maze-dull" rats, that is, those that were the opposite—slow to learn the maze; and (3) an intermediate group, which was discarded.

After identifying his groups, Tryon permitted only bright males to breed with bright females and allowed dull males access only to dull females. The second generation produced by these crosses was tested in the same maze. The progeny of bright rats ran the maze with significantly fewer errors than the offspring of dull rats. (Each was given the same number of learning trials.) In fact, the average performance of both groups of offspring was more extreme than that of their parents (Figure 8). Tryon repeated the procedure over subsequent generations. By interbreeding the very best of the bright line and the very dullest of the dull line, he was able to produce brighter and brighter rats as well as duller and duller ones. This is evidence that among the rats in the original population there were heritable differences related to maze-running. The breeding program concentrated the alleles that promoted this activity in one group and different alleles in the other group.

Tryon called his lines "maze-bright" and "maze-dull," terms that imply that he was selecting for intelligence and stupidity in laboratory rats. However, when the same rats were tested in another kind of maze, which emphasized visual cues (psychologists have invented a wide variety of mazes), the bright rats performed no better than the dull ones [644]. This result suggests that Tryon was not necessarily selecting for degrees of intelligence in his animals. Nor was he creating a population of brilliant maze runners and others deficient in this ability. Instead, he was selecting animals with traits that facilitated learning how to get to the end of the special type of maze he used for his tests.

The specificity of selection has been illustrated over and over again with artificial selection experiments with fruit flies. It is possible to select for flies that are unusually socially tolerant and for those that will kick and butt their neighbors [230]; for flies that will move toward light and for those that move away from it [202]; for flies that will mate quickly when the sexes are brought together and for those that are unusually slow to mate [461]. In the species *Drosophila melanogaster* (there are vast numbers of species of fruit flies), females typically mate once a few days after they have reached adulthood and then once more five to six days later. Donald Pyle and Mark Gromko took a population of 75 females that had mated once when they were three days old and on each day thereafter for five days; they put the females, one to a vial, in a container with two males for a two-hour period. They checked at intervals to see if any female had mated. The first ten females that copulated a second time were selected, and their progeny made up the next generation of flies. A group of 75 females were taken from this next generation and permitted to mate when they were three days old. These daughters of quick remating mothers were then given

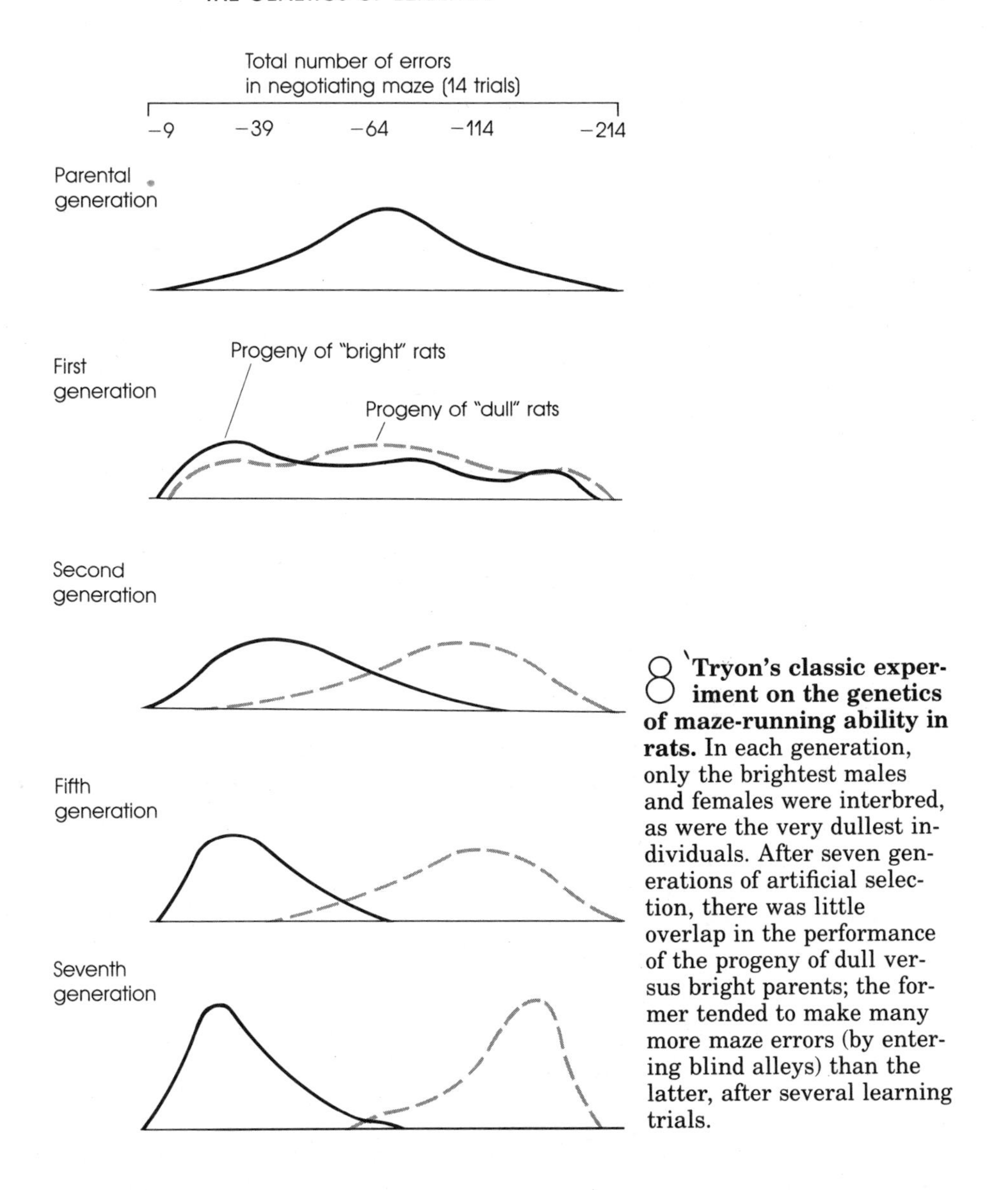

8 Tryon's classic experiment on the genetics of maze-running ability in rats. In each generation, only the brightest males and females were interbred, as were the very dullest individuals. After seven generations of artificial selection, there was little overlap in the performance of the progeny of dull versus bright parents; the former tended to make many more maze errors (by entering blind alleys) than the latter, after several learning trials.

daily opportunities to copulate a second time. The first ten females to accomplish the second mating were used as breeding stock for the third generation and so on.

The results of the selection experiment were clear-cut [592]. The average number of days to the second mating gradually fell from about 5 on the first few generations to about 2 days by generation 10 and to 1.5 days by generation 20. Another way to demonstrate the impact of artificial selection on the time to remate is to examine the percentage of females that had copulated again by a particular day after the initial mating. After ten

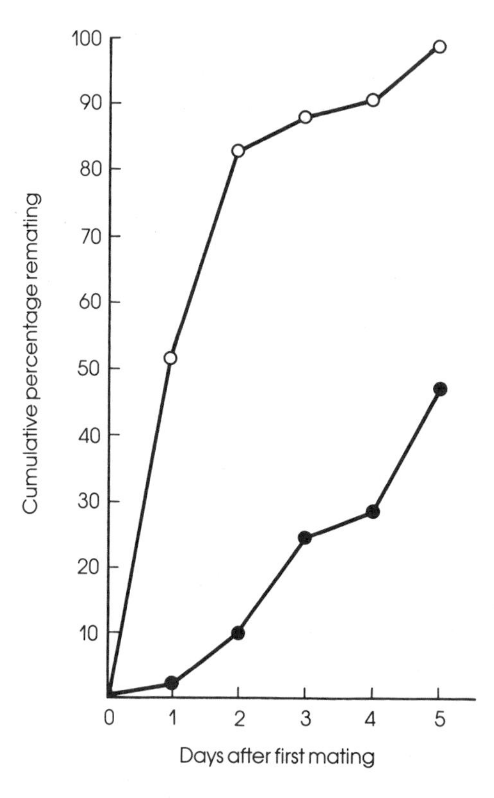

9 Effects of artificial selection. After ten generations of selection for rapid remating, most fruit fly females have copulated again two days after the first mating (open circles). Females in the unselected control line are much less likely to have remated by this time (solid circles).

generations of selection, about 80 percent of the females had copulated a second time by day 2 whereas only 10 percent of the control line remated this quickly (Figure 9).

The experiment shows that there was genetic variation in the founding population that influenced how quickly a female became receptive again after once mating. Artificial selection favored those alleles that contributed to rapid remating, changing their frequency in the population from low to high over time, and changing the average response of individuals in the selected population with respect to acceptance of another male. There is no reason to assume that natural selection does not do exactly the same thing—produce populations of animals with specific behavior patterns, tendencies, and capabilities that enhance individual reproductive success.

Genes, Physiological Mechanisms, and Behavioral Differences

The evolution of behavior requires that genetic changes induce behavioral changes in populations. But genetic information is only very indirectly linked to the development of a behavior pattern because, as we noted, genes do not themselves code for particular behavioral traits. Instead, each gene codes for a different protein whose production may influence the development of an individual's structural and physiological features. If we are to understand the genetics of behavior more fully, we must discover something about the genetic information and the physiological attributes of an animal that enable it to behave in well-defined ways.

Tracing the link between genes and physiology and human behavior is an unusually difficult task because of the complexity of the human brain and the inability of researchers to perform controlled genetic experiments with human subjects. There has been, however, some progress made from studies of the physiological effects of naturally occurring mutations. There are now known to be over 150 metabolic disorders caused by single-gene defects in humans [86]. In many of these, the mutant allele produces a defective enzyme incapable of breaking down a particular substance found in nerve cells. As a result, the substance reaches unusually high concentrations and disrupts the normal functioning of the cells. The unfortunate individual with these abnormal cells may become mentally retarded or may suffer progressive neural degeneration and death, even though he or she is missing just one normal enzyme.

By working with mutants of other species, additional insights have been gained into the gene–physiology connection. A traditional way to tackle difficult biological questions has been to look first at a relatively simple system and then use the results of this research as a guide to understanding more complex cases. With this goal in mind, work is proceeding on the genetic basis of the behavior of single-celled organisms, especially bacteria [386] and paramecia [197, 399]. Protozoans have some interesting and sophisticated behavioral abilities. For example, a bacterium is able to track concentration gradients of many substances, moving toward useful resources and away from toxic materials. A paramecium is a thoroughly competent swimmer, able to maneuver rapidly through its watery universe. One of the rules that organizes the swimming behavior of a paramecium is, if you bump into something, back up for a short time and then start forward again. When the anterior end of the animal collides with an obstacle, the cilia that cover its body reverse their beating stroke for a few seconds. The animal backs up during this time before the cilia resume their forward drive stroke and the paramecium starts off again. Usually its orientation has changed somewhat, and it moves past the obstacle that stopped it before. If not, it switches into reverse and repeats its avoidance response (Figure 10).

By screening large numbers of paramecia, observers have found some mutant individuals of various sorts, among them the slow-swimming "sluggish," the extremely rapid swimming "fast," and two avoidance mutants, "paranoiac" (which swim backward for a much longer time than is normal) and "pawn" (which is named for the chess piece that is forbidden to back up at all). The appropriate set of genetic crosses revealed that each of these behavioral mutants was linked to its own single-gene mutation.

The normal avoidance response is mediated by an effect of a tactile stimulus on the electrical charge gradient across the membrane that surrounds the paramecium. It is possible to place minute recording wires within and without an intact paramecium and measure the charge differ-

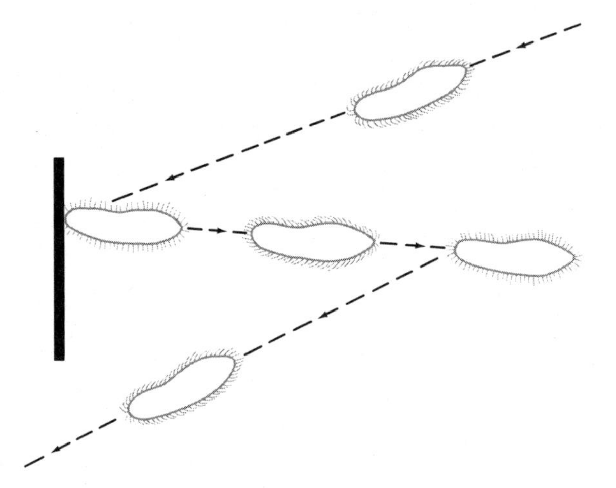

10 **Avoidance response of the paramecium.** When the anterior end of the organism collides with an obstacle in its path, the beating stroke of its cilia is reversed for a short period. The cilia then resume their forward stroke.

ential across the membrane (the membrane potential). This tiny, but significant, charge difference is caused by the way charged particles (ions) are distributed inside and outside the membrane. When the paramecium touches something with its anterior end, this normally changes the permeability of the membrane so that calcium ions in the water can enter the cell. As the positively charged calcium ions enter, they change the membrane potential of the protozoan. In some unknown fashion this causes the cilia to reverse their beating stroke and the animal backs up. Within the cell, enzyme systems begin expelling the calcium ions almost at once; and when the original membrane potential is achieved (after a second or two), the cilia start beating in the forward position and the animal moves forward.

We are now in a position to understand the effect of the "pawn" mutation. In animals with this mutant gene, the membrane does not respond to a tactile stimulus in the normal fashion. There is no influx of calcium ions and thus no membrane signal to the cilia. The protein defect caused by the "pawn" allele may interfere with proper membrane construction in a number of ways. For example, there may be something wrong with an enzyme that helps change the permeability of the membrane to calcium ions; or the components of the membrane may be put together in an abnormal fashion, a condition that makes it incapable of a response. (You should be able to make an educated guess about the nature of the enzymatic defect caused by the "paranoiac" mutation.)

Genes and Nervous Systems

Comparative physiological studies of mutant and normal individuals are also being pursued with animals that have nervous systems [297], animals including *Drosophila* fruit flies. Seymour Benzer and his colleagues, whose screening techniques revealed the fruit fly mutants "stuck" and "coitus interruptus," have gone on to explore the physiological bases of certain mutant behaviors [63]. By examining the flight muscles of flies that held their wings in an unusual position, Benzer discovered that these muscles have an aberrant structure that was apparently responsible for the altered behavior of the flies. The mutant gene possessed by these individuals had interfered with the normal development of just those muscles controlling their wings. As a result, the animals could not fly but could walk about perfectly normally. Other mutant individuals have been shown to possess abnormal or degenerate elements within their nervous systems. One particular mutation has been called "shaker" because when a fly with this allele is anesthetized it will continue to move its legs rhythmically, unlike normal flies. The nerve cells controlling the legs have been shown to produce abnormal signals. Each message has longer lasting effects than normal, a condition that results in hyperstimulation of the muscles linked with these neurons [297].

A different technique for pinpointing the physiological substrate of a particular behavior comes from studies of GENETIC MOSAIC flies [63]. There is a procedure for experimentally producing flies that are combinations of genetically different cells (Figure 11). One can create mosaics that consist partly of cells with two X chromosomes and partly of cells with just one. Cells with two X chromosomes develop into female tissue and cells with a single X chromosome develop into male tissue. Using sophisticated breeding experiments that produce flies with male and female tissues of different colors, persons working in Benzer's laboratory can tell which parts of a fly have one or two X chromosomes. They then permit different kinds of mosaic flies to interact with one another to determine the effects of the genetically different male and female tissues on the reproductive abilities of the flies.

A key discovery was that a fly whose upper brain lies within a zone of male tissue will pursue any other fly whose posterior abdomen is chromosomally female [296, 345]. The pursuer will also initiate courtship with the wing-waving phase. This behavior will occur even if the rest of the courting fly is composed of two X chromosome tissue so that it has female antennae, eyes, wings, and genitalia. These results show that there are no overriding hormonal influences from the female's reproductive tract that regulate behavioral decisions made by the brain. Moreover, female antennae, like male antennae, must be capable of detecting the key olfactory cues that receptive females provide to attract males. But the female's brain must operate differently from that of the male, because only brains whose nerve cells carry a single X chromosome will order the body to move after a female and wave its wings in courtship. Thus, through mosaic fly studies one can localize the specific effects on the development of the nervous system that result from having one or two X chromosomes.

Genes and Geographic Variation in Behavior

Although studies of mutant and mosaic flies have helped identify the physiological effects of genetic differences among individuals, there is a question about how much we can learn about the normal genetic influences from the studies of abnormal individuals who would not exist in the natural environment. Work of this sort is a bit like throwing a wrench at random at a

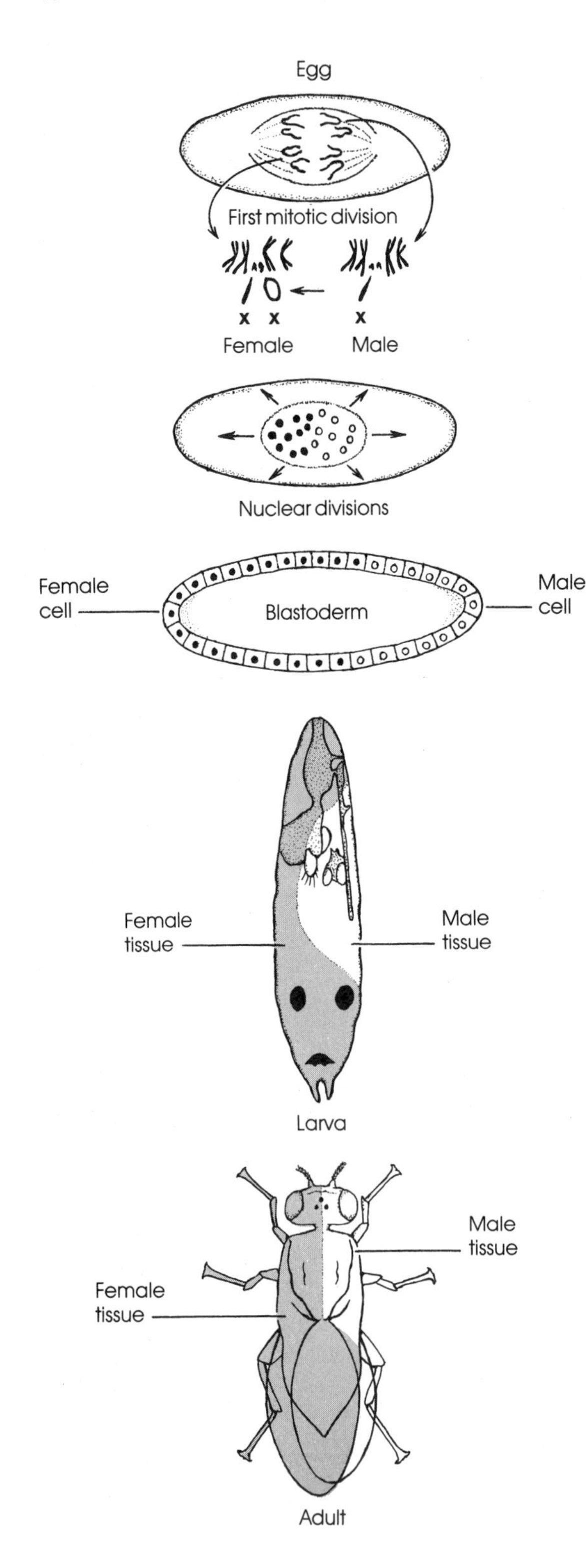

11 **Genetic mosaic fly.** It is possible to produce a fruit fly whose body is composed of both male and female cells. The genotype of some eggs is such that upon an early cell division, one of the daughter cells produced has only one X chromosome. As it divides further, this cell gives rise to a line of genetically male descendant cells. Any cell that retains its two X chromosome complement is the founder of a line of cells that are genetically female. With the appropriate techniques, male and female cells can be identified in the adult fly.

delicate engine and then attempting to deduce from the severely damaged machine how an unviolated engine works. An approach that avoids this complaint is to examine naturally occurring genetic variation that leads to adaptive differences in the behavior of animals. Stevan Arnold has conducted a beautiful piece of research of this sort, a study that makes evolutionary sense of the genetic, physiological, and behavioral differences between two populations of a garter snake (*Thamnophis elegans*) [27]. The snake occurs over much of western North America in a wide variety of habitats including both foggy, wet, coastal California and the drier, elevated, inland areas of that state. There are marked differences in the diets of the populations of snakes living in the two areas (referred to hereafter as "coastal" and "inland"). The coastal snakes search about in humid areas where they find their major prey, slugs. Their ability to consume these creatures will arouse either bewildered admiration or disgust in anyone who has handled these repulsively slimy mollusks (Figure 12). Slugs do not live in inland northern California, and not surprisingly the inland snakes find something else to eat, primarily fish and frogs, which they capture while swimming in lakes and streams. But will snakes from inland locations eat slugs if they are given the opportunity?

One could present slugs to adult snakes from the inland population, but if they failed to eat the prey this might be because their long feeding experience with fish and frogs had biased them against a strange food. The more interesting experiment would be to give newborn baby snakes from inland and coastal areas a chance to eat slugs, because this test would eliminate the possibility that prior feeding experience had shaped their

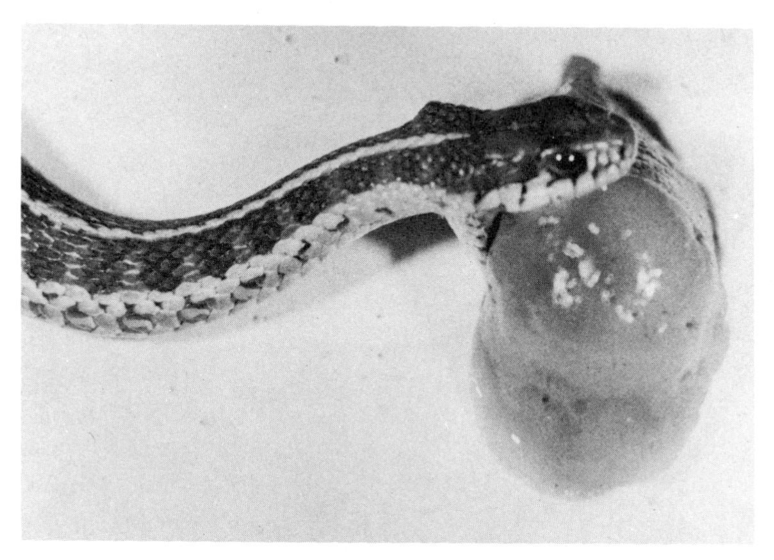

12 **Garter snake with slug prey.** Although slugs secrete mucus, garter snakes from coastal California consider them a preferred prey. Photograph by Stevan Arnold.

dietary preferences. Arnold conducted the more interesting experiment. He captured pregnant females from each location and took them to a laboratory. After they gave birth, he placed each young animal in a separate cage away from its littermates and mother to remove this possible environmental influence on its behavior. Some days later he offered each baby snake a chance to eat a small chunk of freshly thawed banana slug (*Ariolimax*) by placing it on the floor of the young snake's cage. Naive young coastal snakes usually ate all the slug hors d'oeuvres they received; inland snakes usually did not (Figure 13). In both populations, slug-refusing snakes did not even make contact with the slug food but ignored or avoided it completely. The response was very stable; animals that refused slugs as newborns almost always rejected them again when tested a year later, with no intervening experience with slugs.

To delve further into the basis for the variation among individuals in their response to slugs, Arnold took another group of isolated newborn snakes that had never fed on anything and offered them a chance to respond to the *odors* of different prey items. He took advantage of the readiness of newborn snakes to flick their tongues and even attack cotton swabs that have been dipped in fluids from some species of prey (Figure 14). By counting the number of tongue flicks that hit the swab during a one-minute trial, one can measure the relative stimulation provided by different odors to an inexperienced baby snake. (Snakes have an organ in the roof of their mouth that analyzes odors carried to it by the tongue. When the animal's tongue touches a source of chemicals, it carries some molecules back to the organ, which plays a role in prey recognition.)

Populations of inland and coastal snakes respond about the same to swabs dipped in toad tadpole solution (a prey of both groups) but react very

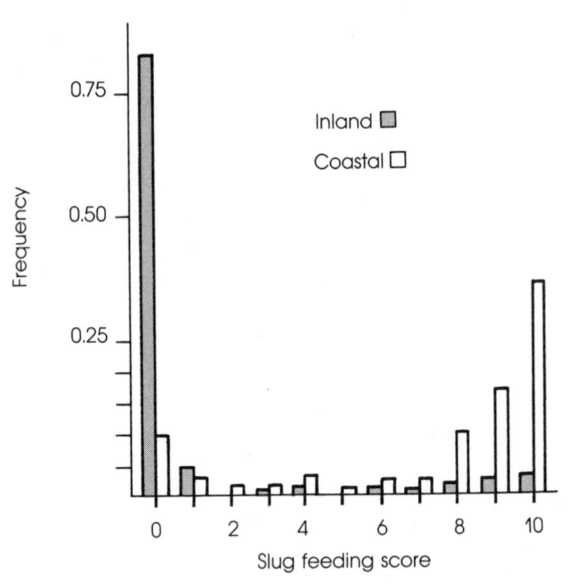

13 **Response of naive, newborn snakes to cubes of slug.** Snakes from an inland California population usually did not eat the cube on any of ten consecutive days on which a slug cube was offered to them (feeding score = 0). In contrast, most coastal snakes consumed the item on eight to ten days (feeding score = 8–10).

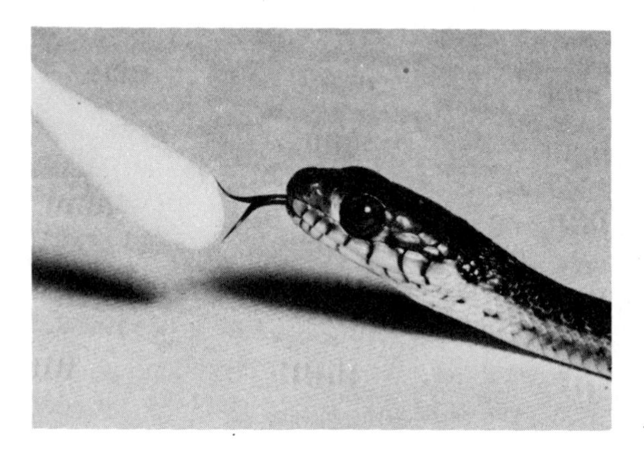

14 **Tongue-flicking** by a naive, newborn garter snake can be triggered by a cotton swab soaked in slug extract. Photograph by Stevan Arnold.

differently to swabs daubed with *Ariolimax* scent (Figure 15). Within each group there is some variation in response, but most inland snakes are unresponsive to slug odor whereas most coastal snakes react strongly to it. By comparing the tongue flick scores of siblings *within* each population, it is possible to determine what percentage of the variation among individuals is due to genetic differences and what percentage is traceable to environmental differences among them. In both populations only about 17 percent of the differences in chemoreceptive responsiveness to slug odor is due to

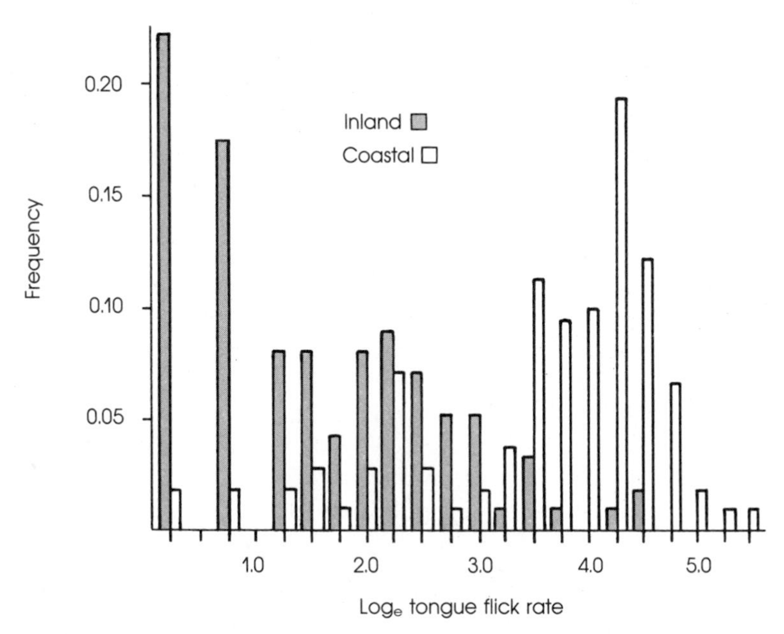

15 **Response to odor of slug** by naive, newborn garter snakes from two populations. Coastal snakes usually tongue-flicked much more frequently to this odor than did inland snakes.

genetic factors. This shows that almost all the genetic variation related to reactivity to slug odor has been eliminated within each population. Most coastal snakes have the allele or alleles that enable them to sense slugs and attack them. In contrast, snakes of the inland population have alternative alleles that result in a low rate of tongue flicking and a low attack probability when presented with a slug. If one crosses snakes from the two populations (Arnold did that, too), there is much more variation in the resulting group of offspring than in either parental population, although the majority of the snakes are slug-refusing. These results confirm that the differences *between* populations have a strong genetic component, and they also indicate that the allele or alleles that contribute to the development of slug-refusing are dominant to the alternative(s) that promote acceptance of slugs as food.

A reasonable scenario for the evolution of the feeding differences between inland and coastal populations is that among the original colonizers of the coastal habitat were a very few individuals that carried the then rare allele(s) for slug-accepting. (There are reasons for thinking that coastal California was inhabited by *T. elegans* more recently than inland western North America.) These slug-eating individuals were able to take advantage of an abundant food resource in the new habitat. If, as a result, their reproductive success was as little as 1 percent higher than that of their slug-rejecting fellows, the coastal population could have reached its present state of divergence from the inland population in less than 10,000 years.

It is easy to imagine why slug-accepting alleles might enjoy an advantage in coastal populations. But why have they been actively selected against (and nearly eliminated) from inland populations? Arnold had shown that the alleles that promote slug-eating also enhance the acceptance of aquatic leeches. These blood-sucking animals are absent in coastal California but plentiful in inland lakes. Although it has not yet been fully proved, it is possible that snakes that try to eat leeches (which slug-acceptors will do) may be damaged by their "victims." Leeches can live even after they have been swallowed by a garter snake, and if they attach themselves to the wall of the snake's digestive tract, they might injure their consumer seriously. Even if this happens only rarely, it would exert a negative selection pressure on snakes with leech-accepting alleles. Because these alleles also lead to the development of the ability to detect and attack slugs, the ability to eat slugs would be disadvantageous in inland populations.

Thus, the geographic differences in the feeding behavior of the snakes can be explained in terms of their genetic basis (different alleles predominate in the two populations), their physiological basis (the chemoreceptors that detect certain molecules associated with both slugs and leeches are more prevalent in coastal populations), and ecological basis (inland snakes must contend with potentially dangerous leeches whereas coastal snakes are exposed only to edible slugs). This case study shows how behavior can evolve both at the genetic and physiological level in response to ecological differences that affect the fitness of individuals with different behavioral abilities [27]. Future research in behavior genetics that integrates genetics, physiology, and ecology will lead to a much improved understanding of the interplay between the proximate and ultimate causes of behavior.

Summary

1 The behavioral differences among individuals may be the result of genetic and/or environmental differences among them. Behavior genetics research suggests that numerous behavioral differences among humans have a genetic component. The same is true for many other animal species.
2 To say that a particular allele contributes to the development of a behavioral characteristic is not to say that the trait is "genetically determined." The statement, "There is an allele for IQ score or brood cell opening or time until acceptance of a second mate," is shorthand for the following: "The presence of a particular allele in an individual's genotype provides information for the production of a distinctive protein whose contribution to the chemical reactions within cells may influence the development of the physiological foundation for a behavioral ability."
3 Because behavioral differences among individuals may be caused by a single genetic difference and because artificial selection experiments are so often effective in altering the behavior of laboratory populations, there is every reason to believe that behavior can rapidly evolve under natural conditions.
4 A study of wild garter snakes in two geographically separated locations confirms this hypothesis. The differences in feeding behavior in the two populations stems in part from their genetic differences. The distinctive genetic features of the two populations affect the development of the chemoreceptive system of the snakes, which in turn affects their perception of prey. Individual selection has favored different perceptual and behavioral attributes in the two areas because the ecological pressures in the two areas differ. This affects the relative reproductive success of snakes with and without the ability to detect and attack slugs.

Suggested Reading

William Dilger's study of the behavior of hybrid lovebirds is an excellent research story [181]. Stevan Arnold's comprehensive study on the genetics, physiology, and ecology of garter snake feeding behavior [27] should be read by everyone interested in behavior genetics. The field of behavior genetics is reviewed in books by Lee Ehrman and Peter Parsons [202], and by Jeffrey Hall, Ralph Greenspan, and William Harris [297].

Suggested Films

DNA, Blueprint of Life. Color, 18 minutes. A highly professional film showing how genes code for proteins with reference to species' differences in genetic blueprints for development.

The Fruit Fly: A Look at Behavior Biology. Color, 21 minutes. A film that shows behavior genetics research in Seymour Benzer's laboratory with the classic animal of genetics, the fruit fly.

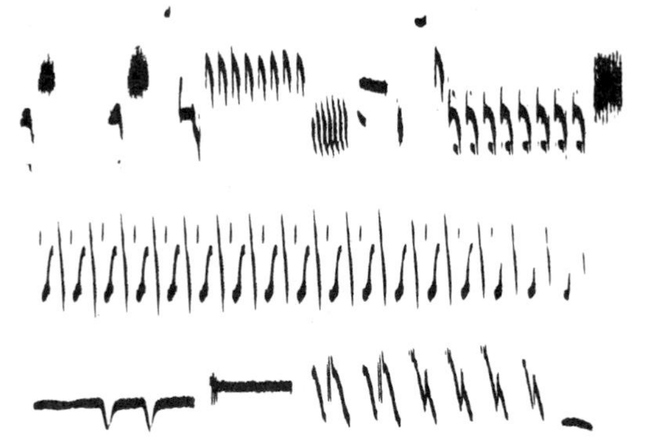

The preceding chapter established that genetic differences among individuals can contribute to differences in their physiological and behavioral attributes. But we left unexamined the problem of precisely how the physiological and behavioral characteristics of an individual develop as the animal grows and matures. This is a problem that requires an analysis of the complex interactions between a growing organism, with its battery of genetic information and developing structures, and the environment in which it is located. Unfortunately, few areas in biology are more poorly understood than the developmental process, because it is so staggeringly complicated. The song of an adult male white-crowned sparrow, for example, is the product of the integrated action of hundreds of millions of nerve and muscle cells. Tracing the pathway followed by even one of these cells as it developed from a single fertilized egg would be a monumental achievement, and yet the accomplishment would barely begin to contribute to an understanding of the totality of developmental events that are the basis for the bird's song.

Despite the difficulties of this research, some useful things have been learned about the interactions between genetic systems and the environment during the development of behavior. Here we shall center our attention on how the differences between the sexes, which are often dramatic and usually biologically significant, come about in some species. We shall define key environmental influences on behavioral development of the sexes and at the same time show the resiliency of the process in overcoming some environmental deficits. This apparent paradox—the sensitivity of behavioral development to some experiential influences *and* its capacity to achieve an adaptive end point in a broad range of different environments—will be the focus of the concluding section of the chapter.

CHAPTER 3

The Development of Behavior

The Development of Sexual Differences in Behavior

Because white rats (the domesticated laboratory variant of wild Norway rats) breed prolifically in the lab, can be kept in small spaces, and have relatively short generation times, they have been used extensively in experimental research in psychology and physiology. Thanks to their sacrifices, much is known about all aspects of their biology, including the development of their behavior.

Not surprisingly, adult male and female rats behave differently. Females, for example, have a specific copulatory position that they assume at the proper time in the ovulatory cycle (when mature eggs are ready to be fertilized) (Figure 1). This position is given in response to stimulation provided by a sexually active male, who has the capacity to mount a female, insert his penis into the vagina, and perform the copulatory response leading to ejaculation. Males almost never assume the female copulatory position; females relatively rarely mount other rats and do not exhibit the full male copulatory performance [52].

Females also differ from males in their parental behavior. Immediately after giving birth, the mother rat remains in very close association with her young for several days, keeping them warm and permitting them to suckle when hungry. If her pups are moved off some distance, the female will quickly retrieve them. In contrast, male rats are rarely solicitous toward their progeny [612].

These differences between the sexes have a proximate basis in the hormonal makeup of male and female rats [417]. If one removes the ovaries from an adult female, she will no longer copulate *unless* she receives injections of estrogen and progesterone (two hormones produced by the ovaries). Likewise removal of the male testes eventually eliminates male copulatory behavior, unless one gives the castrated individual injections of testosterone (the major hormone produced by the testes).

But males and females that have had their hormone-producing, reproductive organs removed when they were adult generally cannot be induced

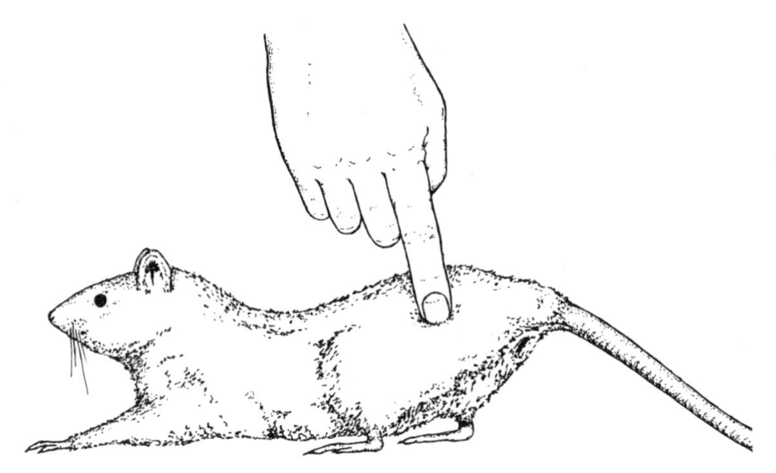

1 **Lordosis.** A female Norway rat adopts the precopulatory posture of a receptive individual when she is touched on her flank.

to behave like members of the opposite sex by giving them the hormones of the opposite sex. Injecting testosterone into a female whose ovaries were removed past puberty does not lead her to engage in male copulatory behavior with receptive females placed in her cage. This suggests that the nervous systems of adult males and females have differentiated in such a way that their brains are not capable of responding to hormonal signals of the opposite sex. In fact, if one examines certain regions of adult rat brains, one finds structural differences between males and females. For example, in the preoptic region of the brain, there are marked differences between males and females in both size and number of nerve cells. In addition, preoptic cells in the male become more active when testosterone is present, but there is no such response to the hormone in the similar region of the female's brain [457].

How do males and females come to develop brain tissue with different responsiveness to hormonal signals? This question has been explored experimentally by surgically removing the ovaries from a newborn female [417]. If one then injects testosterone into the animal, the hormone will be carried by the circulatory system to her developing brain. There it has effects that alter the course of neural development so that when the rat is an adult it behaves like a male when given additional injections of testosterone. (The experimental rat, although genetically female, has neither ovaries nor testes, but its brain has the capacity to respond to testosterone as a result of the early influences of the hormone on brain development.)

The corollary experiment is to extract the testes from a neonatal male rat. With the loss of the testes goes the ability to produce testosterone; and in the absence of this hormone, the rat will develop a brain with female characteristics. Regions of the adult's brain will respond to estrogen treatment when the castrated male becomes an adult, and he will adopt the female copulatory pose. His brain will not, however, respond to testosterone at this stage of development (Figure 2).

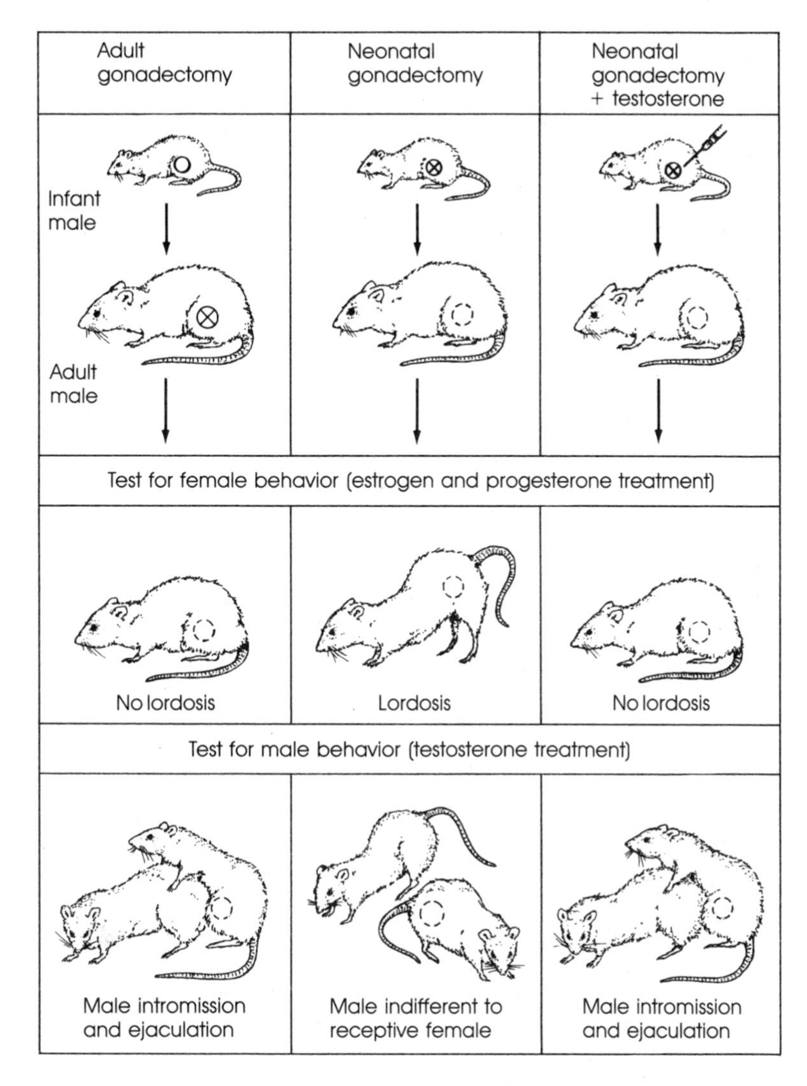

2 **Hormones play a key role** in the development of copulatory behavior in male rats. If testosterone is present in the infant male rat (see columns 1 and 3), the adult rat will develop the neuronal circuitry needed for performance of intromission and ejaculation.

The developmental effects of testosterone on the central nervous system can be demonstrated to take place in a brief CRITICAL PERIOD [457]. If one removes the gonads from a newborn rat and then waits for a week before injecting testosterone into the animal, it will be too late for the hormone to trigger the events that lead to development of a "male" brain. There is a short interval between 18 and 27 days after conception when the brain of the developing rat is sensitive to testosterone in the bloodstream. (A rat is born after a gestation period of 22–23 days.) Certain cells within the young brain have the capacity to bind with testosterone molecules. At this early stage both male and female brains possess these cells. If testosterone is present, either naturally because the animal is an intact male whose testes are producing the substance or unnaturally through experimental injection, the hormone will be drawn into its target cells. Once there the hormone is carried to the nucleus, where its presence affects the genetic activity of a nerve cell. New proteins that were not present before are produced. These molecules affect the biochemical activity within the cell and others in contact with it, influencing the course of cellular development. Thus, testosterone molecules in the bloodstream act as a trigger for a distinctive pulse of chemical activity within certain components of the immature brain [499].

The question then becomes, Where does the original testosterone come from? Until the fetal rat is about 16 days old, there is no testosterone in its circulatory system; and, indeed, the gonads of male and female fetuses prior to this stage are superficially identical in appearance. But at about the 16-day stage, dividing cells in the internal reproductive organs of the male take on a distinctive form. These developmental units of the testes are probably responsible for the production of testosterone, which is released into the bloodstream and carried to the brain, where it has its critical developmental effects shortly before and after the birth of the male [417].

The development of the testes and production of testosterone are also correlated with the inhibition of the development of the ovaries. In an embryonic mammal, cells that have the potential to become ovarian or testicular tissue are both present at a very early stage of development. But in a male embryo, the cells that give rise to the female reproductive organs will not undergo their program of multiplication and differentiation in the presence of the metabolic products of the early-stage testis. Fetal female rats do not develop testosterone-producing cells. As a result, male hormone is not available to initiate the male developmental pattern in the brain and gonads. In the absence of testosterone, the progenitor cells of ovaries will give rise to millions upon millions of additional cells, which form the mature ovaries. These structures produce estrogen and progesterone in specific patterns related to ovulation and so afford the hormonal basis for adult female reproductive behavior.

But if the key to the development of male or female behavior can be traced back to the presence or absence of a few testosterone-producing cells in a early-stage embryo, what is the reason for their existence in male embryos but not female ones? The bottom line is that males and females differ genetically from the moment of conception. In mammals, rats included, there are identifiable chromosomes whose presence or absence is critical for the sexual differences between males and females [312]. A male mammal typically has one X and one Y chromosome, whereas a female has two X chromosomes. Maleness arises from the presence of the Y chromosome. (Very rarely human beings are formed with only a single X sex chromosome. These individuals develop into females. There are also some rare XXY genotypes, which produce males with testes and other secondary masculine characters.)

The Y chromosome of some mammals has been shown to carry very little genetic information. But among the few genes located on the Y chromosome of humans are one or more that code for a specific protein (H-Y antigen) and one or more that regulate the production of this substance. The H-Y antigen in the human embryo is believed to play a central part in activating the development of the cells that give rise to the male testes [312]. Perhaps in rats, too, there is a similar substance that is produced in early embryonic cells of males and then is transported to other cells whose receptor molecules bond with the antigen. The receptor–antigen complex may migrate to the nucleus of the target cell, where its presence can activate previously

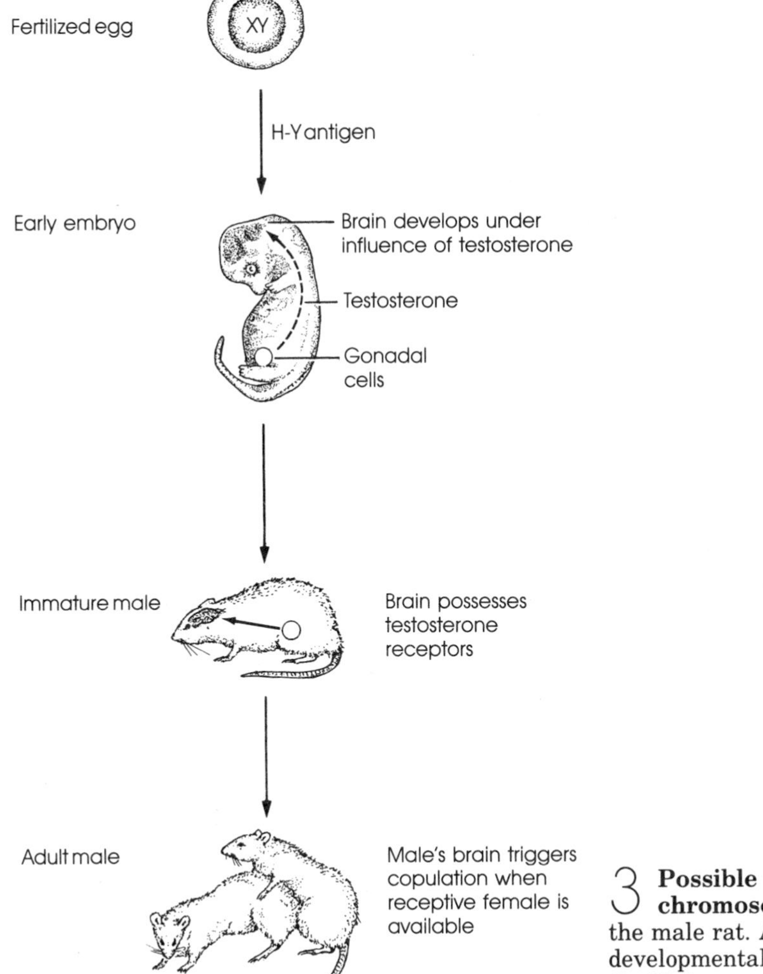

3 **Possible relation between the Y chromosome and behavior** of the male rat. A complex sequence of developmental events is initiated by the Y chromosome.

inactive genes. This will lead to new chemical reactions that trigger the orderly cell divisions and differentiations that take place as the male gonads develop. Some of the first gonadal cells produce testosterone, which signals additional changes in a host of target tissues including the brain. The continued production of testosterone by the mature testes later activates brain tissues that developed under the influence of embryonic testosterone. This enables the adult male to respond appropriately when confronted with a receptive female (Figure 3).

Hormones, Sensory Stimulation, and the Development of Maternal Behavior

The interaction between environment and genotype does not end with the birth of an animal, as can be shown through an examination of the development of maternal behavior in rats [612]. As noted already, a female rat treats her newborn pups in a distinctive manner, keeping them close to her body and retrieving them if they are displaced. These attributes normally first appear when the female is a sexually mature adult and has just given birth to a litter. But one can induce guarding and retrieval behavior in nonpregnant females, indeed even in sexually immature female rats, by repeatedly exposing the female to one- to two-day-old pups taken from other females. If the "sensitization" process is maintained for a week, many females will change from avoiding or even attacking strange pups to "adopting" them as their own.

The striking thing is that adoptive behavior in a nonpregnant female appears more quickly and reliably if she *cannot* smell (which can be arranged through surgical or chemical treatment). Females evidently have an initial aversion to baby rats that do not smell like themselves. If the female cannot detect the odor of the pups, this enables her brain to be positively stimulated by the sounds and tactile cues provided by neonatal rats; this stimulus leads to the appearance of maternal behavior in the female. The nonpregnant female's brain has certain properties that normally develop by the time she gives birth and that would be used in natural conditions to enable the mature female to direct her maternal care exclusively to her own pups. Baby rats provide stimulation that helps maintain maternal behavior in the early days of their lives. (If one removes pups from their mother for five days, provides them with a foster parent, and then returns the pups, a large proportion of the mothers will no longer respond maternally.) It is likely that the stimulation pups provide activates certain regions of the brain and induces hormonal changes that facilitate the maintenance of maternal care.

The role that hormones play in the development of female retrieval behavior has been studied by removing the ovaries from immature rats [612]. If they are later given "sensitization" training, they are only one-third as likely to retrieve foster pups as intact females of the same age that have gone through the same procedure. If, however, one injects the surgically treated females with an estrogen hormone, they respond to their foster pups in much the same way as sensitized nonpregnant females with ovaries.

This experiment establishes that the ovarian hormones circulating in the blood of nonpregnant females helps prepare them for eventual motherhood by promoting the development of behavior patterns that will assist them in the care of their young.

Figure 4 summarizes the developmental sequence that leads to normal maternal behavior in the rat. Note again that the development of brain cells with estrogen receptors is dependent upon an absence of testosterone in the early-stage embryo. No testosterone is produced by the embryonic gonads if the fertilized egg lacks the distinctive genetic information carried on a Y chromosome.

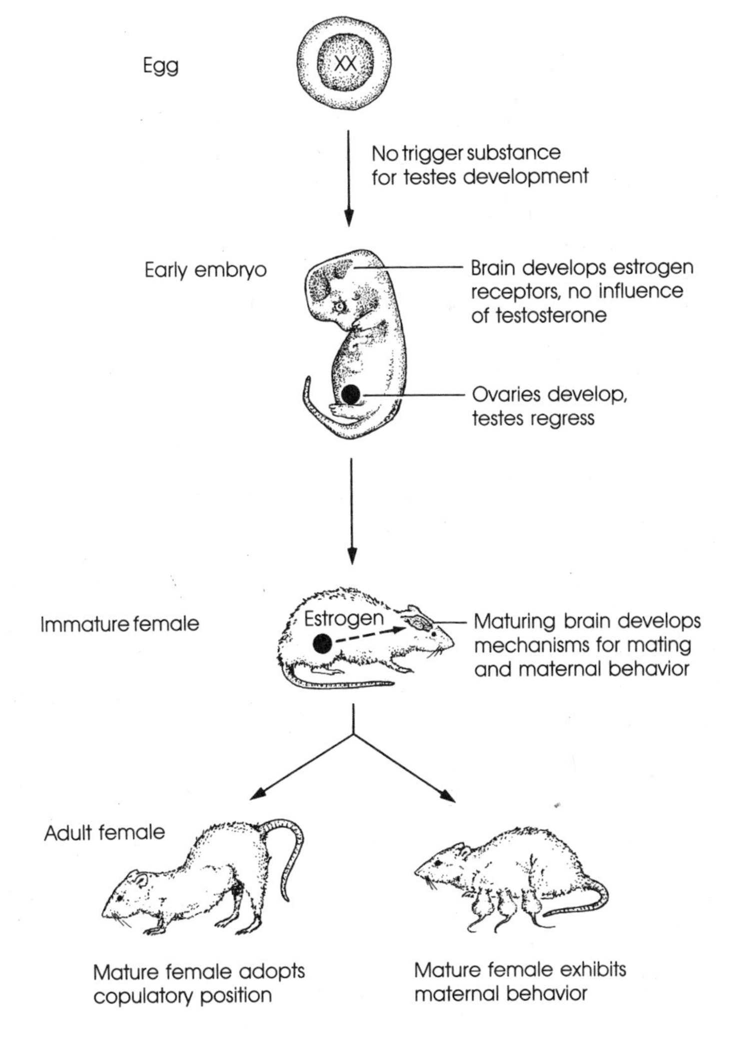

4 **Chromosomes, hormones, and reproductive behavior** of the female rat. In the absence of a chemical signal to activate testes development, the rat develops ovaries and a brain that promotes behavior appropriate for females.

Development as an Interactive Process

The rat example illustrates the inescapable interdependence of genotype and environment in the development of a trait. A small genetic difference, that is, the presence of a Y chromosome with its relatively few genes in a fertilized egg with many other chromosomes and tens of thousands or perhaps millions of other genes, initiates a cascading series of events whose ultimate result is the development of the often striking differences between male and female mammals. Note that the Y chromosome does *not* code for a complete set of masculine characters. Instead, it carries information that shifts development from one track to another. Once certain cells begin to produce testosterone, their options are limited, and the same applies to the many other cells that are eventually affected by the hormone.

In a way, development consists of a sequence of restrictions on the potential capabilities of a cell. An early brain cell in a male or female rat has the flexibility to develop along male or female lines because in each case the cell has receptor molecules that can bind with testosterone. These receptor molecules exist because of an interaction between genetic information in these brain cells and the cytoplasmic environment of the cell that contained the amino acid building blocks necessary for the construction of the receptors. If a cell with testosterone receptors occurs in an embryonic environment in which testosterone is present, it becomes committed to a state that can interact with environmental influences in a way that promotes the eventual development of maleness. But if this decisive environmental stimulus is absent, the cell will tend to move along a pathway that produces a cell with female properties.

The Development of Singing Behavior in Birds

In a few birds, as well as a few mammals, enough is known to identify some of the environmental factors that influence the development of certain differences between the sexes. In zebra finches, for example, sexually mature males sing a complex, species-specific song that differs from the songs produced by males of other finches [356]. One type of song is used to court females, who listen to the singing male and may assume the precopulatory position, an action that enables the male to mount and fertilize his partner. Adult females never produce the courtship song in nature nor do males assume the female's precopulatory pose. Even if one implants testosterone in an adult female, she will not sing. This suggests that by the time a zebra finch has become adult its brain has become sexually differentiated and is limited in its abilities.

Differences in the structure of male and female brains have been discovered in zebra finches, differences that relate to song production [291]. There is a chain of distinctive neural elements, which runs from the front of the brain to its union with the spinal cord. These units constitute the song system, which connects with the neural pathways that lead to the syrinx (the organ that produces vocalizations). The components of the song system are all much larger in males than in females.

The developmental basis of this difference can be traced back to the chromosomal differences between the sexes. (In birds, it is the female that has the Y chromosome, whereas males are XX.) In the absence of the Y chromosome, an embryonic male's gonads apparently produce estrogen, which acts as the critical signal for the masculinization of the brain. If one implants a newly hatched male with pellets containing either estrogen or testosterone, neither has an effect on the development of the song system and singing behavior. But either estrogen or testosterone treatments applied to a nestling female enlarges the song system of the female's brain. Interestingly, estrogen has large effects on three major components of the song system, whereas testosterone treatment results in an increase in the volume of brain cells in only one component. This finding indicates that estrogen is the ORGANIZING SUBSTANCE that activates the development of a male-type song system by directly or indirectly stimulating growth and differentiation in critical portions of the brain [291].

A female that has had an estrogen implant as a nestling will not sing when she reaches adulthood unless she also receives testosterone treatment as an adult. Although her brain has been masculinized, the song system requires a male hormonal signal to prime it for song production. Cells within the song system of males (and masculinized females) have receptors that bond with testosterone, when it is present in the blood circulating through the brain. In nature, males do not sing in all seasons of the year but only during that period when reproduction is likely to be successful. During this time, the adult male testes are stimulated to increase production of testosterone, and a higher testosterone level has cascading effects that result eventually in singing behavior.

Figure 5 shows a sonogram of the song of a female zebra finch that received estrogen as a chick and testosterone when adult. The same figure shows courtship songs produced by two adult males that were reared without hormonal manipulations. One of the two males was the female's father, to whose song she listened in the early days of her life. Close inspection reveals that the songs of father and daughter share certain components in common. This, and other evidence, indicates that the experience of listening to songs produced by particular individuals can shape the course of development of a species-specific song [356]. In other words, learning plays a role in the acquisition of this behavioral trait in the zebra finch.

Experience and the Development of White-Crowned Sparrow Song

Zebra finches are not the only birds in which opportunities to listen to other singing members of their species affect the development of adult song [385, 539, 702]. The white-crowned sparrow is similar in that it too has a species-specific song that it acquires in part through learning [466]. Under natural conditions, nestlings and young birds recently out of the nest begin their lives in an environment in which adult males, including their fathers, will be singing for several months. Then with the end of the

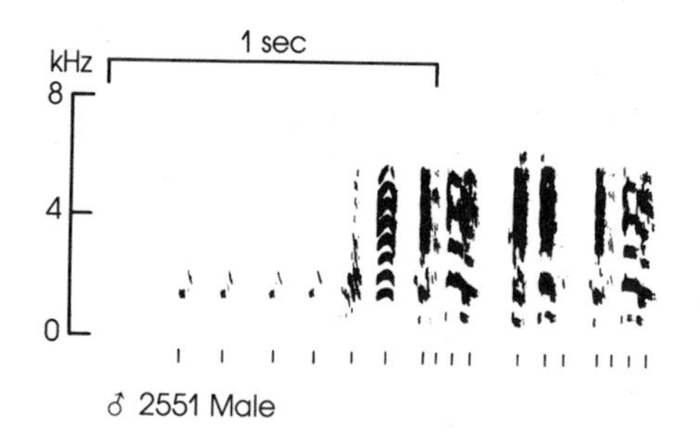

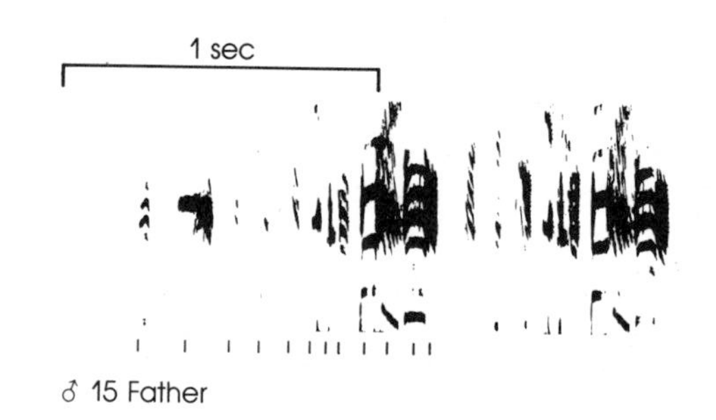

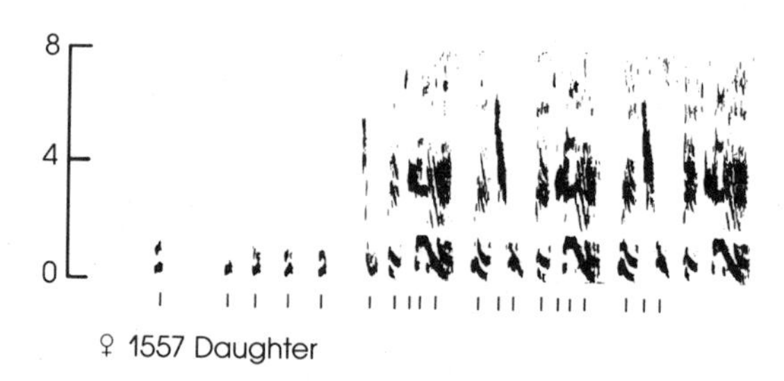

5 **Zebra finch song.** Sound spectrograms of songs produced by two male zebra finches (*top*) and by a female that received estrogen hormone early in her development (*bottom*). Note that elements of the female's song resemble those of one of the males (her father whom she heard while a young bird). Courtesy of Gunvor Pohl-Apel, Roland Sossinka, and Klaus Immelmann.

reproductive season, adults cease singing for the most part. The young male does not begin to sing until it is about five months old. At this time its first efforts are not very promising, a twittering SUBSONG that has only the vaguest similarity to a mature male's FULL SONG. But over the next two months, the song of the juvenile male becomes more and more complete until, by the start of the breeding season, it closely resembles a normal, full song.

The discovery that white-crowned sparrows have geographically stable dialects [469] is consistent with the hypothesis that learning plays a role in song acquisition in this species (Figure 6). Sometimes males in two populations separated by only a few miles have their own easily recognizable song patterns. These dialects retain their distinctive properties from year to year as the young recruits to the population sing the song of their area. Although in theory song differences between populations might be entirely due to genetic differences among them, there is clear evidence that young males learn their dialect by listening to the adult males around them.

This evidence comes from experimental tests of the hypothesis that acoustical experience influences the course of song development [384]. The first step was to find the nests of wild sparrows from which the eggs were stolen and removed to a laboratory incubator. Once the infant birds hatched, they were hand-reared and maintained in sound-proof chambers in which it was possible to control exactly what sounds reached a young male. One group of isolated sparrows were not permitted to hear the songs of white-crowned sparrows as they matured. Another group of young birds listened to tapes of their species *before* they were surgically deafened at five months

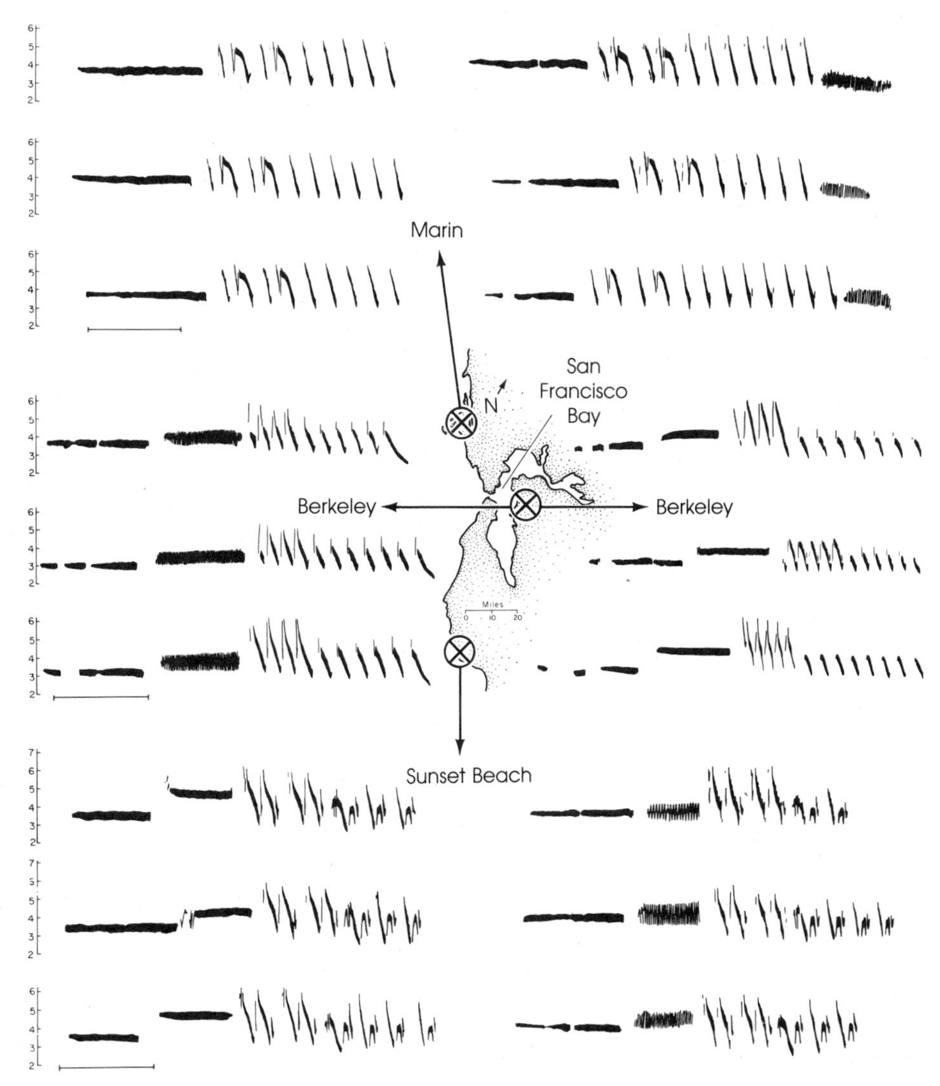

6 **Sound spectrograms of songs** of 18 different male white-crowned sparrows whose songs were recorded in three separate localities around San Francisco Bay. The males found in any one place sing songs with the same distinctive dialect, which differs from the dialect of songs produced by birds of another population. Courtesy of Peter Marler.

of age. The results of these manipulations are shown in Figure 7, which has examples of the songs produced by a wild bird, by a bird with no experience with white-crowned sparrow song, and by a deafened sparrow.

These experiments indicate that a male white-crown must hear other males singing if it is to produce a detailed full song when it is mature. The isolated male, like the male zebra finch with only a foster mother, will eventually sing a song with certain properties of the normal song, but the signal is incomplete. If, however, one plays tapes of white-crown song to a hand-reared bird in its acoustical chamber, it will come to adopt the dialect of that song—provided it has not been deafened before it begins to vocalize at 150 days of age.

With these experiments we can identify a number of "rules" of song development based on the bird's neural responsiveness to acoustical influences:

1. The male white-crowned sparrow is not capable of acquiring the song of any bird species except its own. Isolated birds subjected to tapes of song or swamp sparrows develop a song that is similar to those of birds that have had no songs of any sort played to them. But if the bird hears tapes with both song sparrow and white-crowned sparrow songs on it, it will develop normal white-crowned sparrow song, incorporating white-crown elements from the tape while ignoring the song sparrow sounds.
2. Moreover, the male must hear white-crown song in the period from 10 to 50 days posthatching if it is ever to sing a normal full song. This is the critical period for song learning, and if the bird is not

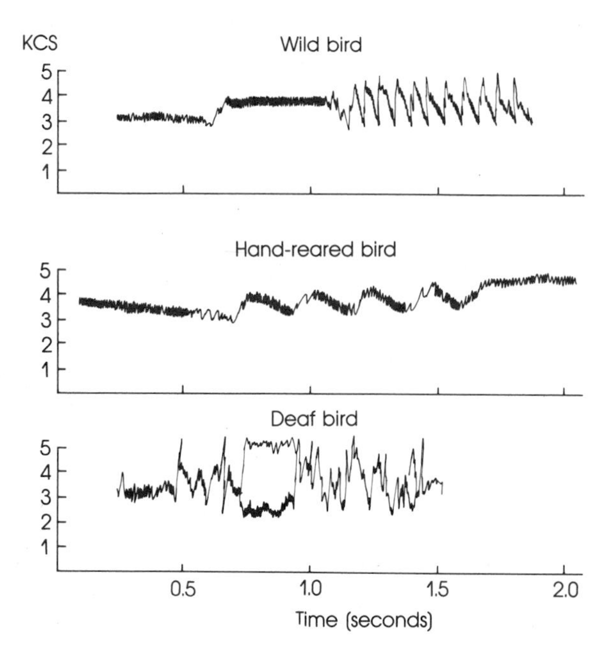

7 **Sound spectrograms of white-crowned sparrow songs.** The song of a wild bird (*top*), the song of a male reared by humans in isolation from other white-crowns (*middle*), and the song of a bird deafened before it reached 5 months of age (*bottom*). Photograph courtesy of Mazukazu Konishi.

exposed to white-crown song during this time it will never sing a complete full song no matter what its later acoustical experiences. Thus, the bird's brain at the 10- to 50-day stage can only be affected by sensory stimulation provided by white-crowned sparrow songs. After 50 days, the brain continues to undergo growth and differentiation that restricts its developmental options. By this time the cells in the song system have lost their flexibility and can no longer store the relevant information for the development of normal song.

3. During the interval from 150 to 200 days after hatching, when the young bird begins subsong, it must be able to hear itself sing if it is ever to produce normal full song. If it is deafened before subsong begins, it cannot match its vocal output with the memory of its species song acquired when it was 10–50 days old. As a result, development of the song is unguided, and the end product remains a twittering, variable vocalization without species-specific properties.

Figure 8 summarizes the key experimental results and provides an interpretation based on the hypothesis that the song system of a juvenile male white-crowned sparrow is specially suited for receipt of information from the sounds of adult white-crown song. The system contains a **TEMPLATE** for storage of this information during a critical sensitive period. The memory offers a model that the bird will eventually match by listening to its own subsong and eliminating discrepant syllables and patterns that do not conform to the model. If a young male does not acquire an appropriate model, its template will not accept a substitute such as the memory of song sparrow song. Therefore, it never develops the song of another species [366].

In more recent work with another bird (the swamp sparrow), Peter Marler and Susan Peters have shown that a young male can remember much more information about songs it heard early in life than it actually uses in its own final full song [468]. Hand-reared sparrows were played tapes containing a variety of different syllables (Figure 9) sung by different adult males of their species. Wild swamp sparrows generally have three different song types, each composed of a different syllable, which is repeated over and over for a period of about two seconds. Hand-reared birds that have heard swamp sparrow song develop a similar number of song types, which usually resemble those they heard as juveniles. Prior to settling on its final repertoire, a first-year bird at about age 300 days (months after the training program is over) will sing a much wider variety of songs, many of which contain syllables that were copies of those on the training tapes. For example, one bird had two final song types that it eventually settled on, both based on syllables learned from the tapes. But during the plastic phase of song development that preceded the final "choice," the sparrow's songs incorporated six additional syllables that appeared on the tapes (as well as five other components that were either poor copies or personal inventions of the bird). This work shows that the swamp sparrow male learns considerably more than one would imagine by comparing the crystallized adult song with the syllables on the training tapes.

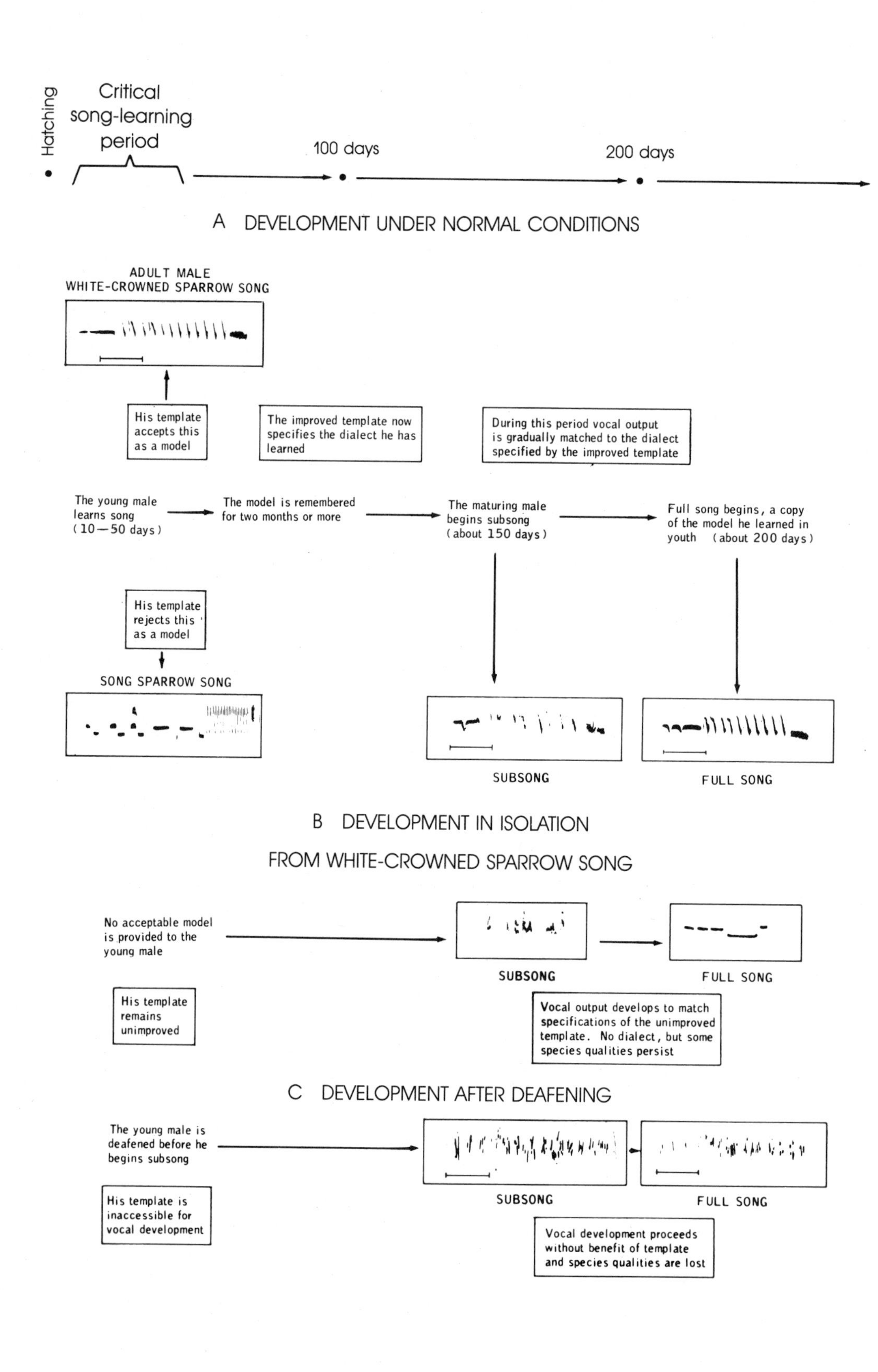

Hatching
Critical song-learning period
100 days
200 days
A DEVELOPMENT UNDER NORMAL CONDITIONS
ADULT MALE WHITE-CROWNED SPARROW SONG
His template accepts this as a model
The improved template now specifies the dialect he has learned
During this period vocal output is gradually matched to the dialect specified by the improved template
The young male learns song (10—50 days)
The model is remembered for two months or more
The maturing male begins subsong (about 150 days)
Full song begins, a copy of the model he learned in youth (about 200 days)
His template rejects this as a model
SONG SPARROW SONG
SUBSONG
FULL SONG
B DEVELOPMENT IN ISOLATION FROM WHITE-CROWNED SPARROW SONG
No acceptable model is provided to the young male
SUBSONG
FULL SONG
His template remains unimproved
Vocal output develops to match specifications of the unimproved template. No dialect, but some species qualities persist
C DEVELOPMENT AFTER DEAFENING
The young male is deafened before he begins subsong
SUBSONG
FULL SONG
His template is inaccessible for vocal development
Vocal development proceeds without benefit of template and species qualities are lost

◀ 8 **Analysis of song learning by white-crowned sparrows.** The key hypothesis is that the young male has a neural template designed to acquire information about the song of males of its species only. If the young bird does not receive this information (B), it cannot develop a typical full song. The juvenile male must also hear itself sing (C); if not, the bird will be unable to match its vocal output with stored information in its template and so will develop a very aberrant full song. Courtesy of Peter Marler.

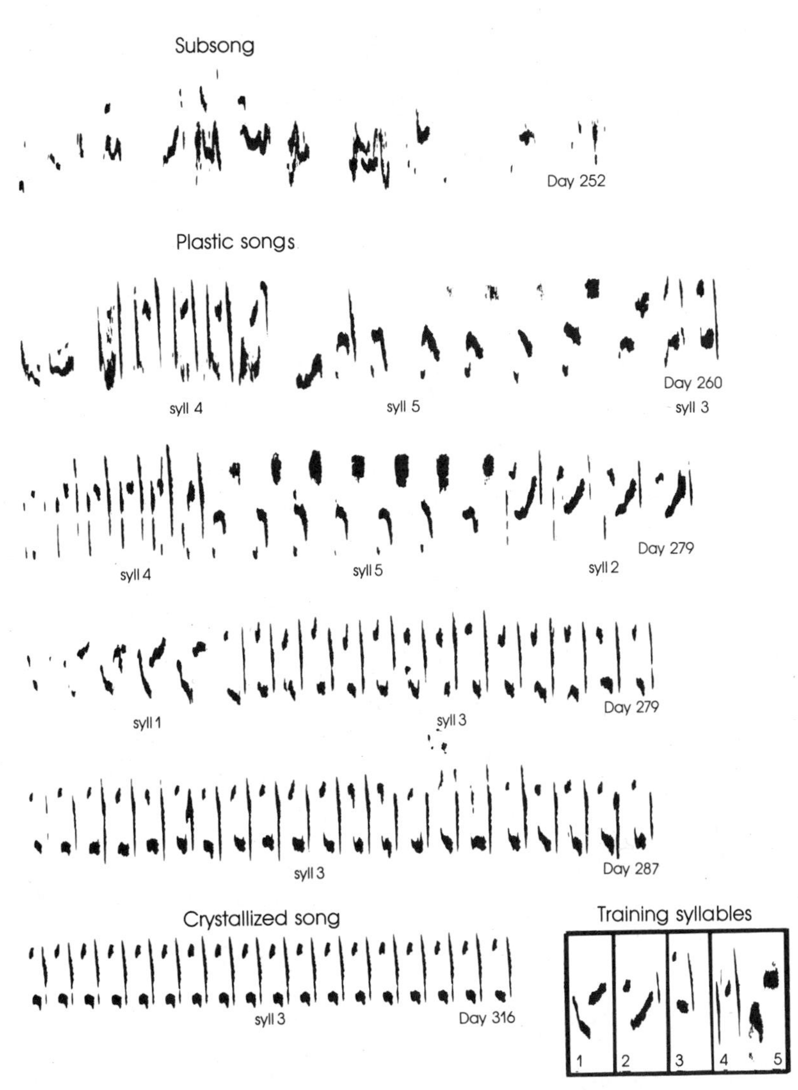

9 **Song syllables used by swamp sparrows.** After listening to a large number of different syllables in training tapes, young birds at first sing plastic songs with many of these elements. But eventually each bird produces a crystallized song with only one syllable type; it will have a total of two or three different crystallized songs. Courtesy of Peter Marler.

Developmental Homeostasis

The development of singing behavior by male birds is a classic illustration of the complex interplay between chromosomal information, hormonal signals, neural differentiation, and acoustical stimulation. The discovery of the subtle and multiple ways in which the environment can alter the course of behavioral development has been a major achievement of this research. But it is also impressive that in nature there are so few birds or mammals that are part male and part female, structurally, physiologically, or behaviorally. The number of female sparrows that sing a male's song is infinitesimally small. The male white-crown that fails to develop a perfectly functional territorial song is an extreme rarity and so is the male rat that never acquires the ability to copulate with receptive females. The fact that we take "normal" development for granted does not make it any less remarkable.

The acquisition of the "correct" behavior in the proper sex becomes all the more puzzling when one considers that each white-crowned sparrow and each wild Norway rat has a genotype that consists of a unique combination of genes drawn from the sample contained in its parents' bodies. Moreover, no two sparrows (and no two rats) eat exactly the same foods, experience identical climatic conditions, hear the same sounds, or encounter identical social situations. The development of each individual is therefore the result of an interaction between a unique genotype and a unique set of environmental influences. It is more than mildly surprising, therefore, that large numbers of individuals within a species come to share many of the same biochemical, physiological, and behavioral characteristics.

The paradox that development is altered by a host of variables and yet animals usually develop the abilities needed for survival and eventual reproduction has been a source of much confusion and argument. Historically this is best exemplified in the debate between Konrad Lorenz [449] and Daniel Lehrman [410]. Lehrman, a comparative psychologist, has written an interesting review [411] of his clash with the zoologist Lorenz, one of the founders of ethology. The review reveals that their dispute had multiple causes, some purely semantic in nature, others having to do with a misunderstanding of the differences between proximate and ultimate explanations of behavior. Still others stemmed from a difference in the scale of their approaches (Lehrman interested in the origins of relatively subtle behavioral differences among individuals; Lorenz ignoring these in order to focus on why there are broad general similarities in the attributes of the members of a species). Rather than rehash their argument, let us see if it is possible to reconcile the fact that development is susceptible to many diverse influences, both genetic and environmental, with the consistent appearance of adaptive traits in developing animals.

The Repeatability of the Developmental Process

To illustrate that development follows a predictable course in different individuals of the same species, one need only consult an embryology textbook. Embryologists can predict with some confidence when certain structures will first appear in a rat or chick embryo and can chart with precision the sequence of changes in these structures over time. There is variation in the course of development among individuals, but it is usually modest and restricted.

In a few cases it has proved possible to follow the fate of identifiable single cells, and this work, too, has shown that development is predictable and repeatable [720]. For example in the grasshopper *Schistocerca nitens,* there is a set of distinctive cells that can be located in essentially every five-day-old embryo when one removes the limb buds from the thorax of the embryo and examines the exposed tissue with a microscope. These cells occur singly in the center between two tissue aggregates that give rise eventually to components of the thoracic ganglia (a part of the central nervous system of the insect). By studying a large number of embryonic grasshoppers of different ages, one can trace what happens over the course of development of one of these central cells. A regular pattern emerges [265]. At a particular stage, the cell begins to fission, a process that produces new cells. These daughter units gradually change their shape, growing into specific neurons with their own predictable schedule of growth, structural change, and alterations of electrical activity. Eventually one particular descendant of the early-stage "mother cell" becomes a fully differentiated motor neuron running between certain cells in the thoracic ganglia to muscle cells in the limb bud destined to become a leg. Figure 10 shows the typical pattern of development of these cells, a pattern that occurs despite considerable genetic diversity in the species and the great environmental differences affecting its members.

It is almost as if the nerve cells of the grasshopper "know where they are supposed to go." This is not unique to insect neurons because workers with amphibians [179] and with birds and mammals [113] also appear impressed by the capacity of neurons to migrate to and make appropriate connections with the proper target tissues. How this happens is still largely mysterious, but it is *not* because the pathway of each cell's development is "genetically determined" in the simplistic sense discussed in Chapter 2. Instead, it may be because the genome of each cell has certain regulatory properties that are triggered by particular chemical signals (such as testosterone molecules) when they appear in the cell's environment. At each stage of development, an embryonic nerve cell is subjected to a barrage of chemical products, some of which it produced itself, others of which have come from near and distant neighbors. The cell's prior development and position in the growing embryo determines the concentration and variety of chemical substances that reach it. The cell's past genetic activity determines what genes are available to be activated or deactivated by these environmental cues. Some patterns of genetic activity are possible, other patterns are not. Changes in which gene products are being made precede developmental changes by the cell and result in the export of materials from it that may affect the cell's neighbors. This can trigger new biochemical activity on their part, with reciprocal effects on the developing nerve cell. To say that the process is dynamic is an understatement of the first degree.

But again the regularity with which functionally effective cells, tissues, and whole bodies develop shows that some developmental outcomes are more probable than others. The interactions between genotype and environment that occur in the development of an individual must in some way be structured to restrict certain options and facilitate those that will lead

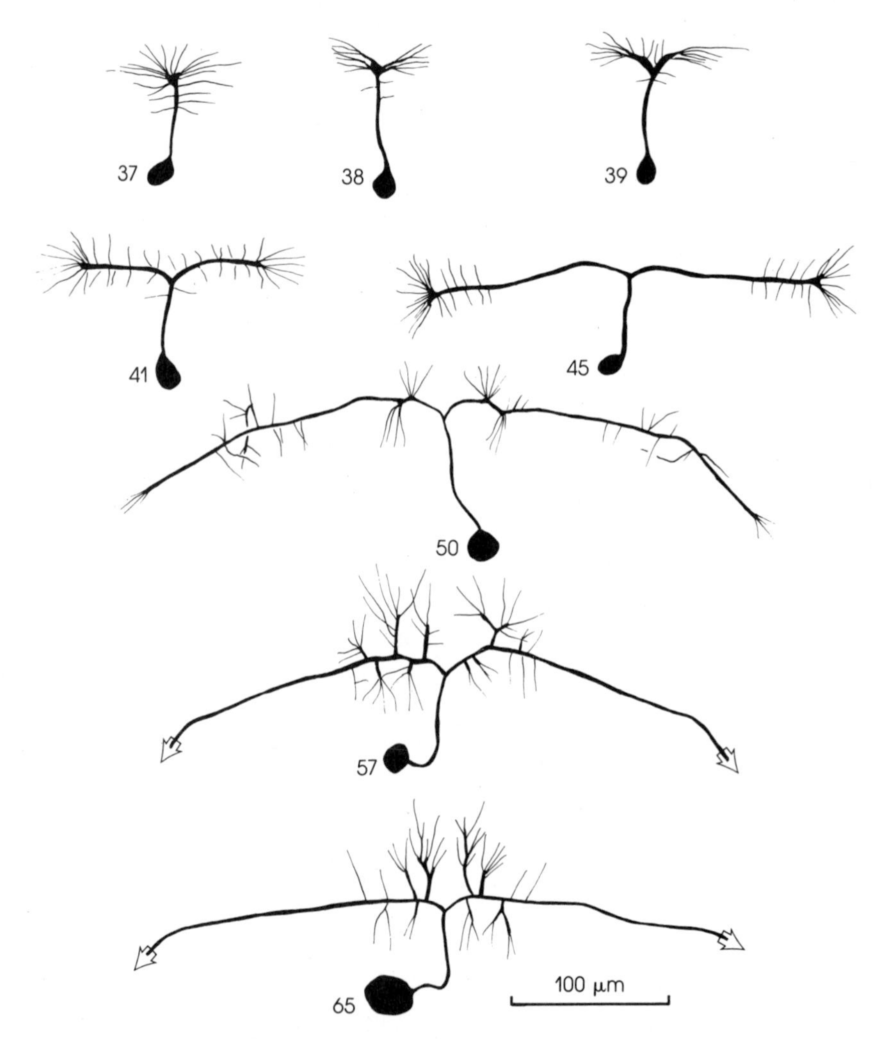

10 **Development of a particular nerve cell** during the embryonic growth of the grasshopper *Schistocerca americana*. One identifiable cell is shown at various stages of embryonic development.

to the construction of a reproductively successful adult. This leads us back again to the regulatory mechanisms of developmental systems. Perhaps genotypes have a number of different regulatory genes that are focused on the same component of development. Just as the designers of manned space vehicles provide these machines with a number of backup systems so that one failure is not fatal, so, too, the genotype may have the capacity to activate different elements within it to cope with different problems confronting a developing cell [48]. Should a mutant gene's product interfere with the manufacture or degradation of a material, a regulatory mechanism might detect this event and trigger other genes to provide substitute enzymes for the task. Should the animal occur in an environment deficient in a key substance, a regulatory gene could activate a combination of genes whose products will lead to the synthesis of the missing compound. The maintenance of metabolic homeostasis within cells should promote the development of functionally useful cells. By adjusting the biochemical responses of its cells to the particular conditions encountered, the organism would have the flexibility to direct its future development toward the same conclusion in a range of different environments.

Backup systems that enable the developmental process to contend with varied conditions have been discovered in some animals. For example, Alain Ghysen performed a surgical maneuver on mature larvae of the fruit fly *Drosophila* [254]. Ghysen misdirected a portion of an incompletely developed nerve cell so that it was placed in a region in the fly's thoracic ganglion where it normally would not occur. The misrouted nerve cell grew into the ganglion at the "wrong" spot, but it then developed in a way that differed dramatically from the typical pattern. (This is clear evidence that the form of the nerve is not "genetically fixed" but is responsive to certain environmental influences.) The cell grew a much longer projection than usual, a projection that made its way through the "foreign" region of the central nervous system until it reached its usual target point. Then it stopped growing and made the normal connections with the cells in this area.

Behavioral Development in Abnormal Conditions

The ability of a nerve cell of a fruit fly to compensate for its physical displacement is proof that the physiological foundation for normal behavior can sometimes develop in an abnormal (to say the least) environment. This speaks to the resilience of the developmental process, a phenomenon well documented in behavioral studies. For example, chickens without feathers (there are various featherless mutants) come to flap their wings every bit as much as chickens with normal feathered wings, despite the fact that featherless wings offer atypical sensory feedback when moved compared to normal wings [587]. Young pigeons do not even need to flap their wings in order to develop the ability to fly; birds whose wings have been restrained from fledging will be able to fly when their wings are freed, at the same time as siblings that have had the opportunity to exercise their wings [285].

Likewise, social deprivation does not always lead to behavioral abnormalities. Male crickets that have spent all their lives isolated from their fellows will sing a normal species-specific song despite their severely restricted social and acoustical experiences [62]. Common European squirrels reared in isolation in a cage with a bare hard floor will nevertheless perform acorn-burying movements, when first given a supply of acorns, in the same way that wild animals store food for later use [203].

These cases show how difficult it is to disrupt the development of certain behavior patterns. Even in the absence of some experiences that might be thought to be useful for the maturation of a trait, the response appears on schedule. Again, experiments of this sort do *not* demonstrate the complete independence of physiological and behavioral development from environmental influences, only the capacity of a genotype in interaction with a limited, abnormal environment to produce the development of certain functional abilities.

The difficulty that experimenters often encounter in their attempts to disrupt normal development can be shown by an examination of the classic experiments of Margaret and Harry Harlow. The Harlows wished to discover what effects early social experience had on the behavioral development of the rhesus monkey [308, 309]. Their basic experimental technique involved separation of the infant rhesus from its mother shortly after birth. The baby was placed in a cage with an artificial "surrogate" mother (Figure 11), which might be a wire cylinder or terry cloth figure with a bottle from which the baby was able to nurse.

The young rhesus gained weight normally and developed physically in the same way that nonisolated rhesus infants do. However, it soon began to demonstrate signs of behavioral abnormality, which might eventually develop into continuous crouching in a corner, rocking back and forth, and various forms of self-mutilation. If confronted with a strange object or another monkey, the isolated baby was especially likely to react fearfully. It would withdraw as completely as possible from the object or playmate by huddling in a corner or by clinging tightly to a terry cloth surrogate mother. Monkeys reared in total isolation for the first six months of their lives sometimes exhibited permanent disturbed behavior in later life, when rocking, head banging, and self-biting were almost the only activities of the adult animal.

The Harlows stressed the significance of receiving maternal care for normal behavioral development. But they also pointed out that young rhesus monkeys reared alone with their mothers did not develop truly normal sexual, play, and aggressive behavior. When they encountered animals their own age later in life, they were likely to behave very inappropriately, reacting with excessive fear or violence. Evidently contact and interactions with peers as well as a mother early in life are essential ingredients for normal behavioral development in the rhesus monkey. These experiments show the importance of early social experience for normal social development, but they also demonstrate how difficult it is to alter the course of

11 **Infant rhesus monkey** with a choice of two surrogate mothers. It prefers the terry cloth model, even though it can receive milk only from the wire cylinder "mother." Photograph courtesy of Harry Harlow.

behavioral development through environmental manipulations. In order to produce a rhesus with behavioral disorders, the Harlows had to isolate an infant in a small wire mesh cage at birth or to keep it in a similar environment with just its mother. Neither event has ever occurred under natural conditions for rhesus infants. An abandoned (i.e., isolated) baby would be dead in a short time; and because rhesus monkeys always live in groups, a young animal is never entirely alone with his mother for prolonged periods of time, to say nothing of six months. Thus, no animal in the history of the species has had to cope with the bizarre environmental situations devised by the Harlows and other researchers in this field. It is not surprising that the animals' developmental systems break down under these conditions.

What is surprising and revealing is that otherwise totally isolated infants developed relatively normal behavior if they were introduced into a cage with three other infants for just 15 minutes each day [308]. At the outset, the young rhesus monkeys simply clung to one another (Figure 12), but later more time was spent in play. In their natural habitat, rhesus babies, beginning at about one month, start to play more and more with others about their age. By six months they spend practically every waking moment in the company of their peers [678]. Yet in the laboratory, despite

12 **Two baby rhesus monkeys.** They have been separated from their mothers and are clinging to one another. Photograph courtesy of Harry Harlow.

the barren existence of isolated infants, a few moments of playtime a day is sufficient to counteract the effects of extreme deprivation. Thus, not only is it difficult to produce abnormal behavior in the rhesus monkey, but a mere fraction of what would be normal social experience is adequate for the development of certain traits. The monkeys exhibit developmental homeostasis—the ability of developmental systems to compensate for environmental (and genetic) deficits—and so produce key adaptive traits in animals exposed to less than optimal conditions.

The compensatory resilience of the developmental system of this primate is further illustrated by more recent studies on the effects of early experience. The surrogate cloth mothers in the Harlows' experiments did not move about; a normal mother spends a good deal of time traveling with her infant. To test the possibility that the sensory stimulation provided by the moving mother influences the course of behavioral development, William Mason devised an experiment in which one group of monkey infants was reared with a mobile cloth surrogate that was moved several times a day from one spot to another in the infant's cage [477]. The behavior of these individuals was contrasted with that of members of a group raised with a stationary surrogate. In general, the mobile-mother group developed much more normal responses. They did not engage in the bizarre body-rocking behavior so common in monkeys raised with a nonmoving mother substitute. More importantly, when members of the mobile-mother group were introduced to other monkeys for the first time at age 14 months and for a second time at 4–5 years of age, their social behavior was far more normal than that of the other group. When mature, several females and one male with moving, surrogate mothers exhibited normal copulatory behavior. The barest minimum of "social" stimulation was sufficient to activate developmental pathways leading to functional social and sexual behavior.

Even more recently, Mason has reared some rhesus monkeys with mongrel dogs as surrogate mothers (Figure 13). This mother substitute is far more active and interactive than a plastic moving mother. Infant rhesus monkeys quickly adopt a dog as an attachment figure and through their social interactions with this being come to develop close-to-normal social behavior, despite having an extremely atypical "mother" [478].

The intellectual development of the monkeys reared with and without dog mothers has been tested by presenting the monkeys with an opportunity to receive a food reward through performance of a simple task, such as pulling or pushing at a novel object. The group with canine companions

13 **Rhesus infant with an "adoptive mother."** Infants reared with mongrel dogs as mother substitutes develop nearly normal social behavior, despite the abnormal parenting provided by their companion. Photograph by W. A. Mason.

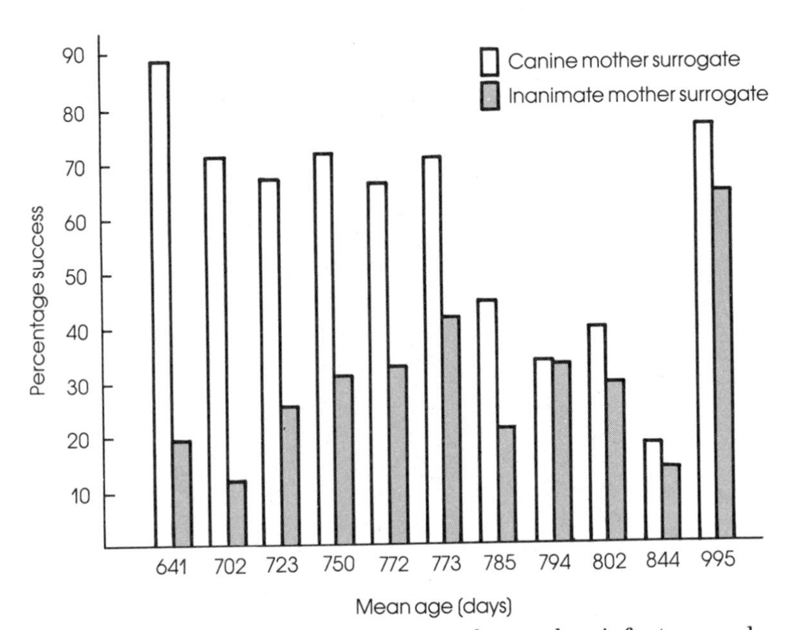

14 **Capacity to solve learning tests** by monkey infants reared with canine and inanimate mother substitutes. Note that the "deprived" group catches up with the dog-raised group by the eighth test.

initially did much better at this learning task (Figure 14), but over time the differences disappeared between the two groups in their willingness to contact the object and in the percentage of problems actually solved. The monkeys with inanimate mother substitutes (a wheeled baby horse with a rug saddle) caught up with the other group after only a few trials scattered over a period of about a year. The recovery of the problem-solving ability of the monkeys with inanimate mother surrogates is further evidence for developmental compensation that enables experimentally deprived animals to develop equal skills with their more socially "enriched" fellows.

Developmental Homeostasis and Human Behavior

Naturally one wonders about the relevance of these studies for human beings and the development of their behavior. A widely held hypothesis states that the experiences very young children have will greatly affect the development of their intelligence, personality, and other attributes. But there is growing evidence that the development of human behavior is also buffered to some extent against some environmental shortfalls, even if these occur early in life [190]. Consider, for example, the results of a study of the mental performance of a group of Dutch teenagers who were born or conceived at a time when their mothers were being starved as a result of the Nazi transport embargo during the winter of 1944–1945 [683]. The embargo prevented food from reaching the large

Dutch cities during this time. Deaths from starvation were common, and many persons lost one fourth of their body weight. For most of the famine period, the average caloric intake was about 750 calories per day, although at times it fell lower still (1500 is the absolute minimum needed to sustain an individual on a long-term basis [812]). Thus, pregnant women living in cities were subjected to famine conditions for months. In contrast, rural women were less dependent on food transported to them. As a result, the birth weights of rural babies born or conceived during the embargo were much greater than those of infants born at the same time to women living in the cities.

One commonly reads that poor nutrition during pregnancy will inevitably result in intellectual damage to the offspring because of the critical nature of this period for the development of the brain. However, the famine babies did not exhibit a higher incidence of mental retardation at age 19 than those born or conceived at the same time in nonfamine areas (Figure 15). Nor did these deprived babies score more poorly than relatively well-nourished infants on the Dutch intelligence test administered to men of draft age. The children were in some way protected from this environmental insult and, despite being born at a lowered weight, suffered no permanent intellectual damage [683]. These data indicate that our developmental programs have the capacity to overcome some deficient prenatal environments. This makes sense. Humans have, in all probability, been regularly exposed to food shortages throughout evolution. Genetic-developmental systems in embryos that were highly sensitive to food deprivation would be likely to produce aberrant individuals with lowered fitness. Those women whose embryonic offspring were relatively insensitive to nutritional deficits would tend to produce normal functional individuals with normal to superior fitness.

I am not saying that humans are impervious to all environmental influences or that it is a good idea for pregnant women to starve. Still, just because there is an environmental factor that potentially could affect development is no guarantee that it will, in fact, have an irreparable effect. Our developmental systems as a general rule can withstand a good deal. Jerome Kagan, a child psychologist, recently went to Guatemala, expecting (according to his own account) to find that the child-rearing techniques practiced in rural towns would permanently stunt the intellectual growth of children [22]. The expectation was based on the knowledge that some Guatemalan villagers keep young infants from birth to about 12 months of age in conditions that strike American observers as amounting to extreme deprivation. The young baby is kept in a dark hut, experiences only limited social interactions (it is fed but seldom talked to), and has few toys. Despite all this (Kagan says that if he had seen American children treated this way he would have called the police), the children's developmental program compensates for the first year or so of confinement. By 11, they score as well on Kagan's test as middle-class Americans of the same age. They are socially well-adjusted, capable human beings [370].

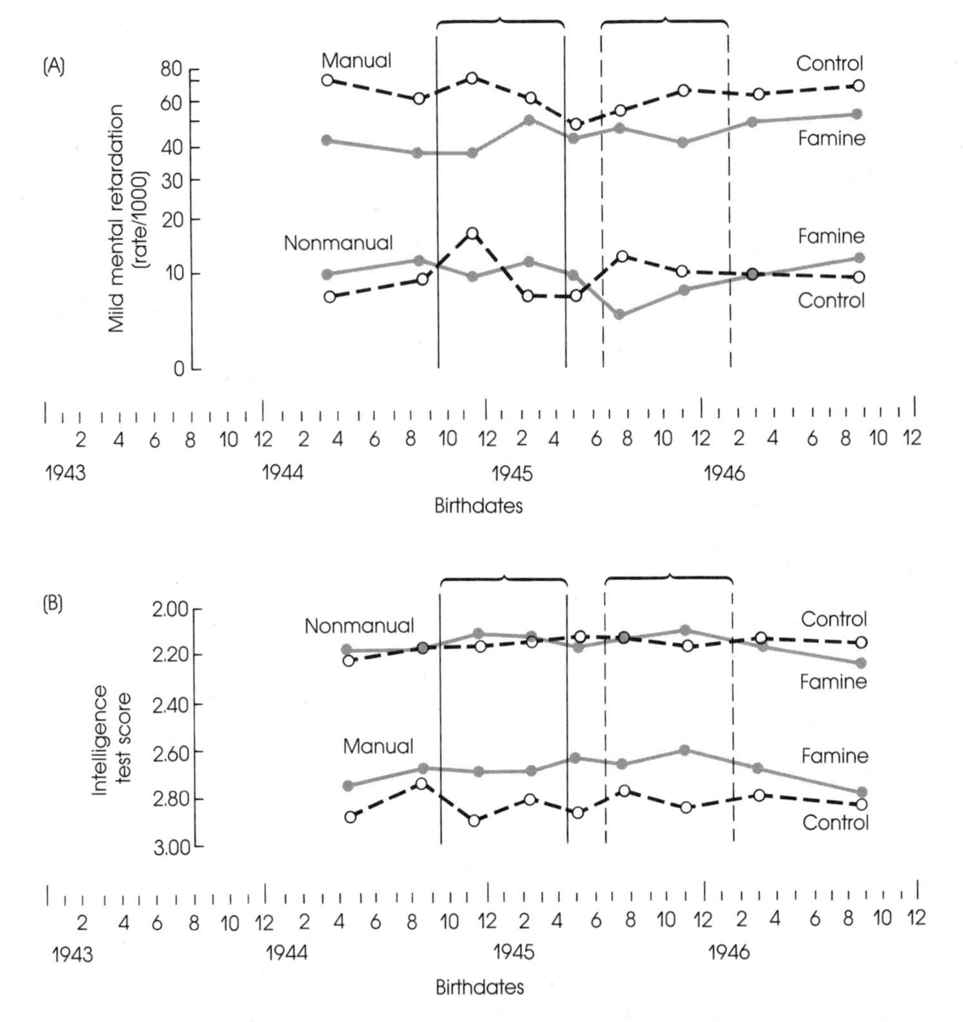

15 **Evidence for developmental homeostasis** in humans. (A) Rates of mild mental retardation in 19-year-old Dutch men divided into two groups on the basis of their father's occupations (manual or nonmanual). All the subjects were either born during the period of urban famine or conceived during this time and born later. The famine group was born in urban areas, where starvation occurred; the control group was born in rural areas, where food shortages were much less severe. (B) The intelligence test scores of 19-year-old men from control and famine groups. Infants born or conceived under famine conditions were no more prone to mental retardation or low intelligence test performance than those conceived or born at the same time in rural, nonfamine conditions.

The Adaptive Value of Developmental Homeostasis

We have seen that in humans, as well as in other animals, the developmental process is a dynamic interaction between genetic information and the environment in which the individual exists. But not all interactions are possible nor are all possible interactions equally likely. We can visualize the course of development in a manner suggested by C. H. Waddington with a fertilized egg or a developing tissue represented as a ball at the top of an inclined plane (Figure 16) [749]. As time passes, the ball rolls down the landscape, which is not a perfectly level slope but is instead highly contoured. The position and shape of the valleys constrains the path followed by the ball and represents the regulatory forces that restrict development. Thus, the presence of testosterone in the environment of brain cells in the very young rat acts in concert with a genetic switch mechanism to channel the development of these cells down a valley leading toward an end point that consists of cells with the physiological characteristics typical of mature male rats. Further influences along the way will determine which of various likely pathways (valleys) the growing tissue will travel along.

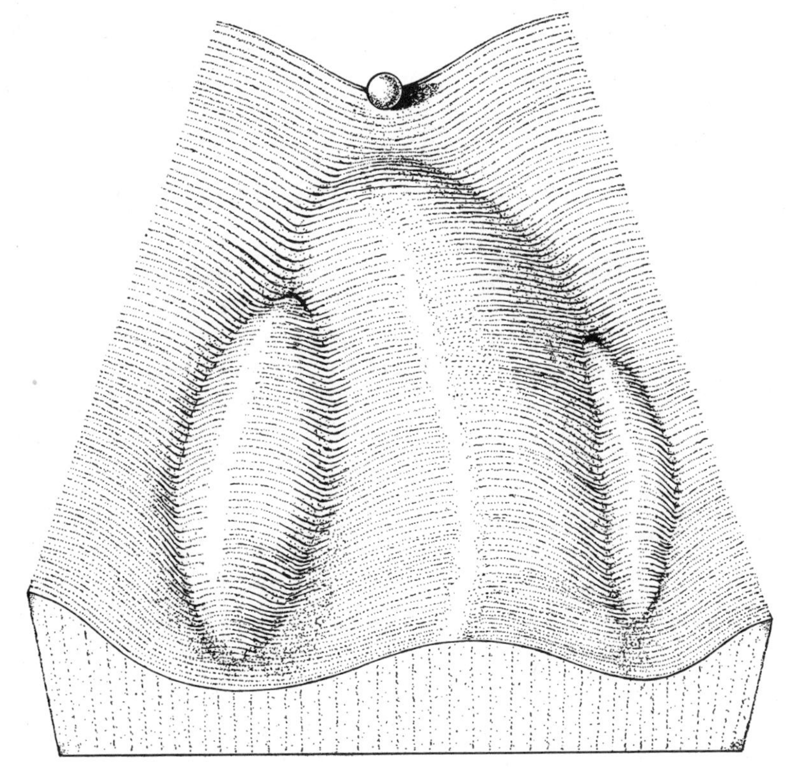

16 **Developmental "landscape."** A diagram of the course of development followed by a cell, tissue, or individual. The diagram represents the guiding constraints on the process of development, constraints that steer it to an adaptive end point.

But only very extreme environmental or genetic forces can cause the descending ball (growing tissue) to depart from a valley once it has become committed to a particular path.

We can extend Waddington's metaphor further to describe the ability of individuals to compensate for deficits in their background by suggesting that many of the valleys on the developmental landscape must come together at various points (i.e., there are many routes to the same end point). Without getting too carried away with this image, it helps remind us that a developing organism is not a passive billiard ball capable of responding equally to every possible environmental influence or peculiarity of its genotype. Such a system would often lead to reproductive disaster. Individuals would run the risk of being permanently warped by a transitory environmental deficit, such as a temporary shortage of food or lack of social stimulation. They might fail to develop the traits that historically have led to survival and reproduction. The genes of these developmentally "sensitive" individuals would surely be less likely to survive than those that had the capacity to guide development around obstacles in a developmental path that had in the past produced animals with high fitness. Although a great deal remains to be learned about the proximate basis of developmental homeostasis, there can be little doubt that it is a widespread phenomenon and that in an ultimate, evolutionary sense individuals benefit by being able to withstand disruptive influences on their behavioral development.

Summary

1 The development of any trait requires a dynamic interaction between the genotype in a fertilized egg and the environment of the developing organism. For example, the sexual differences between males and females of some birds and mammals have their roots in relatively small chromosomal (genetic) differences between males and females. These lead to differences in the hormones produced by embryonic gonadal tissues. The presence of a key hormone acts as an environmental trigger for other cells, altering their genetic and biochemical activity with cascading effects throughout the body over a long time. The result is a spectrum of physiological and behavioral differences between males and females.

2 The environment of a developing organism consists, not only of the metabolic products of its cells and the food materials it receives, but also of its sensory and social experiences. All of these factors can potentially

act as cues (sometimes only if they occur during a restricted critical period) that have long-term developmental consequences (as in the acquisition of bird song in the white-crowned sparrow).

3 Despite the great variation in the genetic and environmental influences operating on the individuals of a species, most animals develop into functionally competent, reproductively capable creatures. Regulatory mechanisms must make certain developmental outcomes more likely than others and confer upon the developing organism a certain protection against environmental or genetic perturbations that might prevent the development of adaptive abilities.

4 The channeled or buffered aspect of development (developmental homeostasis) is seen in (1) the constancy with which individuals of a species pass through certain species-specific stages and (2) the development of normal physiological and behavioral characters in animals placed experimentally in highly abnormal environments. The resilience of the developmental process helps individuals develop the key characteristics that promote genetic success.

Suggested Reading

The March 20, 1981 issue of *Science* magazine has a collection of superb review articles [529] on the relation between genes, hormones, and sexual differentiation in animals. Peter Marler's article [466] on song learning by white-crowned sparrows is excellent.

Suggested Films

A Rhesus Monkey Infant's First Four Months. Color, 32 minutes. A film on the developmental pattern of the rhesus monkey made on a captive group that lives under conditions approximating natural ones. The film shows the normal pattern, which can be contrasted with the disrupted sequence produced by experimental manipulation.

Why Do Birds Sing? (F2047). Color, 27 minutes. One of the best of all currently available films on animal behavior. It shows in detail research on the proximate (and ultimate) basis of bird song, with coverage of Peter Marler's studies.

Historically, students of animal behavior have found it useful to develop a classification scheme for behavioral traits. The traditional major categories, instinctive behavior and learned behavior, were created to acknowledge supposed differences in the proximate mechanisms underlying the two kinds of behavior. Instincts were often held to be genetically controlled, whereas learning was believed to be largely (or entirely) dependent on experience (the environment). It is now recognized, however, that *all* behavioral traits are the product of a dynamic interaction between genetic information within the individual and environmentally supplied materials and experiences. A genotype without environmental building blocks would remain a genotype and nothing more. Molecular building blocks from the environment, in the absence of genetic information to regulate the development of a body, would remain an unorganized collection of substances. The development of every aspect of an individual—its appearance, its physiological mechanisms, its behavior, its everything—involves both genes and environment. Because the influence of heredity and nurture are integrated in development, one cannot legitimately separate behavior into genetically controlled versus environmentally determined categories. At the same time, the developmental program that underlies the baby garter snake's ability to strike at a slug-soaked cotton swab differs from the mechanisms that enable a male white-crowned sparrow to acquire the song dialect of its area. In this chapter we shall attempt to salvage a classification system that acknowledges the diversity of proximate foundations of behavior while avoiding the ill-conceived "*either* genetic *or* environmental" approach so often associated with the instinct–learning dichotomy. The eventual goal of the chapter will be to ask why different kinds of developmental systems underlying behavioral traits have evolved. We begin by analyzing the classical categories of behavior: instincts and learning.

CHAPTER 4

Categories of Behavior: A Proximate Classification

Instincts and Learning

A very young baby cuckoo of the parasitic European species is able to perform a complex behavior pattern. It flops about the nest until it lodges an egg of its foster parents in the cup formed by its outstretched wings and back (Figure 1). The nestling cuckoo then maneuvers itself backward up the wall of the nest higher and higher until the egg topples over the edge. The cuckoo then slides down into the cup of the nest, having removed a competitor for the meals its foster parents will provide (see Figure 5). This is the kind of behavior that has been variously labeled "an instinct," "innate," "inherited," and "genetically determined."

This seemingly preprogrammed ability is different from many other behaviors. For example, if one takes a toad into a laboratory, it can be trained to *modify* its feeding behavior as a result of certain (unpleasant) experiences. An experimenter can offer the toad various insects; and a hungry, cooperative animal will oblige by opening its mouth, flipping out its tongue, and snapping up flies, mealworms, and other edible creatures presented to it. If the experimenter then places a toxic millipede in the enclosure, the toad may take this bait as well. The millipede responds by exuding a substance from pores in its body; the toad finds the exudate violently nauseating and spits out the prey (Figure 2). Later the amphibian will refuse to attack the millipede even though it may be hungry and willing to take edible prey quickly. This is the kind of behavior that has been labeled "learned, not inborn," and "environmentally determined."

Although the differences between the toad's behavior and that of the baby cuckoo are real and significant, at the same time the words *instinct* and *learned* carry with them various connotations, some of which are ambiguous, misleading, or downright wrong. As a result, the use of these terms has led to confusion and debate (often tinged with acrimony) [31]. As noted in the introduction to this chapter, the belief that an "instinct" can develop without environmental influence is incorrect. In order to exhibit its egg-ejection behavior, the baby cuckoo requires environmental contributions of food (from the yolk of its egg and from its foster parents) and sensory stimulation (from the eggs of its host). It is also entirely possible

1 **A newly hatched parasitic cuckoo** can roll the eggs of its host out of the nest. Photograph by Eric Hosking.

that subtle experiential factors, such as tactile contact with the nest cup, contribute to the development of the nestling's ability to orient properly once an egg is nestled in its back.

Equally at odds with reality is the widespread notion that learning can take place without genetic contribution. The toad's ability to remember the visual properties of a noxious millipede and to associate this character with a memory of the disgusting taste of the prey reflects properties of its brain. The toad's nervous system would never have developed at all, let alone developed in a way that permitted information storage and recall, without the genetic information that is in all its cells and that it inherited from its parents.

To try to separate and quantify the relative contributions of genes and environment to the development of an "instinct" or a "learned behavior" is also a misguided endeavor. Genes and environment have totally different effects on development: the one regulating and structuring the process, the other providing the construction materials and the cues that trigger changes in genetic activity. Therefore it is meaningless to say the cuckoo's behavior is 90 percent genetic and 10 percent environmental or that the toad's aversion to millipedes is 71.5 percent environmental and 28.5 percent genetic. It is possible to quantify the contribution of genetic and environmental *differences* among individuals to the behavioral *differences* among them (Chapter 2). But the attributes of any one individual are the developmental outcome of an inextricable blend of genetic information interacting with environmental influences.

I believe that the terms "instinct" and "learning" can be given valid definitions. One meaning for "innate behavior" is an action that is performed in a functional manner the first time an animal of certain age and motivational state encounters the appropriate environmental cue for the behavior. The first time a baby cuckoo tries to maneuver an egg from the nest when it is a few days old, it is likely to succeed. The first time a baby toad ventures onto land after having metamorphosed from a tadpole, it will be able to feed itself when it encounters certain small prey items. The first time an adult female cowbird hears the male courtship song of her species, she is able to adopt the appropriate precopulatory pose (see Figure 21). A useful addition to this definition is that the performance of an instinct can only change *over generations* as individuals with different first-time responses are selected for or against. Thus, the feeding instinct of newborn

2 **Toad spitting out a millipede** (see bottom photograph) that it has found to be repellent and noxious. After a single experience with this and other poisonous prey, the toad will no longer attack them. Photographs by Thomas Eisner.

garter snakes has changed in coastal populations over evolutionary time as individuals with a tendency to attack certain odor cues have enjoyed a reproductive advantage.

In contrast, learning can be defined as adaptive behavioral changes that occur within the lifetime of one individual as a result of that individual's experiences. A toad that has had a mouthful of millipede alters its behavior because of the information it receives from this experience. The change is adaptive for obvious reasons, and it is enduring (the refusal to attack certain species of poisonous millipedes may last the lifetime of the once-educated toad).

As defined in this way, the distinction between instincts and learning implies nothing whatsoever about the relative importance of heredity and environment to the development of behavior. The key is whether the behavior in question appears in functional form the first time it is employed by an animal or whether the behavior represents an adaptive modification of an earlier appearing trait as a result of an experience that the animal has had. But even this definition is open to criticism for at least two reasons. First, there are practical difficulties in eliminating the possibility that what appears to be an instinct is actually derived from earlier responses that have been modified by subtle experiences in the animal's lifetime. It is conceivable that very young cuckoos make primitive or incomplete egg-ejection movements from which they derive sensory experience that modifies the timing of the performance of these movements or other aspects of the behavior. What looks like an unpracticed first-time performance of egg-ejection might be shaped by past experiences of the bird. One could perform a deprivation experiment and prevent a young cuckoo from ever experiencing sensory stimulation from a host's nest and eggs for several days before introducing it into a nest with eggs. But even if the deprived cuckoo performed the ejection behavior, this would still not exclude the possibility that some other experiential factors modified the behavior [325].

A second difficulty is that the term "instinct" carries such strong historical connotations of genetic determinism that carefully qualified definitions of the word tend to be ignored in favor of a persistent attachment to the old idea that instincts mean "genetically fixed" behavior. Therefore, a person who wishes to avoid misunderstanding may choose to avoid the term altogether (difficult though that is).

Because of the problems with the instinct-learning classification, I present here another classification scheme that focuses on the proximate developmental basis of behavior patterns. This classification system is based on the argument that the way in which behavior develops varies from action to action, and from species to species, over a wide and continuous range. This continuum can be arbitrarily divided into three categories: (1) behavior that is based upon a restricted developmental program; (2) behavior that develops as a result of a semirestricted (but somewhat flexible) program; and (3) behavior whose basis is a flexible developmental program [450, 493]. In the next three major sections that follow, I shall illustrate each of the three categories with a variety of examples.

Restricted Developmental Programs

Some behavior patterns are the developmental outcome of a process that appears to be highly channeled, requiring a minimum of sensory experience. The ability of baby garter snakes just out of the egg to perceive slug odor and to strike at objects endowed with the odor falls into this category. The hatchling's functional feeding response suggests that in the course of its development within the egg the molecular building blocks present in the egg yolk were employed in a restricted manner to construct a nervous system with specialized properties. The young coastal garter snake's brain detects and responds to particular sensory cues in a well-defined and useful manner. Moreover, the behavioral product is developmentally stable, as a snake will continue to perceive and attack slugs throughout its lifetime.

The European ethologists, led by Konrad Lorenz and Niko Tinbergen, have provided many other examples of behavioral characteristics that reliably appear at a particular stage of an animal's life, that are triggered by simple sensory cues, and that are difficult to alter once they have appeared [451, 704, 709]. In the jargon of ethology, these responses are FIXED ACTION PATTERNS that occur when an animal senses a SIGN STIMULUS (the effective sensory stimulation provided by an object). Thus, the newborn garter snake will open its mouth wide and lunge forward (the FAP) after having detected certain volatile chemicals (the sign stimulus) associated with a slug.

Sign stimuli that are communication signals used by the member of a species to activate social responses are called RELEASERS. A male blackbird can be induced to copulate with a few feathers plus the tail of a female blackbird provided the tail is raised in the position females customarily adopt prior to mating (Figure 3). Clearly the male is not attending to all the stimuli provided by a female of his species when she assumes the precopulatory display. The visual cues provided by the raised tail act as a releaser of mounting behavior by a sexually active male.

There are numerous other examples in which the highly mechanical nature of the relationship between a very simple cue and a complex behavior leads to a biologically inappropriate use of the behavior (Figure 3). Parent birds of many species react to the shiny whitish fecal sac produced by a baby bird by picking up the sac and carting the waste material away from the nest. If a human bird-bander places a shiny metal ring about the leg of a nestling, its parent may treat it as if it were a fecal sac, sometimes even to the point of actually throwing the baby out with the bathwater, despite its pathetic cries of distress [769].

The simplicity and stability of the rules governing some behavior patterns have been exploited (unconsciously) by certain organisms that have evolved mimetic releasers for their own benefit [781]. Possibly the premier code breaker is a rove beetle, *Atemeles pubicollis,* which lays its eggs in the nest of the ant *Formica polyctena* [328]. The larvae that emerge possess special glands at the tip of the abdomen. These produce an attractant pheromone that releases brood-keeping behavior in their hosts, causing the ant workers to treat the parasitic larvae as if they were their own young.

(A)
(B)
(C)
(D)

3 **Complex responses by birds** to simple stimuli. (A) A territorial male European robin will attack a tuft of red feathers while ignoring a more complete model of a robin that lacks a red breast. (B) Willow warblers are attacking the stuffed head of a cuckoo. (C) A male red-winged blackbird is copulating with a mount consisting of the tail of a female raised in the precopulatory position. (D) A group of jackdaws are attacking the arm of Konrad Lorenz as he carries a pair of black bathing trunks in his hand. These birds react indifferently to a human who is holding a nestling in his hand—provided the nestling has not yet acquired its first black feathers.

They groom the beetles and keep them in their brood chambers, where the larvae feast on ant eggs and ant larvae. Not content with raiding this larder, the parasites also mimic the food-begging behavior of ant larvae. They tap a worker ant's mandibles with their own mouthparts and so trigger regurgitation of liquid food. The larvae eventually metamorphose into adult beetles, which mimic the food-begging behavior of adult worker ants by tapping a worker with their antennae just as a worker ant would tap a nestmate to signal it to stop moving about. *Atemeles* then touches the ant's mouth with its forelegs and feeds on the regurgitated droplet presented by the ant (Figure 4).

In the fall the adult rove beetle leaves the *Formica* nest and moves to a nest of *Myrmica* ants. Species belonging to this group, unlike *Formica* ants, continue to raise brood and feed young throughout the winter, which assures a continuing food supply for the beetle during this period. (The *Atemeles* female returns to a *Formica* nest in the spring to lay eggs.) The beetle locates a *Myrmica* nest by its characteristic odor, being sensitive to this smell only for a brief time in the fall when the beetle is nest moving. When the beetle approaches the nest of its new host, it would normally be attacked as an invader or unwanted guest. However, as the workers rush to inspect the beetle, it presents the tip of its abdomen to them. There, a specialized appeasement gland secretes a substance that inhibits hostile behavior in the inspectors. A subdued ant then moves to the side of the beetle, where other glands exude a chemical communicator of unknown properties. It may be that this pheromone mimics *Myrmica* ant odors (just as it has been shown that another rove beetle species that lives in termite colonies produces the same distinctive hydrocarbons in its cuticle that are present in its species of host termite [349]). When the ant detects the chemical cue offered by the adoption glands, it grasps the hairs about the gland and dutifully hauls the parasite into the nest, where it will enjoy ant provender throughout the winter.

It is even possible for a code-breaking species to provide a more effective mimetic sign stimulus, or supernormal stimulus, than the biologically correct object. Several bird species, among them the European cuckoo and the North American cowbird, are BROOD PARASITES [305, 781]. The female of these species foregoes the time and energy expenses involved in building a

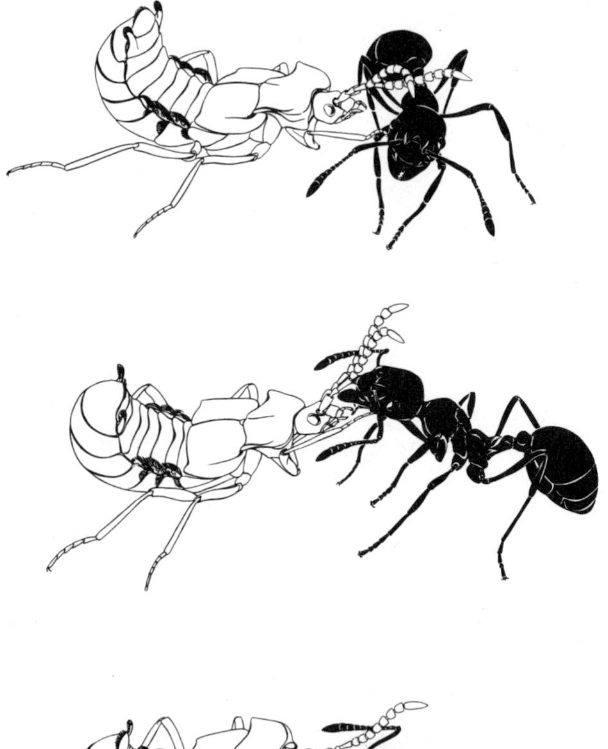

4 **Rove beetle adult** first taps a worker ant with its antennae, then touches the ant's mouthparts with its forelegs, and then consumes food regurgitated by the ant.

nest, incubating the eggs, and feeding the offspring. Instead, she locates the nest of some other species, generally smaller than herself. When the owner of the nest leaves during a pause in egg laying or incubation, the parasite that has been waiting nearby slips into the nest, quickly lays an egg, and disappears. When the owner returns, it often accepts the addition to its clutch, incubates the egg, and hatches the parasite. The young cowbird or cuckoo generally requires less incubation time and develops more rapidly than the host's offspring. This has a number of advantages. First, the newly hatched bird may have an opportunity to push some of the other eggs in the nest out onto the ground (Figure 1). Second, thanks to its larger size and more rapid development, it may also be able to eject its nestmates if they have hatched. Third, it provides, by virtue of its large size and its great demands for food, a supernormal releaser of parental feeding. The relevant cues that determine which of several birds in a nest will be fed by a parent returning with a food item are how high the bird stretches out of

5 **Tree pipit** attempts to meet the voracious demands of a fledgling cuckoo it has reared. Photograph by Eric Hosking.

the nest cup, how noisy its begging calls are, and how energetically it bobs its head and body. A large voracious nestling cowbird or cuckoo exaggerates each of these key cues and is therefore more likely to be fed than the host's own nestlings (Figure 5).

Innate Releasing Mechanisms

In order to account for mechanical responses to simple sensory cues, Lorenz and Tinbergen proposed that the nervous system of animals must have special units that detect sign stimuli and activate an appropriate reaction. They labeled these hypothetical units INNATE RELEASING MECHANISMS. They hypothesized that, given adequate nutrition for the full development of the nervous system, the appropriate "wiring" would emerge. These neural circuits would enable the animal to perceive special sensory cues and then play out a programmed series of instructions to the muscles that would generate the "correct" response. This hypothesis, although the object of much debate over the years since its introduction, has in its general form been supported by the findings of neurophysiologists in recent years [45, 325]. For example, A. O. D. Willows has shown that special cells within the brain of the sea slug *Tritonia* do have a programmed score of motor messages that they play back without guiding sensory feedback when other cells detect the presence of chemical substances from their arch-enemies, predatory seastars [790, 791]. Willow's model of nerve action has been modified slightly as a result of later discoveries [252], but it is clear that the slug's brain has special units that help it perform a highly patterned set of thrashing, swimming movements that make up a FAP (Figure 6).

Likewise, a male cricket's nervous system can produce a complete calling song even if the sensory components of its neural machinery have been severed. The cricket does not have to hear itself in order to sing a perfect calling song. Moreover, the "motor tape" that generates the species-specific calling pattern of a male cricket has been located in some species in a particular fiber consisting of the axons of a very few nerve cells [62]. The fiber runs from a particular region of the brain of the insect to a site in a

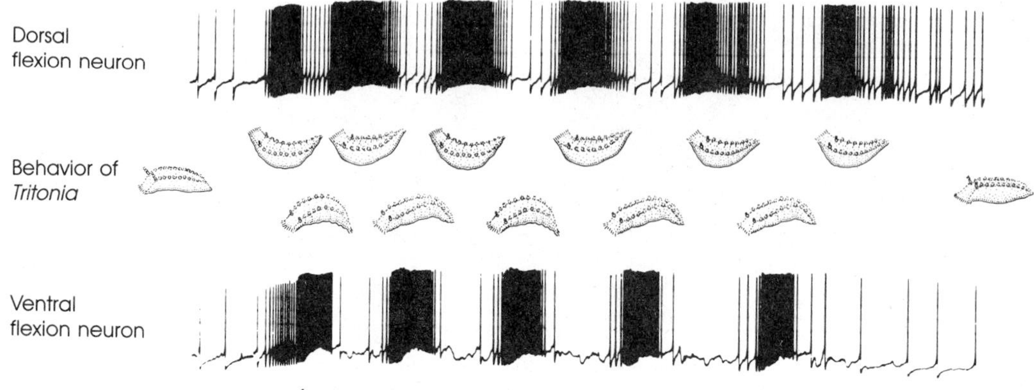

6 **Sea slug escape response,** a fixed action pattern. Recordings are shown from a dorsal flexion neuron (DFN) and a ventral flexion neuron (VFN). The DFN contributes to the contraction of the slug's back muscles and the VFN helps contract the slug's belly muscles. The alternation of contractions causes the slug to "swim" to safety. The system that contains and controls the DFN and VFN cells can be considered to be an innate releasing mechanism.

thoracic ganglion (a mass of neurons in the middle of the body that controls the legs and wings of the animal). It is the precisely timed movements of the wings that produce the song, the temporal pattern of which identifies the species-membership of the singer [753] (Figure 7). (The song is produced as the wings open and close, rubbing the thickened edge of one wing (the scraper) over the row of ridges (the file) on the underside of the other wing cover. Each movement of the scraper across the file generates a pulse of sound.)

If one stimulates the motor tape fiber with a weak electrical current from an electrode, it produces a train of impulses with the correct timing.

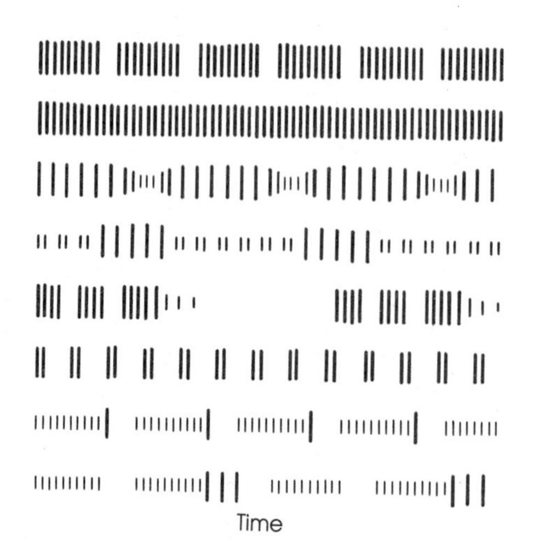

7 **Diversity in calls of crickets.** Each line represents the mate attraction call type of a different species. Each black bar symbolizes a pulse of sound; the height of the bar represents the amplitude of the pulse, and the width represents the duration of the pulse. Crickets achieve species-specific messages by varying the temporal aspects of sound pulse production.

These signals reach neurons in the thoracic ganglion, causing them to relay the right temporal pattern of orders to the wing muscles. The result is a perfect calling song, even in a brainless cricket. In an intact cricket sitting by its burrow, cells in the brain integrate information about things such as the time of day [443] and the weather conditions. If the setting is correct, brain cells fire, a response that turns on the interneurons in the fiber; signals travel to the appropriate battery of cells that contract and relax those muscles that control the calling song (Figure 8).

What is the developmental basis of this system? When a cricket hatches from a fertilized egg, it is a miniature version of the adult except that it lacks wings. If it lives long enough to undergo ten molts, it will become a much larger, mature adult capable of reproduction. The immature stages (or instars) never attempt to sing. Without wings, they could not produce the calling song anyway. The evidence suggests that early instars lack a completely developed, calling-song, neural network. These cells gradually form and are complete by the final nymphal stage. By selectively damaging a certain region in the last instar's brain (which normally suppresses attempts to sing), it is possible to cause the immature male to generate the normal calling-song pattern. The young cricket's wing buds (the two small pads of tissue that will eventually give rise to the wings during the final molt) move in the temporal pattern of the singing adult [62].

In an intact cricket, the male's brain never activates the calling-song fiber until some time after the final molt. But when it does, the first song

8 **Male cricket calling** at its burrow entrance. Note the moving wings. Photograph by E. S. Ross.

the adult sings will have perfect timing. This is true for males raised under a wide range of conditions. If a male receives sufficient food, its maturational process is regulated so that the appropriate interneuron(s) and other elements of the song network are in place and ready to operate when the adult male emerges from the cast cuticle of the last nymphal instar.

Male crickets do not alter the structure of their songs as a result of experiences they have while becoming mature or after they are adult. Developmental stability is a widespread feature of the courtship and copulation behavior of animals [493]. A mature male spider that locates a receptive female of his species mechanically performs a series of behavior patterns in response to sign stimuli. If he is fortunate, copulation follows. Depending on the species, courtship signals may include an elaborate set of leg-waving movements or the use of silk to tie down the female in a special way or the presentation of wrapped prey in a particular manner [90] (Figure 9). In some species the male runs the risk of being eaten prior to copulation if he does not provide the correct courtship messages. This, needless to say, acts as a strong selective force favoring males whose nervous systems enable them to generate a complete and accurate courtship sequence the first time they attempt to mate. If the male is eaten by his mate following copulation or if his chances of encountering another receptive female are almost nil, it is unlikely that males of these species would gain by being able to modify their reproductive behavior as a result of mating experience.

Semirestricted Behavioral Development

If at one end of the spectrum are behavior patterns that require a minimum of sensory experience for their full development, in the middle range are a variety of traits that will not take on their final form until the animal has received some special information in the course of its interactions with its environment. The territorial song of a male white-crowned sparrow is an example of a behavioral trait whose developmental program has some, but not much, flexibility. The young male while in the egg almost invariably develops neural circuitry that will enable it to begin producing a song when it approaches maturity. In addition, there are other neuronal systems (the "template"; Chapter 3) whose function it is to store information of a specific nature (memories of white-crown song) during a defined time period. Without this information the bird's song performance will never fully develop. But with it, the bird will sing a particular dialect for the rest of its life, with no further modification.

The sparrow's singing behavior shows how the instinct-learning dichotomy breaks down in some cases. One might label the twittering subsong of the juvenile male an "instinct," but as the bird engages in this behavior it receives sensory feedback that, in conjunction with its memories of the full songs it heard as a nestling and fledgling, serves to alter the initial behavior. The many restrictions on the developmental process guide the changes in singing behavior safely toward a typical full song of its species. It is not unreasonable in a case like this to speak of the young bird as being "pro-

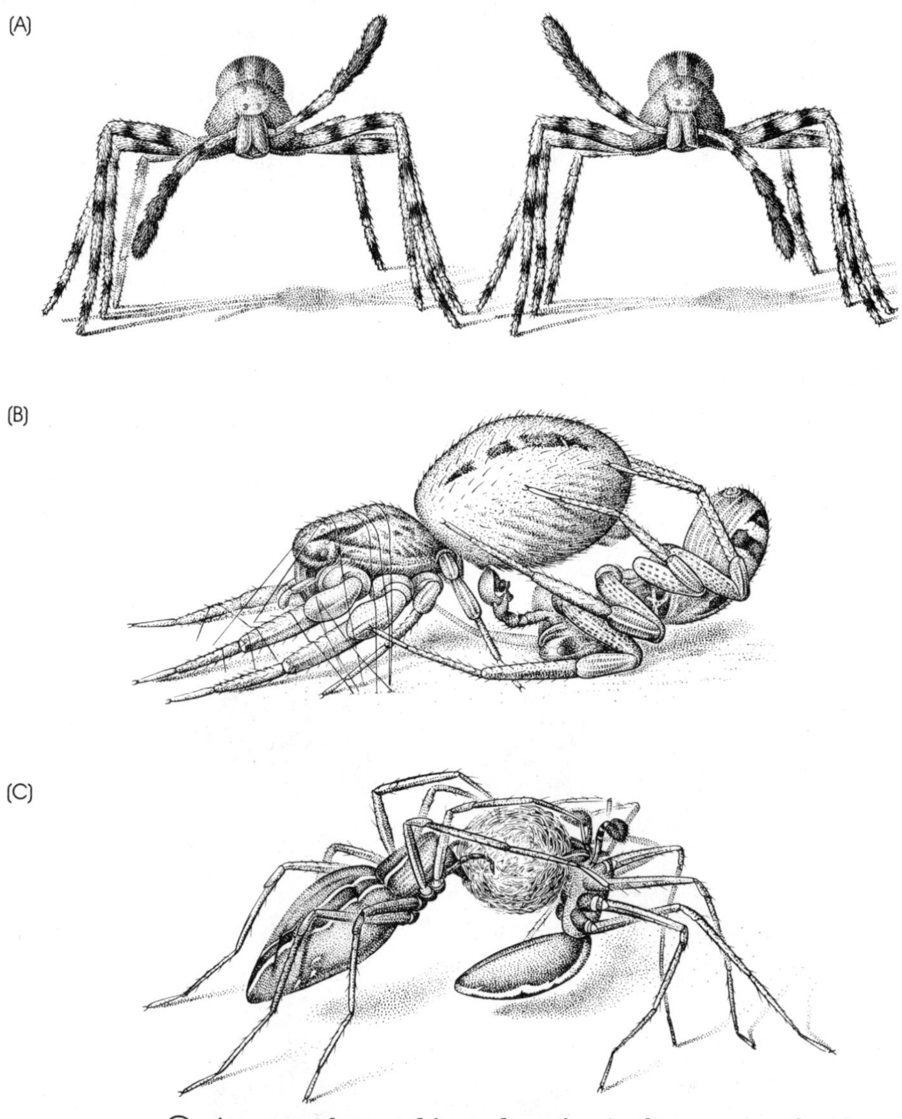

9 **Aspects of courtship and mating** in three species of spiders. (A) The courtship signals of a male wolf spider (*Lycosa amentata*) are being given before approaching the female. (B) The male of *Xysticus cristatus,* having fastened his mate to the ground with silk threads, is about to place sperm in her genital aperture. (C) The male of *Pisaura mirabilis* (on the right) is presenting wrapped prey to the female before mating.

grammed" to learn certain things. At the physiological level, song development reflects the way in which the bird's brain is "wired," which in turn represents the developmental outcome of the way in which food available to the embryo and young hatchling was organized in building the sparrow's nervous system.

There are many additional examples of behavior that are modified in a very constrained way as a result of the sensory experience an animal has while performing its original responses. Perhaps the classic example is the development of food-begging of young gulls. Niko Tinbergen showed that a baby herring gull, only a few hours out of its shell, will peck at the red dot on the tip of its parent's beak when the bill is swung back and forth in front of the chick (Figure 10). The pecks that hit the bill induce the adult to regurgitate a mass of half-digested fish, clams, and refuse from the local garbage dump—a rather unsavory, but much appreciated, first meal for the baby gull. Tinbergen used cardboard models of beaks and gull heads to determine exactly what stimuli triggered begging by the young gulls [710]. By offering various kinds of models to newly born gulls and counting the rate of pecking the models released, he and his associates determined that (1) long, pointed objects were more effective than short, stubby ones; (2) red was preferred over all other colors; (3) the greater the contrast of a dot on the model bill, the greater the rate of pecking; and (4) a moving model elicited more begging than a stationary one. But an elaborate three-dimensional model or a stuffed herring gull head was no more effective than a

10 **Great black-backed gull with young.** One of the chicks is about to peck at the dark red spot on the lower mandible of the adult bird. Photograph by Eric Hosking.

two-dimensional cardboard cutout in releasing pecking in a newborn baby gull.

Thus, shortly after hatching herring gulls have within their brains the neural circuitry that enables them to perform a functional begging response when they perceive a relatively simple sign stimulus. The begging behavior could be called an instinct. The begging action is, however, altered by experience, as Jack Hailman showed in his experiments with young laughing gulls [294]. He found that older chicks were much more selective in their response to models than younger ones. Although one- or two-day-old chicks would peck equally at almost any pointed object offered them, they soon began to refuse models that did not closely resemble laughing gull heads. As a result of their social interactions with their parents, the young birds acquired information, both auditory and visual, that led them to identify their parents as individuals. Eventually they begged only from their mother and father.

Thus, a young herring gull or a young laughing gull, like a young white-crowned sparrow, has a specialized ability to acquire information from its environment, information that alters an initial behavior pattern. The phenomenon of restricted learning is equally apparent in another consequence of parent–offspring interactions in some species—imprinting.

Imprinting

Konrad Lorenz, in his classic studies of imprinting in waterfowl, showed that baby ducks and goslings that normally follow their mother away from the nest within a day after hatching, could be induced to follow a substitute instead [452]. If Lorenz reared the birds himself, they would waddle after him when they were old enough to leave the hatching box. By virtue of this experience, the baby birds formed an immediate attachment to Lorenz and would follow him everywhere for days thereafter instead of following an adult female of their own species or another human being. They had evidently learned to recognize him as an individual just as they would normally learn to identify their mother. Even more remarkably, when they were mature many months later, the male birds would court human beings (including but not specifically limited to Lorenz) in preference to members of their own species.*

*The exploration of imprinting sometimes has rather exacting consequences [448]. A male jackdaw that had become sexually imprinted upon Lorenz insisted upon engaging in courtship feeding with his human "partner." "Remarkably enough, he recognized the human mouth in an anatomically correct way as the orifice of ingestion and he was overjoyed if I opened my lips to him, uttering at the same time an adequate begging note. This must be considered as an act of self-sacrifice on my part, since even I cannot pretend to like the taste of finely minced worms, generously mixed with jackdaw saliva. You will understand that I found it difficult to cooperate with the bird in this manner every few minutes! But if I did not, I had to guard my ears against him, otherwise, before I knew what was happening, the passage of one of these organs would be filled right up to the drum with warm worm pulp, for jackdaws, when feeding their female or their young, push the food mass, with the aid of their tongue, deep down into the partner's pharynx. However, this bird only made use of my ears when I refused him my mouth, on which the first attempt was always made."

11 **Mallard female** with her offspring clustered around her. Because they are imprinted on her, the ducklings will follow their mother wherever she goes. Photograph by Jon E. Cates.

Presumably a baby goose or duck can do these things because its nervous system is "primed" to be altered in a carefully defined way during the first few hours of its posthatching life. Under natural conditions, following a moving object (the mother) during this sensitive period has two distinct developmental consequences: (1) the formation of a social attachment to a specific individual, its mother (Figure 11); and (2) the eventual recognition of suitable mating partners by a male.

Imprinting and imprintinglike behavior occur in some mammals as well as some birds. For example, young shrews of an European species can be stimulated by their mother (when she wishes to lead her brood from one place to another) to hold onto the fur of another shrew (either the mother or a sibling). The mother then sets off with a conga line of babies trailing behind her (Figure 12). The role of imprinting in the attachment behavior of the young shrews has been explored with experiments [820]. When the baby shrews are 6 or 7 days old, they will form a "caravan" by grasping a cloth and many other substitute mothers. However, in the period between 8 and 14 days, they become imprinted on the odor of the individual that is nursing them. Usually, of course, this is their mother, and she alone after this period can induce caravan formation by the young animals. However, if 8-day-old shrews are given to a substitute mother of another species, they will become imprinted upon her and when returned to their biological

12 **Caravan of European shrews** with the mother leading the way. The young shrews learn the identity of their mother through olfactory imprinting.

mother at 15 days of age will not follow her or any siblings that had been left with her. They will follow a cloth impregnated with the odor of the foster mother, a behavior that proves they had learned the identity of the female that nursed them when they were 8–15 days old.

Language Development

Sexual imprinting, song learning by birds, and the learned recognition of parents are all products of developmental pathways that have special restrictions on what information can be acquired from experience. Moreover, the end product is highly stable. For example, once a sexual preference has been imprinted, it is extremely difficult to alter it, as zoo administrators know too well from the refusal of prized hand-reared animals to breed with a member of its own species in preference to attempts to copulate with a human attendant.

At first glance the acquisition of language would not seem to fall into the general category of behavior with a semirestricted developmental basis. Language superficially appears to be a highly flexible behavior, with more than 3000 languages recorded from around the world [525]. Moreover, we all know that humans can acquire new languages throughout their lives. On the other hand, all languages share certain things in common. Although the human vocal apparatus can produce a vast array of sounds, a total of only about 40 speech sounds (phonemes) are used to construct the thousands upon thousands of words in human languages; many of the same phonemes appear in different languages. Even more significantly, all languages employ their vocabularies of words in lawful (grammatical) ways to construct meaningful sentences.

Recent discoveries have shown that the brain of the human infant has properties analogous to those of young white-crowned sparrows, properties that facilitate the acquisition of a language. Six-week-old infants from

English-speaking households are able to discriminate among many consonant sounds including such similar pairs as "ba" and "pa." Peter Eimas discovered this by giving his infant subjects a pacifier wired so as to record the sucking rate of the baby. He then played tapes of artificially synthesized sounds that would be perceived by English-speaking adults as "ba," "ba," "ba," . . . . With the first few "ba" sounds, the infant's sucking rate typically went up; but with repetition of the sound, the attention of the infant declined and the sucking rate fell. When a new consonant appeared, such as "pa" or even a novel, but very similar, consonant sound that Thai speakers use but that does not occur in English, the infants became aroused and their sucking rate promptly accelerated. This work shows that specific perceptual mechanisms involved in the subtle discrimination of speech sounds have already developed in the infant's brain about the time it is born [204].

Soon after birth all young humans go through a stage in which they babble, producing a variety of unstructured sounds [414]. They seem to match the sounds they make against their stored memories of the language sounds they have heard. Their ability to make fine discriminations among similar consonant and vowel sounds surely aids in this matching process. If a young child cannot hear spoken language, it has no store of information about speech sounds and it will not develop a spoken language; it will go through the babbling phase but eventually will stop producing sounds altogether. The parallels with song development in sparrows are numerous and obvious (Table 1).

Vocabulary development occurs remarkably rapidly in children with normal hearing. At 18 months the average child has command of 50 words; just a year and a half later the youngster will be using about 1000 words

Table 1
Parallels in the Development of Territorial Song in Sparrows and Language Development in Humans

White-Crowned Sparrows	Humans
Selective attention by young animal to species-specific song syllables	Same, but to speech sounds
Critical period for storage of song	Some indication that humans must hear spoken language early in life if they are to acquire speech
Relatively unstructured subsong produced by young animal	Same, babbling
Full development of song depends on ability to hear self-generated sounds	Same requirement if spoken language is to develop fully
Population dialects arise through imitation by young animals of sounds produced by older ones	Same
Once full song is acquired, it is resistant to modification	Language acquisition skills decline with age

and understanding many more (in fact, at 30 months children appear to comprehend, if not obey, everything said to them [414]). The ability to process dozens of words in short periods of time and to understand their totally symbolic content is an astonishing human trait.

Moreover, the development of sentence structure of young children in English, Russian, Chinese, Finnish, and Zulu-speaking cultures, to name a few, follows a channeled pattern [525]. The child first uses one-word sentences ("No."; "Go."; "Mama."), then *always* enters a two-word stage ("Dada come."; "Bring me.") that endures for some time before the child enters the third phase: TELEGRAPHIC SPEECH, in which sentences consist of nouns without plurals, verbs without tense endings, speech stripped to its essentials. Yet even at the two-word stage, grammatical rules are employed; and as language development proceeds, the young human shows great skill in extracting (unconsciously) the rules of language use. Without a knowledge of these rules, language communication would be impossible. (Language knowledge these a communication without impossible be of rules.) My ability to communicate with you occurs because we agree (in largely automatic ways) that there are only certain ways subjects, verbs, predicates, verb tenses, and prepositions can be used. The young child acquires the necessary grammar skills without ever receiving formal instruction in them. In fact, correcting a young child is far less useful to it than providing conversational practice (this is not surprising given the difficulty professional linguists have in defining just what the grammatical rules of language really are).

Let me illustrate what I mean with an anecdote modeled after one reported by Ursula Bellugi [58]. Rule extraction by a youngster is shown in the following dialogue:

> CHILD: My mommy buyed me some toys today daddy.
> FATHER: Did you say that mommy bought you some toys today?
> CHILD: Yes.
> FATHER: What did you say she did?
> CHILD: She buyed me some toys today!
> FATHER: Did you say she bought you some big toys?
> CHILD: No, she buyed me some little toys.

This child and all English-speaking children learn automatically by listening that when one wants to talk about something that has happened in the past, the way to do is to add the suffix *-ed* to the verb in the sentence. None of us had to sit down and have this rule explained to us. Nor did we figure it out in the sense of having a conscious revelation about past tense. We just did it. Confusion arises when there are exceptions to general rules; in this case the irregular verb *to buy* has a past tense that violates the standard rule. The fact that the child makes the error *buyed,* although he or she had never heard any adult say *buyed,* shows that rule extraction involves more than simple imitation of adult speech. In some way the brain of a human child acts to acquire grammatical rules and proceeds to use these rules in the construction of novel sentences. The great virtue of language lies in the infinite number of meaningful combinations of words that can be formed if the combinations are grammatical. Children learn the appropriate rules easily and by the age of five years are completely competent language users. Although we take this for granted, the monumental nature of the achievement becomes clearer if we accept the estimate that ten linguists working full time for ten years could not program a computer with the language skills of the average five-year-old [525].

Language and the Brain

That the human brain's capacity for language far outperforms that of the most sophisticated computer program reflects its own remarkably specialized properties. Language ability requires certain regions of the human cerebral cortex [244, 251, 378], a fact known since the middle of the nineteenth century when Paul Broca and Karl Wernicke established that damage to two regions of the temporal lobe of the cerebral cortex (usually the left lobe) resulted in specific language deficits (Figure 13). Injury limited to Broca's area is associated with a nearly or totally complete inability to speak. However, although the affected individual may only be able to produce simple two-word sentences, he or she can read and can understand spoken language with no difficulty.

Damage restricted to Wernicke's area does not disrupt speech, which occurs with its normal rhythm. However, the sentences that are spoken lack real information (e.g., "I was over in the other one and then after they had been in the department I was in this one and . . ."). Similarly, although the person sees and hears perfectly well, written or spoken language is incomprehensible to him.

Development of these two areas of cerebral cortex, the one controlling verbal output, the other regulating understanding and message content, occurs in nearly all human beings. This suggests that they are the evolutionary product of natural selection for neural systems that facilitate language learning. If this hypothesis is correct, animals other than humans

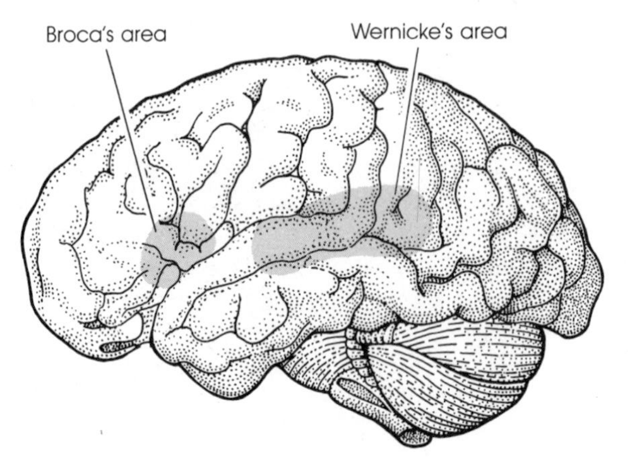

13 **Language centers of the human brain.** Two areas that appear to play key roles in human speech perception and communication are indicated.

ought to be incapable of acquiring language because there is no reason why they would have evolved the special semirestricted mechanisms necessary for the development of language.

The test of this hypothesis is available in the results of attempts to teach language to chimpanzees and gorillas [585, 621]. There is no question that neither species is capable of learning a *spoken* language. Although our fellow primates can make many complex sounds, no one has ever succeeded in teaching a chimp more than three spoken words, despite heroic efforts to do so. But researchers that have raised chimpanzees in a homelike atmosphere and have used American sign language, a language used by deaf humans, often (but not always) claim success in their endeavors [248]. There seems little doubt that young chimps can acquire a substantial "vocabulary" of signs (Figure 14) and that they use them in sequences. But other workers in this area are skeptical that this is equivalent to human language [474, 633, 694]. They argue that what looks like meaningful signing may often be cued (unconsciously) by the human teacher. Moreover, there is no doubt that chimps are relatively "intelligent" in the sense that they can learn to perform complex maneuvers to receive a reward, such as a piece of banana, or social stimulation, such as tickling. Thus, some observers have suggested that the chimps have learned mechanically to string

14 **Chimpanzee using sign language.** Nim Chimpsky is signing "me" with his left hand and "hat" with his right hand in response to his trainer's use of the sign "me". Photograph courtesy of H. S. Terrace.

together certain gestures for the reward they receive from success in this task.

Perhaps the strongest evidence in favor of the ability of chimpanzees to learn language is their occasional construction of *novel* sequences of sign gestures that obey the rules of grammar [248]. I think it is fair to say, however, that these events are extremely rare, especially in comparison with their occurrence in the speech of young humans. The enormous difficulties encountered in demonstrating persuasively that chimps can use a human language is evidence in itself that chimpanzee brains operate very differently from those of young human beings. It is easy to show that a two-year-old human infant can understand and creatively use a language. But doubts persist that even the most thoroughly trained chimpanzee is capable of a rudimentary language. The extraordinary training programs required to teach chimpanzees a battery of signs are not needed by language-learning infants. Human babies appear to find language rewarding in itself and do not require constant food rewards or positive social interaction for "correct" vocalizations. Even neglected or abused children will develop effective speech. In this context it is significant that some deaf children, unable to learn a spoken language and not taught a standard sign language, developed a sign language *of their own invention* that exhibited the grammatical properties of human language [263]. Our developmental systems apparently make it very difficult for us *not* to acquire the ability to use language. This is not surprising from an evolutionary perspective, given the enormous importance of language to the reproductive success of human beings.

Plastic Developmental Systems

In our review thus far we have seen that even apparently flexible and variable behavior, such as language learning, may rest on a developmental system that seems primed for the acquisition of specific experiential information. There are other behavioral traits that seem still more open to continuing modification than language, and these characteristics presumably have their foundation on more plastic developmental systems. But the distinction is one of *degree,* not kind, for even the most flexible of behavioral patterns requires a proximate basis that is structured to guide development of the characteristic toward productive ends. This can be seen through an analysis of the developmentally flexible behavior that is categorized as trial and error learning or operant conditioning. The feeding behavior of the toad is an example. Its dietary choices are to a considerable extent influenced by its feeding experiences and can change in an open-ended way as new food items appear in the animal's surroundings over the course of the seasons. This ability reflects properties of the toad's brain. Elements of its central nervous system generate rewarding sensations in response to sensory stimuli provided by nutritious edible prey and punishing sensations in response to cues offered by toxic, stinging, or poisonous prey. Nowhere is it written that the chemical substances secreted by a millipede *must* taste unpleasant and *must* be remembered to enable a consumer to avoid that species after one experience with it. It is conceivable that the millipede has a special predator capable of coping with its chemical defenses; such a predator might well find the taste of its prey rewarding, just as certain sea slugs feed with impunity on the stinging tentacles of some jellyfish.

The specialized nature of the mechanisms that promote flexible behavior can be illustrated by that most generalized of foragers, the Norway rat. The domesticated laboratory version of this animal has figured prominently in the analysis of OPERANT CONDITIONING [665]. In this form of learning, an apparently random response (an operant) that is not automatically released by a stimulus becomes associated with that stimulus after positive reinforcement (a reward). Imagine a laboratory rat in a Skinner box (named after the famous psychologist B. F. Skinner, who has devoted his life to the study of modifiable behavior). After it has been introduced into the box, the rat will wander about exploring its surroundings (Figure 15). Sooner or later, it accidentally pushes a bar on the wall, perhaps as it reaches up to investigate the side of the cage. As the bar is pressed, a mechanism releases a rat chow pellet into the food hopper in the box. Customarily some time passes before the rat (which has been deprived of food) discovers that there is something to eat in the food hopper. After this happens and the pellet is eaten, the rat may continue to explore its rather limited surroundings for some time before again pressing the bar in the course of its apparently aimless inspections. Out comes another food pellet. The rat may find it quickly this time and then turn back to the bar and press it repeatedly. It has learned to associate a particular activity with the availability of food. It is now operantly conditioned to press the bar.

Because, in their view, the control of behavior is simple (any positively reinforced behavior becomes more frequently practiced), Skinnerians believe that one can condition virtually any response. They point to the success of shaping experiments in which, by first reinforcing behavior that vaguely resembles the goal of a researcher and then rewarding better and better approximations, pigeons can be conditioned to waltz in circles or play table tennis with their beaks or "communicate" symbolically with one another [220, 664] (Figure 16). Skinnerian techniques have also been used to condition humans and other animals to regulate internal activities, which were previously thought to be entirely unconscious and automatically regulated [73]. Given appropriate reinforcements, a human can be taught to alter his heart rate while sitting in a chair or his brain waves (which are being recorded by electrodes taped to his head).

Constraints on Conditioning

Without question there are many impressive demonstrations of the power of reinforcement theory. But the Skinnerian approach, with its emphasis on the extreme flexibility of behavior, tends to overlook the fact that even operant conditioning is subject to a number of restrictions. By the mid-1970s psychologists had discovered numerous cases that showed

15 **Rat in a Skinner box.** The rat approaches the bar (*top left*). The rat presses the bar (*top right*). The animal awaits the arrival of a pellet of rat chow in the food dispenser (*bottom left*). The pellet is consumed and the rat is reinforced for having pressed the bar (*bottom right*). Photographs by Larry Stein.

that not all actions could be equally easily conditioned, counter to Skinnerian theory [652, 659]. For example, one can condition a rat to do some things like running in a running wheel by sounding a warning noise and giving it an electric shock unless it is engaged in the proper operant. After a rat has had some experience with hearing the sound and receiving a shock while standing in the wheel but not while running, it will make the appropriate association and start running whenever it hears the warning cue [77].

Another operant is standing upright in the running wheel. This is something rats do frequently, yet they cannot be conditioned in the manner just

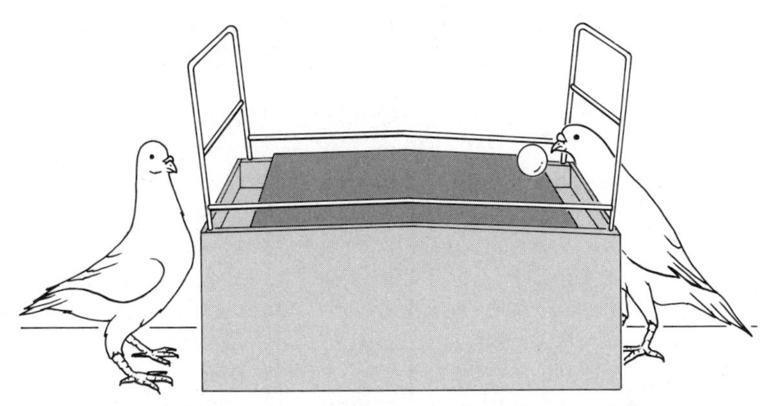

16 **Pigeon table tennis match.** If the ball falls into the trough on either end of the table, it trips a switch that produces a pellet of food for the pigeon that won the point.

described to stand in response to the warning sound. Rats that happen to rear up just after the sound are not shocked; but despite this, they fail to make the association between this behavior and avoidance of electrical punishment. The frequency with which they perform the standing response when they hear the warning signal actually declines over time (Figure 17). An evolutionary hypothesis to account for this result is that running is an adaptive response to a threatening sound whereas rearing up, which is an exploratory behavior, is not [77]. Therefore, at the proximate level rats are endowed with the neuronal circuitry that facilitates the formation of learned associations between auditory cues and running but not between

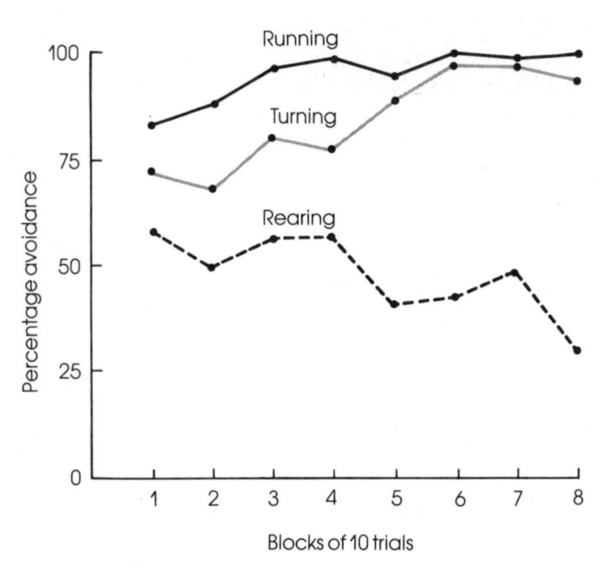

17 **Learning curves** for three operants (running, turning, rearing) that were equally rewarded. A shock was not given if the animal made the correct response. Note that the rats could not learn to rear on their hindlegs to avoid a shock.

auditory cues and rearing up when the reward is avoidance of painful skin sensations (such as those provided by shock and by biting rival rats).

The psychologist who probably has done most to advance the argument that there are evolutionary constraints on the learning process is John Garcia [246, 247]. He and his co-workers have examined the ability of white rats to learn to avoid various sensory cues that are associated with punishing consequences. One punishment, X-ray radiation, mimics the effects of ingestion of toxic substances. Laboratory rats are extremely sensitive to radiation, which disrupts cell metabolism and leads to the buildup of toxic products in tissues and body fluids. If a rat consumes a food or fluid with a distinctive taste and then is exposed to minute amounts of radiation, it will refuse to eat this material at a later time. Even tiny doses of radiation induce mild nausea, which the animal associates with any unusual tasting liquid or food it has recently consumed. Studies of the effects of illness-inducing treatments on the learned aversion to foods and fluids in rats have revealed the following rules that constrain the learning process. The degree to which the food or fluid that is linked with a treatment is avoided is proportional to (1) the intensity of resulting illness, (2) the intensity of the taste of the substance, (3) the novelty of the substance, and (4) the shortness of the interval between consumption and illness [247].

These rules make biological sense, given the food-sampling behavior of wild rats. Under natural conditions, a Norway rat becomes completely familiar with the area around its burrow, foraging within that area for a wide variety of foods, plant and animal [447]. New plants and insects are constantly coming into season and then disappearing. Some of these organisms are edible and nutritious; others are toxic and potentially lethal to the rat. A rat cannot clear its digestive system of toxic foods by vomiting. Instead, the animal samples novel food items in very small amounts; if they cause internal illness, the rat avoids them thereafter. If the rat becomes even slightly ill after consuming a new food item, it should avoid this food or liquid.

It is especially significant that even if there is long delay (up to seven hours) between eating a distinctive food and exposure to radiation and consequent illness, the white rat is still able to link the two events and use the information to modify its behavior. An aversion to the food is not formed as readily as when the illness follows immediately, but learning can still occur under the conditions of long delay. It is far more difficult to teach a rat to press a bar for food if the reinforcing food pellet is not delivered within three seconds of the bar press. In nature, if a rat manipulates an object with its paws and that object does not produce food quickly, the chances are high that it never will. The rat requires instant gratification if it is to learn to perform a manipulation in order to get food. On the other hand, poisonous foods rarely have an instantaneous effect, especially if eaten in tiny quantities. The rat's brain has evolved the ability to accommodate long delays between gustatory stimulation and illness because this accommodation helps individuals learn to avoid dangerous foods [247].

The moderately specialized nature of this form of conditioning is further shown by the rat's complete failure to associate a distinctive sound (a click) with internal illness. A rat is, however, perfectly capable of learning that a click sound means a shock is coming and will learn to leave a chamber in its cage if its departure prevents it from getting the shock. The reinforcement here is the avoidance of a shock punishment that is signaled by a click. Yet if the punishment is nausea, the rat cannot learn to leave an area and so avoid the sickening treatment. In addition, rats have great difficulty in making the association between a distinctive taste as a signal that a shock is about to be delivered. If, after drinking a sweet-tasting fluid, the rat receives a shock on its feet, it often remains as fond of the fluid as it was before (as measured by amount drunk per unit time), and this is true no matter how many times it is shock-punished after drinking sweet liquids. Thus, the nature of the cue and the consequence determine whether a rat can learn to modify its food and liquid consumption patterns (Figure 18).

Ultimately, only certain associations are likely to be biologically meaningful to rats. Sounds do not precede nausea in natural conditions any more than an unusual taste normally causes pain receptors in the skin to fire. The rat's genes guide the development of a brain designed to cope with real world events for rats: the sickening effects of eating a poisonous substance (linking taste with subsequent nausea) and perhaps the painful skin wounds that follow certain threatening vocalizations from aggressive fellow rats (linking sound with subsequent skin pain).

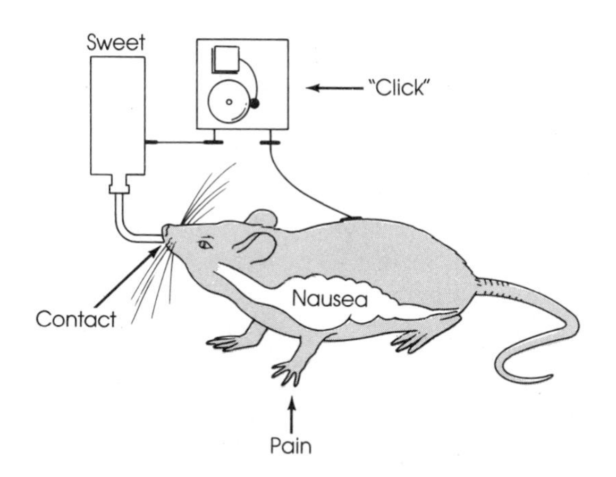

		CONSEQUENCES	
		Nausea	Pain
CUES	Sweet	Acquires aversion	Does not
	"Click"	Does not	Learns defense

18 **Selective learning** by the white rat. Rats easily make certain associations between cues and consequences, but they have great difficulty with other associations.

More recently, Garcia and co-workers have conducted experiments in which rats were permitted to drink a novel liquid with a mild odor (almond) and a highly distinctive taste (saccharin). Immediately after drinking, the rats were poisoned with a dose of lithium chloride. After a period of recuperation, the animals were allowed to consume unpoisoned water with a strong saccharin taste or with an almond scent. For many days after being poisoned, rats offered the scented water were more reluctant to approach it and drank far less than those given water with the distinctive taste [623]. Again, this feature of rat learning makes biological sense. Through their ability to pair odor cues with a memory of severe internal illness, rats need not taste a potentially toxic liquid or food in order to remember that they should avoid it. This may save them from repeated contacts with dangerous foods. Although the feeding behavior of the lab rat is remarkably plastic and modifiable, the things the animal can learn most easily are limited in ways that appear to increase individual reproductive success.

Adaptation and Constraints on Behavioral Development

Although even the most flexible of conditioned responses depends upon regulated developmental systems that impose restrictions on how experience can influence the course of conditioning, there is variation in the extent to which behavioral development is constrained. Why do animals vary in the degree to which the sensory stimulation they receive as juveniles or adults can alter the development of particular behavior patterns? Perhaps restricted, semirestricted, and flexible developmental systems offer a different mix of advantages and disadvantages to individuals in terms of their ultimate effects on gene-copying success. The ratio of benefits to costs varies depending on the ecological circumstances confronting an individual, so that under one set of conditions a behavior based on a rigidly structured developmental system may confer greater fitness than one based on a more plastic system. In other situations a more flexible response based upon a developmental system capable of incorporating individual experience may be the superior route to genetic success.

Consider how animals develop responses to potentially lethal foods. The Norway rat possesses physiological mechanisms that store information about the illness that follows consumption of such foods and leads the animal to avoid them after a single experience. The stable nature of this learning suggests that its developmental basis could be categorized as semirestricted.

A more constrained system is exhibited by fledgling motmots (tropical birds related to kingfishers), which do *not* require experience with coral snakes in order to develop an aversion to them. These birds eat lizards and small snakes. Even young fledglings are attracted to thin snakelike objects—unless the objects happen to have some of the visual characteristics of coral snakes (Figure 19). Susan Smith gave young motmots, which she had hand-reared from the nestling stage, opportunities to peck at painted wooden models that were about eight centimeters long [672]. The birds

19 **Coral snake.** The color pattern of this animal consists of bright red, yellow, and black bands. Photograph by James D. Jenkins.

readily approached and pecked at almost all models *except* those with alternating red and yellow rings, which they completely refused to touch (Table 2). Although inexperienced motmots avoided the color pattern of coral snakes, they readily pecked at models with red and yellow longitudinal stripes and at models with green and blue rings. The developmental difference between Norway rats and motmots is instructive. The dangerous foods that rats are likely to encounter are primarily poisonous plants (or

Table 2
Reaction of Motmots to Models of Various Sorts: Evidence for an Aversion to Yellow-and-Red-Ringed Pattern in Young, Inexperienced, Hand-Reared Birds

Model	Model Colors	Reaction: Number of Pecks	Reaction: Comment
	Yellow and red rings	0	
	Green and blue rings	89	
	Yellow and red stripes	60	
	One end yellow and red rings; other end plain	79	Only 15% directed at yellow-and-red-ringed end
	One end yellow and red stripes; other end plain	90	47% directed at yellow-and-red-striped end

Source: Smith [672].

baits). An interaction with one of these foods is not likely to be lethal if the rat only eats a very small portion (which it does). In contrast, the life expectancy of a motmot fledgling that picks up a red-and-yellow-ringed coral snake is extremely limited because of the deadly neurotoxin possessed by these snakes. Coral snakes have provided a selection pressure favoring nestling motmots that happened to be able to recognize and avoid the cues associated with these snakes by the time they had fledged and were beginning to feed on their own.

The Development of Predator Avoidance

There is a similar basis for the evolution of variation in the developmental systems underlying predator recognition and avoidance. A young sea slug, when it is touched by a predatory seastar that is entirely capable of and willing to kill it, has the perceptual mechanisms in place to detect key chemical cues from the enemy's sucker feet [790]. Sensory signals from the chemoreceptors activate cells in the slug's brain and "turn on" the appropriate motor tape that causes the animal to flop away from danger. If the slug were to perform its escape response imperfectly on its first encounter with a hungry seastar, it would become an eaten slug in short order. Therefore, it could not benefit by having to acquire information through personal experience that it should not permit itself to be grasped by a deadly predator.

But vervet monkey infants do learn certain things about their predators. In this species adults have four different alarm calls, one each for leopards, martial eagles, pythons, and baboons—all potential monkey killers [654]. Infant vervets also employ the same four calls, but the range of objects that elicits each signal is initially much greater than in later life [653]. For example, a harmless pigeon flying overhead may trigger the "eagle alarm call" in a young animal. Moreover, the reaction of the baby vervet monkey to an alarm call given by others is sometimes inappropriate (as when an infant walking on the ground will look overhead when another individual gives the "snake alarm"). Over time these kinds of mistakes become less frequent. Although it has not been proved, it seems likely that young animals learn through observation of their mothers and other older animals in the band what the truly dangerous enemies are and what responses are more effective. Because the young animal is partly protected by its mother and because it tends to err on the side of being too easily alarmed rather than too difficult to frighten, the chance that an initial error will be lethal is much less for the vervet monkey than for the sea slug. The young monkey has the time and opportunity to fine tune its responses to its particular environment through its experiences so that it avoids predators more and more economically and effectively.

The Development of Courtship Signals: Song Dialects

There is also diversity with respect to the role social experience plays in the development of courtship signals. For a cricket or spider whose life is brief and often largely solitary, there is a premium on the ability to produce the correct species-specific signal on the male's first attempt to attract a mate. This favors a restricted developmental system that increases the probability that a male's communication signals will be stable and effective.

A bird like the white-crowned sparrow differs from the cricket because it begins its life in close association with its singing father and other nearby males. It has the time and opportunity to acquire information about the details of the local song dialect. But what does it gain from its investment in the neuronal circuitry required to store this information? If the sole function of its song were to convey a message of what species it belonged to, a preprogrammed signal of the sort crickets use would seem entirely adequate. Perhaps, however, the dialect of a male's song conveys additional information that females may find attractive. Fledgling female white-crowned sparrows learn to recognize their population's dialect at the same time their brothers are storing the acoustical information they will need for the later development of full song. A recent study has proved that captive mature females (that have been primed with an estrogen implant) often adopt the precopulatory position upon hearing the dialect of their natal region but rarely do so when the tape plays songs of males with different dialects. By choosing a male with a familiar dialect, a female might gain a mate that had demonstrated his capacity to survive in the locale where the female's offspring will also live. Her selection of a locally adapted mate might improve her offspring's chances of acquiring the adaptations that are most effective for the region in which they too will live and breed [32–34, 306].

Mate choice by female white-crowns might be even more subtle. A female might prefer a male with the local dialect whose song was familiar but not *too* familiar. In this way she could avoid mating with her father or brothers and so avoid inbreeding depression (the production of congenitally defective offspring is often increased when close relatives mate) while at the same time accepting sperm from a locally adapted individual. Some birds apparently do use information acquired during early imprinting to avoid mating with parents or other close relatives (Figure 20) [49, 50].

Ecological Correlates of Song Dialects

We can test the hypothesis that female preferences for locally adapted males favors the evolution of song dialects by comparing species with and without such dialects. Donald Kroodsma has shown experimentally that the long-billed and short-billed marsh wrens (two closely related species) differ greatly in their ability to incorporate acoustical experience in the development of their songs [395]. The male long-billed marsh wren is like the white-crowned sparrow in its ability to copy the song types that it hears as a juvenile. In contrast, hand-reared short-billed marsh wrens that are exposed to a tape with various song types of their species do not develop copies of these sounds but instead improvise a considerable variety of songs in adult life. Why the difference?

Although these species are generally ecologically similar (males may

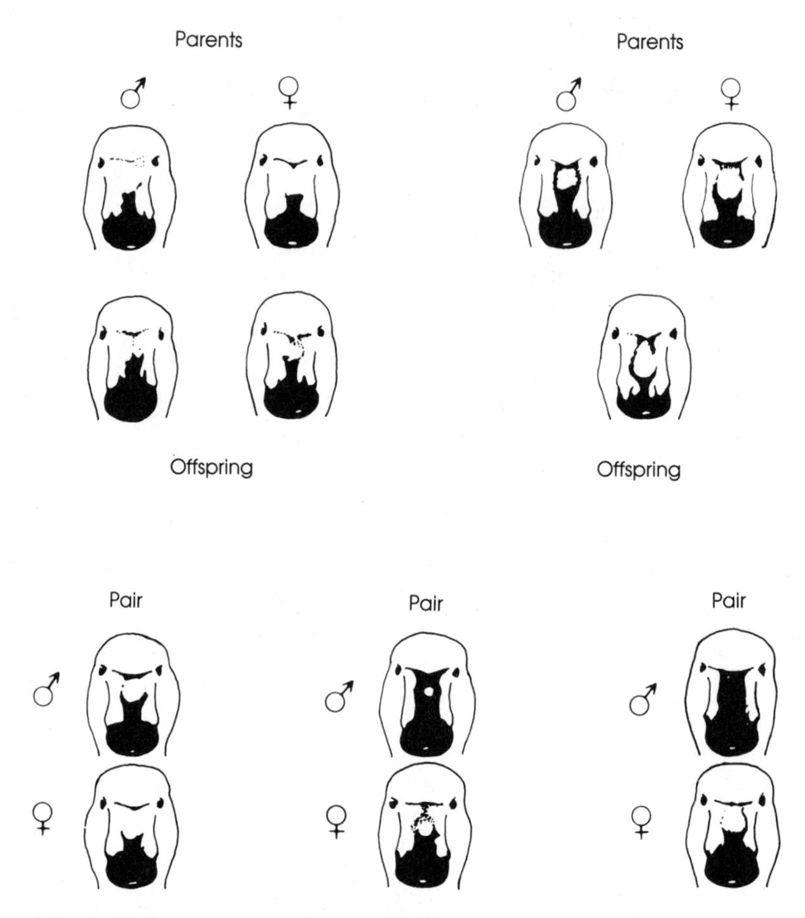

20 **Facial patterns in Bewick's swans.** (*Top*) Offspring resemble their parents. (*Bottom*) Adults often choose partners that are different in appearance than they are, suggesting that swans avoid inbreeding by choosing mates who do not resemble their parents.

acquire more than one mate and they nest in dense populations in wet meadows and marshes), there is one major distinction. The long-billed marsh wren nests in wetter areas that are less likely to dry up and become unsuitable for breeding. Short-bills nest in less stable habitats (flooded meadows) that may or may not be available in successive years. Long-bills typically return season after season to the same marsh, but short-bills are highly nomadic. There are distinctive song dialects in the long-billed marsh wren but none in the short-billed species, whose song everywhere begins with a stereotyped, nearly identical, introduction and concludes with a highly variable segment. This enables males to communicate effectively with short-billed marsh wrens born in many different areas. Because in this species there is no reproductive advantage to be gained from possession of locally adapted traits (because of the nomadic nature of the birds), females would gain nothing by selecting mates on the basis of a song dialect that identified them as members of a particular population.

The Adaptive Value of Restricted Song Development

Nomadic behavior and populational mixing are not the only environmental factors that can favor a restricted song development system, as the cowbird illustrates. This is a parasitic species whose females lay their eggs in nests of a wide variety of host species. Thus, young male and female cowbirds are reared in environments in which they are very likely to hear a singing red-eyed vireo or bluebird but not necessarily the songs of adult male cowbirds. Meredith West and her colleagues have shown that if one isolates baby cowbirds the males nevertheless develop functionally effective cowbird song and the females develop the ability to recognize their species' song [379, 772]. When one plays tapes of songs of various species to isolated females, they adopt the precopulatory position (Figure 21) only when hearing cowbird song (even without any previous experience with this song).

In the course of this research, West and her co-workers discovered that female cowbirds were much more likely to respond sexually to the songs of isolated males than to the songs of wild-caught, naturally reared individuals! This suggested that the apparently rigid developmental system underlying singing behavior in the cowbird has some flexibility. To identify what experiential factors might modify song development, West reared some birds in a screened flight cage in which the isolated male could hear other cowbirds but could not see (or interact) with them. These *socially* isolated males sang songs as potent as those of cowbirds that were both socially and acoustically isolated (Figure 22). Males that were reared in their own cage from which they could see into an adjacent enclosure where there were other males developed much less potent songs. But how can it

21 **Precopulatory response** of female cowbird. The response can be triggered by hearing the song of a dominant male. Photograph by Andrew King.

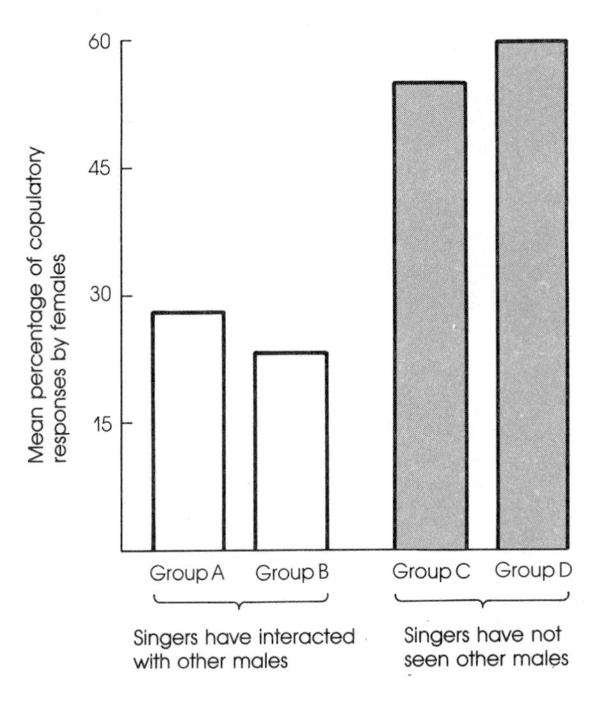

22 **Social experience and the effectiveness of male song** in cowbirds. Songs recorded from males that have interacted with other males (Groups A and B) are less effective in triggering the precopulatory position in females than the songs of males that have not been "inhibited" by experience with other males (Groups C and D). Males in groups A and B had at least visual contact with males in another cage, whereas males in groups C and D lived in partial or complete social isolation from other cowbirds.

be advantageous for a young adult male to sing a song that is less likely to induce a female to permit him to fertilize her eggs?

The answer to this question came from a revealing experiment in which naive, socially isolated cowbirds were introduced as young adults into cages containing a number of other males that had been living together. Typically such groups had established a DOMINANCE HIERARCHY in which one male dominated his cagemates and could displace them from their perches upon his approach. The other males ranked second, third, fourth, and so on in their ability to dominate their fellows. The top-ranking male in a group was generally the only bird to sing high-potency songs and as a result was able to monopolize the receptive females in the flight cage. Naive, isolated birds, after being introduced into the group cage, continued to sing their sexually effective songs, which stimulated the resident males to attack. Male cowbirds are capable of killing one another. But when birds are reared in a social group, they learn to suppress their potent songs as a result of interactions with more dominant males. A socially isolated male does not have the chance to acquire the experience that will teach it to restrain its potent song until it becomes dominant and can *safely* sing the kind of song that females find most sexually stimulating [772].

Reproductive Success and Developmental Flexibility

Because humans are developmentally flexible, capable of using personal experience to modify our behavior in so many ways, there is a tendency to think that this must be the evolutionarily superior tactic. This self-congratulatory conclusion is not justified. As our review demonstrates, a rigid developmental program can be highly effective under certain conditions. Because juvenile cowbirds may not hear other conspecific males singing, selection has favored those individuals whose song system can develop without benefit of this information. The male that relied on acoustical experience for completion of the development of his singing behavior might sometimes fail to acquire his species-specific signal and so fail to persuade females to copulate with him.

On the other hand, the social environment of young cowbird males is not completely predictable. There may be many or a few or no other more dominant competitors in the area in which it lives. Information about the nature of social competition is critical for determining the risk of dangerous attacks from its rivals. The young adult male acquires this information through its aggressive interactions with its fellows and uses experience to adjust the potency of its courtship signal to reduce the risk of assault. This flexibility means that the male is not locked into giving an impotent song for any set period but can adjust its mate attraction behavior to its particular circumstances. As soon as he becomes the dominant bird, a male will employ his most potent song.

Not all ecological conditions favor developmental flexibility. The ability to acquire, store, and use experiential inputs costs something, including the energy needed to manufacture and maintain the physiological substrate required for these processes. If the large brain size of humans is related to our great behavioral flexibility, as many people have argued, then the physiological expenses of human learning are great indeed. Although the brain makes up only 2 percent of our total body weight, it demands 15 percent of all cardiac output and 20 percent of the oxygen captured by the lungs. The brain is extremely sensitive to oxygen deprivation, a fact that has been responsible for the shortening of many lives. Moreover, it is also developmentally vulnerable to genetic accidents [86]; many of the known single-gene effects on human behavior involve severe brain damage resulting from a single mutant allele (see Chapter 2). Finally, our reinforcement networks, although often rewarding or punishing us in biologically appropriate ways [619], may malfunction, an event that causes various kinds of debilitating mental illnesses. A less massive brain, more simply constructed, might be less expensive to maintain and less susceptible to injury and operation failure.

Simple rules of behavior are probably cheaper in physiological terms, and they are highly effective in situations in which there is a reliable relationship between a particular cue and a specific, productive response. It is true that restricted developmental systems may produce behavior prone to exploitation by parasites and others, as in the case of ant-robbing rove beetles. But the evolutionary maintenance of a simple, cued response requires only that the behavior be more genetically advantageous for an individual on average than any alternative trait. Ants may occasionally feed a parasitic beetle, but almost invariably they succeed in feeding a fellow colony member. Feeding the beetle is a small disadvantage that presumably is outweighed by the reliable performance of a behavior that is nearly always genetically appropriate. An ant that happened to have a beetle detector in its brain would have to pay a physiological cost to construct and maintain these neural units; their cost might exceed any benefit because the probability of having to use the detector to avoid beetles is so slight.

The risk of exploitation or misapplication does not apply only to developmentally restricted behavior. It is entirely possible for an animal to modify its behavior as a result of an experience in a way that lowers, not raises, its fitness. A human being or a Norway rat may mistakenly learn to avoid an edible food if the first time he eats the food it happens to be spoiled or if he consumes it and becomes ill for some other reason [67, 135]. By associating illness with the novel item, lifetime aversions to perfectly nutritious foods have become firmly established.

The curiosity and intelligence that are trademarks of human beings have been genetically advantageous in the past, but they have inadvertently (again from the genotype's perspective) led to the discovery of highly effective methods of birth control, which may be used to reduce individual fitness, not raise it (see Birth Control in Chapter 15). Likewise, the reinforcement mechanisms of human brains may often teach us to do things that are linked with gene-copying success, but they are also used by alcoholics, extremely obese persons, and drug addicts in ways that are harmful to survival and reproductive success.

Learning abilities not only may be misapplied but also may be exploited by conspecifics and members of other species (Figure 23) [98]. Humans regularly practice deception, providing false information for others to learn. This trait, however, is not restricted to human beings. For example, a pregnant female langur sometimes undergoes a false estrus after a takeover has occurred in her troop (Chapter 1) and will mate with the new band owner even though she cannot be fertilized. The female may be attempting (unconsciously) to deceive the new resident male into accepting the infant when it is born as one of his own instead of killing it as a competitor's offspring [350]. Males may store information about which females they have inseminated in order to learn which infants not to kill, but their mates may exploit this learning ability for their own ends.

Therefore the superiority of a trait is not determined by its developmental flexibility but by its costs and benefits relative to alternative traits. The ability to learn from experience does help if the individual can adjust to variable conditions that could not have been completely anticipated prior to its birth, such as the location of food, hiding places, and paths in its living space, or precisely what edible prey species are available, or exactly what its parents look like, or what language is spoken in its home. It is the variety of ecological problems facing individual animals that sets the costs and benefits of the different degrees of developmental flexibility and thereby generates the great diversity of animal behavior.

23 **Four edible flies** that resemble stinging insects. Exploiting predators' ability to learn to avoid noxious prey, these insects mimic a yellowjacket wasp (*top left*), an eumenid wasp (*top right*), a honeybee (*bottom left*), and a paper wasp (*bottom right*). Photographs by the author.

Summary

1 The traditional classification of behavior into the proximate categories—instinct versus learning—led to the widespread, but mistaken, belief that some behavior patterns are more genetically determined or more environmentally determined than others. Genes and environment play different, but complementary, roles in the development of all behavioral traits.
2 Another proximate classification scheme recognizes that behavioral development may be more or less dependent upon information acquired through experience by the individual following its birth, hatching, or emergence. At one end of the continuum is the restricted development category in which the neural circuitry required for stable, functional traits such as the copulatory behavior of certain spiders or female cow-

birds emerges with a minimum of experiential shaping. These behavior patterns appear to be governed by a simple rule: when cue X appears, perform response Y.

3 Other traits such as sexual imprinting and the singing behavior of some birds seem to be the product of a developmental program with some (limited) flexibility. Here key experiences provide information critical for the completion of the development of the animal's nervous system and behavior. Still other behavioral characters such as trial and error learning or exploratory learning are even less constrained and may be modified repeatedly throughout an animal's life. But these traits too depend upon specialized physiological mechanisms capable of incorporating information from a variety of experiences.

4 No one kind of developmental system is intrinsically superior to another in terms of its contribution to individual fitness. Simple rules of behavior can be more adaptive than complex learning abilities under the appropriate conditions. The benefit-to-cost ratio of a behavior pattern depends upon such things as the costs of constructing the physiological foundation of the trait, the danger that the behavior will be exploited by parasites and others, the cost of making an initial mistake in an interaction with a predator or toxic food, and the degree of variation in the social environment that cannot be anticipated prior to the birth of the individual. These factors vary considerably from species to species. The result has been the evolution of variety in developmental systems and behavioral abilities.

Suggested Reading

Readers interested in the approach and findings of traditional European ethologists should see Niko Tinbergen's *The Herring Gulls' World* [707], the anthology of Tinbergen's major studies [709], or Irenäus Eibl-Eibesfeldt's textbook, *Ethology, The Biology of Behavior* [203].

Konrad Lorenz has written a good article on the ethological approach to learning [450]. John Garcia's review of his controversial research on white rat learning is also recommended [247].

Bert Hölldobler's *Scientific American* article on ants and their nest parasites makes excellent reading [328], as does the report by West, King, and Eastzer on song development in the cowbird [772].

Suggested Films

Birth of the Red Kangaroo. Color, 21 minutes. A beautiful film that demonstrates the complex behavior of the newborn infant kangaroo, behavior that is the product of a restricted developmental program.

Complex Behavior: Chaining. Black and white, 7 minutes. A film illustrating what can be accomplished through conditioning.

Development of the Child: Infancy and *Development of the Child: Language Development.* Color, 20 minutes each. Two good films on different aspects of human child development; these films illustrate the behavioral abilities of newborns and the programmed nature of language learning.

The First Signs of Washoe (F2021). Color, 56 minutes. A superior film on teaching sign language to a chimpanzee.

Imprinting. Color, 37 minutes.

Konrad Lorenz: Science of Animal Behavior (No. 05909). Color, 14 minutes. A look at the founder of ethology and the phenomenon of imprinting.

Token Economy: Behaviorism Applied. Color, 20 minutes.

To Alter Human Behavior: . . . Without Mind Control. Color, 22 minutes. This film and the preceding one deal with the application of Skinnerian techniques to the modification of human behavior.

The three previous chapters on the proximate causes of behavior were based on the proposition that genes have their influence on behavior indirectly by affecting the development of physiological mechanisms within individuals. We have already discussed the profound consequences on behavior that endocrine mechanisms have in some species. Thus, in the rat, hormonal physiology regulates the development of the rat brain and primes mature animals to respond to certain stimuli. Here our focus will be on how nerve cells operate to produce behavioral responses to stimulation from the environment. Nerve cells are highly specialized for the perception of certain events in an animal's external and internal environment and for ordering rapid responses to this information. Although one can make some generalizations about how nerve cells carry information throughout an animal's body, these units are tremendously diverse, both within individuals and across species, in their structure, electrical properties, sensitivity to different kinds of stimulation, and effects on other components of neural networks. If this diversity is the product of natural selection, the variety of perceptual and behavioral abilities of animals should be related to the solution of obstacles to reproduction by individuals. This chapter describes a series of examples of different kinds of neural mechanisms in different species that appear beautifully adapted to particular ecological problems. These examples should persuade the reader that our own perception of and response to the world around us are not at all representative of animals generally but, as in every other species, reflect a unique nervous system with adaptive limitations, biases, and distortions.

CHAPTER 5

Nerve Cells and Behavior Patterns

How Do Moths Evade Bats?

We shall illustrate many of the principles of this chapter with an analysis of how the nervous system of certain noctuid moths helps them avoid a deadly enemy and thereby improves the adults' chances of surviving long enough to reproduce. The enemy consists of nocturnal insectivorous bats. On a summer evening, one can sometimes watch bats hunting moths over open grassy areas. As Kenneth Roeder notes, all that is required is "a minimum amount of illumination, perhaps a 100-watt bulb with a reflector, and a fair amount of patience and mosquito repellent" [608]. A patient observer would sometimes see a bat catch a moth in its tail membrane and fly off with his catch. But if he or she were acute enough, the bat watcher would also see some flying moths turn abruptly even before a bat came rushing into view. Moreover, the observer might see moths dive or cartwheel out of the grasp of an approaching bat. These observations indicate that some moths have the ability to detect a bat at some distance and can make themselves difficult to capture.

But how do they do this? It is hard to believe that they can see bats far off at night, and there are for human observers no other cues that could alert the moth to approaching danger. But this is because we cannot hear high-frequency vocalizations, which bats produce in abundance as they fly along; most of these sounds are so high pitched that human auditory receptors fail to respond to them. But these sounds are precisely what moths can hear.

First, however, why do bats vocalize as they cruise the night sky? In the 1950s, Donald Griffin proposed that bats employ a form of sonar, that is, they emit pulses of high-frequency sound and then listen for the weak echoes reflected back from objects in the animal's flight path [282, 283]. He tested this hypothesis by placing the little brown bat, a common New England species, in a room in which fruit flies had been released and wire obstructions had been strung from the ceiling to the floor [284]. Some bats were cooperative and flew successfully about the room, uttering their cries and gobbling up the flies. Griffin then proved that a bat's calls were critical for its navigational abilities by turning on a machine that generated high-frequency sounds above 20,000 hertz, in the range of those produced by the predators themselves. As soon as these extraneous sounds began to bombard them, flying bats began to collide with obstacles and crash to the floor, where they remained until the jamming device was turned off. In contrast, noisy low-frequency sounds of 1000–15,000 hertz had no effect because these sounds did not mask the high-frequency echoes that the bats require if they are to fly safely and find food in the dark.

The bat's reliance on a sonar system has favored individual moths that can detect the pulses of sound produced by bats navigating at night [608, 609]. The detection device consists of two "ears" located in the thorax of the moth (Figure 1). A moth ear has a tympanic membrane on the outside of its body; to the membrane are attached two sensory receptor cells, A1 and A2 fibers. When airborne vibrations strike the moth, they may cause the tympanic membrane to vibrate. The mechanical energy in these vibrations reaches the receptors and may induce them to respond. The response is manifested in a change in the permeability of the membrane of the cell to sodium ions. The positively charged ions enter the cell at a point near the tympanum and this alters the charge differential across the membrane at a neighboring point. Sodium ions enter at this site and repeat the effect at a neighboring section of membrane, initiating a cascading series of

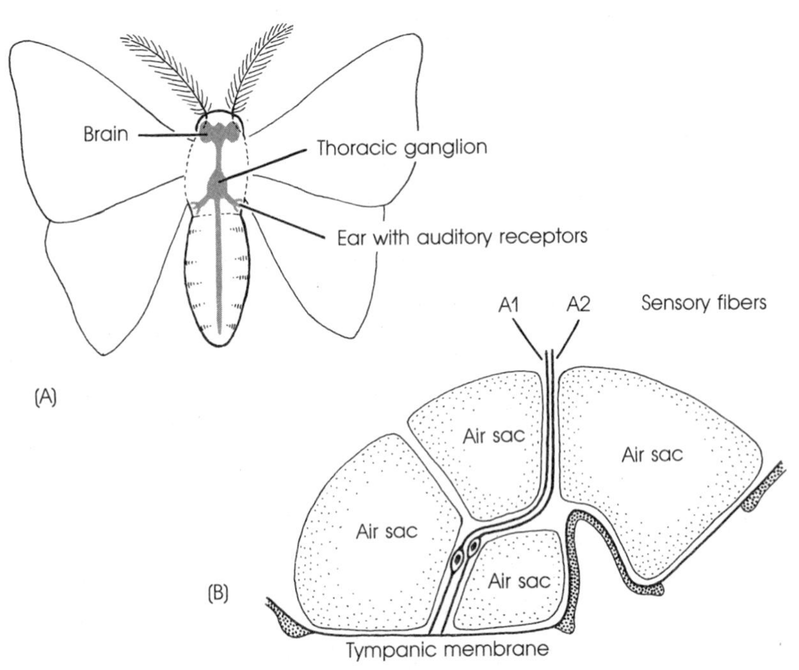

1 **Ear of the noctuid moth *Agrotis ypsilon.*** (A) The location of the ear and (B) its design. Sounds of sufficient intensity that strike the tympanic membrane will induce the auditory neurons to fire.

changes in membrane permeability that sweeps along the cell toward and around the *cell body* until it reaches the proximal end of the *axon* (Figure 2). Depending upon the intensity of the activating stimulus, the change in cell permeability to sodium ions may be sufficiently great to induce the cell to fire. A neural message, or ACTION POTENTIAL, is a brief, standardized change in membrane permeability to sodium ions that travels the length of the axon to the point of near contact (the SYNAPSE) with the next cell in the network. Nerve cells (or NEURONS) communicate with one another in a number of ways. A common method is for the arrival of an action potential at a synapse to cause the release of TRANSMITTER SUBSTANCE by one cell that reaches the neighboring cell(s). These substances may affect the membrane permeability of the next link in the network in ways that increase *or* decrease the probability that the cell will produce its own action potential(s). (See Kuffler and Nicholls [398] and Ewert [229] for a detailed description of the elements of neural action and interaction.)

The cells that relay receptor information to the brain (or its equivalent) are called SENSORY INTERNEURONS. Their messages can change the activity of neural activity in cells in the central nervous system, which in the moth consists of the aggregates of neurons in the thoracic ganglia and other

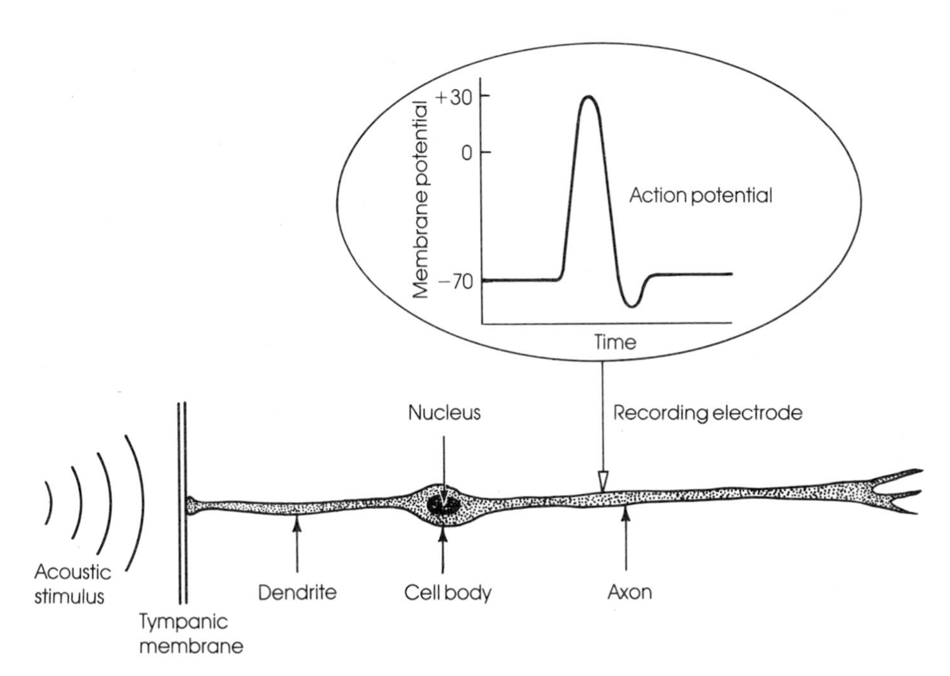

2 **Diagram of acoustical receptor.** The electrical activity of the membrane of the dendrite of the cell can be altered by certain acoustic stimuli. These changes, if sufficiently great, can trigger an action potential that starts near the cell body and travels along the axon of the receptor toward the next cell in the network that processes acoustical information.

ganglia in the head. These cells can be thought of as decoders, analyzing sensory inputs and making "decisions" about which responses to order. Certain patterns of activity in the thoracic ganglia affect MOTOR INTERNEURONS, whose action potentials in turn reach nerve cells that are connected with the wing muscles of the moth. When a motor neuron fires, the transmitter substances it releases at the synapse with a muscle fiber induce complementary changes in membrane permeability in muscle cells. These regulate the contraction or relaxation of the muscle, with consequent effects on the movements of the wings and the moth's behavior.

An action by a moth (or other animal) can, therefore, be viewed as an integrated series of changes in cell chemistry, initiated by receptor cells and carried on by sensory interneurons, brain cells, motor interneurons and motor cells, and muscles (Figure 3). Because these changes can occur with remarkable rapidity, an individual can react to changing stimuli in its environment almost immediately and can make fine adjustments in its behavior in fractions of a second.

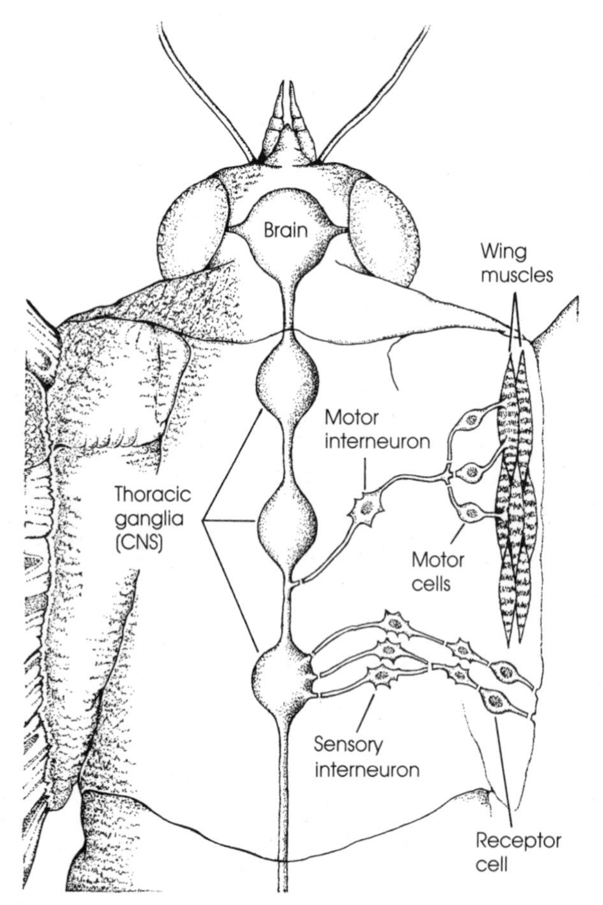

3 **Basic elements of nervous systems,** illustrated with a noctuid moth. Receptors detect sensory information. Sensory interneurons relay receptor messages to the central nervous system (CNS), where cells command motor interneurons to activate motor cells. These neurons regulate muscular responses and so generate behavior.

Receptor Cell Design and Adaptive Information Gathering

Nowhere in this scheme are genes obviously present. But their information indirectly influences an animal's behavior by their impact on the development of each cell of the nervous system, its connections with other cells, the properties of the cell's membrane, the nature of its transmitter substances, and its reactivity to various kinds of stimulation. This is evident in the different responses of the two auditory receptors of noctuid moths, whose capabilities were defined in simple, but elegant, experiments performed by Kenneth Roeder [608, 609]. He attached recording electrodes to each sensory fiber in a living, but restrained, moth and projected a variety of sounds at the ear. The electrical activity that resulted was relayed to an oscilloscope, which can convert action potentials into a visual pattern on a screen and on a paper reel. The result is a permanent record of the activity of a nerve cell in response to different kinds of acoustical stimulation. Roeder made the following discoveries about the two receptors (Figure 4):

1. The A1 cell is sensitive to low-intensity sounds. The other receptor is not and begins to produce action potentials only when a sound is loud.
2. As sounds increase in intensity, the A1 neuron fires more often and with a shorter delay between arrival of the stimulus at the tympanum and onset of a series of action potentials.
3. The A1 fiber fires much more frequently in response to pulses of sound than to steady uninterrupted sounds.
4. Neither neuron responds differently to sounds of different frequency over a broad ultrasonic range. A burst of sound of 20,000 hertz elicits basically the same pattern of firing as an equally intense sound at 40,000 hertz.
5. The receptor cells do not respond at all to low-frequency sounds. The moths are evidently deaf to sounds that we can easily hear.

Although each ear has just two receptors, the amount of information about the presence and location of bats that the receptors can relay to the central nervous system for analysis is impressive. Note first that the A1 fiber is most sensitive to *pulsed, ultrasonic* sound. Bat orientation cries consist of repeated bursts of sound, four to five per second in the case of the little brown bat. The highly sensitive A1 fiber begins firing to cries from a bat that is about 100 feet away, long before the bat could detect the moth. Because the rate of firing in this cell is proportional to the loudness of the sound, the insect has a system for determining whether the bat is getting closer.

In addition, the moth's ears gather information that can be used to locate the bat in space. For example, if a hunting bat is on the right, the A1 receptor on the right side will be stimulated sooner and more strongly than the A1 receptor in the left ear, which is shielded from the sound by

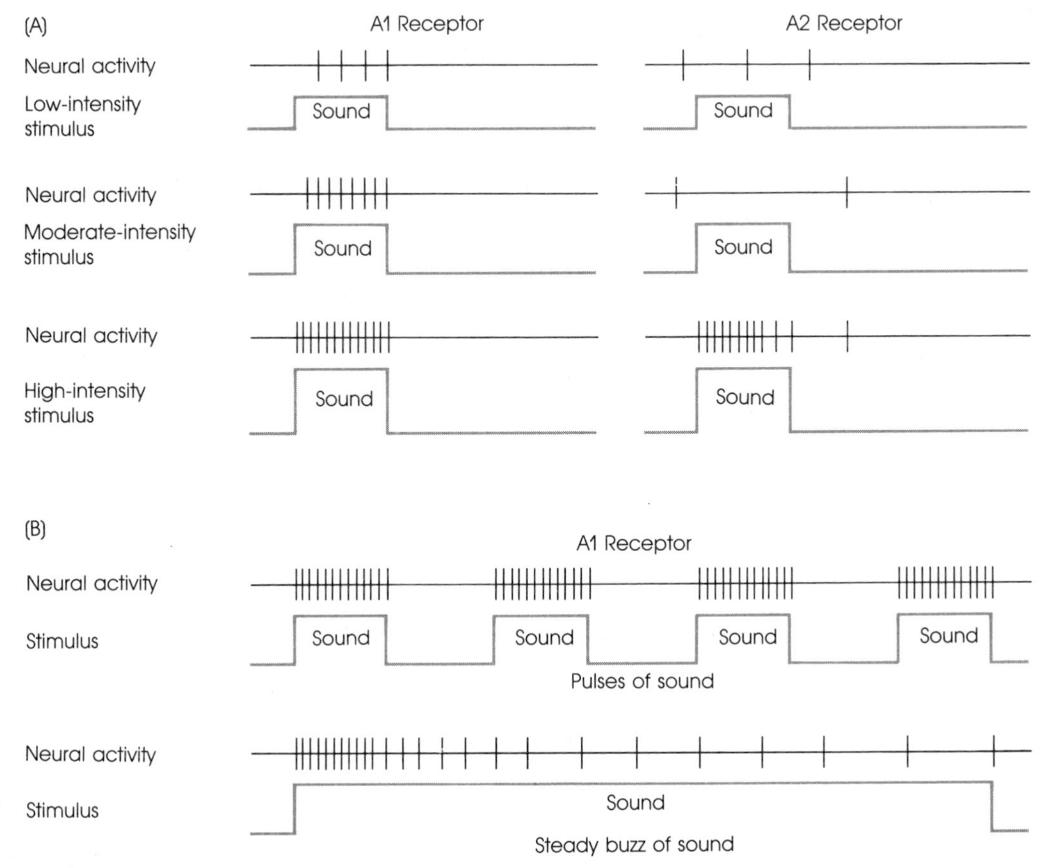

4 **Response of moth's auditory receptors** to sounds of various sorts. (A) The A2 receptor does not react to sounds of low or moderate amplitude. The A1 fiber fires sooner and more often as sound intensity increases. (B) The A1 receptor reacts strongly to pulses of high-frequency sound but ceases to fire after a short while if the stimulus is a steady hum of sound of the same frequency and intensity.

the moth's body. As a result, the right receptor will fire sooner and more often than the left receptor. (If the bat is directly behind the moth, both A1 cells will be equally active at the same time.) Thus, the brain's decoder neurons, by comparing sensory inputs from both ears, can locate the bat in the horizontal plane.

The moth's nervous system can also determine whether the bat is above it or below it. If the predator is higher than the moth, then with every up and down movement of the insect's wings there will be a corresponding fluctuation in the rate of firing by the A1 receptors when exposed to and then shielded from bat cries by the wings. If the bat is lower than the moth, there will be no such fluctuation. A summary of

A1 activity in response to stimuli coming from various points in space is provided in Figure 5.

As the waves of neural activity initiated by the receptors sweep through the moth's nervous system, they may ultimately generate two basic patterns of activity in motor neurons. One battery of motor messages causes the moth to turn and fly directly away from a source of ultrasonic sound [611]. This may help it avoid detection by a bat. When a moth is moving away from a bat, it exposes less echo-reflecting area than if it were flying at right angles to the predator and presenting the full surface of its wings to the bat's vocalizations. If a bat receives no echoes from its calls, it cannot detect a prey. Bats rarely fly in a straight line for a long period, and therefore the odds are good that a moth will remain undetected if it can stay out of range for a few seconds. By then the bat will have found something else within its eight-foot moth detection range and will have veered off in pursuit of it.

In order to employ its *antidetection* response, a moth need only orient so that the auditory stimulation of each ear is equalized, thus synchronizing the activity of the two A1 fibers. Differences in the rate of action potential production by the receptors in the two ears are monitored by the brain, which relays neural messages to the wing muscles via the thoracic ganglia and allied motor neurons. The resulting changes in muscular action steer the moth away from the side of its body with the ear that is more strongly stimulated. As the moth turns, it will reach a point where both A1 cells are equally active. If it then maintains this condition, the insect will be flying in the same direction as, and away from, the bat.

Although this reaction is an effective one if the moth has not been detected, it is not useful if the bat has come within the eight-foot detection range because the predator could easily overtake a moth that tried to fly away from it. Moths close to bats do not engage in futile antidetection behavior but instead employ evasive responses, including wild loops and power dives, that make it relatively difficult for bats to intercept them. A moth that executes a successful power dive and reaches a bush or grassy spot is safe from further attack because echoes from the resting place mask those coming from the moth itself [611].

Roeder has speculated that the physiological basis for the erratic flight of the moth lies in circuitry leading from the A2 fiber to the brain and back to the thoracic ganglion [610]. Whenever a bat is about to collide with a moth, the intensity of sound waves reaching the insect's auditory receptors is high. It is under these conditions that the A2 cells are stimulated sufficiently to fire. Their messages are relayed to the brain, which in turn may shut down central steering mechanisms in the thoracic ganglion that regulate the activity of motor neurons (Figure 3). When the steering mechanism is inhibited, the moth's wings begin beating out of synchrony or irregularly or not at all. As a result, the insect does not know where it is going, but neither does the pursuing bat, whose inability to plot the path of its prey may permit the insect to escape.

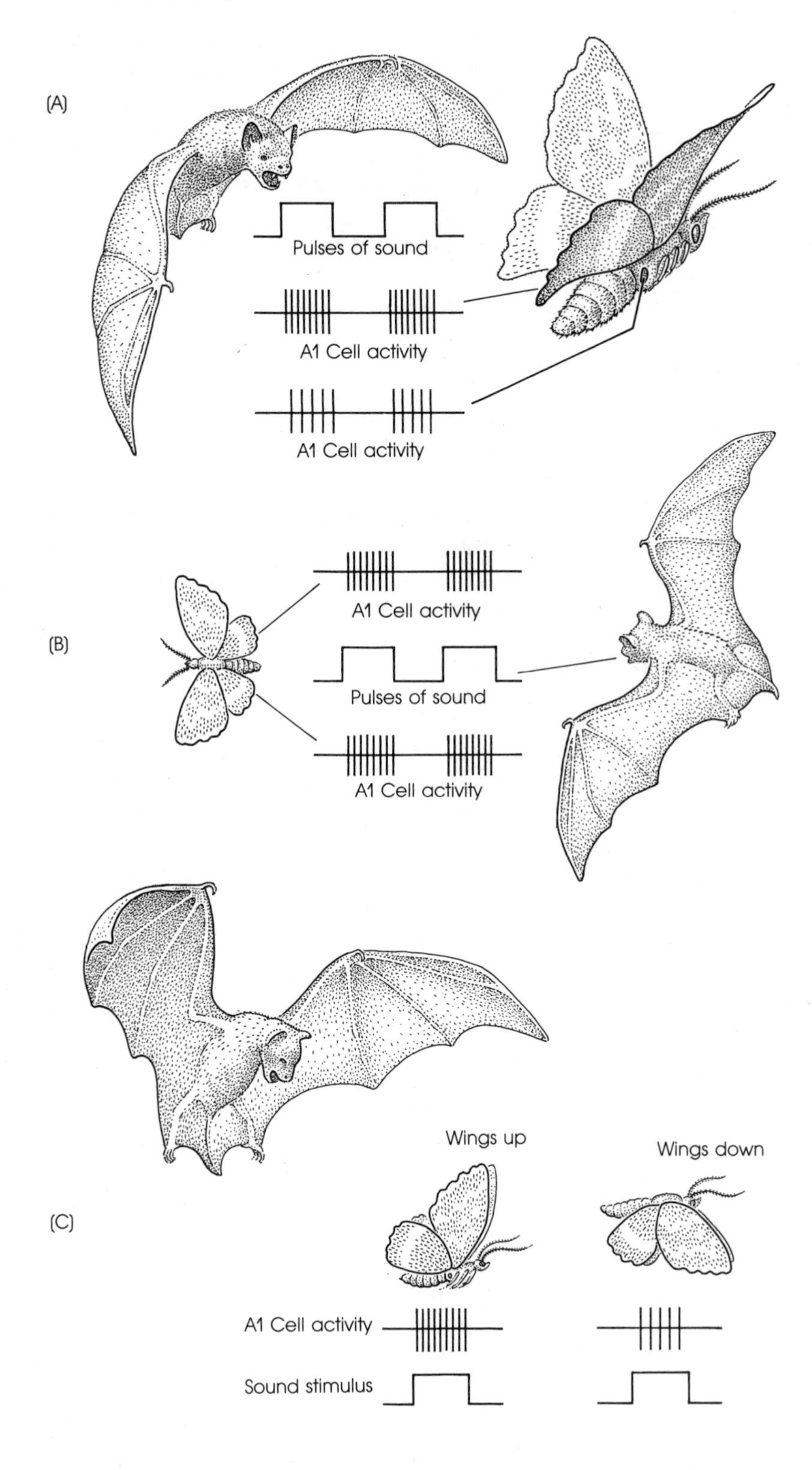
(A)
Pulses of sound
A1 Cell activity
A1 Cell activity
(B)
A1 Cell activity
Pulses of sound
A1 Cell activity
Wings up
Wings down
(C)
A1 Cell activity
Sound stimulus

Stimulus-Filtering and Behavior

The noctuid moth's auditory system and antipredator behavior provide a classic example of the relation between stimulus filtering and adaptive programmed responses to simple sensory cues. The moth's ear does *not* relay information about a host of acoustical stimuli in its environment. Low-frequency sounds audible to us do not trigger membrane changes and action potentials in the A1 and A2 fibers. Ultrasonic sounds of different frequency do not elicit different patterns of activity by the receptors and so cannot be discriminated perceptually by the moth (whereas humans can easily tell the difference between C and C sharp). Prolonged steady sounds are quickly ignored by the receptors. Thus, the moth's auditory system sacrifices a great deal of potential information. The ear appears to have one task of paramount and overriding importance—the detection of cues associated with its nocturnal archenemies. To this end its auditory capabilities are distorted and limited, sensitive to pulsed ultrasonic sound at the expense of most other sounds. Likewise, its behavioral repertoire of responses is constrained and simple. It turns away from low-intensity ultrasound and dives, flips, or spirals erratically to high-intensity ultrasound. The simplicity of perception and response has multiple advantages. By filtering out potential inputs of little biological significance, the moths avoid the need to build and maintain a complex battery of receptor and decoder cells (each of which carries with it a physiological cost). Moreover, the simplicity of the system increases the probability that the moth will not be distracted or confused by extraneous acoustical stimulation but will detect and respond in a discriminating fashion to just those events critical to the propagation of its genes.

Selective perception and specialized responses to acoustical stimuli are not limited to noctuid moths but are characteristic of many other species. For example, consider the lacewing, a creature wholly unrelated to noctuid moths (it is an insect that belongs to the order Neuroptera rather than Lepidoptera). Its ear is not in its thorax but lies within an enlarged vein in each forewing. Working with totally different raw material, selection has favored individuals with a tympanum in the vein that is linked with receptors (25 cells in modern lacewings) that can respond to airborne ultrasound [513, 514]. The electrical messages from stimulated receptors are rapidly relayed through the lacewing's body, eventually triggering motor neurons attached to muscles that fold the wings over the insect's back.

5 **Activity in the A1 receptors** of a moth's ears on detecting the cries of bats located in different points in space. (A) A bat is to one side of the moth; the receptor on the side closest to the predator fires sooner and more often than the shielded receptor. (B) A bat is directly behind the moth; both A1 fibers fire at the same rate and time. (C) A bat is above the moth; activity in the A1 receptors fluctuates in synchrony with the wing beat of the insect. Figures are not drawn to scale.

Intense orientation cries of bats (which are about 50 centimeters from the insect) are sufficient to stimulate the receptors; and, as a consequence, the flying lacewing closes its wings and dives downward at two meters per second (Figure 6).

This response unmodified might save some lacewings, but bats can track falling objects and at 50 centimeters they can readily detect the descending lacewing. However, in studies with caged bats and flying lacewings, Lee Miller found that only 30 percent of the power-diving insects were caught by hungry hunting bats. This led him to take a closer look at the interaction between predator and falling prey. By examining sequences of strobe-flash photographs, he often found that 50–100 milliseconds before a bat at-

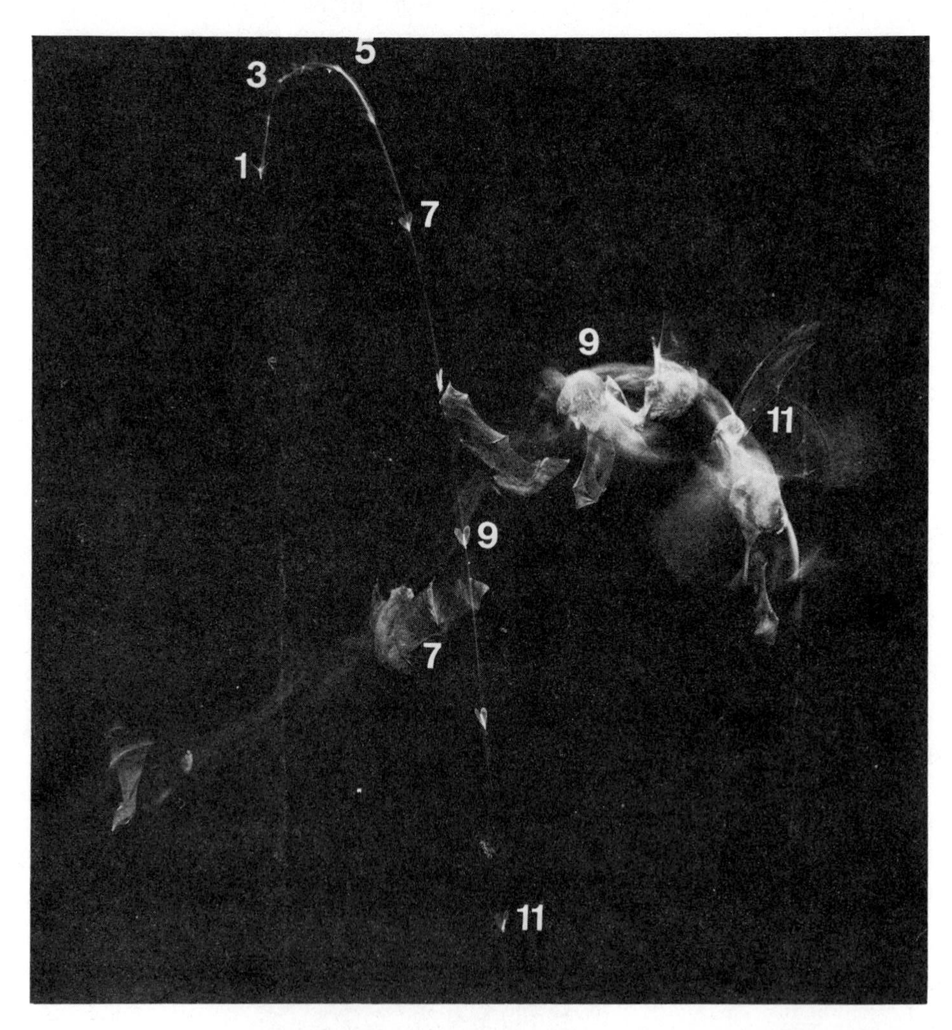

6 **Bat–lacewing interaction.** A multiple exposure photograph whose numbers show the relative position of the bat and lacewing over time. The bat missed the power-diving lacewing, despite performing an aerial somersault. Photograph by Lee Miller.

tempted to sweep up a lacewing, the insect performed a "wing flip," sharply beating its wings once. This interrupted and altered its descending path. The bat, therefore, often missed the lacewing because its attack was based on where the lacewing appeared to be going. Miller believes that the wing flip is triggered by the intense terminal buzz of cries made by a bat that has tracked a prey and is on the verge of capturing it. Although the receptors responsible for detecting the buzz have not been precisely identified, the point is that the lacewing, like the noctuid moth, has a two-part response to bat vocalizations, a reponse that includes a specific antiinterception behavior pattern and that stems from the unique design of its nervous system [513, 54].

Lacewings and moths need not have the vaguest mental image of a bat nor any idea of the danger they represent. It is sufficient that they have inherited the neural "wiring" that automatically helps them avoid predators better than individuals with different nervous systems. The neural wiring patterns that have survived to the present provide their owners with a most incomplete picture of their world and with a highly limited behavioral repertoire—but one that works in the competition to survive and reproduce.

If nervous system capabilities are related to the ecology of a species, animals that are not confronted with bat predators should not have evolved special neural responses to bat vocalizations. If some bats had wingspans of four meters and enjoyed humans for dinner, I imagine that we would have evolved the capacity to detect these hunters in the dark. There are, however, no such bats, and we lack the ability to hear ultrasonic sounds. But perhaps a better demonstration of the correlation between ecology and perception comes from examination of what a noctuid moth *larva* can hear, as contrasted with an adult moth.

The cabbage moth larva, bane of backyard gardeners, is a small green caterpillar that spends its days feeding voraciously on the leaves of cabbage and other related plants. It is obviously under no special risk of attack by aerial hunting, nocturnal bats. Yet it has an "ear" of sorts, which is capable of detecting airborne sounds [462]. The ear consists of eight thin hairs located in little sockets on its anterior segments (Figure 7). If one projects noises at the caterpillar, some frequencies will cause the grub to stop feeding; it may then begin to squirm on its leaf, regurgitate a greenish fluid, and eventually drop to the ground. The frequencies that have this effect are all *under* 1000 hertz, well within our hearing range and well outside the 20,000–100,000 hertz produced by aerial hunting bats.

The larva's hearing ability helps it avoid its enemies, which are not bats but are instead paper wasps of the genus *Polistes*. These wasps hunt for moth larvae, which they sting, rip apart, and feed to their brood within a nest (see Figure 12, Chapter 14). When the paper wasp flies, its wings vibrate and produce a low-frequency buzz of about 150 hertz. These sounds are detected by the caterpillar because the larva's filiform hairs are so designed that they readily vibrate in response to sounds between 100 and 700 hertz (Figure 8). Information from the stimulated hairs in some way

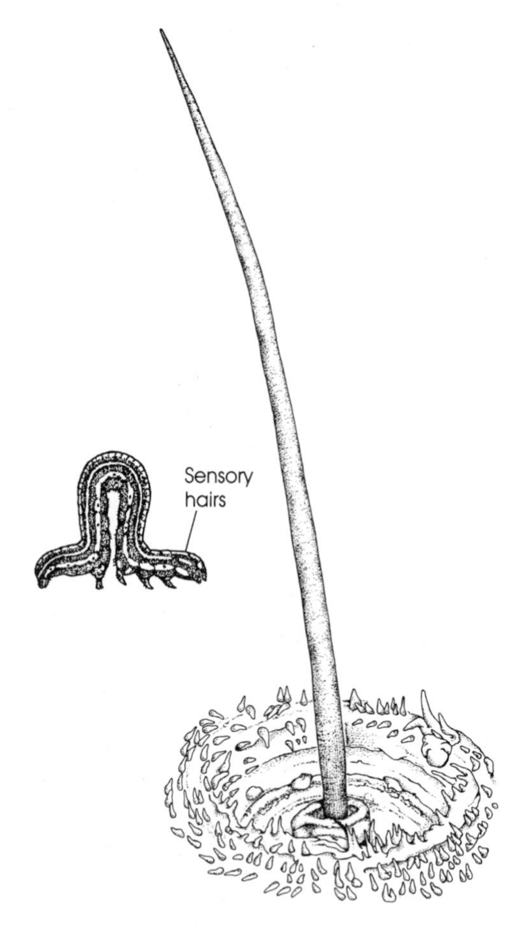

7 **Acoustical hairs** of a noctuid moth larva. The larva has eight sensory hairs on each side. The enlarged drawing shows how the base of a hair is poised in a socket. Airborne vibrations can cause the hair to move.

activates the set of behavioral responses described earlier. The reactions are effective in reducing the risk of predation. If one removes the hairs from a larva, it does not detect wasps at a distance and so does not begin reacting until the wasp touches it. In one experiment, "deaf" caterpillars had a 30 percent higher probability of being captured by wasp hunters than intact individuals [692].

Thus, even members of the same species may at different stages of their life cycle possess different specialized sensory mechanisms and motor responses. Animals are adept at perceiving sensory cues that are of special significance to their survival and reproductive success.

Stimulus-Filtering and Selective Perception in Vertebrates

A skeptic might say that it is all well and good to show that insects, with their almost ridiculously simple nervous systems, exhibit extreme stimulus filtering, but the perceptual abilities of vertebrates, which have much larger and more complex networks, should be essentially complete. This argument is not without merit. After all, the human ear has millions of receptor cells, not just two, and we are able to make many auditory perceptions that a noctuid moth cannot. But even vertebrate systems have their limitations and specializations. This becomes immediately clear upon comparison of ecologically distinct species, like humans and nocturnally active bats. Our acoustic system does not respond to ultrasonic sounds that are routinely detected by a bat's ear and brain. Thus, our auditory system exhibits stimulus filtering. In addition, if we were to analyze any one of our auditory receptors, we would find that it was specialized to some degree to detect certain stimuli much more readily than others. These two general properties of nervous systems—stimulus filtering by the system as a whole and specialized feature detection by certain of its component cells—are also evident in the way bats perceive sounds.

Superficially a bat's auditory receptor is similar to our own. Most bats have an impressively large external ear. Sound waves in the environment pass through the ear openings and along the middle ear canal to the inner ear. At the end of the ear channel, they strike a membrane, causing it to vibrate. These vibrations are transmitted to the auditory sensory unit, the cochlea, which contains the primary receptor cells. The cochlea is a fluid-filled organ with a thin basilar membrane running through the middle of the entire structure. Vibrations reaching the cochlea cause the fluid to

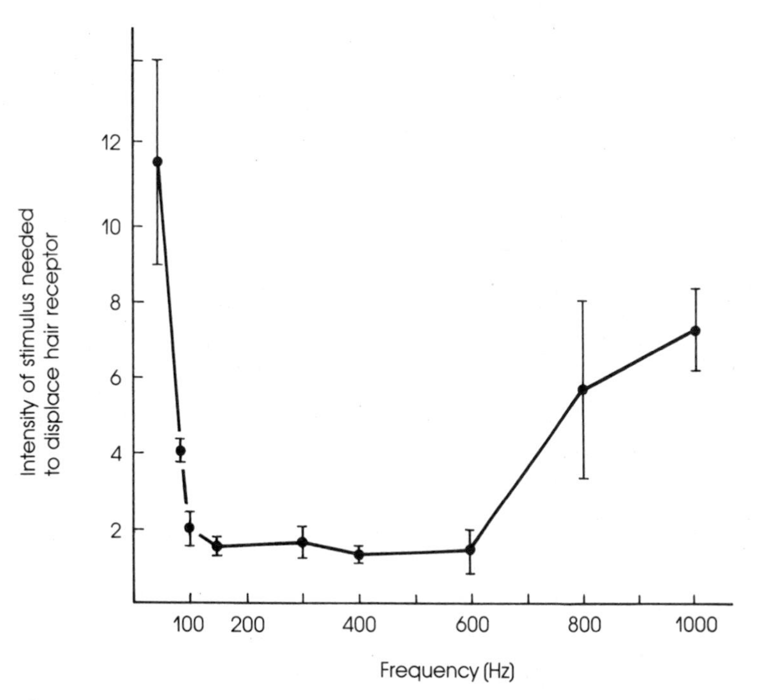

8 **Tuning curve** of an acoustical receptor of the noctuid moth larva. The receptor is most sensitive to sounds within a narrow range of relatively low frequencies. Other sounds must be very loud to generate any response by the receptor.

oscillate back and forth; this motion in turn moves specific portions of the basilar membrane, the position depending upon the frequency of the sound waves entering the ear. The mechanical energy present in these movements deforms receptor cells attached to the membrane; this energy is transformed into a receptor signal that is relayed to sensory neurons associated with each receptor. When these neurons fire, their messages are carried away from the cochlea along the fibers that constitute the auditory nerve. This cable runs to the lateral lemniscus and inferior colliculus in the bat's brain (Figure 9). The brain cells analyze input from the auditory sensory system and make decisions that control the bat's movements, which determine its success in capturing prey [322].

The Detection of Ultrasonic Echoes

The auditory differences between humans and echo-locating bats are not only evident in the relative size of the external ears but also in the proportion of brain devoted to auditory analysis (much larger in most bats) and in the capacity of ultrasonic sound waves to stimulate the basilar membrane in a bat's cochlea (but not ours). If, however, a bat is to catch a flying insect, it is not enough that it possess a cochlea whose membrane design enables high-frequency sounds to stimulate auditory receptor cells. Its nervous system must be capable of detecting *very weak* ultrasounds reflected from prey items. The trouble is that these faint echoes return to the bat within milliseconds after it has produced an orientation vocalization about 2000 times as intense as the echo is likely to be [283]. An auditory cell that has just been exposed to a loud sound becomes temporarily insensitive to soft noise. But bats preserve the sensitivity of their auditory receptors in the following way. Just before cells in the bat's brain order muscles controlling the larynx to produce an orientation cry, they also send messages that cause a muscle in the middle ear to contract. The muscle damps the vibrations of the middle ear bones that transmit the energy in sound waves to the cochlea. The muscle relaxes two to eight milliseconds after an orientation pulse, an action that permits a full response to the weak reflected sound waves coming from objects in the bat's auditory field. Thus, during the relatively quiet interval between cries, the receptors are not recovering from a recent vocalization blast but are adjusted for and can detect sounds of very low intensity [366].

The middle ear muscle decreases the loudness of sounds passing down the ear canal to 1 percent of their original intensity. But this is not the only mechanism for protecting the auditory system against self-generated sounds. In the lateral lemniscus, there are neurons that block transmission of auditory messages to higher regions of the brain during the period of a vocalization. Nobuo Suga and his associates placed an electrode in cells in the lateral lemniscus and recorded the cells' activity to cries the bat produced itself. They found that these relay cells are inhibited by other neurons when the animal is vocalizing [687]. This attenuation system further reduces the response of a bat's brain to the orientation cries it produces. The

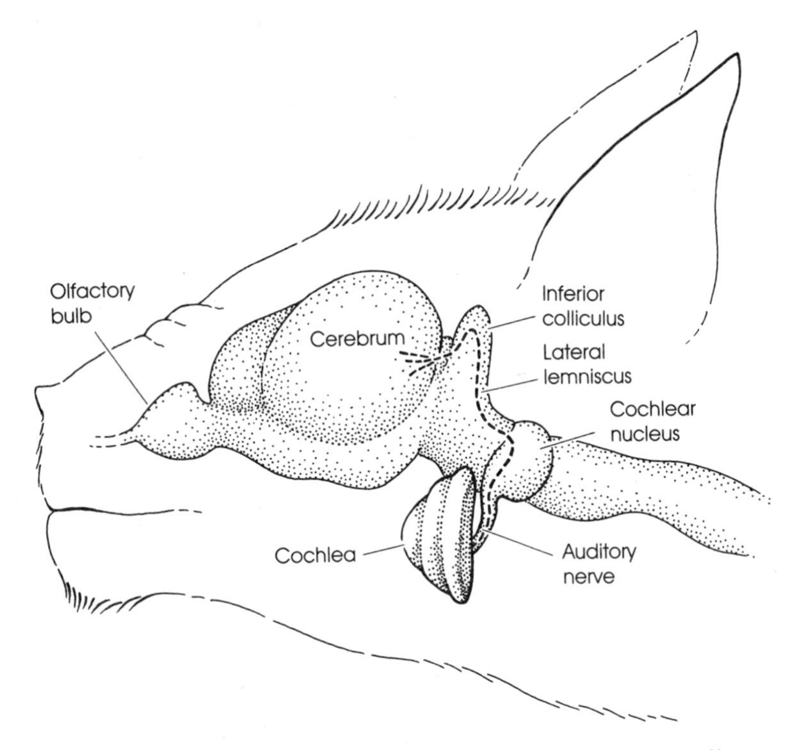

9 **Auditory system of the bat *Hipposiderus commersonii.*** Sound waves entering the ear are relayed to the cochlea, where they may stimulate receptors within this organ. Auditory signals are transmitted via the auditory nerve to the cochlear nucleus, lateral lemniscus, inferior colliculus, and then to regions of the cerebrum for analysis.

combined effect of the middle ear muscle contraction and the inhibition of neurons in the lateral lemniscus may be as great as a 10,000-fold reduction in the transmission of auditory impulses from self-produced sounds. (If a call generated 10,000 signals in the auditory nerve, cells in the inferior colliculus might receive as little as one action potential.)

In addition to these two mechanisms designed to protect the sensitivity of the auditory system to sound wave echoes, bat brains have within them cells that can be called echo-detector units [582]. Within both the lateral lemniscus and inferior colliculus of some bats, neurons have been discovered that respond more intensely to the *second* of two separate pulses of sound, one following the other in close succession, just as an echo would follow the vocalization pulse. Of special interest are certain nerve cells in the inferior colliculus of the Mexican freetail bat. These cells respond to sounds of extremely low amplitude at the very threshold of bat hearing *if, and only if,* the soft sound has been preceded by a loud burst of sound, preferably of the intensity and frequency of a typical Mexican freetail orientation pulse

(Figure 10). The specifications of the stimuli needed to excite these cells clearly match the properties of echoes coming from relatively distant objects following soon after an orientation cry.

The Identification and Interception of Prey

Echo-detector cells have been discovered in the mustache bat as well. These cells do more than merely detect reflected ultrasound. Some units are "tracking neurons" that remain active over time only if there is a continuing *decrease* in the interval between the bat's emitted orientation sound and the echo. Continuing activity in these cells would inform the bat that it was getting closer and closer to an echo-reflecting source. Needless to say, this is what the bat "wants" to do if the object is a moth or other edible flying insect [548].

Still other cells in the mustache bat's brain are "range-tuned." These cells respond most actively to one constant interval between a vocalization and an echo. For example, one of these cells would fire at its highest rate only when the delay between orientation pulse and returning echo was three milliseconds; another might be tuned to a six-millisecond delay; a third might fire most rapidly when the interval was nine milliseconds. These nerve cells are arranged in linear sequence, running from short- to

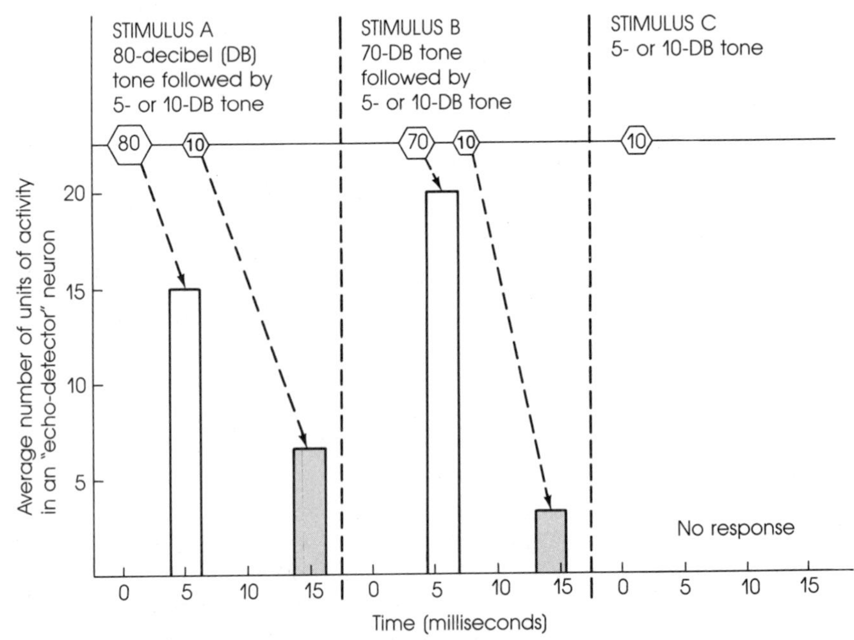

10 **Neural activity of an echo-detector cell** of a bat in response to three kinds of auditory stimulation. Although the unit does not respond to isolated low-intensity tones (stimulus C), it will produce neural impulses when these same tones are preceded by a very loud burst of noise (stimuli A and B).

long-delay neurons on the bat's cerebral cortex. Presumably still other higher-order cells that receive inputs from many of these range-tuned neurons integrate the information they provide and order the adjustments in flight path needed to keep the predator moving toward its targeted meal [686].

There is no doubt that the auditory system of the bat is more complex and provides much more information than the ear of the moth. In part this is because the bat relies much more on auditory inputs to control its behavior and because the tasks that it undertakes are exceptionally demanding. (Under these conditions, it may be advantageous to have numbers of cells monitoring the same stimuli to provide cross-checks on behavioral decisions. Although tracking neurons and range-tuned brain cells may appear redundant, they may permit multiple plots of the path to an elusive, evasive prey.)

But despite the differences, there is a critical similarity between bat and moth audition. In both animals the biochemical and functional properties of the components of the auditory system are exceedingly specialized. A bat's nerve cells help it focus on and interpret biologically vital information, especially echoes from pulses of sound it emitted, while ignoring less relevant sounds (such as weak noises unrelated to its vocalizations). Its nervous system makes discriminating decisions on every level. The behavioral consequences are adaptive. Bats have little difficulty avoiding tree trunks, nor do they mistake leaves for moths. In captivity, little brown bats can fly safely through a forest of wires 0.1 millimeter thick strung in a laboratory room. Captive bats can also quickly learn to discriminate between very similar potential prey items, such as mealworms and round plastic balls dropped from the ceiling of their enclosure [608]. Members of one species have even been trained to discriminate between two plastic blocks with holes drilled part way into them; the only difference was that in one block the holes were 1 millimeter deeper than in the other. Thus, bats are able to analyze extremely subtle differences in complex patterns of echoes, an ability that in nature may help them discriminate between prey of very similar shape but different palatability.

Needless to say, these are abilities that we lack altogether; our "deficiencies" reflect the many proximate differences in the nervous systems of humans and bats, differences that are based on evolutionary differences between us. To reinforce the point that the perceptual world of each species has its own unique features unshared by other animals, let us consider electrical perception and its behavioral uses in certain fish.

The Neurophysiology of Electric Fishes

Just as we find it difficult to imagine what it would be like to navigate through the air in complete darkness by listening to the echoes from our vocalizations, so too it is humbling to realize that there are animals who can interpret their environments in terms of distortions in a self-generated electrical field. Two unrelated groups of fishes, the gymnotids of South America and the mor-

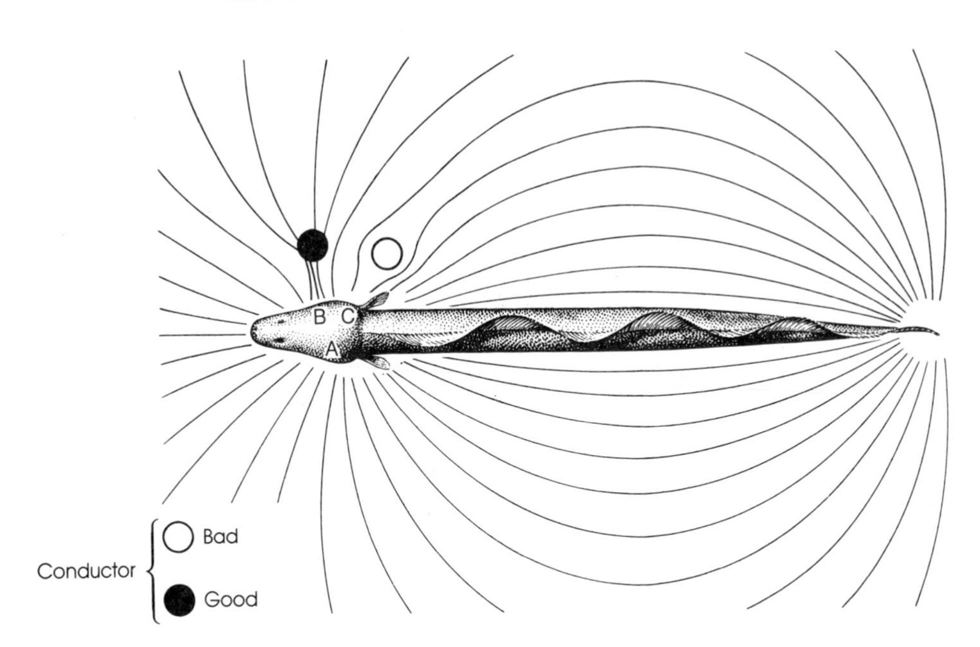

11 **Electric field** that the fish *Gymnarchus* generates from an organ in its tail. Objects near the animal distort the field and are detected through changes in the relative activity of electroreceptors in the fish's skin. Receptor activity is normal at A, elevated at B, and depressed at C.

myrids of Africa, have independently evolved electric organs whose discharges create a weak electrical field about the body [342, 432, 433]. Objects in the field distort the lines of electrical force, either concentrating the current flow (if the object is a better conductor of electricity than water) or dispersing it (if the object is a poorer conductor) (Figure 11). These fishes have rows of electroreceptors in pores along the length of the body. They respond to changes in the electrical field by *changing* their rate of firing. Cells in the central nervous system of the fish monitor the altered activity of receptors in the zone affected by the "shadow" cast by an intruding object. An increase or decrease in the activity of a group of neighboring receptors results in new patterns of activity in certain brain neurons. This could cause a behavioral change, with the fish orienting to or moving away from the item in its electrical field. Electric fishes are remarkably skillful in their detection of alterations in their electrical fields and in their use of this information. *Gymnarchus niloticus* can sense a change in electric current as small as 0.03 microvolts and can be trained with food rewards to discriminate between glass rods that are 0.8 or 2.0 millimeters in diameter [433].

In many respects the electroreception system of these fishes is similar to the sonar apparatus of the bat at both the ultimate and proximate levels. Functionally it is the kind of sensory system that helps an animal navigate and search for prey in the dark. Both the South American and African species tend to live in turbid, muddy waters or to be active primarily at night. They are both predators whose prey do not advertize their position, as it is not to their advantage to be eaten. A bat's sonar and a fish's electrical field "forces" prey items to give up information about their location that they would not otherwise provide.

How To Avoid Electrical "Jamming"

At the proximate level, the extraordinary abilities of the fishes reside in the specialized nature of their receptors and decoders. Just as insectivorous bats devote a large portion of their brains to analysis of echoes, so too the brain of an electric fish has an unusually well-developed region whose major task is to respond to information from the electroreceptors. Electric fishes and bats also have achieved a similar solution for a shared problem—how to discriminate between the sensory stimulation that their signals have created and the stimulation that nearby individuals have generated. Bats have evolved brain cells capable of responding selectively to echoes from each individual's own cries while ignoring ultrasound produced by other vocalizing bats. Many electric fishes live in moderately high densities, and therefore their nervous systems potentially could be overwhelmed with input from electrical fields generated by their neighbors. A common solution to this potential risk is illustrated by fish of the genus *Eigenmannia* [482]. Two electrically isolated individuals often generate signals at about the same rate (say, 370–375 discharges per second). But if they are placed together so that their electric fields overlap and interfere with one another, the fish have a JAMMING AVOIDANCE REFLEX (JAR) in which they automatically adjust the rate of electric organ discharge up or down away from their neighbor's frequency. When each has its own private channel, brain cells can selectively monitor distortions in the distinctive baseline activity of its receptors associated with the fish's personal frequency. The fish can remain informed about the nature of its surroundings.

But there is one weakly electric fish (*Sternopygus*) that lacks a JAR and yet is able to locate objects in its environment, even in the presence of a discharging neighbor. In its brain are cells that are insensitive to the large-scale distortions in its electric field caused by a neighbor's electric activity [482]. When such cells receive inputs from many receptors whose firing rate has been altered, they do not respond. But if a few receptors fire among the many that feed their signals to type 3 brain cell, it will then become active (Figure 12). This neuron is designed to detect small local distortions and to ignore irrelevant widespread changes in electrical inputs; the small-scale changes are likely to be caused by biologically significant objects.

Here again we see an example of a specialized decoder cell that receives messages from many receptors, integrates their information, and responds maximally to a few patterns while ignoring many others. This selectivity of reaction helps the fish perceive critical cues in an electrically complex environment.

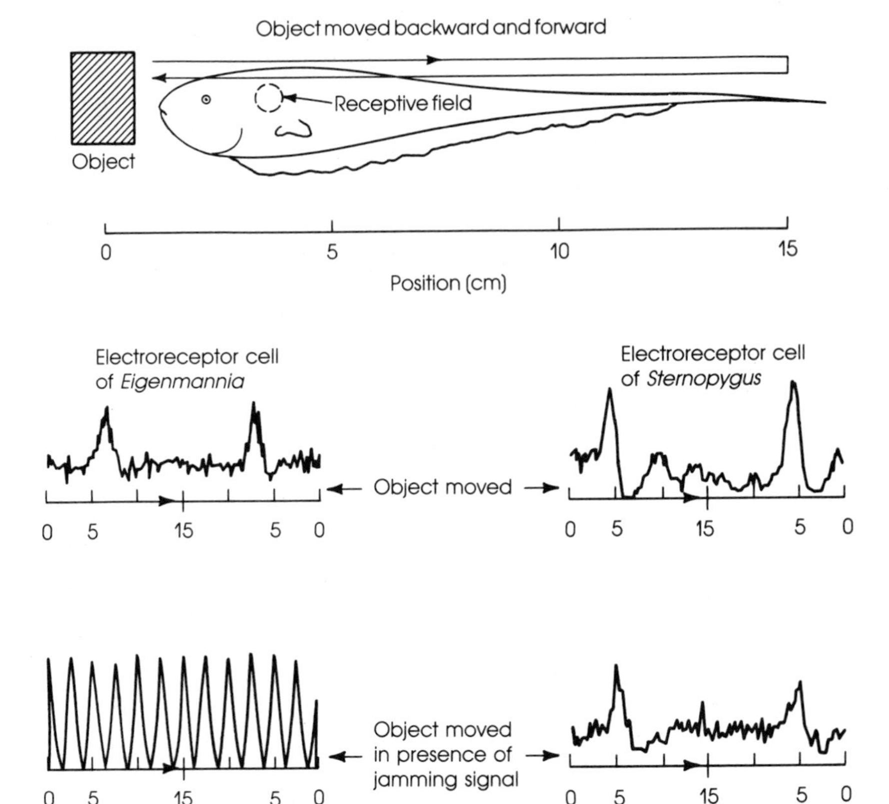

12 **Nonjammable electroreceptors** of the electric fish *Sternopygus*. Recordings were made from a cell whose receptive field is five cm from the fish's head. As the object passes this point, the cell's activity changes. If a jamming electrical signal is turned on as the object is moved, some cells of *Sternopygus* are unaffected because they ignore widespread background electrical activity while continuing to respond to small localized changes in their electrical field.

How To Communicate Electrically

The same capacity to avoid potentially distracting, irrelevant stimulation through possession of selective decoders is apparent in the sexual recognition systems of these fishes. In many species, sexually mature males and females each have their own distinctive electrical pulses that vary from species to species and from sex to sex within a species (Figure 13) [341, 343]. Thus, for example, the female of *Brienomyrus brachyistisus* produces a characteristic, positively charged pulse that lasts for 0.4 milliseconds and then dips sharply down to become briefly negative at its conclusion. During the breeding season, males answer signaling females with a burst of electrical activity. These messages, when converted into a signal audible to humans, have a sound rather like a rude "raspberry," but nevertheless they are presumably appealing to receptive females.

The male's response provides a way to identify what properties of the female's signal enable a male to recognize a potential mate. Males have a

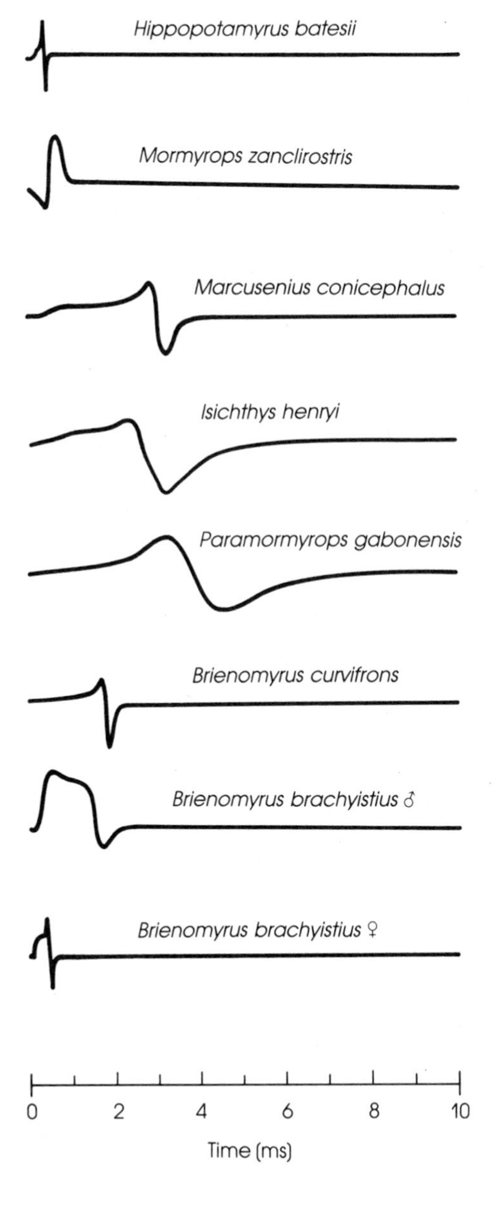

13 **Communication signals** produced by the electric organs of various African electric fishes. Note that a male and a female of the same species may have different electric organ discharge patterns.

significant problem because they may live in areas with other males and immature fishes of their own species as well as with members of many other signaling electric species. Carl Hopkins explored electrical analysis by males of *B. brachyistisus* by playing various electrical signals through electrodes placed in the water near a male and by recording his responses. The rasping response can be elicited by many different kinds of electrical patterns provided that the stimulation lasts precisely 0.4 milliseconds. The positive onset of the normal signal activates receptors on one side of the male's body and the negative terminal portion stimulates receptors on the other side. Certain cells in the fish's brain receive inputs from receptors on both sides and are tuned to the key feature: a 0.4-millisecond delay in the change in activity of the spatially separated receptors. This delay stimulates activity in the selectively tuned brain cells; the male calls to the female and presumably sometimes induces her to mate.

The Visual World of The European Toad

One reason for considering perceptual worlds that are so different from our own, like echolocation and electroreception, is that we cannot take them for granted. Instead, we immediately recognize them for what they are, a highly specialized way of gathering and processing information from the environment. In contrast, vision is so natural to us that it is easy to assume that what we see must be exactly what other animals see. We can disabuse ourselves of this notion by examining visual perception in the common toad of Europe. It teaches us that what an animal sees and how it responds are every bit as much a reflection of specialized discriminating cells as the sonar system of a bat or the mate detectors of an electric fish.

The toad possesses a visual system the structure of which has many similarities to our own [228, 229]. The creature has two large eyes with a retinal surface exposed to the environment. The retina contains a layer of receptor cells that specifically detect the energy in the light reflected from objects. Changes in membrane permeability of these receptors generate messages that are relayed to the next link in the network, bipolar cells, which in turn feed their output to ganglion cells. The long axonal projections from the ganglion cells run together to form the optic nerve, which carries action potentials to regions within the optic tectum and thalamus of the toad's brain. There the information that is received provides the basis for decisions that the animal makes in responding to prey or predators (Figure 14).

Jorg-Peter Ewert has conducted a remarkably comprehensive research program aimed at defining the properties of nerve cells throughout the visual system and their relation to the toad's behavior [228, 229]. He has, for example, devised a sophisticated apparatus that can record the activity of single cells in the optic tectum of a toad. The apparatus can be mounted on a living toad that is free to move about and respond to stimuli in its (laboratory) environment. This device informs us that, when a toad is stitting still (something toads do a great deal of) and there is nothing

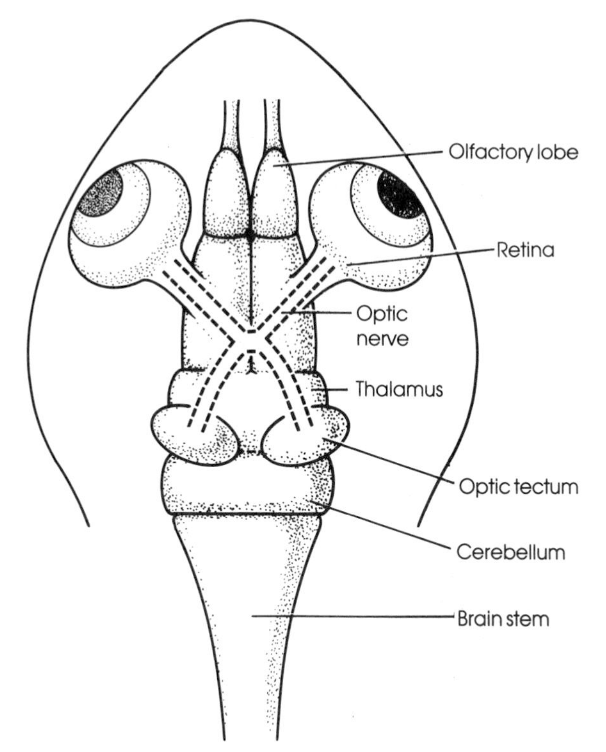

14 **Visual system of a toad.** This view from above shows the toad's brain with the optic nerve carrying visual input to the optic tectum.

moving in its surroundings, the animal apparently does not see a thing! Photons of light are striking its retinal receptors, but the optic nerve does not relay receptor messages to the optic tectum—a clear example of stimulus filtering. The ganglion cells, which receive signals from many receptors via many bipolar cells, are not passive instruments that record and transmit everything they receive but instead are active filtering agents.

There are three classes of ganglion cells, all of which fire *only* in response to moving stimuli that pass through the receptive field of the cells. The RECEPTIVE FIELD of a single ganglion cell is that area of the retina whose receptors feed messages back to the cell. In the toad and numerous other animals, each ganglion cell in effect monitors a small elliptical portion of the retina; its receptive field typically is organized into a central excitatory region surrounded by an inhibitory ring (Figure 15). Imagine a brown beetle moving over dark soil in front of a toad. The beetle will cast a very small image on the surface of the retina as the light waves reflected from the beetle enter the lens and are focused on the back of the eye. The image made by the beetle moves over clusters of receptors. Some of these clusters will constitute the excitatory central area of the receptive field of one (or more) ganglion cells. These cells will fire and information will be sent to the brain. The same receptors may constitute the inhibitory surrounding ring of the receptive field of other ganglion cells. These cells will not be

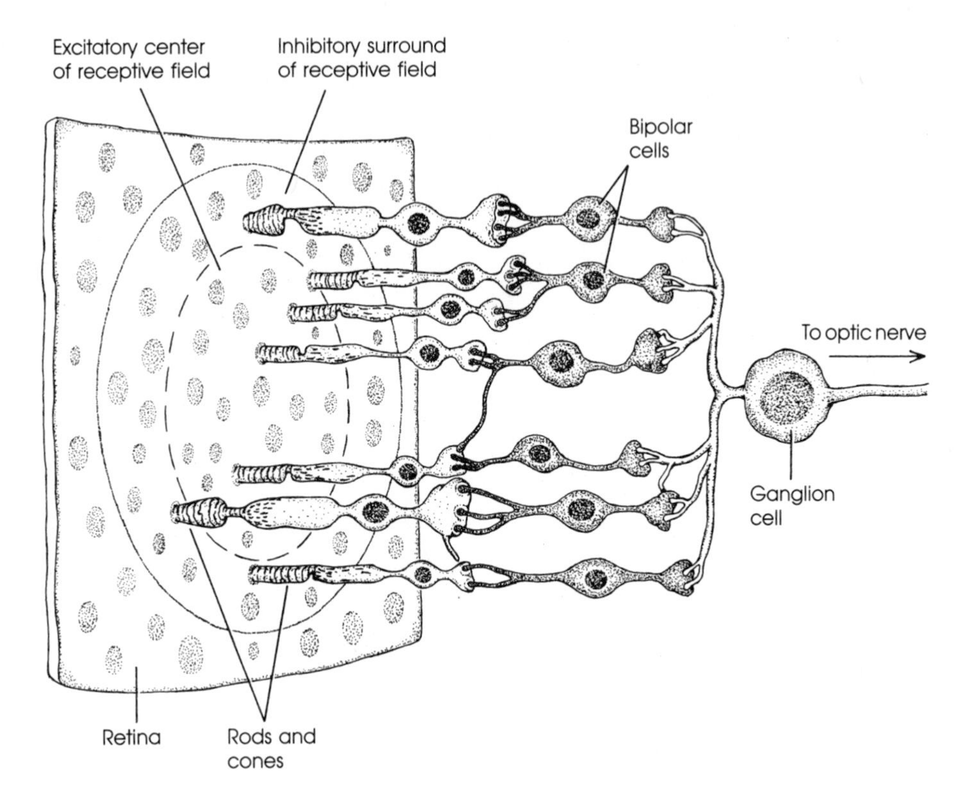

15 **Receptive field of ganglion cell.** The ganglion cell receives input from many bipolar cells, which in turn are linked to many receptor cells. As a result, the ganglion cell integrates information from a portion of the surface of the retina. Objects that cover the entire receptive field elicit little response. Small objects that move through the center of the field may generate many action potentials from the ganglion cell.

stimulated by the passage of the beetle's image across the retina and will remain relatively inactive.

Imagine a large object such as a human hand passed close to a toad's eye. The hand will cast a relatively large image on the retina, an image that will cover the entire receptive field of most ganglion cells. For these neurons, the inputs received from the inhibitory surrounding ring will largely or entirely cancel any messages from the excitatory central zone. These cells are not likely to fire.

The three classes of ganglion cells can be defined by their different responses to small moving images that enter the excitatory centers of their receptive fields. One type produces a burst of action potentials when the image enters the center (the ON-CELL). Another responds only as the image moves out of the center (the OFF-CELL). The third class generates two clusters of action potentials, once when the stimulus enters the center and again when it leaves (the ON-OFF CELL) (Figure 16).

Thus, the ganglion cells are movement detectors that respond to different aspects of changes in light intensity that reach the retina. From a toad's perspective, small moving images on its retina are much more likely to be caused by its prey (nearby beetles and worms) or its enemies (distant herons or hedgehogs) than by objects that cast large stationary images on its receptor surface.

The selective analysis of visual stimulation does not end at the level of the ganglion cells. Neurons within the optic tectum receive messages from many neighboring ganglion cells. Each brain neuron has, therefore, a receptive field of its own, a field consisting of that area of the retina monitored by the ganglion cells linked to it. The properties of the receptive field of a tectal cell can be studied by attaching a recording electrode to the cell and then moving various objects in front of the eye of a live, but immobilized, toad. Some stimuli will elicit a considerable response; others will generate a few action potentials; and still others will have no effect on the activity of the neuron.

Some cells in the European toad's tectum respond most to long, thin objects that move horizontally across the toad's visual field (Figure 17). These cells have a roughly circular receptive field that consists of an excitatory central strip lying horizontally in an inhibitory surrounding region. Objects that happen to move primarily through the excitatory area of the receptive field will cause the cell to fire. Objects that move through both the excitatory and inhibitory areas produce few or no neural impulses because the excitatory effect of the stimulus is canceled by its inhibitory

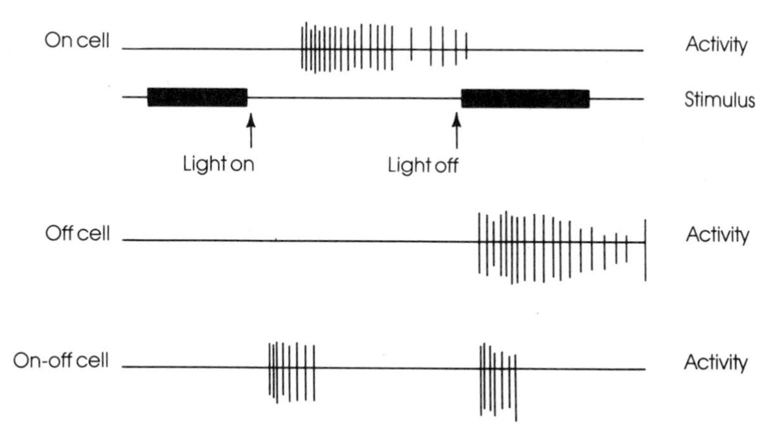

16 **Activity of some ganglion cell types.** Each three cell types shown has its own characteristic response pattern, firing upon the onset of a light stimulus on the cell's receptive field, *or* upon removal of light from the field, *or* reacting with action potentials to both the onset and cessation of the stimulus.

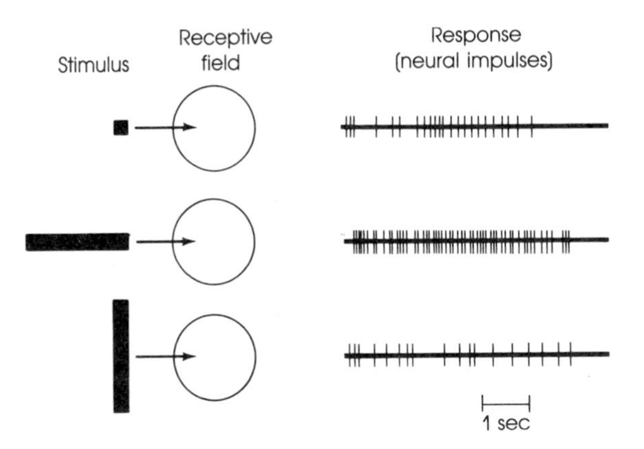

17 **Worm detector** of the European toad. A cell in the optic tectum of the toad responds differently to different kinds of stimuli that are moved through the cell's receptive field on the retina. Note that long, thin objects moved horizontally elicit the maximum response.

or blocking effect. In particular, objects that are oriented perpendicularly to the direction of movement elicit no reaction from these cells.

The consequences of the design of this class of tectal neurons becomes clear if one imagines how they would respond to a worm moving within toad attack range. The worm would be represented by a small, five- to ten-millimeter image on the retina of the sitting predator. This horizontally oriented image is small enough to pass through the excitatory central strip of the receptive fields of many tectal cells, causing them to fire rapidly. These messages travel to yet another group of brain cells that also receive inputs from other components of the visual system that are located in the thalamus. Certain thalamic cells respond strongly to moving objects that cast a *perpendicular* image on the retina, such as a hunting heron or stork. (What does the receptive field of these cells look like?) If both the worm- and stork-detector cells are active, the decision-making cells in the toad's brain either tell the toad's muscles to hold the toad's present position or command the toad to crouch down so as to become less conspicuous. If, however, the input is largely from the worm-detector cells, a different set of muscular commands are issued. The toad begins to turn toward the object. In the optic tectum there are three other kinds of cells that sequentially provide the motor commands that cause the toad (1) to stop turning when both eyes are fixed on the potential victim; (2) to lean closer within tongue range; and (3) to open the mouth, flip out the tongue, and snap up the worm [523].

Visual Perception in Other Animals

The toad's visual system is a marvelous example of the integration of physiology, behavior, and ecology. Toads benefit reproductively if they are able to detect and capture worms and detect and avoid their predators. They possess visual mechanisms that emphasize those stimuli most likely to be worms and predators. They possess the capacity to respond effectively to these categories of objects. To what extent is the visual system of other vertebrates similarly selective?

Although cats lack specialized worm-detector cells, nevertheless in the visual cortex of their brains there are nerve cells that actively accentuate certain patterns of stimulation at the expense of others. These feature-detector cells have been studied intensively by the Nobel prize-winning team of David Hubel and Torsten Wiesel [353, 354]. Figure 18 provides a summary of the activity of one kind of "hypercomplex" cell in the brain of a cat. This cell has a receptive field consisting of two inhibitory rectangles sandwiched about a thinner, central, excitatory strip. In order to stimulate this cell maximally, a dark bar must move downward at an angle so that the stimulus enters and then stops completely within the excitatory central strip. The relative activity of thousands upon thousands of cells of this sort is integrated within the brain to provide a visual perception in which the key features of objects in a cat's visual field are emphasized. This helps the animal categorize the things it sees, alerts it to stimuli of special biological importance, and provides the basis for adaptive discriminating responses to its environment.

Human beings also possess a visual system that is designed to detect certain stimuli in preference to others. The optical illusion illustrated in Figure 19 reflects the specialized features of cells within our retina. We see gray patches at the corners of the black squares because our receptor system

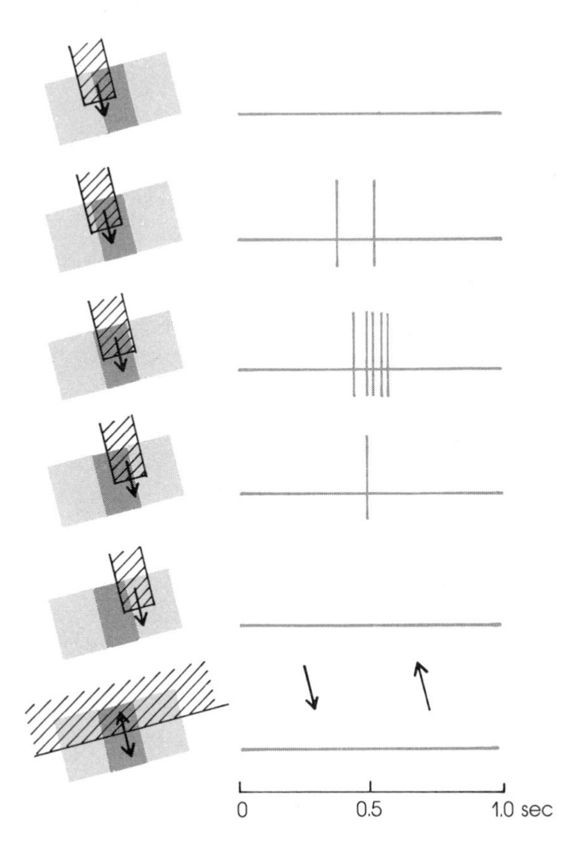

18 **Responses of a hypercomplex cell** in the visual cortex of the cat's brain. This cell responds solely to dark objects of restricted dimensions that move downward at a particular angle.

19 **An optical illusion.** Illusory gray spots appear at intersections of the white bars. This is because our brains exaggerate the contrast wherever there is an edge (black is bordered by white) but not where there is no edge.

accentuates the contrast between black and white edges through LATERAL INHIBITION. Our receptor cells have both excitatory and inhibitory effects on the array of relay cells that carry receptor messages to other regions of the nervous system (Figure 20). When receptor C is activated by light energy, it excites relay cell C but partially inhibits the neighbors of C, relay cells B and D; this partial inhibition reduces the frequency with which they fire. Thus, in a retinal area that is evenly bathed in light, the receptor outputs partially cancel one another. But in the edge area with a dark region abutting a light one, there will be a receptor (H in Figure 20) that receives reflected light while its neighbor I is not. The unstimulated cell will not produce signals that reduce the activity of relay cell H. Relay cell H, therefore, will fire relatively more often than cells whose activity is dampened by two active receptors. The greater rate of production of action potentials by H ultimately is converted into the perception that the white area is exceptionally bright in the areas just adjacent to the dark edge. There are no more photons coming from this point than from white areas farther from the black squares, but our brain cells provide us with this sensation because of the "false" information received from the retina.

Lateral inhibition produces an optical illusion that is adaptive [269]. Contrast enhancement enables us to detect subtle patterns in our environment that would otherwise be missed. As a result, we are especially adept at seeing the outlines of objects and can better detect the body of an animal partly hidden in vegetation, a sharp twig projecting into a path we are traveling, the subtle change of expression in a companion.

Throughout this chapter I have emphasized the particular biases of each species' perceptual systems to overcome the anthropomorhpic tendency to assume that all animals must share our perception of the world around us. My emphasis should not be interpreted to mean that there is no perceptual overlap among species nor that perception is purely arbitrary and idiosyn-

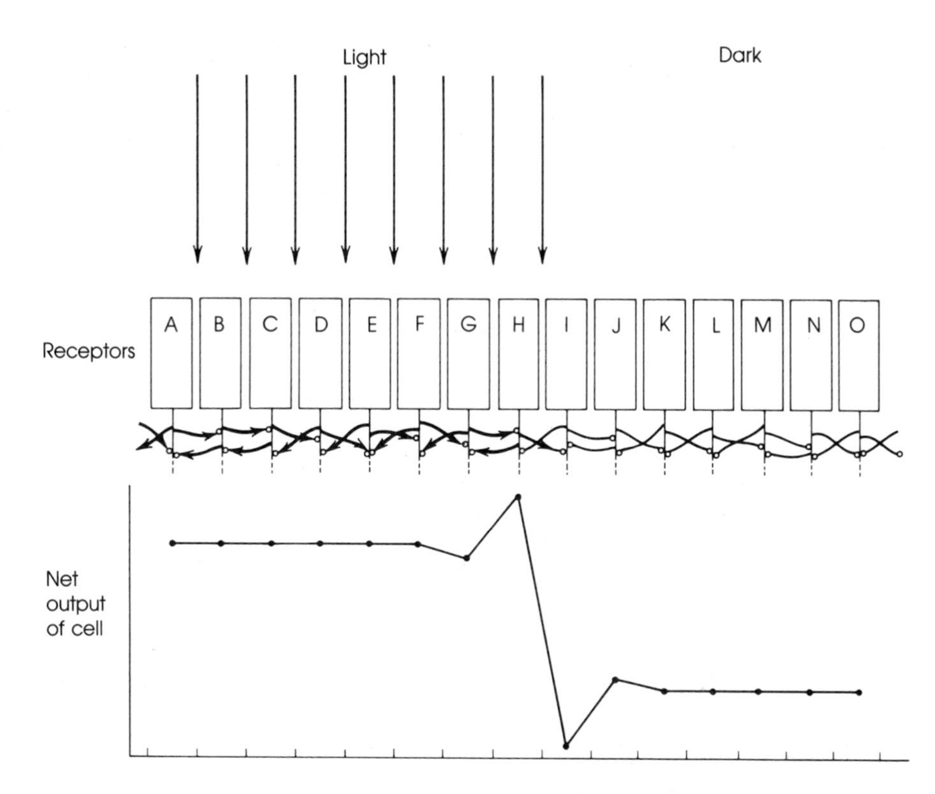

20 **Lateral inhibition.** At the edge of the light spot, one cell is being stimulated by radiant energy but is not being inhibited by one of its neighbors (which is in the dark and hence inactive). As a result, it is more active than its other light-receiving neighbors. This ultimately creates the perception of an artificially bright contrast between the light and dark areas of the visual stimulus.

cratic. Many species do doubtless share similar perceptual abilities, especially those that experience similar ecological pressures and possess complex nervous systems. Accurate representations of some components of the same real world are adaptive whether one is a honeybee or a human or a cat. Nevertheless, the discovery of slight perceptual exaggerations, subtle illusions, and peripheral screening emphasizes that nervous systems actively interpret the "real world" in ways appropriate for the ecology of the species.

Summary

1 The sensory and behavioral abilities of different animal species are often highly distinctive, as demonstrated by bats that can navigate by echolocation and fishes that can perceive weak electric fields.

2 One can define the special properties of nerve cells that contribute to the unique behavioral capacities of a species. An auditory receptor in a noctuid moth, a cell in the optic tectum of a toad, a neuron in brain of a bat each has its own characteristic properties that result in its sensitivity to a few patterns of stimulation and insensitivity to others.
3 STIMULUS FILTERING is a universal attribute of perceptual systems. The design of an animal's receptors limits what information they can collect and relay for further processing. Furthermore, neural systems typically consolidate complex patterns of sensory information into simpler patterns (as when a visual ganglion cell integrates inputs from many receptors into a few action potentials of its own). The result is that animals screen out many biologically insignificant stimuli. FEATURE DETECTION is a major aspect of many sensory cells, which are highly specialized to respond primarily to a particular stimulus of biological significance to an animal.
4 Behavioral responses, as well as sensory perceptions, are specialized and restricted because of the way an animal's nervous system operates. Animal species often have discrete reactions to key sensory cues, reactions that may be as simple as the power dive of a noctuid moth or as complex as the prey-tracking behavior of a rapidly flying insectivorous bat.
5 There is a correspondence between selective neural units (e.g., worm detectors in the toad), specialized behavioral responses (e.g., the actions that are involved in prey capture by toads), and the ecology of a species (e.g., European toads live in an environment in which worms are an abundant, edible, and vulnerable prey). This correspondence supports the prediction that natural selection will favor individuals whose proximate neural mechanisms contribute to the ability to detect and respond preferentially to those events in the environment most likely to have an effect on individual reproductive success.

Suggested Reading

Kenneth Roeder's *Nerve Cells and Insect Behavior* [608] should be required reading for all students of behavior. It is an understandable, entertaining, and exciting account of how to conduct research on the physiology of behavior. The book contains material on bats and moths. Donald Griffin's *Listening in the Dark* [282] is the classic book on bat sonar. Jorg-Peter Ewert's research on the neurophysiology of the European toad is nicely summarized in his textbook, *Neuro-Ethology* [229]. This book is also a valuable up-to-date survey of the study of the neural basis of behavior. *Mechanisms of Animal Behavor* by Peter Marler and W. J. Hamilton [467] provides many examples of the adaptive relation between physiological mechanisms and animal ecology.

Suggested Films

In a Frog's Eye. Black and white, 30 minutes. On visual perception of the leopard frog, an animal with a selective sensory system similar to that of the European toad.

Survival and the Senses. Color, 25 minutes. Something of a smorgasbord, with short segments on the sensory capabilities of a variety of invertebrates and vertebrates.

The previous chapter showed that animal species possess nerve cells that enable individuals to detect key stimuli and to respond appropriately to these cues. However, the proximate mechanisms that regulate an animal's behavior are even more sophisticated. They can adjust an individual's responsiveness to its environment in adaptive ways. A toad that has just eaten eight earthworms in a row is not committed to eat the ninth one like an automaton. Even though its worm detectors may be firing enthusiastically, feedback signals from its full digestive tract presumably inhibit its feeding response. Likewise, a lacewing that is flying during the day and is bombarded experimentally with recorded bat cries does not dive to the earth as it would at night but continues undeterred on its way [514]. This makes sense because bats do not pose a threat during daylight hours. The lacewing's nervous system must be able to integrate sensory inputs from its auditory receptors with information about time of day, adjusting its behavioral reaction to acoustical stimulation accordingly. Thus, moving worms or high-frequency sounds can evoke very different responses in the same individual, depending on its internal state, the time of day, and competing stimuli in the environment. These conditions can change within minutes, days, months, or years. As a result, individuals alter their priorities over time as their biological "needs" change or as environmental circumstances fluctuate. This chapter examines the proximate endocrine and neural mechanisms that help animals adjust their responsiveness to stimuli in a manner that ultimately improves their chances of reproductive success.

CHAPTER 6

The Organization of Behavior

The Physiology of Behavioral Changes in *Anolis* Lizards

The green anole, *Anolis carolinensis,* of the southern United States does not behave the same way the year round. Thanks to work by David Crews we now know a fair amount about the hormonal and neural basis for this lizard's changing reactions to the stimuli in its environment [145–149]. Anoles can do a great many things: stalk a fly, dart for cover when a sparrow hawk swoops, go dormant for months, fight with rivals, lay eggs, and copulate. All of these actions are *potentially* useful to the animal in terms of successful reproduction. But in order to be *actually* useful, it is vital that the animal be able to set priorities.

For example, during the summer in South Carolina, sexually mature females will regularly encounter sexually mature male anoles. A male will usually respond with a courtship display in which he extends his dewlap (Figure 1) and bobs his head up and down, flagging the female with the red dewlap. These signals may induce the female to arch her neck, in which case the male grasps her behind the head, crawls onto her back, and inserts one of his two penises into her cloacal opening.

But females often ignore a male's invitation to mate. This is adaptive because each copulation consumes 5–20 minutes and the immobilized female cannot feed or accomplish anything else. Moreover, a mating female is probably more vulnerable to attack by predators, particularly because pairs often copulate on exposed limbs and other open surfaces. Female anoles avoid the costs of superfluous matings by refusing to copulate when courted *except* during a particular stage of development of an egg. In *A. carolinensis,* only one egg develops fully at a time (over a period of 10–14 days). When it is mature and the female is about to ovulate, she becomes receptive; only then will she exercise her precopulatory neck-arching behavior in response to a territorial male's courtship signals. Immediately after mating, the female becomes unreceptive again and her rejection of males persists for the next week or two while a new egg develops. She eventually regains her willingness to copulate, and the cycle is repeated (Figure 2).

1 **Extended dewlap** of a male *Anolis opalinus* from Jamaica. This species has one of the more striking displays involving a spread dewlap. Photograph by Thomas A. Jenssen.

What regulates the female anole's reproductive cycle? How does she know when to respond and when not to accept a partner's advances? The regulatory mechanisms that control receptivity are complex. As an egg develops, cells in the ovary communicate hormonally with the brain of the anole by releasing a sex hormone, estradiol, into the bloodstream. (In ad-

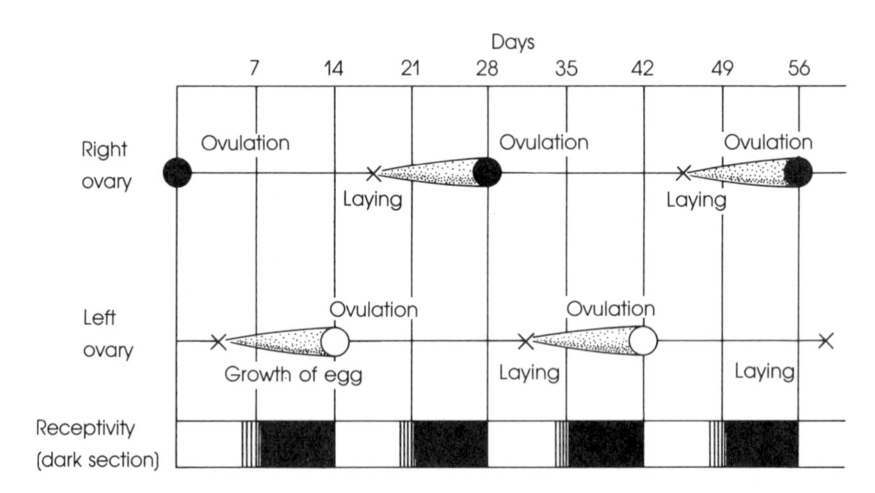

2 **Physiological and behavioral cycling** during the reproductive season of female American anoles. One egg matures at a time. As the egg is completing its growth, the female gradually becomes receptive so that she is sure to have been inseminated by the time the egg has been released into the oviduct. Receptivity temporarily ceases after copulation, to be regained as the next egg matures.

dition, another hormone—progesterone—also has a particular pattern of ovarian production and release, but its role is less well understood.) Estradiol has specific target cells within particular regions of the female's brain. When ovarian hormones enter these cells, they set in motion a chain of biochemical events that ultimately leads the pituitary gland on the underside of the lizard's brain to secrete certain hormones of its own. These travel back to the ovaries, where they help maintain the schedule of production of estrogen and progesterone. The ultimate impact of this feedback loop is to alter the responsiveness of those brain cells that react to courtship stimuli (although precisely how this happens is not known). At the time when ovulation is about to occur, the female's hormonal state permits her to arch her neck when she sees the bobbing dewlap of a male.

One can experimentally demonstrate the importance of estrogen and progesterone in the control of female receptivity by removing an anole's ovaries [504, 711]. The ovariectomized female will never respond sexually to male courtship again unless she receives compensatory hormone injections. An effective combination in inducing readiness to copulate is to give the anole a shot of estradiol 24 hours before she receives an injection of progesterone. Given the proper concentrations of hormone doses, almost all the females will be ready to mate in about 24 hours, despite their eggless condition.

The Control of Receptivity

Once a copulation is completed, one can arrange (experimentally) to have a new male presented to an intact female. If this is done within a minute or two after her first male dismounts, a female may copulate once more. But if five to seven minutes have passed, she will religiously refuse to respond to courtship signals and her unwillingness to mate persists for 10–14 days, until she is about to ovulate again. The transition from receptivity to unreceptivity is therefore extremely abrupt. It is unlikely that a change in estrogen levels in the blood is responsible for the switch to unreceptivity because it is hard to imagine that a female could alter estradiol concentrations in the space of a few minutes [148]. What then is the basis for the change in her behavior?

The onset of nonreceptivity is closely related to intromission by the male. If a female is courted and then mounted by a male anole whose double penis has been surgically removed (the life of a laboratory lizard has its unpleasant moments), she will retain her willingness to copulate after the male dismounts. Thus, there is something about the tactile stimulation provided by the penis or some material in the male's ejaculate that quickly induces the female to become unreceptive. Richard Tokarz and David Crews guessed that prostaglandins might be involved (because these substances regulate female receptivity in a variety of other animals [712]). They injected minute amounts of a prostaglandin (PG) in females whose receptivity had been tested; the females had permitted a mature male lizard to grip them by the neck, after which the male was removed. (The females were ovariectomized and had received estrogen–progesterone treatment to prime them to mate.) Shortly after PG injection, the females were no longer receptive (Figure 3).

Perhaps in a normal female that has mated, mechanical stimulation of the genital tract induces certain cells in this region to release PG into the bloodstream. The substance may be carried to target cells in the female's brain. Prostaglandin may inhibit (block) those units responsible for ordering the female to remain still and arch her neck in response to visual stimuli from a courting male. Therefore, when a female sees a bobbing dewlap, she does not wait to be mounted but either departs in a flash or bites the male, treating him as an aggressor rather than a sexual partner.

The effects of PG injection last no more than six hours in females whose ovaries have been removed; after this induced period of unreceptivity, these individuals regain their readiness to mate. This finding suggests that the long-term maintenance of unreceptivity in normal females depends upon hormonal changes produced by the ovaries. It would not be surprising if the PG signal set in motion the release of hormonal messages from the pituitary gland at the base of the brain, messages that had two effects: (1) an increase in the likelihood of egg laying and (2) a decline in estrogen released from the ovaries; estrogen release would fall further once the mature egg was actually laid. Low estrogen production would continue until a new egg became mature. The rise in estrogen levels, coupled synergistically with progesterone releases, would lead once again to copulation and fertilization of the egg, after which oviposition and a new cycle of egg maturation and receptivity would occur.

Seasonal Changes in Anole Behavior

The cyclical pattern of receptivity in the female green anole persists only during the "breeding season" of March to August. As is true for many, but not all, animal species, the anole follows an annual cycle

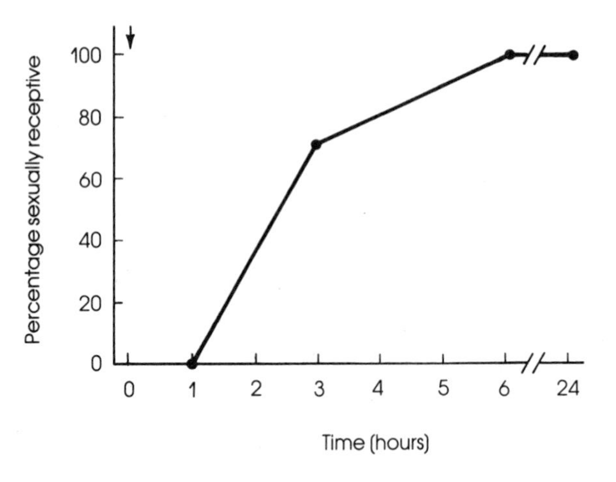

3 **Effect of prostaglandin** on the sexual receptivity of female American anoles. When injected into seven fully receptive females, prostaglandin temporarily abolished the readiness of the anoles to accept courting males. The time of injection is shown by the arrow.

characterized by numerous changes in responsiveness to various stimuli [145]. A female anole that has been alternately avoiding males and then permitting them to copulate will lay a final egg in August. After this time, she will not undergo a new cycle of egg maturation and will not regain her receptivity. Instead, she hunts insects only. Likewise, male courtship readiness declines precipitously at this time. They too feed for some time until, with the approach of winter, they slip under a sheltering log or mass of palm fronds and become dormant until February. Then they rouse themselves and begin to claim territories, repelling other males with bobbing display threats and occasional physical combat (activities that are correlated with a sharp rise in testosterone levels in the bloodstream and brain tissue). Females emerge in March and reproduce for the first time in late April. They then enter a new summer breeding season and, if fortunate enough to survive until August, will once again enter the fall feeding and winter dormant periods (Figure 4).

The annual behavioral cycle of the anole may be partly founded on an internal annual clock mechanism that primes males and females to behave in particular ways at different seasons of the year [146]. Indirect evidence for an annual clock comes from studies of the sharp decline of responsiveness by females to courtship stimuli in the fall. This is due to the failure

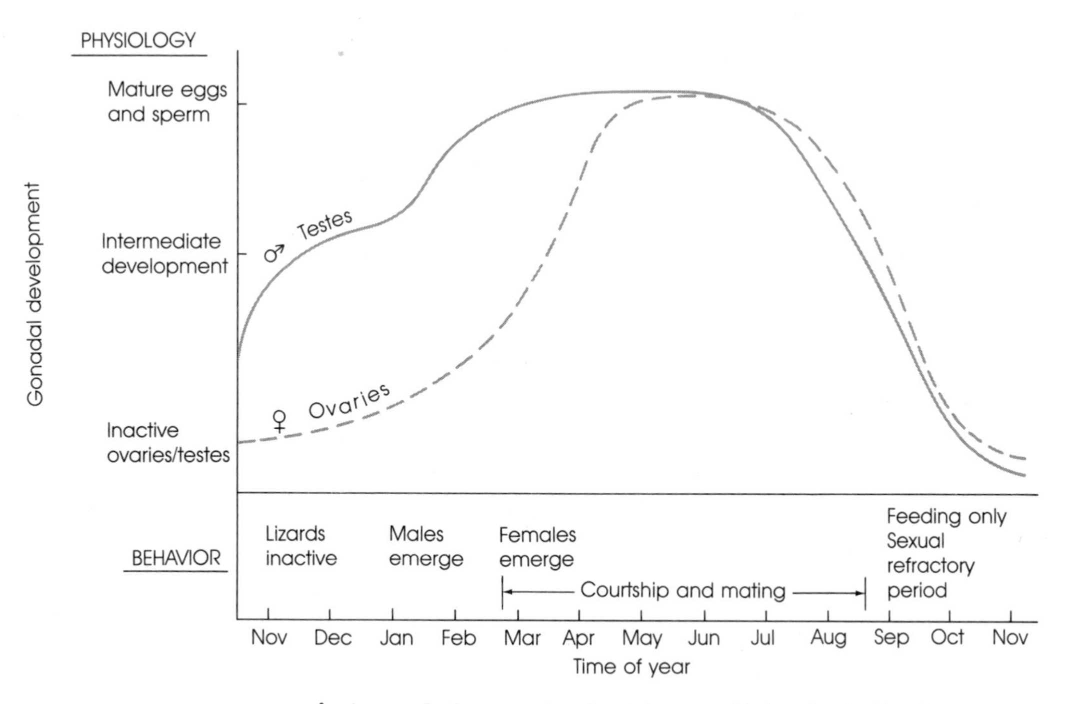

4 **Annual changes** in physiology and behavior in the American anole. The changes in behavior of the lizard over the year are linked to changes in the development of the ovaries and testes of females and males.

of the ovaries to produce a mature egg (and the associated estrogens) during the fall SEXUAL REFRACTORY PERIOD. One can show that external temperature cues are not solely responsible for onset of the refractory period. Female anoles taken into the laboratory and exposed to warming temperatures are largely unaffected by these conditions at this time. In contrast, females taken prematurely from their overwintering sites and exposed to increasing temperatures, respond readily with ovarian growth and eventual sexual receptivity. Likewise, if one injects pituitary hormones in females, there is little ovarian response if the experiment is done in September, but there is pronounced ovarian development in February [149]. Whatever the basis, the lizard does not create mature eggs in the fall, even if an unusual warm spell happens naturally or even if the lizard is experimentally exposed to conditions that would induce egg production in the spring.

One can go further still (and David Crews has) by removing the ovaries completely from females captured in the fall. If one injects these animals with estrogen and then offers them courting males, they still refuse to copulate. One gets a very different response from estrogen-treated, ovariectomized females that are captured in June and tested two weeks later (Figure 5). Thus, cells in the brain as well as in the ovaries apparently have an annual cycle of sensitivity to hormones involved in the reproductive process. In the summer, estrogen treatment activates sexual receptivity via

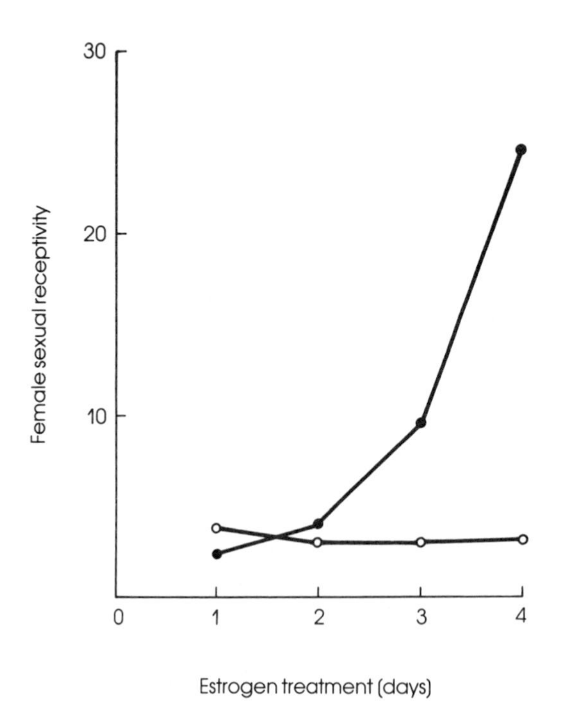

5 **Behavioral sensitivity to estrogen** in the anole. Sensitivity to estrogen depends on the season of the year. Females were collected in June (•) and September (○), ovariectomized, and then treated with estrogen. Only the June females regained sexual receptivity.

altered target cells in the brain. It is possible that in the fall the properties of the target cells have changed so that the same hormonal signal has a different outcome [149].

Climate and Social Interaction

The schedule of behavioral activities of a green anole has some flexibility despite the overall seasonal pattern of territoriality, breeding, feeding, and winter dormancy. Crews and his associates have shown that individual lizards are sensitive to many environmental factors that can either accelerate or inhibit ovarian development and reproductive activity. For example, temperature is very important in determining precisely when a female will begin breeding in the spring [420]. Anoles possess temperature receptors whose information is integrated in the brain. Increasing temperatures in the early spring are correlated with releases from the pituitary gland of gonadotropic hormones that stimulate the growth of the ovaries (note that the schedule for the growth of the male gonads differs from that of the female; Figure 4). There is a limited period during which the anole's brain is primed to be sensitive to inputs from its temperature receptors in order to reach a decision on beginning reproduction. The precise pattern of temperatures in any one spring, however, will determine exactly when an individual emerges from its hibernaculum and begins to engage in activities appropriate for the season. Colder weather will delay gonadal growth and emergence; warmer days will facilitate the cessation of dormancy.

However, reproduction is regulated not only by temperature but also by an anole's social environment. Males of *A. carolinensis* compete for foraging territories on vegetation that will attract females. Females that have left the overwintering site will move to a suitable living–feeding place where they are likely to encounter a territory owner and perhaps other males as well. Their observations of these males can have a profound influence on the onset of reproduction in the anole female, as the following experiment illustrates.

Crews removed dormant females from their winter hiding places, moved them to the laboratory, and placed them in cages in groups (1) with no males, or (2) with castrated males only (which will not court), or (3) with males whose dewlaps had been surgically removed, or (4) with intact males, one of whom had established territorial dominance over the other males present [145]. Females in the first three groups did not have the opportunity to observe the courtship display of a territorial male, whereas females in the fourth group regularly had dewlaps spread before them by an "ardent" male. The members of the fourth group experienced much more rapid ovarian development than those in the "dewlap-deprived" groups. The experiment demonstrates that visual stimulation from a displaying male does more than simply trigger the neck-arching response in a receptive female. It actually prepares recently emerged females to become sexually receptive, presumably because some visual stimuli activates nerve cells in the brain that relay messages to the pituitary gland. Females that are courted frequently secrete greater quantities of pituitary gonadotropins than uncourted anoles, speeding ovarian development and the onset of mating readiness in the spring.

The Adaptive Value of the Anole's Regulatory Mechanisms

A sensitivity to social feedback from courting males leads female anoles to adjust their sexual readiness in ways that increase the probability that they will mate with a dominant territory owner. This is almost certainly an adaptive result of the proximate mechanisms that regulate the female's reproductive behavior. Producing a large, nutrition-packed egg is a physiologically demanding process for a lizard (an anole manufactures no more than 15 per season). Moreover, carrying a large ovum may slow the female as she races for cover with a sparrow hawk close behind. Therefore, it is reasonable that a female should (1) not begin ovulating until a mate was available and (2) exercise care in the selection of a partner. In species in which male dominance varies considerably, females may gain by mating with a territorial male in preference to a subordinate individual (see Chapter 12). Because a female anole's control mechanisms link ovarian growth to regular courtship stimulation, she is unlikely to become receptive until one male in her environment has achieved sufficient dominance to be able to court her without interference from rivals.

One can develop hypotheses about the ultimate value of the other features of the anole's changing responsiveness to certain stimuli (Table 1). The control systems that regulate these changes appear adaptive in the seasonally and socially variable environment of the anole. This environment differs for males and females, which in any case have different routes to reproductive success (Chapter 11), with the result that the control mech-

Table 1
Hypotheses on the Adaptive Value of the Changing Behavioral Priorities of Male and Female Green Anoles

Characteristic	Hypothesis
WINTER DORMANCY: Inactivity takes precedence	Cool winter weather makes activity more difficult and less productive for a cold-blooded predator that feeds on insects that are common in the summer
EMERGENCE: Males appear before females	Male fitness is tied to acquisition of territory; the early-emerger gets the better sites
RECEPTIVITY: Female sexual behavior is cyclical	A female's eggs mature singly at long intervals; one mating provides sufficient sperm to guarantee fertilization of the ripe egg; other matings are superfluous and risky
FALL REFRACTORY PERIOD: Feeding claims top priority	Reproduction in the fall would generally fail because hatchling anoles face declining food supplies and cold weather at this time; the adults gain future chances to reproduce by building up fat reserves to carry them through the winter

anisms differ for the two sexes. The end product is that individuals are able to alter their behavioral priorities in a dynamic fashion over the course of a year.

The anole example was chosen to illustrate three adaptive aspects of behavioral control. First, individuals confronted with conflicting choices (to mate or to feed or to escape) manage to select one action while suppressing others. Second, many behavioral priorities change over the short-term in ways that advance reproductive success, as in the relation between female receptivity in the anole and maturity of the ovum. Third, an animal's behavior may be organized in an annual cycle, with the onset of certain activities precisely timed through monitoring of various climatic and social factors. The remainder of the chapter will be devoted to an examination of these three components of behavioral regulation.

Coping With Competing Stimuli

One can view the organization of anole behavior as the outcome of a nervous system that consists of a number of interrelated control modules, each with a primary responsibility. Thus, one might speak of a sexual behavior control system, a feeding center, or an escape unit. Each component monitors a variety of stimuli, responding to hormonal messages or to direct sensory stimulation and adjusting its relation with other units in a lawful manner. Thus, a female with a mature ovum becomes sexually responsive to courtship stimuli. After receiving hormonal signals correlated with the mature ovum, her sexual control center is especially likely to become active and to block the escape and feeding centers while activating the neck-arching reaction. This regulatory mechanism permits copulation to occur even in the face of certain low-level distractions from other visual stimuli in her environment. But immediately after mating or during the entire fall feeding period, the relations between control centers are altered, with other modules blocking the sexual control unit and thereby preventing the female from responding sexually to male courtship signals even if she should encounter a receptive male.

This model, although speculative and primitively simple in detail, does accommodate an important aspect of reality. Anoles (and other animals) rarely try to carry out two incompatible activities at the same time, for good and obvious reasons. To achieve this result, command decisions that order one reaction must directly or indirectly inhibit competing circuits whose messages would interfere with the effective completion of the selected response. Little is known about the rules of operation of nerve cells within the anole's small, but vastly complicated, brain. Inhibitory interactions among nerve cells, however, are very likely to play a critical role in decision-making in this and other animals.

We have already seen an example in the European toad. If the toad is simultaneously presented with objects that equally stimulate a set of worm-detector cells and a set of predator-detector cells, thalamic neurons inhibit those tectal cells responsible for ordering the first step in prey catching (Figure 6). The action potentials arriving from the thalamus have the effect

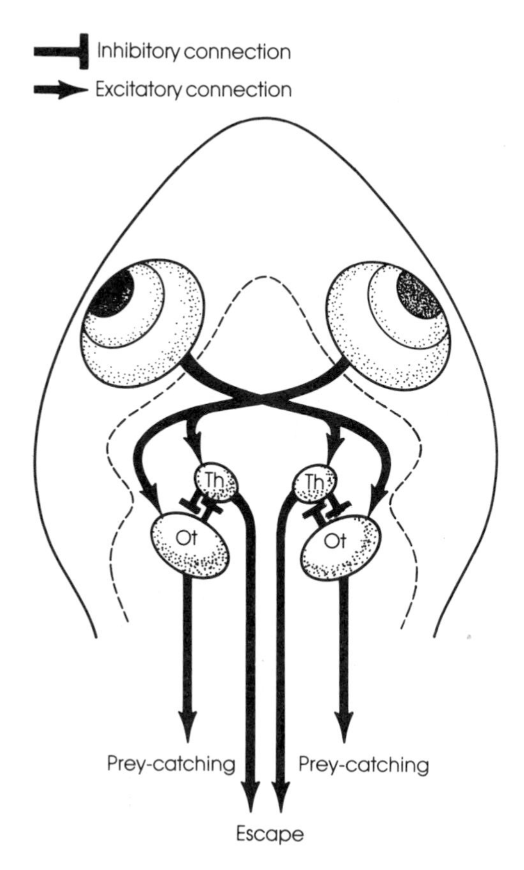

6 **Neural control centers** of the European toad. Interactions between thalamic cells (Th) and cells in the optic tectum (Ot) determine whether the toad will exhibit escape behavior or a feeding response. Simultaneous strong activation of both centers by different visual stimuli will result in the inhibition of the optic tectal cells and the selection of the escape response.

of raising the threshold of response for the tectal cells. They become less likely to fire, and the toad becomes less likely to orient toward the potential prey item and more likely to carry out the crouch avoidance-response. The unequally weighted inhibitory arrangements between cell clusters in the thalamus and optic tectum set behavioral priorities for the toad [229]. All other things being equal, escape takes precedence over feeding. On the other hand, if the stimulation from worm-detector cells vastly exceeds the signals provided by enemy-detectors, prey catching can take precedence. Optic tectal cells can be inhibited and they can inhibit other cells, including the thalamic cells that control the crouch response. Given sufficiently strong stimulation from a worm or beetle moving nearby, the toad will be able to carry out the complete and productive sequence of prey catching actions without interruption. A toad that oriented, then crouched—or one that tried to combine the two activities—would not do well in capturing its prey.

The Control of Mantis Behavior

The utility of inhibitory connections between nerve cells in the central nervous system applies to invertebrates as well as to toads and anoles. Kenneth Roeder's work on the preying mantis offers a classic demonstration of this point [608, 610]. A mantis perched on the leaf is bathed in a rich assortment of stimuli, many of which the insect can detect with its receptor systems. Most of the time, however, it does absolutely nothing. This is adaptive for the mantis, whose feeding success depends on its ability to remain motionless until an unsuspecting prey wanders within striking distance. Once this occurs, however, the insect is capable of making very rapid, accurate, and powerful grasping movements with its front pair of legs. If it does not encounter a meal for a long time, the mantis is likely to move to another waiting site. Moreover, reproductively mature males are able to search out females. The mantis is capable of sorting out these options because of a simple, but elegant, set of controls imposed by the brain and subesophageal ganglion on the thoracic ganglia that control the walking and striking legs.

Roeder explored this system of controls by surgically cutting the connections between the various components of the mantis's nervous system and observing the behavioral effects of different operations. If he separated a single segmental ganglion from all others, the movements in that segment ceased. However, if the ganglion was stimulated electrically, the muscles and any limbs in that segment would make vigorous, complete movements. Therefore, each segmental ganglion is responsible for the patterning of motor output that drives a limb forward or contributes to the movement of the abdomen.

If the segmental ganglia are responsible for telling individual muscles how to contract or relax, what is the brain doing? Removal of the protocerebral lobes produces a mantis that walks and grasps simultaneously. This is a disastrous situation for the insect because eventually its forelimbs find something to grasp firmly while its walking legs pull the animal ahead. This dilemma graphically illustrates the importance of avoiding the simultaneous activation of competing responses. The protocerebral lobes are designed to avoid the dilemma by exerting an either–neither control of the rest of the nervous system. The mantis either walks or grasps, or it does neither.

What is the nature of the commands originating in the brain? Roeder answered this question by removing the entire head of the mantis, a procedure that eliminates the subesophageal ganglion as well as the protocerebral lobes. Under these circumstances the animal is immobile. Single, irrelevant movements can be induced by poking the creature sharply, but that is the extent of its ability to behave.

These experiments provide a picture of the total control system of the praying mantis (Figure 7). Most of the time, environmental information received by the protocerebral lobes does not affect the normal stream of the inhibitory messages that this region sends to the subesophageal ganglion. However, particular patterns of stimulation block the relay of inhibitory messages from protocerebral units to other areas in the subesophageal ganglion; these regions, freed from control, send messages to the thoracic ganglia where new signals are generated that order muscles to take specific actions. Depending on what sections of the subesophageal ganglion are no

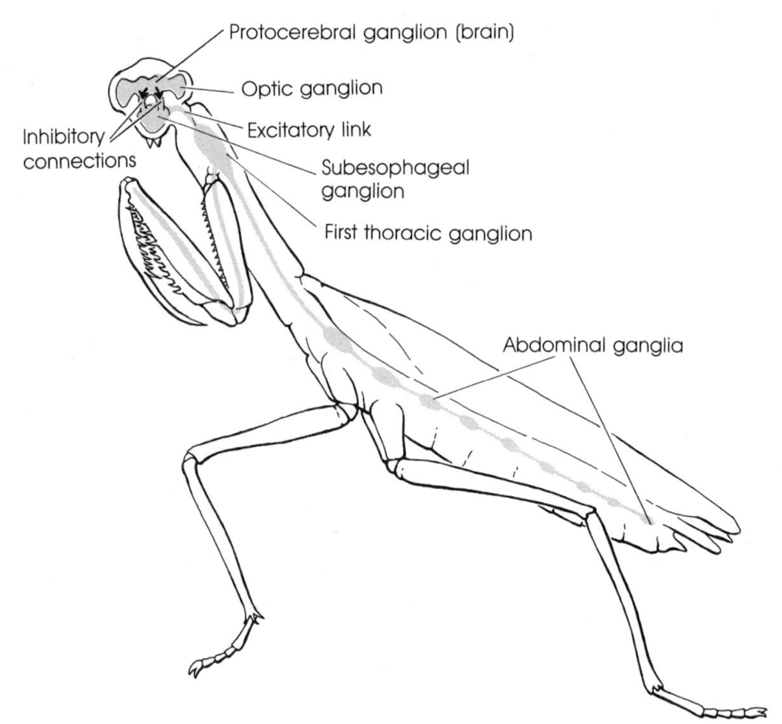

7 **Nervous system of a praying mantis.** The upper brain of the mantis generally inhibits the subesophageal ganglion, preventing the animal from attempting more than one activity at a time. If the connections between the upper and lower portions of the brain are cut, the subesophageal ganglion sends a stream of excitatory messages to the thoracic and abdominal ganglia; the mantis then attempts to do several competing activities simultaneously.

longer inhibited, the mantis walks forward or strikes out with its forelegs.

There is one exception to this scheme. If a mature male's head is removed, the animal, instead of losing its ability to behave, performs a series of rotary movements that swing its body sideways in a circle. While this is happening, the mantis's abdomen is twisting around and down. The behavior of headless males has special significance. A reproductively mature male stalks an adult female until he is close enough to leap upon her and make vigorous downward movements with the tip of his abdomen in an attempt to copulate. Sooner or later the female may reach back over her shoulder and begin munching on the head of the male (Figure 8). Even if copulation has not been effected, the "S bending" of the abdomen continues despite the loss of the brain and subesophageal ganglion. Still more remarkably, if a stalking male is spotted by a female, captured, and partly eaten before he is able to climb upon the female's back, his legs carry him around in a circular path. This brings the male's body against the female's, at which

8 **Female grass mantis** (above, head tipped from lower left to upper right) is consuming the winged male during copulation in Zambia, Africa. The female has largely eaten the male's head at this stage. Photograph by E. S. Ross.

point the headless insect climbs up onto the back of its would-be mate. The male's abdomen probes about professionally, and copulation occurs normally. As a result of the cannibalistic behavior of their mates, selection has favored mature males with a control system that is strikingly different from that of other mantises, one in which the thoracic and abdominal ganglia can independently order the headless animal to copulate.

The nervous system of the mantis, like that of the anole, appears to be functionally organized as a cluster of "centers," each with specific responsibilities for certain activities and decisions. Some mantis neurons manufacture their own output: a stream of inhibitory messages that control the activities of other groups of cells. The inhibition of competing centers is the basis for the mantis's ability to do one thing at a time. However, after receiving selected patterns of sensory information, inhibitory neurons in the brain lobes are themselves inhibited, freeing special circuits in the subesophageal ganglion that relay triggering signals to the segmental ganglia for the performance of a particular behavior pattern [608, 610].

Cyclical Changes in Behavior

The ability of the mantis's nervous system to block all but one selected series of motor commands is obviously adaptive. Equally beneficial is the capacity of animal nervous systems to organize some categories of behavior in repeating cycles. We have already commented on how the pattern of egg production by anoles favors females that copulate only when they possess a mature egg. The carnivorous sea slug *Pleurobranchaea californica* (Figure 9) also produces eggs at intervals as it acquires the food necessary to support egg development (although the slug manufactures a clutch of fairly small eggs rather than one large egg per cycle, as does the anole). During the period of egg maturation, feeding behavior has very high priority for *P. californica,* so much so that if given a simultaneous choice between feeding stimuli (squid homogenate) and mating stimuli (a receptive partner), the slug will extend its mouthparts and strike at the homogenate rather than copulate [165, 166]. Similarly if given a choice between feeding and withdrawing its oral veil in response to a tap on this structure (Figure 10), a hungry slug will feed. Nerve cells that regulate the ingestion of food must also inhibit other cells responsible for copulation and withdrawal. The ability of certain cells that are involved in the feeding response to inhibit other identified neurons that participate in the withdrawal response is biologically appropriate. A feeding animal is likely to have its oral veil stimulated by the thrashings of a victim; withdrawal might lead it to lose contact with a prey.

The ability of feeding neurons to inhibit other components of the slug's nervous system means that feeding usually has high priority in the behavioral hierarchy of *P. californica.* But this changes at the time the eggs are laid. A slug offered a chance to strike at a food stimulus becomes much less likely to do so in the few hours after egg laying. In conjunction with egg development, a hormone is produced and circulates in the bloodstream. At the time of egg laying, concentrations of this hormone peak and suppress the activity of nerve cells associated with the feeding strike response.

W. J. Davis and his co-workers established that hormones were involved in feeding inhibition by taking blood samples from slugs that had just laid their eggs and injecting the blood into other individuals [598]. The transfusions carried substances that induced the recipients to lay their eggs (whether the eggs were mature or not) and reduced their response to squid

9 Carnivorous marine slug, *Pleurobranchaea.* This animal has a well-defined set of interactions among its behavioral options. Photograph courtesy of W. J. Davis.

homogenate. Normally the hormonal suppression of feeding would lower the chance that a slug would cannibalize its own progeny; by the time feeding regained its dominance as hormonal levels fell, the slug would usually have moved away from its clutch. Thus, the slug is similar to the anole in that its cycle of egg laying is monitored internally via hormonal changes that lead to correlated adjustments in behavior.

Feeding Cycles

The interaction between the feeding network and other regulatory units within the sea slug is not only influenced by the schedule of egg development but also by the results of feeding. A slug that has been fed to satiation on strips of squid meat becomes much more likely to withdraw its oral veil when touched, even in the presence of squid juice [387]. Perhaps as the gut becomes distended with eaten squid, mechanoreceptors in the digestive tract are stimulated. Their signals directly or indirectly lead to the production of messages that travel to and inhibit the feeding network. This has two consequences. Because the feeding neurons are blocked, (1) they are less likely to order a feeding reaction to chemosensory cues signaling "squid present" and (2) they are less likely to inhibit the competing network controlling the withdrawal response. The slug will not try to capture food for which there is no digestive space. It will become more "cautious," readily withdrawing its sensitive oral veil from mild tactile stimulation. As the mechanical bulk in its gut declines during food processing, the feeding response gradually regains its usual high-priority position.

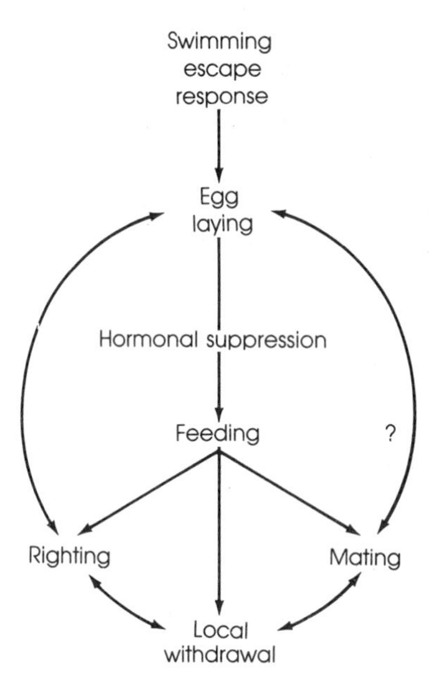

10 **The behavioral hierarchy** of *Pleurobranchaea*. Escape consistently takes precedence over all other behavioral options of the animal. If given a choice among responding to a mate, righting its body, and feeding, the animal will feed. Lowest on the hierarchy is withdrawl of the head region in response to mild tactile stimuli. When the hormone associated with egg laying is present in its bloodstream, the animal's behavioral hierarchy changes and feeding is suppressed.

A similar mechanism for the cyclical control of feeding behavior exists in the totally unrelated blowfly [173, 174], an insect that feeds, not on squid and other marine animals, but on various exudates from plants, juices of liquifying animal corpses, and other savory fluids rich in sugars and proteins. During the night, the nutrients collected in earlier meals are metabolized to provide energy for maintenance of the insect. By morning, a neural monitoring system detects that the fly's blood sugars have been depleted and orders the animal to become active. The fly flies. Upon detection of certain odors, it heads on a zigzag course into the wind until olfactory stimulation from key stimuli becomes so intense that the thoracic ganglion is shut down and the fly alights.

Once on the leaf, the insect walks until it steps into a solution that contains carbohydrates, at which time the tarsal sugar receptors fire. Action potentials from these cells reach the brain, where decision-making neurons command muscles to extend the proboscis. The lower surface of this apparatus comes into contact with the liquid, the labellum is spread (Figure 11), and liquid flows into the oral groove, stimulating still more sugar receptors there. Input from the oral receptors is routed directly to the brain; central motor neurons order the fly to begin drinking. The speed with which fluids are imbibed and the duration of sucking are proportional to the concentration of sugar in the solution. If the liquid is not very sugary, the

11 **Blowfly extends its proboscis** in response to an application of sugar water to its feet. Note also the fringe of sensory hairs around the labellum, the base of the proboscis. Photograph by George Gamboa.

oral receptors adapt (cease firing) quickly and sucking ceases. If sugar concentrations are high, adaptation of the oral receptors may not occur for 90 seconds or thereabouts. After a period of time the sensory cells recover and will respond again to stimulation. This causes reextension of the proboscis and a new bout of sucking, usually shorter than the previous one because the taste receptors adapt more rapidly. Eventually feeding will cease entirely, despite the fact that the receptors have disadapted and will fire on contact with appropriate solutions.

The cessation of drinking happens because of events taking place along the digestive tract of the fly. As liquid is consumed, it passes into the foregut and, from there, a small slug may enter the midgut. Then a valve closes the passage between the two compartments while sugars in the midgut are absorbed. Liquid that continues to pass along the digestive tract is shunted into the crop, a storage area. As the crop gradually fills, liquid is forced back into the foregut, distorting stretch receptors attached to the muscles of this part of the tract. The receptors fire and relay their messages to the recurrent nerve, which runs between the foregut and the brain (Figure 12). While the recurrent nerve fires, extension of the proboscis is permanently inhibited, even after the sugar receptors are disadapted. Cutting this nerve fiber produces a fly that literally blows itself up in an orgy

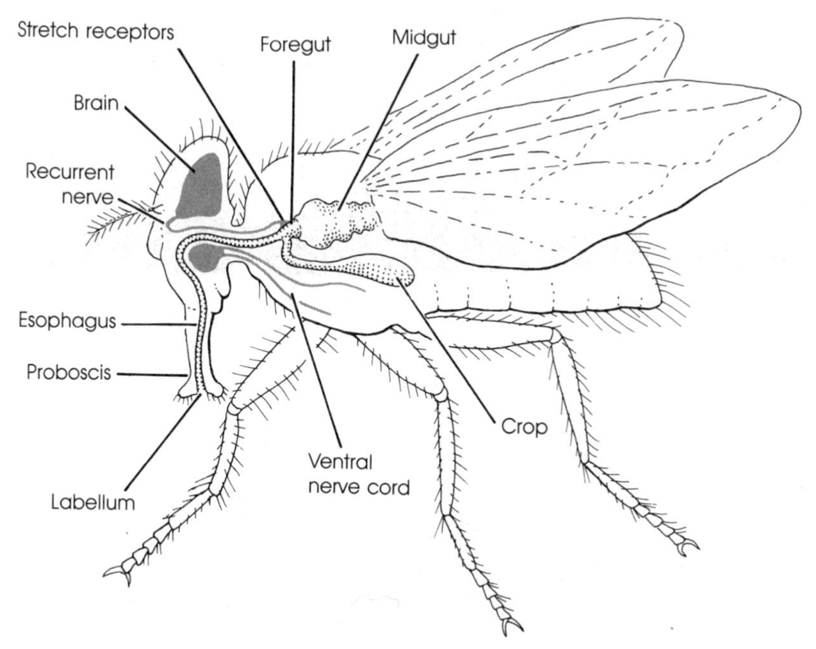

12 **Digestive tract and nervous system** of the blowfly *Phormia regina*. The recurrent nerve plays a critical role in controlling drinking by the fly.

of drinking. In the absence of inhibitory signals from the recurrent nerve, the insect starts sucking up liquids again as soon as its sugar receptors disadapt.

The fly's feeding control system prevents its life from coming to a spectacularly explosive conclusion. Instead, the fly feeds in sensible bouts when it has a nutritional deficit. This requires that it be able to integrate information on its internal state with diverse external stimulation encountered on its travels. Its nervous system contains the appropriate receptors and integrators so that it feeds cyclically in response to its fluctuating needs [173].

Circadian Rhythms

The fly's feeding behavior is organized not only in very short, repetitive bouts during the daylight hours but also on a longer, 24-hour cycle as well. Blowflies are inactive at night; only during the day do they search for food, mates, egg-laying sites, and so on. Because we follow a daily rhythm of a similar sort, we take this ability almost for granted (although most of us are aware that some animals are active only at night, others at dawn, and still others at dusk). Daily rhythms reflect the specialized properties of physiological mechanisms and their interactions with many other components of an animal's body. One might argue that the blowfly's night–day cycle of inactivity–activity is a fairly simple response to changes in temperature and light intensity. A test of this hypothesis would be to hold individuals under controlled laboratory conditions of constant temperature and uniform light intensity. If the fly persisted in becoming active at certain times of the day and inactive during what would correspond to the nocturnal hours, then one would discard the hypothesis that activity was purely externally signaled. Instead, the insect would be said to possess a CIRCADIAN RHYTHM, a daily cycle that persists under constant conditions.

Experiments on circadian rhythms have been conducted with a vast array of animals, including many insects. I have chosen to present the results of a study of a cricket, *Teleogryllus commodus,* whose males and females possess a similar circadian clock but use it for very different purposes [443, 444].

Males of *T. commodus* (and many other crickets) spend a substantial period of each day calling from the entrance of their burrows (Figure 8, Chapter 4). Their signals may attract receptive females, which travel to them in order to mate and receive the sperm needed to fertilize their eggs. These activities are not randomly distributed throughout the 24 hours of a day but are largely restricted to the nighttime. Now this could be because the crickets use declining light intensity as a cue to activate the appropriate responses. But if one takes a population of males into a laboratory and leaves the light on permanently (or places the crickets in continuous darkness), the males will continue to sing only for a limited block of hours each day. Under conditions of constant light, the start of the bout of singing begins about 25.5 hours later than the onset of singing the previous day. If the room is continually dark, the FREE-RUNNING CYCLE has a period of about 23.5 hours (Figure 13). Thus, without the cues provided by nightfall and sunrise, a male cricket will continue to sing in a repeating cycle, but he gradually drifts out of phase with the actual nighttime hours because his circadian rhythm is not quite 24 hours. The slight divergence of a free-running cycle from 24 hours is typical of animal species generally [564, 565].

But if one switches from constant conditions to a regime of 12 hours of light, 12 hours of dark, the crickets receive a cue that they use to ENTRAIN the circadian pattern of singing. In a few days, the males will all be starting to sing about 2 hours before the lights go off and they will continue until about 2.5 hours before the lights go on again in the "morning." This cycle of singing matches the natural one, which is synchronized with dusk; it does not drift out of phase with the 24-hour day but will continue to be reset each day so that it begins at the same time in relation to lights-out [443].

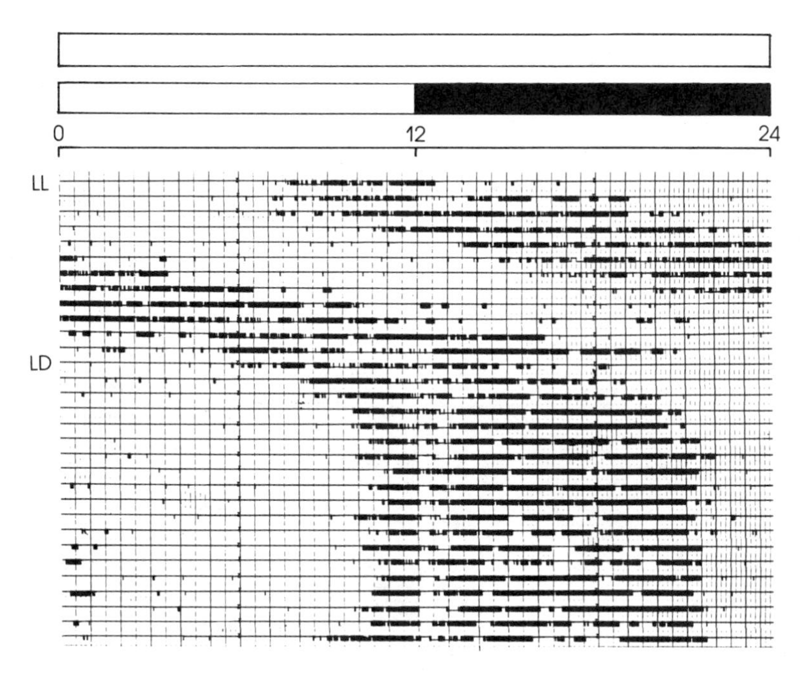

13 **Circadian rhythm in calling** by males of the cricket *Teleogryllus commodus.* In constant light (LL; days 1–12), the crickets do not call randomly, but the start of the calling period begins somewhat later each day. When the day is divided into 12 hours of light and 12 hours of dark (LD; days 13–31), the calling period stops shifting and begins an hour or two before the lights are turned off each day.

Females of *T. commodus* do not sing, but, as noted, they will travel to singing males at night. An unmated female that has recently metamorphosed into an adult will begin moving about just after the lights go out at night in a laboratory that is set on a 12L:12D schedule. Her activity can be monitored if she is placed in a special cage with a running wheel, an apparatus rather like a tiny treadmill whose revolutions activate an electronic circuit and generate a pen mark on a slowly moving paper strip. This creates a permanent record of when the female was walking (Figure 14). The female need not hear a male calling in order to begin or continue her search, which lasts for most of the night. This activity is influenced by an internal timer, as can be shown by holding females in constant darkness. They will enter a free-running cycle with a period of about 23.5 hours, which is identical to the male's free-running song cycle [444]. As in the case of the male, the switch from light to dark, if it is available, will synchronize the female's circadian rhythm so that she begins searching the moment the lights go out each day. This response to darkness ensures that her search will coincide with the period of male calling, and improves her chances of ultimately encountering a mate.

The timer mechanism that regulates singing and searching activity by male and female crickets, respectively, has not been precisely located, and its manner of operation remains largely mysterious. It is believed that the clock resides somewhere in the brain. If one cuts the nerve carrying sensory information from the eyes of a male to the optic lobes of its brain (depriving a male of his vision), he enters the free-running pattern. Visual signals are needed to entrain the daily rhythm, but a rhythm persists in the absence

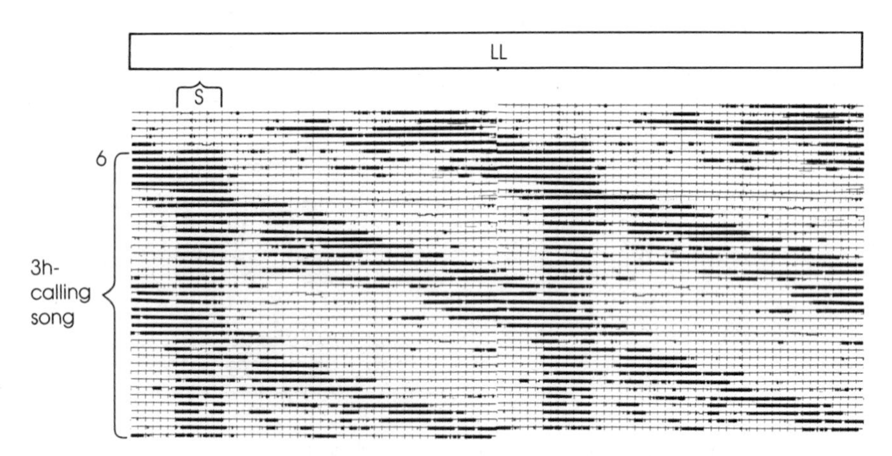

14 **Circadian rhythm in location** by females of the cricket *Teleogryllus commodus*. Under conditions of constant light, females exhibit a shifting daily period of running activity. After day 6, male song was played to the females for three hours twice daily. The calls stimulated female activity but did not abolish their clock-triggered locomotory rhythm.

of this information. If, however, one cuts both optical lobes, separating them from the brain mass, there is a complete breakdown of the singing cycle and a male will sing with equal probability at any time of the day.

The search for the location and operating rules of clock mechanisms in other animals has not been vastly more successful than in the cricket. In general, however, rhythmic behavior in animals often reflects the capacity of certain nerve cells in the central nervous system to *spontaneously* generate their own rhythmic pattern of messages. This is best documented in the case of neurons regulating the cyclical breathing, feeding, or locomotory activities of some animals [171], but the principle should apply as well to circadian rhythms. In the marine mollusk *Aplysia californica,* a distinctive cell that has been found in the abdominal ganglion of the animal will display a circadian pattern of action potential output—even when the ganglion in which it resides is surgically removed from its owner and kept alive in a seawater bath with constant temperature and light! The isolated ganglion no longer receives any sensory inputs nor is subject to hormonal influences, and yet it persists in its programmed rhythm of activity [565].

Examples of this sort have called into question the hypothesis that there is a single biological clock that can be located in one structure within an animal. (Plants have circadian rhythms, too, a finding that shows that nervous systems are not even required for this phenomenon.) Perhaps, however, there is in some animals a master clock that regulates circadian clocks in organs and tissues throughout the body. The best-documented example of such a mechanism has been discovered in the brains of some rodents. In these mammals it appears that the suprachiasmatic nuclei (SCN) contains a major circadian oscillator. Experimental lesions that damage this structurally distinctive region in hamsters and Norway rats abolish circadian rhythms in a host of attributes such as heart rate, hormone secretion patterns, locomotory cycles, and feeding behavior [821, 822]. In addition, electrical stimulation of cells in the SCN can disrupt the circadian rhythm of some rodents [622].

The SCN is in an ideal position to exert its major organizing influence because it is located within the hypothalamus of the brain, which in turn is adjacent to the pituitary gland. The hypothalamus of mammals is known to play a central role in integrating and motivating many activities via the pituitary gland. Furthermore, the SCN receives input by way of a direct neural pathway running from the retina. This pathway is not part of the visual system itself but could provide information about photoperiod and the timing of dawn and nightfall. It is probably an entrainment pathway that resets the master clock each day and keeps it from drifting out of phase with real world conditions [821, 822].

Long-Term Behavioral Cycles

The regular changes in responsiveness to certain stimuli that occur during some portion of a day or over a 24-hour repeating cycle are not the only rhythms of behavioral change in the animal kingdom. We have already encountered a species, the green anole, with an ANNUAL CYCLE of activity. In this section we shall examine some other species with similar biological rhythms. But first let us consider two species whose neural mechanisms enable them to time selected activities by the appearance or phase of the moon. The gravitational fields of the sun and moon act in concert every 14.7 days to generate the greatest variation in high and low tides in marine habitats. The occurrence of these "spring tides" is of the greatest reproductive significance for the grunion, a small silvery fish that mates and lays its eggs on the beaches of California and Mexico. During certain months of the year, the grunion comes ashore in very large numbers a few hours after high tide on nights when there is a full moon or a new moon (Figure 15). Females bury the lower part of their body in the moist sand;

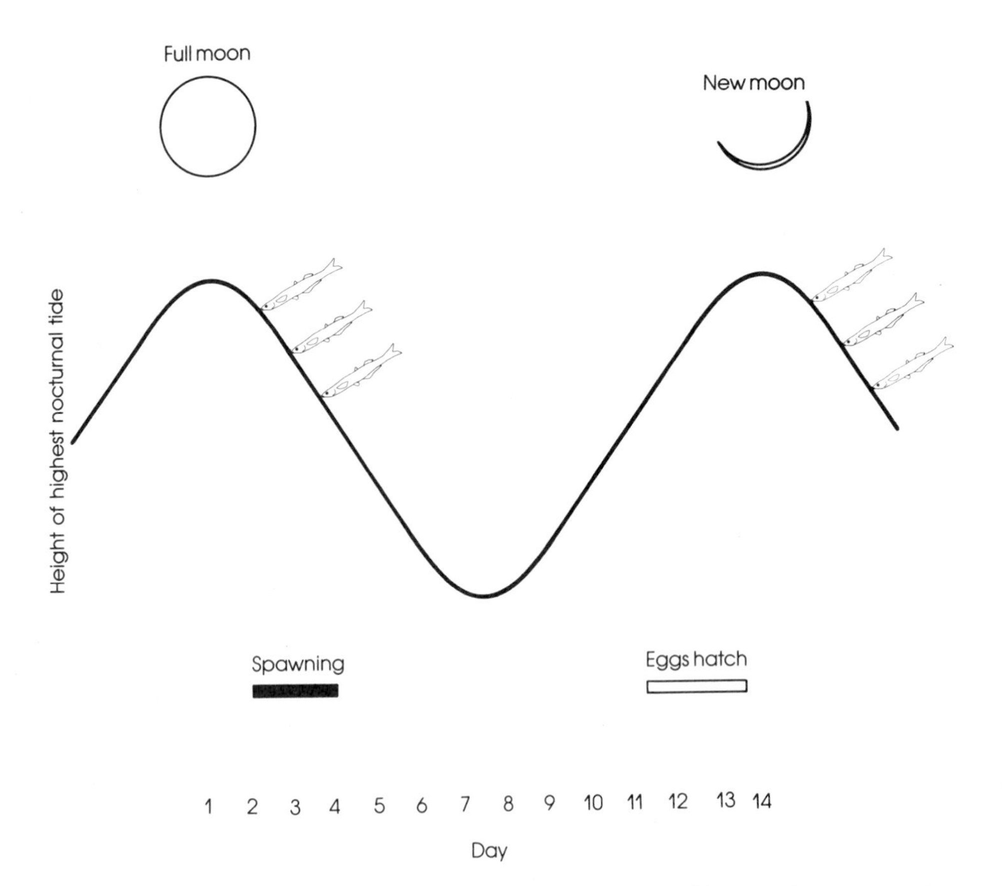

15 **Tidal cycle and grunion behavior.** The grunion spawn, at two-week intervals, when the nightly high tide has just passed its peak. This ensures that when their eggs hatch, the fry are likely to be swept out to sea by increasing high tides.

males release their sperm as the eggs are laid. Waves then carry the fish back to the ocean, leaving behind the buried eggs that are covered with sand deposited by the receding tide. The eggs will not be exposed for about ten days until, in the course of a new tide cycle with high tides of increasing magnitude, the surf uncovers the eggs (a process that stimulates the mature embryos to hatch) and carries the fry out to sea. The precise means by which grunion determine when to come ashore and breed is not known, but it would not be surprising if the fish possessed a fortnightly clock that primed them to become sexually active at the optimal time for successful reproduction [564].

A very different creature whose behavior is also temporally regulated by the phase of the moon is the banner-tailed kangaroo rat. The foraging activity of these rodents for much of the year has a distinct circadian rhythm, with the animals restricting their seed-collecting searches to the night. But the story is more complex, as research on wild kangaroo rats by Robert Lockard and Donald Owings has shown [441, 442]. The activity of individual rats was monitored by an ingenious food dispenser–timer invented by Lockard. The device released very small quantities of millet seed at hourly intervals. To retrieve the seed, the animal had to walk through the dispenser. When it entered, the animal depressed a treadle that moved a pen that made a mark on a paper disk that was turned slowly through the night by a clock mechanism. When the paper disk was collected in the morning, Lockard had a temporal record of all nocturnal visits to the machine. Data collection was sometimes frustrated by ants that perversely drank all the ink or by cows (disgraceful animals that have consumed most of the vegetation in the western United States) that stepped on the recorders. Nevertheless, Lockard's records showed that in the fall when the animals had accumulated a large cache of seeds, the rats were selective about foraging, usually coming out of their underground bunkers only at night when the moon was not shining (Figure 16). Because the predators of kangaroo rats (coyotes and owls) locate their prey more easily when visual cues are available, the banner-tails probably reduce their risk of detection by foraging in complete darkness.

Just as it is biologically sensible for kangaroo rats and grunion to regulate their behavior in accord with lunar cycles, so too it may be adaptive for an animal to adjust its behavior according to an annual cycle, particularly if the species lives in an environment with marked seasonality. An annual clock mechanism is a potential aid in dealing with seasonal changes in conditions because it provides an internal signal that helps an individual prepare physiologically and behaviorally for migration, or reproduction at the appropriate time of the year.

However, the only definitive demonstration of an annual clock involves the golden-mantled ground squirrel [573], which in nature spends the late fall and winter hibernating in an underground chamber. Five members of this species were born in captivity, and blinded, and held throughout their lives in constant darkness and constant temperature while being provided

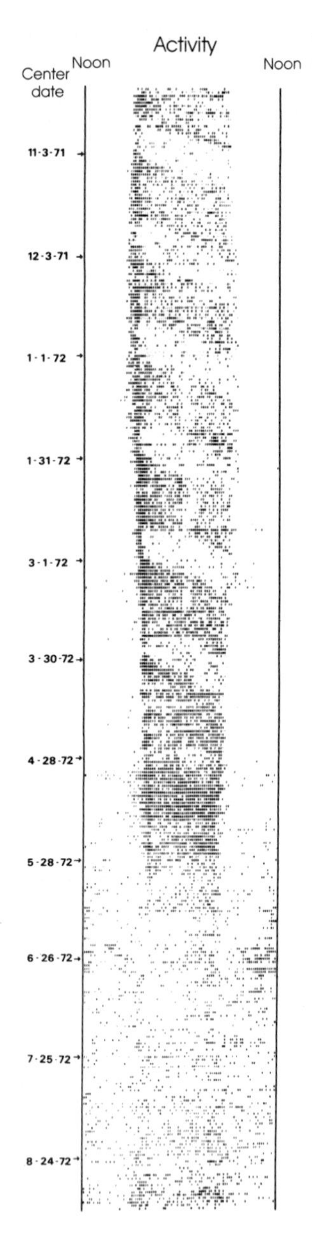

16 **Patterns of activity** of banner-tailed kangaroo rats. Each black mark represents one or more records made by rats feeding at a timer device. In the period from November to March, the rats were active during the dark hours of the night (the white diagonal bands correspond to the hours when the moon was shining). However, a shortage of seeds first led the animals to forage throughout the evening, whether the moon was up or not, and later to forage throughout all hours of the day. Courtesy of Robert B. Lockard.

with an abundance of food. Year after year they entered hibernation at about the same time that their fellows living in the wild were doing the same thing (Figure 17).

The Annual Cycle of The White-Crowned Sparrow

Although white-crowned sparrows may or may not have an annual clock mechanism, they are faced with profound seasonal changes in conditions that make changes in behavioral priorities highly adaptive. Imagine a white-crown from a population that breeds in central Alaska. Food is abundant there in the late spring and summer but not in the winter. Sensibly enough, white-crowned sparrows abandon Alaska in the fall and travel as much as 2400 miles to wintering spots in southern United States and Mexico. But in order to return to reproduce successfully in Alaska, a white-crown must not only navigate many miles north in the spring but also must time its journey so as to arrive in Alaska when conditions are suitable for breeding. There is obviously no direct information on Alaskan weather conditions available to a white-crown wintering in sunny northern Mexico in early April. The bird that begins its spring migration too soon may reach the breeding area in time to be killed or severely weakened by a cold snap. On the other hand, the male that lingers too long on the wintering grounds may arrive to find choice territories already occupied. The female that arrives, builds a nest, mates, and lays eggs too quickly may lose her clutch during a cold

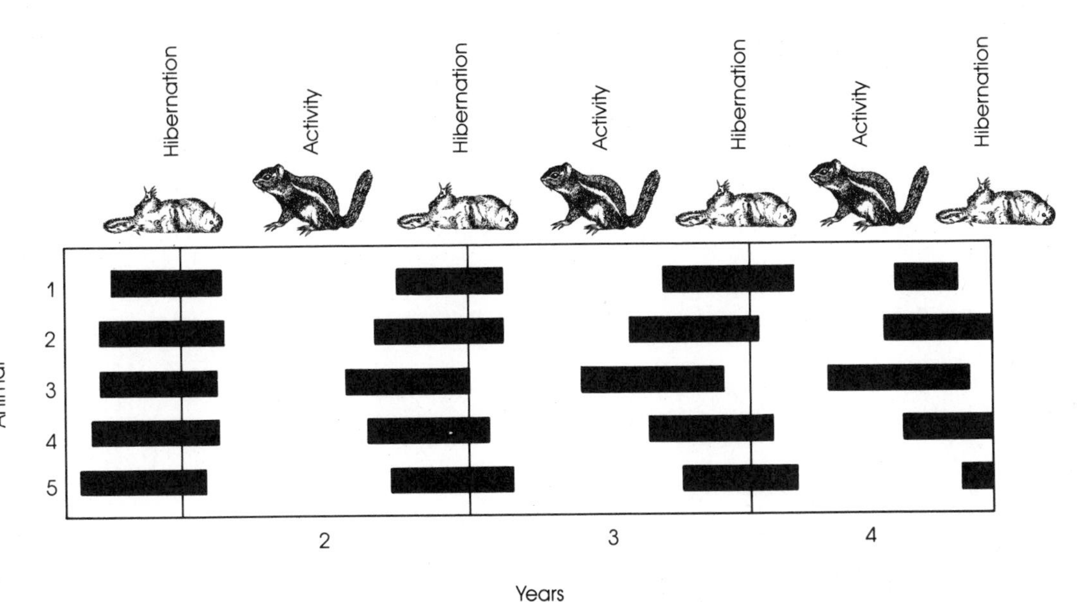

17 **Annual clocks** in five golden-mantled ground squirrels. The annual onset of hibernation persists year after year, even in squirrels that have had their visual system destroyed.

spell. However, one that waits too long may not have sufficient time to rear her offspring before the fall migration. The timing and coordination of reproductive activities has to be precise in north temperate zones because the margin for error is so small. Individual sparrows often succeed because of a complex array of physiological mechanisms that exert their control on reproductive behavior and physiology largely through the regulated release of hormones [234, 800] (Figure 18).

The key to success for the bird clearly has something to do with the ability to determine the time of year. No one doubts that white-crowns migrate north in the spring and south in the fall, but just how they manage this feat is unknown. The control mechanism could be based on (1) seasonal or geographic differences in environmental cues or (2) seasonal changes in the birds' internal physiological state induced by a circannual clock. In the garden warbler, another songbird that migrates long distances, Eberhard Gwinner and Wolfgang Wiltschko have experimental evidence that sup-

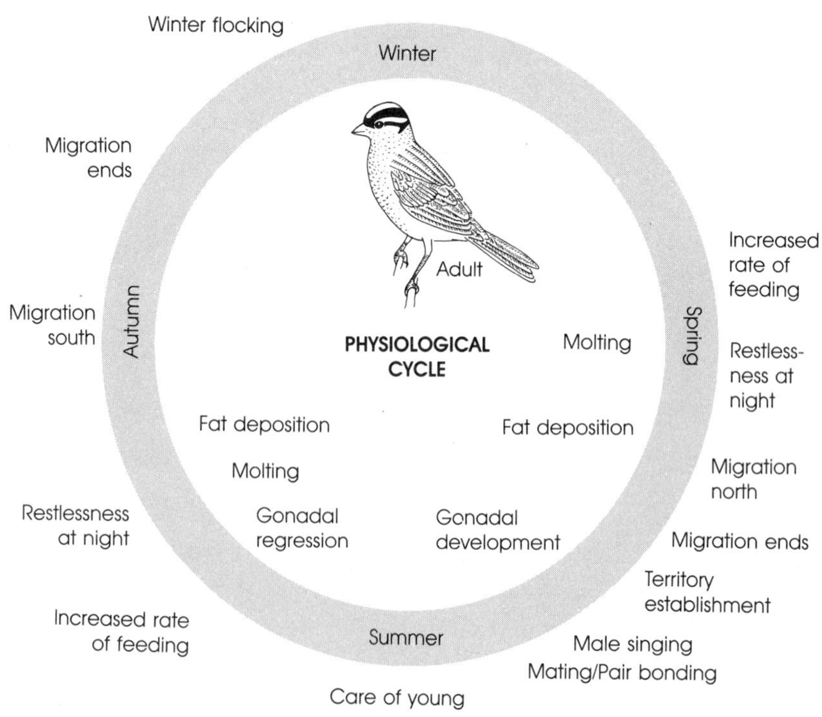

18 **Annual behavioral and physiological cycle** of the white-crowned sparrow. The regular sequence of behavioral changes over the seasons is the product of a structured series of changes in hormone releases within the bird.

ports the second possibility. They hand-reared 11 baby garden warblers. At the start of the fall migratory season in Europe, the birds were housed individually in cages and exposed thereafter to a constant environment (12 hours of light and 12 hours of dark each day under constant temperatures). Despite their constant surroundings, the birds underwent a moult of their feathers on schedule and began to exhibit nocturnal restlessness (jumping from perch to perch in a cage where each hop could be recorded electrically), a behavior that persisted until early spring. (Garden warblers migrate at night; typically, caged birds become restless and active at night during the migratory season, whereas at other times they sleep at night.) As the spring migratory season approached, the birds—still held under 12L:12D conditions—went through another moult and once again resumed their nightly hopping, an action that indicated a desire to migrate at a time when their wild companions would have been heading north on the spring migration [292].

Not only did the birds show migratory restlessness at the right seasons, they also oriented in the correct direction both in the fall and spring. During both seasons the warblers occupied special octagonal cages with eight pairs of perches that electrically recorded the location of the bird within the cage. The record of perch use therefore enables one to determine the directional preference of the caged bird. The warblers could not see the night sky and so had no celestial cues to guide them, but they could sense the earth's magnetic field, as earlier experiments had demonstrated [292]. The lines of weak magnetism running north to south provide a compass cue that garden warblers and some other migrants use to orient their flights. The caged birds in the fall employed this cue to jump toward perches on the south side of their cages; in the spring the same birds selected perches on the north side of the cages. The birds not only "knew" when to migrate, they also knew which direction to go. They did not need any obvious environmental cues in order to behave in a manner appropriate to the fall or spring migratory season. These experiments support the hypothesis that garden warblers can unconsciously "anticipate" the migratory season by producing new flight feathers, fattening up for the trip, and then setting off in the right direction.

The Regulation of Reproduction

A component of the annual cycle of white-crowned sparrows involves the growth and decline of their reproductive organs. When a bird arrives on its breeding grounds, its gonads will have not yet reached their full reproductive dimensions. Because of the high energetic costs of flight, it is advantageous for birds, particularly when they are migrating, to carry as little unnecessary weight as possible. During the nonreproductive seasons, the gonads of songbirds atrophy, often to less than 1 percent of their maximum weight. Functional ovaries and testes are, however, indispensable for reproduction and therefore must be redeveloped at the appropriate time. White-crowns manage in part by using information about photoperiod to regulate the development of their gonads [234].

Information about the length of the photoperiod is collected by light-sensitive receptors present somewhere in the brain itself. If the white-crown's visual system is destroyed or if its eyes are covered by light-proof goggles, the bird's circadian rhythm is not eliminated [811]. It will continue to wake up in the morning, become active for some time, less active during the middle of the day, and more active again in the early evening before going to sleep once more.

The bird's circadian clock may work in the following fashion. Built into its time-measuring system is a daily cyclical change in sensitivity to light, a cycle that is set each morning at dawn. During the initial 13 hours or so after the clock is set, the animal's control system is highly insensitive to light; this insensitivity then steadily gives way to increasing sensitivity, reaching a peak in the 16- to 20-hour period after the starting point in the cycle. Sensitivity then fades very rapidly to a low point 24 hours later, at the start of a new day and a new cycle. If the days are 12 to 13 hours long and nights 11 to 12 hours long, the photosensitive part of the system simply is never activated because there is no light available during the light-sensitive phase of the cycle. However, if the days are 14 to 15 hours long, light reaches the bird's brain during the photosensitive phase. This causes the release of hormones from the hypothalamus that stimulate the anterior pituitary. Cells there release the hormone prolactin and assorted gonadotropins. The gonadotropins are carried by the bloodstream to the gonads of the bird, where they initiate development of the reproductive equipment of the sparrow.

If this model of the clock system is correct, it should be possible to deceive it experimentally. Donald Farner and his colleagues have done exactly that with several elegant experiments (Figure 19). They first set the clock by exposing birds to a regular sequence of 8 hours of light and 16 hours of darkness (8L:16D). They then put a group of birds on an 8L:28D schedule. Because the light periods are now out of phase with a 24-hour cycle, these birds receive light during the time at which their clocks are highly photosensitive. The male birds' testes grow under these conditions, even though there is a lower ratio of light to dark hours than during the 8L:16D cycle, which does not stimulate testicular growth [233]. The same effect can be achieved by switching a bird from an 8L:16D cycle to an 8L:16D cycle that is interrupted by a two-hour light period 17 to 19 hours after the start of the day. This coincides with the time of peak photosensitivity during the circadian cycle. The result is marked testicular development.

Thus, by the time a white-crown arrives on the Alaskan breeding ground, it is primed for rapid gonadal growth [799]. In *both* the male and the female, testosterone levels become much higher than they were during migration. Males establish territories, sing constantly, attack intruders, and attract a mate. The female contributes to defense of the territory after she joins a partner. As she feeds and acquires sufficient supplies for egg production,

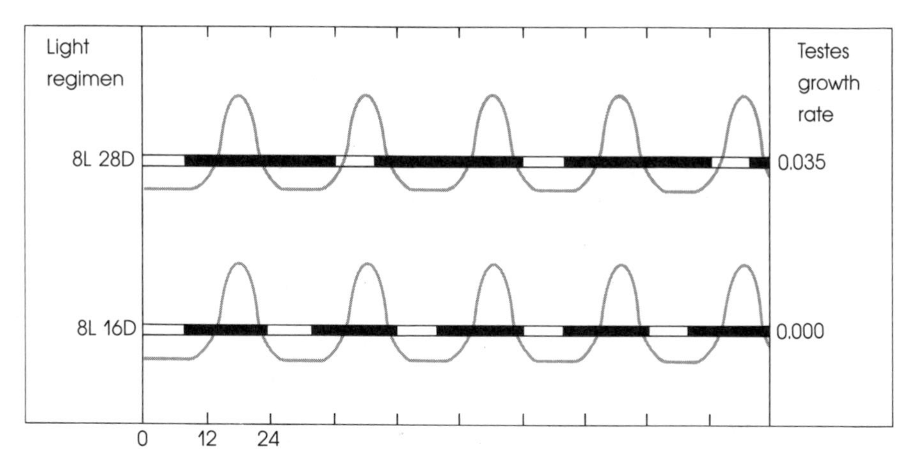

19 **Physiological system regulating testicular growth** in the male white-crowned sparrow has a cyclical component. In the experiment the open and black sections of the two bars show the light and dark periods of two different L:D regimens. The curves represent a hypothetical cycle of "photosensitivity" in the control system. During the 8L:16D regimen, the birds are never exposed to light during the daily photosensitive phase, and gonadal growth does not occur. Sparrows experiencing a 8L:28D schedule are exposed to light occasionally in the photosensitive phase of the cycle, and they respond with testicular growth.

the concentration of estrogen increases in her blood. The female is likely to begin soliciting copulations from her partner. He in turn is influenced by his female's behavior [520]. Males that are experimentally paired with females that have received estradiol implants have higher testosterone levels than males held with hormonally untreated females (Figure 20). As a result of the signals a male receives from his mate, his sexual activity rises and the pair copulates regularly. It is probable that in turn the male's song, courtship, and sexual behavior have reciprocal effects on the female's hormonal state, further fine-tuning the schedule of reproduction.

The biological priorities of a pair change once the eggs are laid. Territorial defense and copulation are no longer the most productive activities; in fact, they would interfere with effective incubation of the eggs by the female. In both sexes it is thought that prolactin secretion becomes dominant as gonadotropin and testosterone levels wane. The gonads decline in weight, the birds become far less aggressive, and the males stop singing [799]. In experiments with the pied flycatcher, Bengt Silverin found that he could reverse the trend of testicular regression in a male by giving him testosterone injections at this time. Treated males devoted themselves to singing and territorial defense but neglected their offspring, with the result that the reproductive success of their mates fell [661].

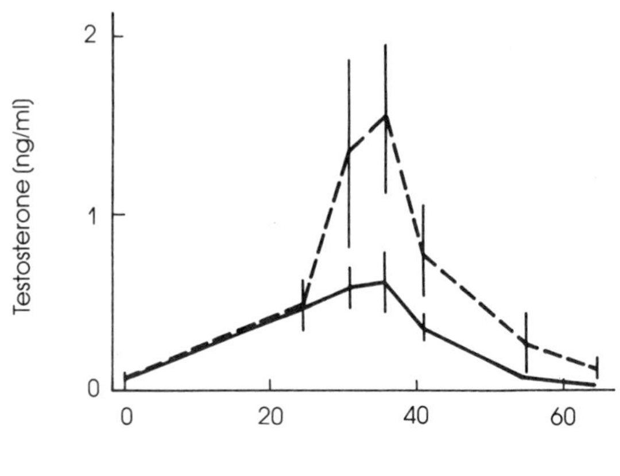

20 **Testosterone levels** in male white-crowned sparrows that were caged with estrogen-treated females (---) or untreated females (——). All birds were exposed to 20L:4D days.

As the eggs of a white-crowned female hatch, the nestlings begin to beg for food and are fed by both parents; they grow up, leave the nest, and soon become independent. The adult's tendency to provide parental care declines correspondingly, perhaps linked with a decline in prolactin secretion. The birds, like green anoles, then enter a sexual refractory period in the fall, during which time captive birds experimentally exposed to long photoperiods simply do not respond with gonadal growth or other components of reproductive activity. The hypothalamus, which in the spring reacts to increasing day lengths by stimulating the pituitary to release luteinizing hormone and other gonadotropins, fails to do so in the fall under the same external stimulation.

Sexually refractory birds feed voraciously, put on weight, molt, and migrate (south) in time to avoid being caught in an early fall snowstorm. If they are able to survive the journey and the winter, their sensitivity to photoperiodic change and their hormonal mechanisms will act in concert to trigger a new series of behavioral changes in the spring, one cascading after the other in an adaptive sequence [800].

Environmental Cues and the Timing of Reproduction

Although our emphasis has been on properties of the physiological systems of songbirds that structure basic seasonal changes in behavior, we have also indicated that the birds use a host of cues from their environments to fine tune their changing behavior. Increasing and decreasing photoperiods are of prime importance in scheduling the reproductive cycles, as are the social interactions between males and females. The capacity to be influenced by aspects of the physical and social environment is true for many other animal species, and in the pages that follow we shall compare how some vertebrates use certain external cues to make fine adjustments in the timing of reproduction.

Let us begin by contrasting the effect of photoperiod and temperature on the reproductive cycles of green anoles and white-crowned sparrows. Although both factors affect the onset of reproduction in these animals, temperature is of paramount importance for the anole, whereas photoperiod is more critical for the sparrow. Because white-crowns are warm-blooded, they are less vulnerable to temperature fluctuations than lizards and so can use photoperiodic cues as a helpful indicator of the season of the year. The anole, a cold-blooded animal hibernating in a dark burrow, gains more by using the immediate temperature of its environment to regulate its behavior.

Other warm-blooded animals that probably respond primarily to photoperiodic cues use this information to generate a very different schedule of seasonal reproduction. For example, the Eleanora's falcon does not lay its eggs until late July or early August, long after white-crowned sparrows have begun to breed in most portions of their range [755]. The distinctive aspect of the falcon's annual cycle stems from its exploitation of migrant songbirds, which it uses to rear its nestlings. The hawk breeds on islands and sea cliffs, primarily in the southern Mediterranean. In the autumn, a flood of migratory songbirds crosses the Mediterranean Sea. Many songbirds arrive in North Africa exhausted after the trip; their condition enables the falcon (which hunts over the water) to capture prey easily. Because the hawk lays its eggs slightly before the songbird migration begins, its young hatch at precisely the time when food is most available for them.

A consistent correlation between a rich food supply and the late summer also occurs in southern Arizona, home of the rufous-winged sparrow, a relative of the white-crown. The rains in the Sonoran desert generally fall in July and August but are much less regular in May and June. Thus, many desert plants and the insects that feed upon them become abundant only in the late summer. Gonadal growth in the sparrow starts in the spring but may not reach its peak until June. Even in birds with mature gonads, nesting generally will not start until the first intense rainfall of the spring or summer. Thus, in a population of ten pairs observed in 1965, all but one female delayed nest construction until the first major storm of the summer on 8 July [545]. It appears that the birds use photoperiodic cues to develop the gonadal condition that permits breeding but do not actually attempt to reproduce until a rainstorm triggers the process (Figure 21). As a result, the rufous-winged sparrow often does not nest until the late summer, long after white-crowned sparrows have begun to do so in areas in which insect prey are abundant earlier in the year.

For the rufous-winged sparrow and Eleanora's falcon, the probability that prey will be available in the late summer is high and these species possess reproductive mechanisms that prepare individuals for this likelihood. But other species live in environments in which there is no such consistent relationship between a particular season of the year and plentiful food. These animals may require access to food itself before coming into reproductive condition rather than anticipating the presence of food by

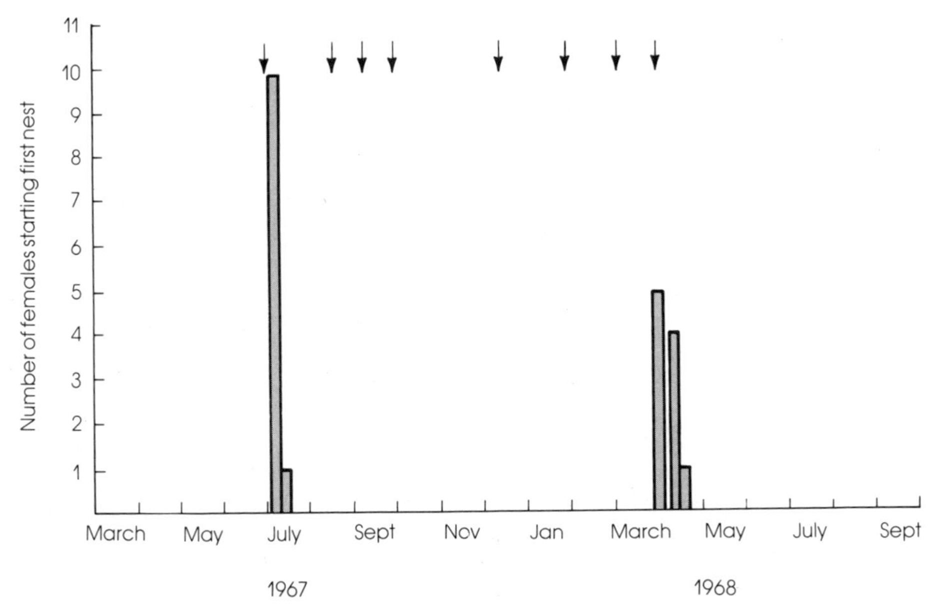

21 **Timing of reproduction in a desert sparrow.** Females in a population of rufous-winged sparrows initiated nesting in July 1967 after the first heavy rainfall of the year. (Arrows indicate weeks in which rainfall of 2.5 cm or more occurred.) In 1968 members of the same population bred in April, after several late winter storms and an early spring rain.

using photoperiodic, temperature, or rainfall indicators. For example, the pinyon jay is an extreme specialist on the seeds of pinyon and ponderosa pine. These trees set seed erratically from year to year. Some years there is a bumper crop of cones; in others, the trees may produce practically no seeds. The jays, therefore, have evolved physiological mechanisms that tie reproduction directly to food supply. In good years the birds begin to court and to build nests much earlier than in years of reduced seed production [422]. Moreover, simply seeing numerous green (unripe) cones of pinyon pines in the summer reverses the gonadal decline typical of the fall, a reaction that prepares the birds to become sexually active in the late winter and early spring, when they will be feeding largely on stored seeds gathered from the then-mature cones.

The direct assessment of food availability also influences reproductive cycles in the California quail [415] and the rodent *Microtus montanus* [65, 535]. Both species live in areas of the western United States in which rainfall and snow melt fluctuate greatly from year to year. The irregular and seasonally unpredictable nature of the water supply means that plant growth, upon which these animals depend, is also unpredictable. In a dry season, new plant growth is reduced, and the older, less nutritious material contains high concentrations of certain chemicals that act as direct inhibitors of gonadal growth in the quail and vole. However, in a wet year, when new plant growth is available, females that feed upon this food do not receive high concentrations of inhibitors; therefore, they come into reproductive condition and produce offspring that can be supported by the available food.

Social Regulation of Reproduction

In seasonal or variable environments, the importance of having the reproductive cycle coincide with the maximum availability of food for egg production or offspring development has led many unrelated species to evolve similar proximate regulatory mechanisms. But food is not the only factor critical to reproductive success. We have already seen that in anoles and white-crowned sparrows certain kinds of social stimulation can advance or retard the onset of gonadal development and sexual behavior at the start of the breeding season, whatever the temperature or photoperiod. There are other examples of this phenomenon, particularly among birds and small rodents.

Barbara Brockway's research on the budgerigar offers a classic example of the influence of the social, as well as the physical, environment on reproductive timing. Contrary to the impression of most Americans that budgerigars live exclusively in department stores, this small parrot does occur in the wild as well, notably in the arid center of Australia. The birds travel in large flocks in search of the rare places in which rainfall has occurred. Upon finding a suitable location, the birds tend to come into reproductive condition promptly. But social factors, as well as climatic ones, are vital to reproductive development. Brockway found that in the laboratory the testicular growth of males was promoted if they heard tapes of the "loud warble," the male territorial call of this species. But unpaired females that heard only this aggressive call experienced *reduced* ovarian growth. Increased ovarian development was stimulated by the male's courtship signals, particularly the "soft warble," which is given when a male perches beak to beak with a female [92, 93].

In some (probably most) species of birds, females do not respond equally to all courting males but instead discriminate among them on the basis of the quality of their courtship. Male canaries vary a great deal in how many song types they have in their repertoire. Young birds typically have fewer than older males [543]. Females that are experimentally exposed to tapes with a reduced number of song types build their nests more slowly and lay fewer eggs than females that have a chance to listen to more diversified songs [394].

The female preference for complex song may be responsible for a remarkable aspect of the neurophysiology of male canaries. Male song is under the control of an identified region of its brain (Figure 22). The neurons in this system undergo an annual cycle of their own. Following each reproductive season, the mass of neurons in this region of the brain declines, only to redevelop at the start of the next breeding season [541, 542]. Fernando Nottebohm has suggested that this may be related to the development of an ever more complex song repertoire as the male grows older and more experienced. Birds with a larger battery of songs develop a larger song system in their brain than males with simpler repertoires.

Social Effects on Mammalian Reproductive Physiology

Certain mammals, as well as many birds, exhibit a number of parallels in the kinds of reciprocal physiological effects that males and females have on one another. These have been particularly well studied in the house mouse, whose females, as noted before, live and reproduce in territories held by a single dominant male who copulates with his harem. Table 2 lists a set of social effects on house mouse reproductive physiology, many of which can be explained in the same way that similar effects were accounted for in anoles. Thus, a dominant male in a stable social setting promotes reproduction in the females he lives with; the absence of a male, or social instability due to the death of a mate, or a population explosion (as evidenced by the presence of the scent of a strange intruder male or the scent of females from a dense population) inhibits reproduction in females. This is probably adaptive because the odds of rearing one's young to adulthood fall sharply in times of a harem takeover (Chapter 1) or overpopulation.

Although the presence of a strange male tends to block female sexual activity, the effect of a strange female on a male is very different. A strange female represents an additional opportunity for the male to pass on his

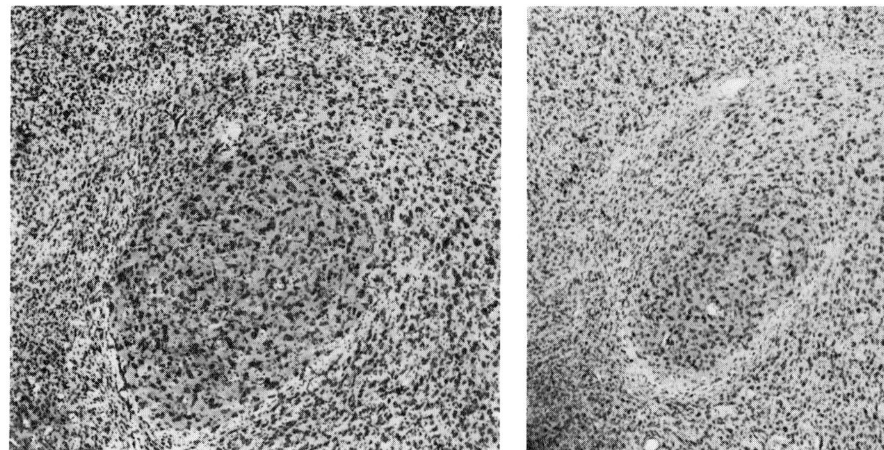

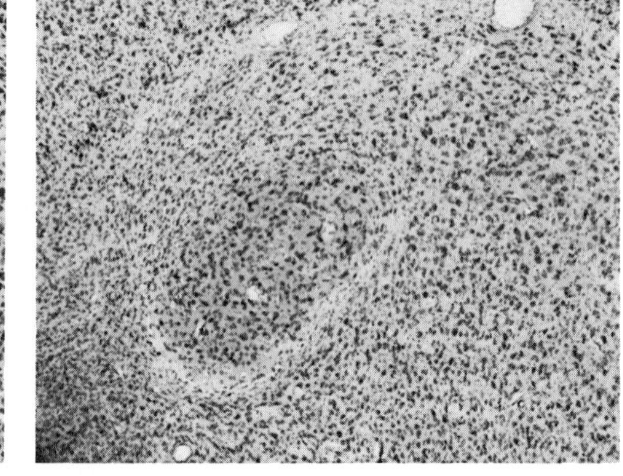

22 **Sexual differences in brain structure** in the canary. A cross section through a male canary's brain (*left*) reveals that the robust nucleus of the archistriatum is much larger than in the female's brain (*right*). Photographs by Fernando Nottebohm.

Table 2
Social Influences on the Reproductive Physiology and Behavior of House Mice

Social Condition	Response
Juvenile females held with dominant male	Juveniles mature more rapidly [446]
Juvenile females held with adult females from dense population	Juveniles mature more slowly [479]
Juvenile male held with dominant male	Juvenile's gonadal development and sexual activity suppressed [458]
Adult females held in group without male	Females stop estrous cycling [780]
Dominant male added	Females resume estrous cycling
Pregnant females exposed to urine of strange male	Females resorb embryos or abort spontaneously [780]
Adult male exposed to strange female	Male's testosterone levels and sexual activity rise [458]
Adult male exposed to urine of pregnant or lactating mates	Male's sexual activity falls [458]

genes. Thus, males become more sexually active in the presence of new females, whereas a resident male's response to his familiar harem of pregnant mates is just the opposite. Having already inseminated these females, there is little or nothing to be gained by copulating with them again until they can rear new litters. Note also that the presence of a dominant male has opposite effects on other males (reducing their gonadal development) and females (speeding sexual maturity). Again, this is biologically sensible because if a subordinate male cannot supplant a dominant individual he will not have opportunities to mate. The superior strategy is to avoid fruitless conflict with the dominant mouse and invest energy in body growth and survival while waiting for a more opportune time to seize a territory and the harem of females associated with it.

The examples that have been sketched here show that the underlying mechanisms that organize an animal's behavior over time are remarkably sophisticated. The time when an individual achieves reproductive maturity and the moment when it exercises its capacity to reproduce may take place within a general time table that matches seasonal variation in the probability of successful reproduction. But the broad pattern of reproductive control can be adjusted in response to a host of factors, helping individuals avoid futile attempts at propagating their genes while hastening reproductive efforts when unusually favorable circumstances apply.

Summary

1 Physiological mechanisms not only provide animals with the ability to perform special behavior patterns in response to key stimuli, they also structure the use of the animal's entire repertoire of behavioral abilities in the adaptive manner. Thus, an anole's behavior is constantly changing

over the course of 24 hours and is organized in a number of short-term to long-term cycles.

2 One aspect of the organization of behavior is the ability of an individual to carry out an action in complete form without interference from other competing behaviors. There is a hierarchical pattern of priorities among control units. Typically, for example, escape behavior takes precedence over all other possible actions, with neural activity in the escape system suppressing activity in other centers.

3 The nature of the inhibitory relations between neural control systems may change in response to changing internal states. This may involve short-term changes in priority, as in the case of a fly with a full digestive system in which feeding no longer has precedence over grooming behavior. Or the changes in control system interaction may follow a regular long-term seasonal or annual cycle, as in the case of anoles, in which reproductive behavior has priority only during a restricted period of the year and is totally suppressed at other times.

4 Although they are still poorly understood, circadian clocks play a significant role in the organization of behavior. They enable an animal to become prepared to take advantage of predictable changes in the environment that are correlated with the day–night cycle.

5 Hormones are exceptionally important in the long-term organization of behavior, especially in the timing of reproductive activities so that the young are produced at the optimal time of the year. Often a series of seasonally related events, such as increasing day length or improving food quality, is detected by neural mechanisms and translated into a sequential release of hormone messengers that activate one previously blocked neural system after another in the biologically correct order.

6 The broad pattern of seasonal changes in behavior that is typically induced by photoperiodic or food changes in the environment is in some species subject to fine adjustment in response to a host of other cues. In particular, seasonal reproductive cycles are often speeded up or inhibited

by particular kinds of social interactions between members of the same species. As a result, animals time their reproductive activity not only to match the optimal seasonal conditions of the year but to coincide with an optimal social environment as well.

Suggested Reading

David Crews and his colleagues are responsible for a beautifully detailed picture of the organization of anole behavior [145–149]. Kenneth Roeder's *Nerve Cells and Insect Behavior* [608] and Vincent Dethier's *The Hungry Fly* [173] have sections on the temporal structuring of the behavior of the mantis and the blowfly. Irving Zucker's review article [821] on biological clocks and circadian rhythms is recommended, as are the reviews of white-crowned sparrow reproductive behavior written by Donald Farner and his colleagues, R. A. Lewis and John Wingfield [234, 800].

Suggested Films

Fish, Moon and Tides—The Grunion Story. Color, 15 minutes. Documents the reproductive cycle of the grunion, a fish whose behavior is linked to tidal and lunar cycles.

Signals for Survival. Color, 50 minutes. See below.

Strategy for Survival. Color, 30 minutes. Although neither film relates physiology to behavior, the first shows in detail the annual cycle of the lesser black-backed gull and the second shows the annual cycle of the monarch butterfly. Both could be used to illustrate the temporal aspects of behavior.

Our attention up to this point has centered on the proximate mechanisms of behavior. This chapter begins a shift to an analysis of the ultimate causes of an animal's actions. Questions about why it is adaptive for an animal to possess certain behavioral abilities are the province of the discipline of behavioral ecology. Unlike studies on the genetic and physiological bases of behavior, which are grounded in a long tradition of scientific research, behavioral ecology is a relatively recent development. Its roots lie in the pioneering work of the European ethologists Niko Tinbergen and Konrad Lorenz in the 1930s and 1940s. The ethological approach to questions about behavioral adaptation, however, has undergone major changes since the 1960s, that is, since the initiation of the debate on the primacy of individual selection versus group selection (Chapter 1). This chapter introduces modern behavioral ecology with its emphasis on the importance of Darwinian natural selection for an understanding of behavioral evolution. We shall begin with a description of the natural history of the black-headed gull and shall then attempt to develop hypotheses on the adaptive significance of its behavioral abilities. These hypotheses will be based on the assumption that individuals are engaged in reproductive competition rather than reproductive sacrifice for the preservation of the species. As we noted earlier, hypotheses of any sort are useful only to the extent that they lead to testable predictions. Some representative individual selectionist hypotheses will be used to derive predictions that we will then test. The goal is to illustrate the range of topics covered by behavioral ecology and the variety of techniques available to test ideas about adaptation in animal behavior.

CHAPTER 7

The Ecology of Behavior

The Behavioral Ecology of Black-Headed Gulls

The black-headed gull is a common seabird of the European coast whose behavior has been studied with exceptional thoroughness by Niko Tinbergen and his associates [705, 706]. The gull has an annual cycle of behavior not vastly different in its general outline from that of the white-crowned sparrow. The adult breeds in the spring and summer, disperses along the coast to feed in the fall, migrates south (to North Africa) in the winter, and then returns to Europe the next spring, generally coming back to the same place where it bred the previous year. Its behavior has been studied most intensively during the reproductive season, and we will sketch the bird's behavior at this time.

The black-headed gull nests in colonies that form on open grassy areas, often on islands (Figure 1). Males arrive at the breeding grounds in the spring before females. They take up a position in the dunes and repeatedly utter a characteristic loud raucous call (the LONG CALL), given as the bird stretches forward obliquely (Figure 2). It is unusual for a male giving the long call to be approached by another male; but if this should happen, the caller may adopt a position called the UPRIGHT DISPLAY, a position in which the gull stretches its neck up, holding its beak ready to stab downward. If the other bird persists, the gull may then perform a FORWARD DISPLAY, with its body held parallel to the ground and its beak pointed directly at the opponent. This display is often followed by a charge in which the bird attempts to strike the other male with the beak and wrist of the wing. Almost always the intruder is driven from the area. Examination of a colony day after day reveals that the same male returns repeatedly to his territory, which he defends against all comers.

A female circling over the colony is attracted by long-calling males. She comes closer and alights in the place a male defends from other males. Initially the male may drive the female off as well, and she appears quick to leave the spot. However, a female may return to a male over and over again. Eventually the two birds may begin to engage in a number of mutual displays, such as the PARALLEL-FORWARD DISPLAY (both birds walk side by side with their bodies stretched forward) and the FACING-AWAY DISPLAY

1 **Ground-nesting gull colony** in typical habitat in northern Europe. This area of rolling, grass-covered dunes is populated chiefly by herring gulls and lesser black-backed gulls. The small structures are blinds for close observation of the birds. Photograph by Niko Tinbergen and Hugh Falkus.

(Figure 2). These occur very frequently early in courtship but gradually fade away and are replaced by food-begging actions by the female. In response, the male regurgitates items he has collected, and his mate feeds upon them. Finally, copulation (Figure 2) is preceded by a mutual HEAD-TOSSING DISPLAY, in which both birds walk about together jerking their heads up repeatedly while uttering soft calls.

If courtship was conducted in a preterritory (an area outside the nesting part of the colony), the pair searches for a place to build a nest in the breeding area. They settle in a spot, defend it vigorously from their neighbors with a great deal of noisy bickering, build a nest, and raise a brood. The nest is a simple loose mound of grasses and straw collected from the area. A pocket is made when the bird settles into the mound. After the female completes her clutch, incubation is shared by both birds.

When an egg hatches and the young bird emerges, the adult on guard at the nest will wait for an hour or so and then pick up the empty eggshell, fly off, and drop it some distance from the nest. The baby gull's feathers are dry by then and soon it can stand and beg for food when a parent, its

2 **Black-headed gulls.** At rest (*top left*). Note that the neck is not stretched out and that the wings are held close to the bird's body. The long call of a black-headed gull male (*top right*). The bird is stretched forward obliquely and the wings are held out a little way from the body. The facing-away display given by a courting pair (*bottom left*). Note the contrast in body shape with the resting gull. A copulating pair (*bottom right*). Photographs at top left and bottom right by Eric Hosking; photographs at top right and bottom left by Monica Impeckoven.

crop engorged with food, returns to the nest. The chick begs in much the same fashion as other gulls (Figure 10, Chapter 4), pecking at its parent's beak until the adult regurgitates a meal. The demands of a brood of baby gulls soon have the parents foraging for food much of the day in alternating trips, with one bird always remaining behind at the nest. The adults gather a great variety of items in many different places and can often be seen flying off in raucous flocks to join others feeding on earthworms in a freshly plowed field, or gathering garbage from a dump, or plunge-diving for fish at sea.

As the young birds grow older, they are watched less closely and often wander about in their parents' territory. Nevertheless, they attend to their parents' calls, running to an adult when they hear the ATTRACTION CALL given by a food-laden mother or father. In contrast, they rarely react to attraction calls given by other adults, and if they do they are likely to receive a pecking. Parents recognize their offspring and vice versa. But if a youngster hears an ALARM CALL given by any bird, it freezes in the nest or dives for cover. All the adults are extremely sensitive to the presence of predators, and the appearance of a large gull, fox, crow, badger, hawk, or human will set off a volley of loud cries. If the predator approaches, groups of gulls will fly toward it, calling noisily and defecating profusely while dive-bombing the enemy.

Eventually the young birds become mature enough to fly and a short time later are left to fend for themselves by their parents, who leave the breeding colony in search of food elsewhere. The surviving youngsters will join the adults on the migratory trek south as winter approaches and with luck will be back in several years to have a go at reproducing.

Ultimate Hypotheses about Black-Headed Gull Behavior

It is not difficult to envision adaptive value in much of what the gulls do. They manage to find relatively safe places in which to breed, navigate successfully hundreds of miles to good wintering places, usually find enough food to survive and feed a brood of rapacious youngsters, and are not helpless in the face of predation. Each element of their behavior can be categorized with respect to its possible ecological significance and assigned an adaptive function (Table 1). But the functional interpretations presented in Table 1 are hypothetical. Each character could have been interpreted in a different way that was consistent with individual selection *or* could have been given a group-benefiting function *or* could have been classified as having no evolved function at all. I have tried to select plausible hypotheses based on individual selection. For example, from my personal experience with attacking gulls, I find it easy to imagine that a crow or hawk beset with a mob of screaming, defecating gulls finds it hard to concentrate on finding a gull nest or a chick. When I have gone walking in a gull colony, some of the upset gulls have slipped around behind me before diving down toward my head. Distracted by the chaos about me, I have not noticed the bombers until they pulled out of their dive at the very last moment, sending the wind roaring through their wings as they sailed over my head. The effect was unnerving to say the least, particularly if the gull applied the coup de grace—a not-so-gentle clip on the top of my head with a trailing foot. But although I find it believable that mobbing deters gull predators, it is a belief that needs to be tested.

Table 1
The Adaptive Value of Some Behavioral Traits of Black-Headed Gulls: Individual Selectionist Hypotheses

Antipredator Traits

Ecological Problem — Gull eggs and juveniles are palatable and can be consumed by a large number of avian and mammalian predators.

Behavioral Solution
1. *Alarm calls* warn family members of danger and alert colony members, providing the basis for group action.
2. *Mobbing behavior* is a mutually beneficial response by those colony members most likely to be damaged by an unchecked predator; through cooperation, the gulls distract enemies that no one individual could repel.
3. *Eggshell removal* leads to the removal of a visual cue (white interior of opened egg) that predators might use to locate a nest with young, vulnerable chicks.

Food-Gathering Traits

Ecological Problem — Gulls and their offspring require large amounts of nourishing food if they are to grow, maintain themselves, and reproduce successfully.

Behavioral Solutions
1. *Broad-based diet.* The birds are able to consume many different food items and so can switch to whatever happens to be the most productive source of nutrition in their variable environment.
2. *Foraging in flocks.* Some individuals can exploit the knowledge of food sources of other birds by flying after them when successful foragers head out to a productive site that they have discovered on an earlier trip.

Reproductive Traits

Ecological Problem — Males and females vary in their reproductive value to potential partners.

Behavioral Solutions
1. *Prolonged courtship with multiple displays* permit the male to assess the reproductive state of the female so as to avoid mating with (and then helping) an already-mated female whose eggs were fertilized by another male.
2. *Courtship feeding* enables a female to build up her food reserves for egg production and helps her assess the foraging skill of a potential partner whose assistance is vital in rearing her offspring.

Ecological Problem — In a dense colony adults run the risk of feeding young birds of other parents, an action that would enhance the reproductive success of rivals.

Behavioral Solution
1. *Individual recognition of offspring.* By learning the visual and acoustical features of their young, parent gulls can restrict their feeding efforts to their biological offspring.

Testing Ultimate Hypotheses

As we noted in Chapter 1, there are a number of ways in which one can test hypotheses about the adaptive function of a characteristic. These techniques all involve using a hypothesis to make a prediction that can then be tested by (1) additional observation, (2) experiment, or (3) comparative data from other species. Evolutionary predictions are based on the premise that an evolved trait helps the individual maximize its lifetime propagation of genes (generally by maximizing its lifetime production of surviving offspring). The problem is that every trait has both negative and positive effects on individual fitness. A gull participating in a group attack on a hawk runs some risk of being killed as well as expending a certain amount of time and energy that could be spent in other activities. All of this decreases to some extent the probability of successful future reproduction by the gull. This is the negative aspect, the EVOLUTIONARY COST, of the action. Set against the costs are the EVOLUTIONARY BENEFITS of mobbing a predator, namely, the potential increase in individual fitness that comes from saving some currently existing offspring.

If we could measure the costs and benefits of a trait in the same currency of fitness (in this case the average number of current offspring saved by a mobber versus the average number of future offspring lost), we could make precise quantitative predictions about many things. For example, we could predict exactly how far a mobbing gull should stay from a hawk or how many minutes it should invest in an attempt to repel a foraging crow. We could then precisely test our prediction through additional observations of mobbing birds or by arranging certain experiments.

The difficulty with an approach based on OPTIMALITY THEORY is that it is often hard to measure the costs and benefits of a trait in the same currency of fitness. In order to predict the optimal distance to stay from a hawk, one would have to be able to translate the probability of dying when mobbing a hawk at various distances into exact measures of lost offspring. We would also need to measure the average number of current offspring saved as a result of mobbing at various distances. We could then determine the distance at which the benefits exceeded the costs by the greatest amount and test whether the gulls actually behaved in the predicted optimal fashion. But at present no one knows how to make the appropriate translation of risk of dying during mobbing into units of fitness lost. In general, application of optimization tests have been limited largely, but not exclusively, to cases, like foraging behavior, in which one can measure both benefits and costs in the same units (e.g., calories gained and expended as a result of various feeding decisions). Here one can more readily make predictions about things such as the kind of food that should be selected by a hungry animal if it is to maximize its net caloric gain. We shall defer detailed discussion of the admirably quantitative optimality approach until the chapter on feeding behavior.

Because quantitative predictions are often hard to come by, evolutionary hypotheses are more frequently analyzed in a qualitative fashion. The general procedure is to establish that there is a correspondence between a particular trait and a solution to an environmental obstacle to reproductive success. If mobbing is truly adaptive for a gull, mobbed predators should be more likely to overlook nests and young than predators that escape the attentions of a group of parents. Moreover, the risk of injury to a mobbing adult gull should be slight, for if it were otherwise the costs of the trait would rise dramatically and almost surely outweigh the presumably modest benefits of mobbing.

The simplest way to test these qualitative predictions is through direct observation of colonies of gulls to document that mobbed predators are repelled and that mobbing adults are rarely captured. Hans Kruuk conducted a two-year study of predation at a black-headed gull colony; during this time he saw many crows and herring gulls attacked while foraging on the ground within the colony [396]. Carrion crows and herring gulls are consumers of gull eggs but cannot capture and eat adult birds, so mobbing black-headed gull adults are reasonably safe. Both crows and herring gulls can avoid dive-bombing gulls, but this requires that they continually face their attackers. As a result, they are distracted and ineffective in their search for nests and eggs while under group attack.

Kruuk also employed experiments to test the prediction that mobbing by black-headed gulls deters some predators [396]. For each experiment he placed ten hen eggs, one every ten meters, in a line running from outside to inside the gull nesting area. Carrion crows and herring gulls hunting for food in the vicinity of the colony often discovered the eggs; and Kruuk was able to determine the order in which the eggs were taken. If the presence of nesting gulls makes egg predators less effective, then hen eggs outside the colony should be more vulnerable than those inside the nesting boundaries. They were. Moreover, as predicted, the success of predators in finding the hen eggs decreased when the attack rate by black-headed gulls

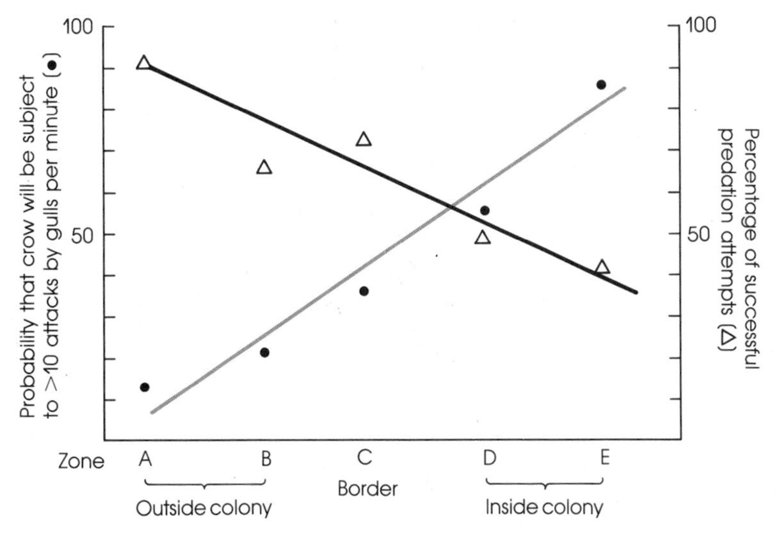

3 **Antipredator function of colonial nesting.** The frequency of intense attacks (●) on egg-hunting crows increases when the predator is within the borders of a nesting colony. This reduces the percentage of successful predation attempts (△) on hen eggs experimentally placed there.

was high. At times the gulls would bombard their enemies constantly, and during such intense group mobbing the egg-finding success of both herring gulls and carrion crows declined precipitously (Figure 3).

Thus, both observational and experimental tests support the hypothesis that group mobbing is an adaptive response to certain egg-hunting predators. But there is still another method of prediction testing—the comparative method—the rationale of which we shall now examine in detail.

The Uses of Convergent and Divergent Evolution

As noted before, the comparative method is a kind of natural experiment in which one first identifies the ecological factors that seem to make a behavioral trait advantageous for the individuals of a species. One then examines other species to test the prediction that the behavior will have evolved whenever the critical ecological factor(s) are present. The species chosen for this test must be unrelated to one another. Species with different phylogenetic histories could be expected to behave differently because they have different ancestors and distinct genetic-developmental characteristics. But if they have been subjected to similar selection pressures, they may have independently evolved similar behavioral traits through CONVERGENT EVOLUTION (Figure 4). If a common ecological factor can be related to an apparent case of convergent evolution, this strengthens the argument that the trait represents an adaptation to this environmental pressure.

There is another comparative test that makes use of DIVERGENT EVOLUTION among related species (Figure 4). Species that have a common ancestor will have "inherited" similar genetic-developmental mechanisms and therefore can be expected to be similar behaviorally unless they have been subjected to unique selection pressures. If behavior evolves in response to special ecological conditions, then related species may diverge behaviorally. If we can show that related species with different ecological problems have different behavioral abilities, we will have evidence to support the hypothesis that the particular traits represent adaptations to the special problems each species faces.

Are there cases of divergent and convergent evolution that support the prediction that mobbing behavior will evolve when a colonial species is subjected to nest predators that groups of adults can repel? The answer is yes. Almost all gulls nest in exposed areas on the ground and are vulnerable to the same spectrum of predators that afflict the black-headed gull; they also exhibit mobbing behavior. But there is a divergent exception—the kittiwake. This species is one of a handful that has evolved the ability to nest on nearly vertical coastal cliffs, where it is all but exempt from nest predation (Figure 5). Mammals cannot safely venture onto the cliffs. The larger gulls and hawks cannot safely land near nests because the cliff ledges are too small to accommodate them easily and the swirling sea breezes make maneuvering by the cliffs difficult. Kittiwakes are relatively small, delicate gulls with clawed feet, and they can land on ledges that few other birds of similar size could use as perches and nest sites (Figure 6). As a result, the kittiwake's eggs and young are not at risk from predators and the adults do not mob predators when the occasional enemy drifts past the colonial nesting site. The absence of mobbing in the predator-free kittiwake supports the hypothesis that mobbing evolves in response to predator pressure [153].

On the other hand, there are a number of cases of species that are distantly or totally unrelated to black-headed gulls that have convergently evolved mobbing behavior. They experience ecological circumstances that

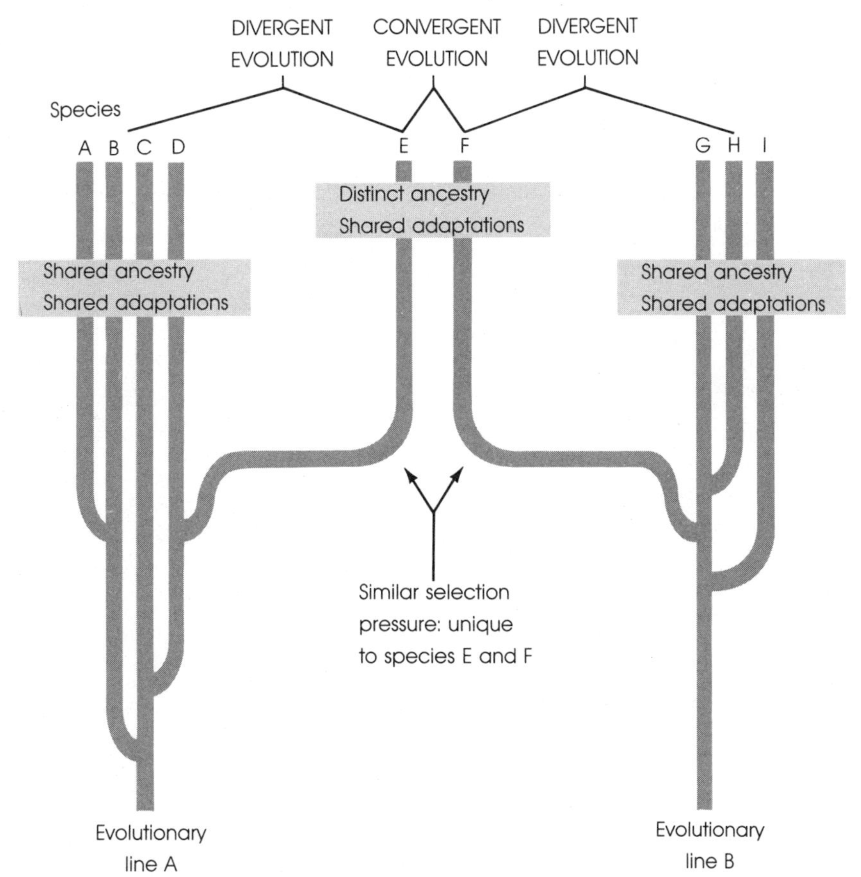

4 **Convergent and divergent evolution.** Species A to D have a number of traits in common because selection has not favored divergence from patterns possessed by a common ancestor. The same is true for species G to I. On the other hand, species E and F have diverged from their relatives because of unique selection pressures operating on them. Species E and F happen to have traits in common, despite distinct ancestry, because the selection pressures acting on them have been similar and have led to convergent evolution.

6 **Kittiwakes** have evolved the ability to alight on the narrowest of cliff ledges. This small spot is evidently often used as a perch, judging from the guano staining the cliff below it. Photograph by the author.

are similar to those affecting black-headed gulls but not the kittiwake. For example, the bank swallow is a colonial nesting species, large numbers of which may gather to breed in a sand quarry or river bank. Their young, like the black-headed gulls, are vulnerable to predators such as blue jays and weasels. The swallows are active mobbers. As expected, (1) they mob only potential enemies, (2) they are very rarely captured by a predator, and (3) they successfully deter certain of their enemies through the harassment and chaos created by a swirling mob of adults [340].

Similarly Tex Sordahl has shown that nesting adults in loose colonies of avocets and black-necked stilts mob their enemies, which range from skunks to gulls and hawks. The birds respond differently, however, to their various enemies at different stages of the nesting cycle [677]. When there are eggs in the nest, encroaching gulls are especially likely to be attacked whereas black-crowned night herons passing overhead are ignored. Gulls eat eggs; herons do not. But when there are nestlings to defend, gulls are less likely to be mobbed whereas herons are vigorously attacked. Herons

◀ 5 **Nesting colony of kittiwakes** on the cliffs at Orkney Island, Great Britain. Photograph by Arthur Gilpin.

eat chicks; California gulls rarely consume them. Hawks, primarily the northern harrier, are a danger both to the eggs and to the chicks. They are mobbed at both phases of the breeding cycle (Table 2). These results support the hypothesis that mobbing serves to distract real enemies from the offspring of avocets and stilts.

A still more distantly related colonial species that engages in group harassment of predators is the California ground squirrel, whose behavior has been the subject of a series of papers by Donald Owings and his colleagues [319, 320, 559]. The squirrel's primary enemies are snakes, which are able to invade nest burrows and capture the young. The adults above ground are relatively safe, and they collectively mob a snake intruder, kicking sand in its face and generally making its life as miserable as possible to "encourage" it to go elsewhere (Figure 7).

Owings has found an interesting difference in the treatment of snakes by California ground squirrels living in different regions. The range of the rodent only partially overlaps the range of venomous rattlesnakes, whose bite can quickly kill a ground squirrel. Young, inexperienced squirrels from the zone of overlap with rattlers are significantly less likely to approach *any* snake than are naive young squirrels from areas without rattlesnakes. Wild-caught adult squirrels from overlap areas will mob rattlesnakes, but they evidently discriminate between them and nonvenomous gopher snakes—keeping farther away from rattlers when harassing them [559].

These findings make sense in the context of a benefit–cost analysis of mobbing behavior. The greater the personal danger in approaching an enemy, the higher the costs of mobbing and the more likely that the costs will exceed the benefits of this trait. In places in which mobbing carries with it higher costs, the smaller, less agile, juvenile squirrels do not engage in this activity at all. Although the adults do mob, they do so more cau-

Table 2
Response to Egg Predators (E) and Chick Predators (C) by American Avocets and Black-Necked Stilts at Different Stages of the Breeding Cycle

		Observations of Mobbing by Parents Guarding	
	Species Mobbed	*Eggs*	*Chicks*
E	California gull[a]	18	5
E	Other gulls[a]	10	5
C	Black-crowned night heron	0	16
C	Other herons	0	8
E, C	Northern harrier[a]	5	10

Source: Sordahl [677].
[a]Gulls only occasionally take avocet chicks, whereas the northern harrier (hawk) is primarily a chick predator.

tiously when dealing with a highly dangerous opponent. This example also illustrates that INTRASPECIFIC VARIATION in behavior that can be traced to an ecological difference between populations provides another useful comparative tool with which to test ultimate predictions.

On the Adaptive Value of Eggshell Removal

Our various tests have supported the contention that group mobbing of predators is an evolved adaptation whose benefits exceed its costs in certain environments. We can apply the same approach to another hypothetical antipredator adaptation: eggshell removal from the nest by black-headed gulls. The benefit of this behavior pattern may be the removal of a conspicious visual cue that certain predators might use to locate a nest and destroy its contents. This benefit, if it exists, seems large enough to outweigh the costs of the behavior, costs that include the energy required to carry off the shell and the danger that an unattended nest will be raided during the parent's brief absence.

The hypothesis has been tested experimentally by Niko Tinbergen [708]. He took some intact gull eggs from a colony and scattered them through sand dunes visited by carrion crows, which have a great fondness for this food. By some of the unhatched eggs (which have a highly camouflaged, mottled exterior of tan and dark brown), he placed some broken eggshells whose white interior was visible; by others, no shells were placed. The

7 **Mobbing behavior of California ground squirrels.** One squirrel is kicking sand at a rattlesnake, while the others are giving a variety of alarm signals. Courtesy of R. G. Coss and D. F. Hennessy.

crows found and ate a much higher proportion of the eggs that had shell cues lying nearby (Table 3).

Another test of the antipredator hypothesis involves a comparison of kittiwake and black-headed gull behavior. What prediction would you make about the kittiwake parent's response to eggshells after the hatching of its babies? You are correct. The kittiwake, which has essentially no nest predators, does not even bother to push the shells over the cliff ledge on which its nest rests [153].

On the other hand, some birds that are unrelated to the black-headed gull and that also rely heavily on nest concealment for the safety of their brood do carefully collect and discard broken eggshells far from the nest (Figure 8). Thus, there appears to be a lawful relationship between the intensity of nest predation and the existence of eggshell removal. This supports the hypothesis that this behavioral trait has evolved in response to the ecological pressures created by nest predators.

A Foraging Adaptation: Tests of a Hypothesis

Observational, experimental, and comparative tests are not restricted to hypotheses about the function of possible antipredator adaptations but can be applied to all categories of behavior. For example, one can ask why it is that black-headed gulls often forage together and sometimes seem to fly out from nesting colonies toward a feeding site in small flocks? This trait is open to multiple interpretations, some of which I list below.

The hypothesis presented in Table 1 is that the followers are gaining information about the location of distant food by somehow identifying and following a successful forager back to a productive spot. Note that this hypothesis does not require the "leader" to be acting for the good of the colony or the species by "sharing" its information about food sources. Indeed a "foraging parasitism" hypothesis is that "leaders" lose fitness by having followers pursue them because the additional gulls will deplete food a leader has found. The followers are, according to this view, parasites that take advantage of a successful forager to locate food that they had not found on their own [756].

Table 3
The Presence of an Eggshell Near a Gull Egg Increases the Risk That the Egg Will Be Discovered and Eaten by Crows

Distance from Egg to Opened Eggshell (cm)	Crow Predation		Risk of Predation (%)
	Eggs Taken	*Eggs Not Taken*	
15	63	87	42
100	48	102	32
200	32	118	21

Source: Tinbergen [708].

8 Convergent evolution of eggshell removal behavior. The secretive European bittern, like ground-nesting gulls, removes eggshells from its nest. Photograph by Eric Hosking.

An alternative hypothesis is that leaders and followers both gain by flying together to a food source. Mutual benefits might include a reduction in the chance of predation (see Chapter 10 on the safety advantages of flocking) or an improvement in foraging success through cooperative feeding. Certain kinds of prey may be caught more easily when subjected to group assault than when confronted by a solitary predator. One might imagine, for example, that when a flock of gulls plunges repeatedly into a school of fish, some of the frantic prey, in attempting to escape from one gull, might fail to see its diving neighbor and so be more easily captured. Thus, the feeding rate might be higher for all the members of a cooperative hunting flock than for a solitary hunter.

A third hypothesis is that traveling in groups is an incidental effect of the fact that many gulls have learned the location of a food source and happen to be flying to it at about the same time. Thus, what looks like leaders and followers may actually consist of a mass of informed gulls all heading to the same spot.

The fact that multiple interpretations of group travel are possible makes us aware of the need to develop predictions that can help us discriminate among the various hypotheses.

A prediction from the "mutual benefit" hypothesis would be that gulls traveling in groups are safer than solitary individuals or that gulls travel together *only* when going in pursuit of certain kinds of food (e.g., schooling herring, but not earthworms or crabs or garbage).

A prediction from the "incidental effect" hypothesis is that the composition of traveling groups should be the same from trip to trip. If it could be shown that the same (color-marked) individuals went repeatedly to the same plowed field or sewage outlet, then it would be unlikely that the "flock" members are trying to learn the location of a new food source.

A prediction from the "parasitism" hypothesis is that highly successful foragers should be followed out from the colony on their way back to a distant, rich, food site. This hypothesis has been subjected to an experimental test in which an abundance of dead fish was placed on a raft out of sight of the colony a kilometer or two offshore. Wandering gulls soon found the food, however, and gorged themselves. An observer near the raft radioed information about the identity of a successful gull back to a colleague by the colony, who could then record what happened upon the return of the bird. The gull landed by its nest with conspicuously distended crop and often with fish in its beak. Thus, a successful forager could readily have been identified by its neighbors even before it began feeding the chicks. But although the opportunity existed for a neighbor to follow the successful forager when it left, no neighbor actually flew behind the forager when it headed back to the raft. Moreover, when the next gull did leave the colony, there was absolutely no tendency for it to fly in the same direction as the successful fish-finder [20].

The failure of the data to support the parasitism hypothesis is useful. It sharply decreases our confidence that black-headed gulls actually possess a sophisticated ability to exploit birds that have found food out of sight of the colony. This encourages us to test other plausible alternative hypotheses for the occurrence of flocking in gulls. This example illustrates that not all the problems have been solved about the function of gull behavior, a fact I find more exciting than depressing.

Parent–Offspring Recognition: Tests of a Hypothesis

We have examined elements of the gull's antipredator and feeding behaviors. Now let us turn to a third aspect of its behavioral ecology—its reproductive behavior. We have noted that adult black-headed gulls learn to recognize their progeny and that the chicks learn to recognize their parents. This information is used by an adult so that it regurgitates food to its offspring alone while handing out blows from its beak to other young gulls that happen to trespass on its territory. One might hypothesize that this ability is adaptive in environments in which parents run the risk of assisting the offspring of genetically unrelated individuals.

The chick-recognition ability is very widespread among ground-nesting colonial gulls [706]. As previously noted, however, the occurrence of a similar trait in related species cannot be used to support a hypothesis on the adaptive value of that characteristic because shared ancestry, rather than shared ecological pressures, could cause a trait to be widely distributed among related species. But *differences* in behavioral abilities among species with shared ancestry are relevant to the comparative approach. One could predict that any gull species in which there were no opportunities for misdirected parental care would not have evolved the capacity for offspring recognition. Again we turn to the cliff-nesting kittiwake for a test of this prediction [153]. Because the birds nest on small cliff ledges, often with sheer drops hundreds of feet to rocks and waves below, there has been strong selection favoring chicks that stay put and do not wander out of the nest until they are capable of flight. This is exactly what kittiwake chicks do. They crouch within their nest, pointed away from the precipice until they are several weeks old and have fully feathered wings. Only then do they leap from the ledge to glide down to the ocean, where they will enter the next phase of their development. In contrast, by the time they are five to seven days old, the chicks of ground-nesting gulls are highly mobile and often move about away from the nest. If one switches chicks this age or older in a ground-nesting species like the herring gull, the parents will not care for the transferred young and indeed may even kill (and eat) them. Similar experiments with kittiwakes yield a totally different result, as nesting adults will calmly accept strange chicks much older and larger or much younger and smaller than their own. In fact, one can substitute a cormorant chick (the very antithesis of a young kittiwake; Figure 9) and the parent gull will feed the newcomer as if it were its own progeny.

Further support for the hypothesis that the risk of misdirected parental care selects for offspring recognition comes from chick transplant tests done with an unusual population of herring gulls that nests on cliffs. As one would predict, these individuals accept chicks other than their own into their nests even when the offspring are three to five weeks old [64]. The danger of moving about on cliff ledges has led to a convergent pattern of chick mobility in these herring gulls and kittiwakes, and this in turn is correlated with an absence of chick recognition prior to fledging. (I predict

9 **Polar opposites.** A kittiwake chick in a nest with its parent (*left*); two young shags (a species of cormorant) in a nest with their parent (*right*). Kittiwake adults will accept young shags that are put into their nests. Photographs by the author.

that in both the cliff-nesting herring gulls and the kittiwakes parent birds do learn who their offspring are in the postfledging phase when the young birds are still partially dependent on their parents for food and protection.)

The capacity exhibited by the black-headed gull and other ground-nesting species for identifying young offspring has some limitations that add still more weight to the hypothesis under scrutiny. Ground-nesting herring gulls, for example, will incubate transplanted eggs (even if these are of a different color than their own) as well as care for foster chicks if the switch is done when their own infants are one to four days old. Because gull nests are well separated and because eggs and freshly hatched chicks do not leave the nest, there is no risk of misdirected parental care until the young become mobile at age five days. Only then does the ability for offspring recognition express itself in herring gulls [707].

Learned Egg-Recognition in Birds

Not all species are incapable of identifying their own eggs. Many birds (but not gulls) have nest parasites that add their eggs to a clutch while the incubator is off the nest. European cuckoos and cowbirds, for example, are highly adept at victimizing a variety of hosts (see Figure 5, Chapter 4). Species parasitized by them are under selection pressure to identify their own eggs in order to discriminate against a parasite's egg. Numerous small songbirds (but not all) have this ability and can remember what their eggs look like. If they detect a foreign egg in the nest, they will either eject it or abandon the nest altogether [305, 618].

Another ecological factor that promotes the evolution of offspring recognition at the egg stage is the possibility of *accidentally* brooding a conspecific's eggs. Although one might imagine that the chance of making this mistake would be nil, it does occur in certain species. Guillemots are oceanic birds that nest in colonies on rocky seaside ledges. In these colonies the density of incubating individuals can reach astonishing levels—up to 34 guillemots per square meter! Under these conditions the birds are packed in shoulder to shoulder (Figure 10). Their "nests" are nothing to write home about—a few pebbles pushed together on the ledge. As a result there is considerable danger of incubating the wrong eggs. But there is great variation in the color pattern of guillemot eggs; each bird learns the distinctive features of its own egg [71].

Another closely related seabird, the razorbill, lives in the same sites as the guillemots. But each pair of razorbills has its own substantial nest well separated from any other nest. T. R. Birkhead predicted that the members of this species would not therefore be able to discriminate between their eggs and those of another bird [71]. To test this, he chased an incubating razorbill from its nest and displaced its single egg a short distance and also placed another egg equidistant from the nest. When the incubating razorbill returned, it had to choose which egg to retrieve and care for. In 40 trials, 20 parents rolled their own egg back into the nest and 17 selected the foreign egg (3 callously refused to participate in the experiment). This result

10 **Colonial guillemots** nest in amazingly dense clusters, a situation that creates the risk of errors in egg incubation. Guillemots have, however, evolved the ability to learn to recognize their own eggs. Photograph by T. R. Birkhead.

shows the razorbills could not identify their own eggs and so chose an egg at random to retrieve. Thus, there has been divergent evolution in egg recognition in two closely related alcids, a divergence that can be ascribed to a difference in the risk of misdirected parental care affecting each species.

Divergent Evolution in Fledgling Recognition in Swallows

Another combination of experimental and comparative approaches has been used to study offspring recognition in bank and rough-winged swallows. These birds are close relatives. They not only resemble one another in appearance but also share the habit of nesting in burrows in clay banks and sand quarries. They differ, however, in that the bank swallow is a highly colonial species (Figure 11) whereas the rough-wing is a solitary breeder, with pairs nesting in isolation from one another. By the time the fledgling emerges from the nest burrow, parent bank swallows have learned the key vocal characteristics of their progeny (Figure 12) and so can discriminate them from other members of the colony [53, 55]. This enables them to feed just their own offspring while treating strange fledglings that happen to land at their burrow entrance in an unkind manner.

11 **Bank swallows examining nest burrows** in a small portion of their colony. Because they live in dense colonies, these birds run the risk of providing parental care for chicks other than their own. Photograph by Michael D. Beecher.

In a large colony with many closely spaced entrances, young birds often wind up in "wrong" nests.

Michael and Inger Beecher predicted that the rough-winged swallow, which through its evolutionary history has never had the chance to feed another's fledglings, would lack the ability to identify their offspring. To test this prediction experimentally, they transferred fledglings between rough-winged burrows and found complete acceptance of the transplants. Indeed rough-winged swallows will even act as foster parents for bank swallows [53].

Offspring Recognition in Other Animals

The comparative method can be extended further to test the misdirected parental care hypothesis still more extensively. To take just two examples, consider a crustacean (the isopod *Hemilepistus reaumuri* [429, 430]) and a mammal (the ground squirrel *Spermophilus beldingi* [658]). The isopod lives in burrows in the barren desert of North Africa (Figure 13). When temperatures permit, it ventures forth in search of food, animal or vegetable, which may be taken back to the shelter of the home burrow before being eaten. Pairs occupy burrows together, and in the spring a male and a female produce a brood of up to 100 youngsters, which stay

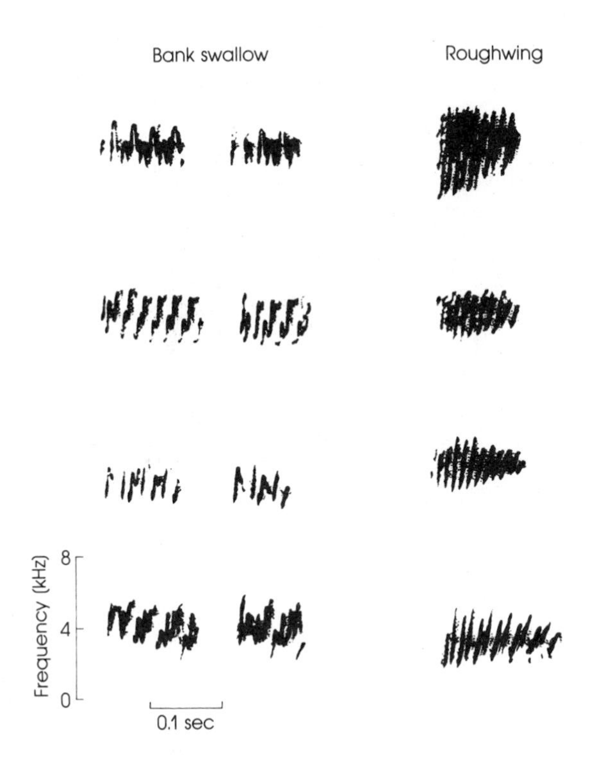

12 Calls of young bank swallows and rough-winged swallows. The calls of four bank swallow chicks on the left are more complex and more individually distinctive than the calls of the solitary nesting rough-winged swallows (four individuals' calls are on the right). Courtesy of Michael D. Beecher.

with the parents for as much as a year, receiving protection and some food during this time. The young animals learn where their home burrow is—an important ability because if they happen to wander into another family's burrow they will be killed and eaten. Isopod infanticide occurs because all the members of a family share a specific odor that all members learn to recognize—the better to identify a stranger who may be eaten. Family members returning from foraging excursions in the desert are permitted safe entry. Although positively prehistoric in appearance, this supposedly primitive creature faces the same ecological problem as herring gulls and bank swallows and has independently evolved a similar solution.

The behavior of Belding's ground squirrel is also informative about the relation between offspring recognition and the potential for misdirected parental care. This animal lives in high mountain meadows in the Sierra Nevadas of California, where it forms loose colonies, each adult with its own burrow. After emerging from hibernation, males seek out receptive females, mate, and then go on their way. The female has a nest burrow in which she gives birth to a litter of three to six babies, which spend three to four weeks within the burrow before venturing to the surface (Figure 14). Prior to the "fledging" stage, mothers can be induced to accept and care for foster young by placing young helpless ground squirrels near a burrow's entrance. When the burrow owner finds these pups, she often carries them into the burrow to join her brood. (Perhaps in nature some

13 **Pair-bonded male and female** of the North Africa isopod *Hemilepistus* at their burrow entrance with one of their juvenile offspring. In this species, parents recognize their young by their distinctive odor. Photograph by K. E. Linsenmair.

predators may bring a female's litter to the surface but not consume them all, favoring the evolution of retrieval behavior that is triggered by the sight of young pups lying near the burrow.) But by the time the young have reached sufficient maturity to come up to the surface on their own, their mothers can identify them and will reject transferred juveniles [336].

In the Belding's ground squirrel, young males soon disperse from the area in which they were born, but females usually remain to live near their mother [657]. Thus, the opportunity exists to test whether mothers will over the long term assist their daughters to a greater extent than distant relatives or unrelated individuals. In order to exploit this opportunity, however, one must have a large marked population of ground squirrels and one must follow their activities and interactions over a period of years. Fortunately, Paul Sherman and his 22 assistants heroically conducted the necessary project by marking more than 2000 individuals in the period from 1974 to 1980 while spending more than 5000 hours recording their interactions. Each squirrel was captured and given an individually numbered ear tag; and many had numbers or letters painted on them with Lady Clairol hair dye (black) to enable an observer to identify them from a distance. Sherman discovered that mothers do know who their mature daughters are and vice versa. An encounter between unrelated female squirrels is four times as likely to lead to a fight as an interaction between a mother and her daughter. Moreover, mother and daughter are more likely to engage in cooperative chases of an intruder in areas about their neighboring nest burrows than any other pairs of individuals (e.g., cousin–cousin,

14 **Female Belding's ground squirrel** and her offspring. A mother is able to identify her young at the time they first emerge from the burrow. Photograph by C. M. Kagarise and P. W. Sherman.

sister–sister, nonrelative and nonrelative). Therefore, mothers and daughters recognize one another for periods spanning several years. They use this ability to moderate their usual aggressiveness and even to cooperate in the defense of their territories (and the young within them), which are subject to attacks by wandering infanticidal females and males (Figure 15).

The Web of Adaptation

The examples in this chapter have been chosen to preview the chapters ahead in which habitat selection, feeding traits, antipredator actions, and the reproductive behavior of certain species will be analyzed to determine their adaptive function. We have

Proportion cooperative chases

0.00 0.05 0.10 0.15 0.20

	Number of pairs
Mother-daughter	59
Littermate sisters	52
Nonlittermate sisters	36
Grandmother-granddaughter	17
Aunt-niece	15
Gtgrandmother-gtgranddaughter	4
Half aunt-niece	9
First cousins	10
First cousins, once removed	3
"Nonkin"	89

15 **Cooperation among female Belding's ground squirrels.** Mothers and daughters are significantly more likely to cooperate in defense of a territory than nonrelatives or more distantly related individuals.

shown here that it is possible to select a behavior of one species, prepare an individual selectionist hypothesis for its existence, derive predictions from the hypothesis, and test them via observation, experiment, and rigorous comparison. The results of these tests sometimes support the argument that a behavioral attribute has evolved in response to a particular selection pressure.

In treating a behavioral characteristic in isolation from the other attributes of an animal, however, we run the risk of overlooking an important point. The evolution of any one component of an animal may be profoundly influenced by its other characteristics. The interlocking evolutionary effects of adaptations can be seen in the kittiwake gull. The switch from ground-nesting to cliff-nesting during the history of the kittiwake has had far-reaching, multiplicative consequences on the evolution of its behavior and structural characteristics. The reduction in predator pressure that presumably made the change in nesting habitat selection advantageous for the kittiwake is nicely correlated with the absence of many antipredator traits, both behavioral and morphological [153]. As predator pressure falls, so too does the benefit of group mobbing and alarm calling; at some point the costs of these traits will exceed their benefits. But on the other hand, nesting on tiny cliff ledges created its own special set of obstacles to reproductive success, favoring mutant individuals with respect to a host of attributes. The birds have evolved exceptionally strong toes and clawed feet, which assist them in landing and perching on precarious ledges. The dangers of losing eggs and young chicks over the cliff created selection for individuals that happen to construct elaborate, deep-cupped nests despite the increased time and energetic costs of nest construction for the kittiwake. Moreover, selection has favored chicks that remain in the nest far longer than is typical for ground-nesting juveniles. With this change has come a reduction in the risk of misdirected parental care in the kittiwake and the loss of the ability to learn to recognize their young during the first weeks of development as well as a loss of the feeding attraction call given by ground-nesters to retrieve their dispersed progeny.

Because kittiwakes are essentially the only gulls to nest on cliffs in the northern hemisphere, there are no benefits to be gained from a species-specific recognition signal and the distinctive long call so characteristic of the ground-nesting species is not employed by the kittiwake. Instead, males announce territory ownership by using a "choking display," a signal derived from nest-building movements, at a potential nest site. In contrast, many ground-nesting species first form pair bonds in one area and then move to their nest sites, which are far more available than on cliff faces where suitable ledges are in short supply.

The catalog of probable changes in kittiwakes from an ancestral pattern could be extended, but the point has been made. Although we can and did consider certain attributes of kittiwakes in isolation from each other, a fuller understanding of the evolution of the gull's behavior comes from an examination of the entire behavioral repertoire of the bird in conjunction with multiple environmental pressures of cliff-nesting. All the behavioral abilities of a species are interrelated to a greater or lesser degree and therefore impose constraints on the evolution of each attribute.

Summary

1 If one accepts an individual selectionist approach, the ultimate function of a behavioral adaptation is to contribute to the gene-copying success of an individual. There are many environmental obstacles that stand in the way of individual reproductive success. Behavioral ecology attempts to discover how behavioral traits overcome these obstacles.
2 All traits have benefits and costs in terms of their effect on the fitness of an individual. The working hypothesis of behavioral ecology is that traits will survive and spread through populations only when their benefits exceed their costs. The benefits and costs of a trait should vary, depending on the environmental pressures acting on an animal. One can develop hypotheses about the net benefits of certain traits in certain environments.
3 A useful hypothesis generates testable predictions, which can be analyzed by gathering additional observational data, by conducting controlled experiments, and by using the comparative approach. Much of this chapter focuses on the comparative method of prediction testing. One can predict that differences in the key environmental factors acting on two closely related species will lead to divergent evolution in behavior. One also can predict that similarities in the ecology of two or more phylogenetically unrelated species will lead to the convergent evolution of a similar behavioral solution to a shared obstacle to reproductive success.
4 The comparative method was illustrated by examination of several cases of apparent divergent evolution, especially the classic comparison of kittiwake gulls with their ground-nesting relatives. The adoption of cliff-nesting by kittiwakes is correlated with a host of novel characteristics. The comparative method also makes use of cases of convergent evolution in behavior. Thus, there are lawful relations between certain behavioral traits and shared ecological pressures in unrelated animals—as in the connection between a parent's ability to recognize its offspring and the risk of misdirected parental care, which occurs in various unrelated birds, an isopod, and a mammal.
5 Each aspect of the behavior of a species is interrelated. Cliff nesting by kittiwakes has affected the evolution of not only the antipredator behavior of the gulls but also many aspects of its reproductive behavior, including nest-building, species-recognition, and offspring-recognition. The habitat selection, reproductive behavior, feeding behavior, and antipre-

dator responses of an animal are part of an integrated behavioral package whose ultimate effectiveness will be determined by the number of genes contributed by the individual to future generations.

Suggested Reading

Intensive research on ground-nesting gulls has been done by Niko Tinbergen and his associates [705–709]. One of his students, Esther Cullen, has written a classic article [153] on the kittiwake. The textbook by John Krebs and Nicholas Davies [392] has a helpful chapter on the application of the comparative method to prediction testing in behavioral ecology.

Suggested Films

Signals for Survival. Color, 50 minutes. One of the very best films ever made on the behavior of an animal species. A beautifully filmed description of the reproductive cycle of the lesser black-backed gull, a ground-nesting species.

The Social Behavior of the Belding's Ground Squirrel. Color, about 20 minutes. An excellent film on the natural history of this mammal, showing the helpful interactions between related females, interactions that are based in part on a mother's ability to recognize her female offspring.

Most animals make four major decisions that influence their fitness: where to live, how to gather food, how to avoid predators, and what tactics to use to reproduce. This chapter deals with the first of these categories of decisions. There is evidence that some animals actively choose to live in certain places rather than others. If individuals can select where to live, they should, according to evolutionary theory, settle in places most favorable for their reproductive success. The results of some illustrative tests of this prediction will be described in the pages that follow. Animals may do more than merely choose suitable places in which to live. Once having secured a site, they may later have the option of leaving it or staying on. Do animals leave when their fitness gains are improved by a search for an alternative living place? Do they stay when costs of leaving exceed any benefits to be gained by dispersing? When are the enormous costs of migratory behavior outweighed by its benefits? If an animal is confronted with an intruder in its living space, should it (a) try to repel the invader, (b) permit it to share the area, or (c) abandon the site and try to find a new spot? There is great diversity in the animal kingdom in the evolved "answers" to these questions. This is helpful because it enables us to search for correlations between particular kinds of behavioral solutions and the environmental pressures that make them reproductively advantageous. This chapter uses the comparative approach and other methods to analyze some possible examples of adaptation in habitat selection, homing, migration, and territoriality.

CHAPTER 8

The Ecology of Finding a Place to Live

Active Habitat Selection

All naturalists know that certain species are reliably found only in certain habitats. A recipe for seeing a great many species of birds in one day, a popular activity in some circles, is to visit a great variety of habitat types. Were there amateur insect-listers or spider-watchers, the rule would be the same. In theory, the association of an animal species with a particular habitat could be achieved (1) passively with the death of all individuals that happened to land in all but the appropriate location or (2) actively through the preferences of dispersing members of a species for a certain environment. Moderately mobile animals should be able to identify places in which their evolved characteristics are most effective. A host of studies have confirmed that animals as different as aphids and the planktonic larvae of some marine invertebrates have the ability to discriminate between habitats, choosing to live in some places while ignoring or avoiding others [286, 703, 777].

Stanley Wecker performed a classic demonstration of active habitat discrimination and selection in his experiment with prairie deer mice, a subspecies of *Peromyscus maniculatus* [762]. As their common name implies, these mice are usually found in grasslands and prairies. But do they actively choose to live there? To test the hypothesis that they do, Wecker built a 100-foot by 16-foot enclosure consisting of a series of compartments divided equally in number between grassland and woodland habitat (Figure 1). At either end of the enclosure was an identical nest box. Each time a mouse left a nest box and every time it passed from one compartment to another, an electric switch recorded the event. The records show the amount of time the mouse spent moving about in each compartment and the amount of time it spent in the field or woods nest box.

Wild-caught field mice exhibited a strong preference for the field habitat, spending more time in the grassy compartments and in the nest box located in this habitat. Moreover, the offspring of wild-caught mice reared either in the laboratory or in the experimental woodland habitat also exhibited a strong tendency to nest in and inspect the field area rather than the woodland enclosure [762]. This choice makes ecological sense if one assumes

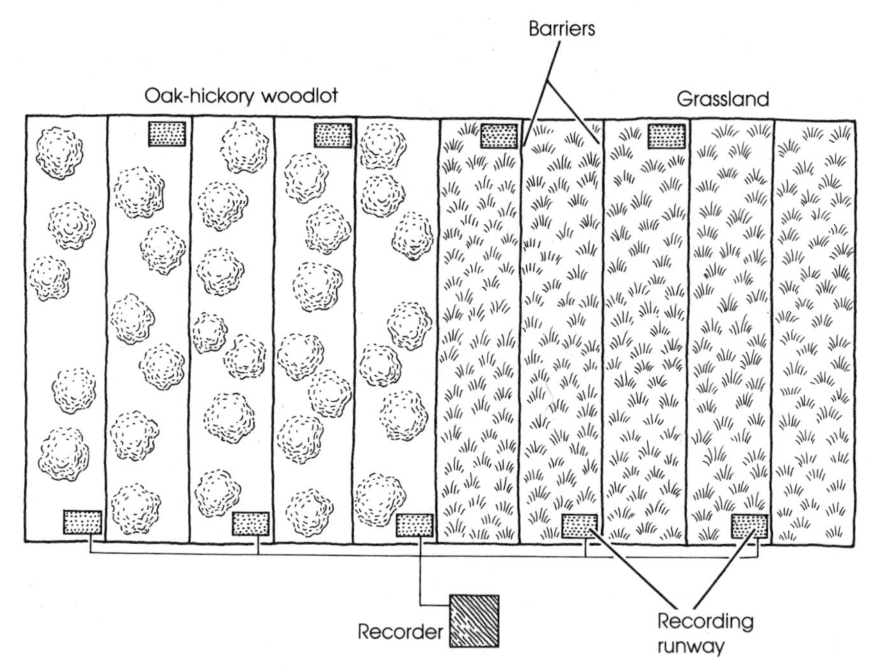

1 **Testing habitat preferences in deer mice.** Five compartments contained grassland habitat and five were built in an oak–hickory woodlot. Mice released in the center of the enclosure could select where they wanted to live by passing through runways between compartments. An automatic recorder monitored the movements of the mice in the enclosure.

that prairie mice are better adapted for prairies than woodlands. In any case, the experiments show that early experience cannot override the relatively closed developmental program that generates habitat preferences in this subspecies of deer mouse.

Discriminating habitat selection is exhibited by animals far smaller than deer mice. One insect whose habitat selection has been well studied is the great golden digger wasp, *Sphex ichneumoneus,* a hunter of katydids, which the wasp paralyzes and feeds to her offspring in a burrow in the ground [91]. Before a female digs her burrow, she spends from one to four hours searching for a suitable site, stopping to gnaw at the earth in place after place before finally selecting a location in which she will excavate a descending tunnel that ends in a brood chamber (see Figure 1, Chapter 1 for a similar nest). The wasp can be watched as she rejects sites in which the soil is very hard or too soft. Eventually she settles on an open sunny location in which the soil is almost always of a certain type (sandy loam) characterized by very fine particle size. The preferred soil is firm, but not too compact, and offers good drainage, but is not dessicated. It is likely that these conditions facilitate construction of a stable burrow that will not collapse half-way through the nesting cycle and that the end result offers a microenvironment well suited for development of the wasp's youngster.

Proximate Cues of Habitat Selection

It is possible that the nervous system of the great golden digger wasp has certain specialized perceptual mechanisms that focus on key cues (soil friability, moisture content) and trigger the onset of serious nest digging. The use of simple sensory signals in decision making is widespread in the animal kingdom (Chapter 4), and this is true for habitat selection as much as for feeding or mating behavior. Coho salmon, for example, imprint on the odor of the stream in which they lived during the early part of their lives. They leave the home stream as juveniles and travel to the ocean (or Lake Michigan, where they are an introduced species) but return many months later. They are able to track the odor plume of their stream by its distinctive chemical bouquet, using this simple cue to determine where they will reproduce [640].

Limited sensory cues are also a sufficient basis for habitat selection in the coal tit and blue tit, two very similar, closely related songbirds of Europe. The coal tit is a bird of the pinewoods, whereas the blue tit is a bird of oak woodlands. Linda Partridge hand-reared some young members of each species and then placed her subjects in laboratory test cages containing a composite "limb" consisting of pieces of oak and pine boughs with oak leaves and pine needles attached [570]. The habitat preferences of the birds were measured by recording the perching time spent by each individual on pine versus oak segments of the artificial branch. Hand-reared coal tits gravitated to the coniferous portions, blue tits to the deciduous tree parts, choices that matched the foraging preferences of wild birds of both species observed in natural mixed woodlands (Figure 2).

Habitat Selection by Honeybees

Habitat choices by honeybees have also been the subject of experimental studies, which have revealed that certain parameters of a potential homesite are of major importance to the nest-searching workers [428]. Nest selection occurs in the spring when a resident queen abandons her old hive to a daughter and flies with half her workers to a nearby spot, where they form a compact swarm (Figure 3). Scout bees leave the group and search for new hive sites. They are attracted to small openings that lead to chambers in the ground, in cliffs, and in hollow trees. Once inside, the worker scout marches up and down, measuring the volume of the potential homesite. As a general rule, only a chamber with a volume between 30 and 60 liters is attractive to a worker [646] and causes her to return to the swarm, where she performs a dance that communicates information about the distance, direction, and quality of the potential new home (Chapter 14). Other workers attend to a dancing scout and may be sufficiently stimulated to fly out to the spot themselves. If it is attractive, they too will dance and send still more workers to the area. The decision

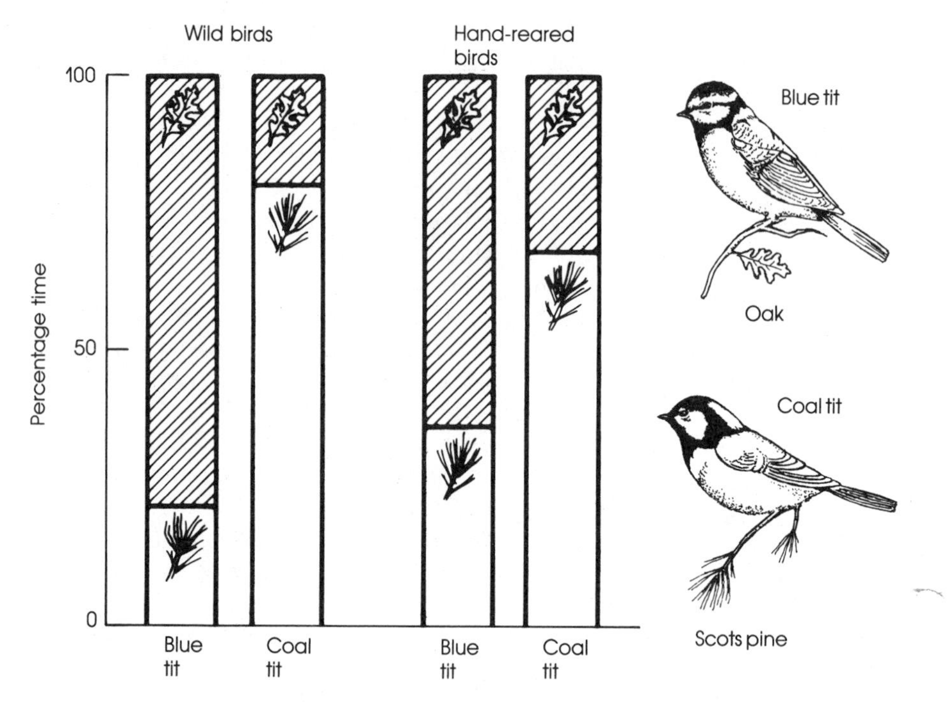

2 **Habitat preferences of two titmice.** Hand-reared, naive blue tits perch preferentially on oak branches, whereas coal tits prefer pine boughs. Wild-caught birds exhibit the same preferences.

to move is based on competition among the various reports. Almost always one site is sufficiently superior to all others so that scouts that had previously been announcing other potential homes cease to display and the great majority dance for just one site. Once agreement has been reached, the swarm flies to the spot to set up housekeeping there. This strategy has costs: the bees are exposed to the elements for several days while the decision is being made; and this time is lost as far as comb building and food gathering are concerned. But there are important advantages to this procedure, the chief of which is that the swarm is provided with information about a large number of potential sites and can choose the best one available.

Martin Lindauer was able, by reading the reports of scouts and observing the reaction of other workers, to determine what qualities of a nest site are the basis for its selection [428]. Perhaps the single most important hive characteristic for bees that live in areas with cold winters is its protective properties. In Bavaria, Germany, the bees generally choose holes in the ground, with hollow trees and straw-basket hives as second and third choices, respectively. The scout bees will reject sites that are either too large or too small to accommodate their colony snugly. Places that are too large will be drafty and therefore dangerously cold during the winter.

3 **Spring swarm of honeybees** settled in a bush. Scouts dance on the surface of the swarm, announcing the location of potential homes. Photograph by E. S. Ross.

Another aspect of a potential hive site is significant. If a swarm is given a choice between two apparently equivalent hives, one close to the original hive (say, about 50 meters away) and another farther away (about 200 meters off), it will choose the more distant of the two [428]. In experiments of this sort, Lindauer has watched the queen bee become exhausted as she struggled to cover the distance (queens are primarily egg layers and only secondarily flying insects). Despite the cost to the queen, the choice of a location far from the other hive presumably reduces competition for nectar and pollen between colonies. Reduced competition benefits both the queen and her daughter's colony, increasing the probability that her daughter will survive and reproduce while enhancing her own chances for another successful season.

Are Habitat Preferences Adaptive?

Having shown that some animals can discriminate among habitats (usually on the basis of relatively simple cues), we are in position to test whether the proximate mechanisms of habitat selection have adaptive outcomes. One can argue plausibly that prairie deer mice, having evolved in grassland habitats, are likely to reproduce better there or that nest site selection by a wasp seems to help her progeny develop in a warm, well-drained burrow. But the plausibility of a hypothesis is no substitute for testability.

Linda Partridge argued that if the habitat preferences of coal tits and blue tits were adaptive, one should be able to show that the birds could gather food more efficiently in the preferred habitat [571]. (Efficiency of food collecting should generally be correlated with reproductive success; see Chapter 9). She tested the connection between habitat type and foraging success by giving hand-reared birds artificial feeding tasks that required each bird to (a) hack through or pull off a cover that concealed a bit of food or (b) hang upside down to get at a prey item. Blue tits in nature rip and pull at broad oak leaves to get leaf-miners and concealed spiders within folded leaves, and they often hang beneath a leaf, inspecting its underside for insect eggs and hidden caterpillars. Coal tits forage on the open clusters of pine needles and pine bark and therefore do not often employ these techniques. As predicted, the hand-reared, untrained blue tits were significantly faster at extracting food during the tests than the naive coal tits (Table 1). This supports the argument that young birds prefer habitats for which their food-gathering abilities are best suited.

The habitat preferences of the two titmice species appear to have diverged in conjunction with selection for different skills in resource gathering. A comparison of habitat choices by different races of honeybees (*Apis mellifera*) indicates that here too divergent evolution has occurred, but

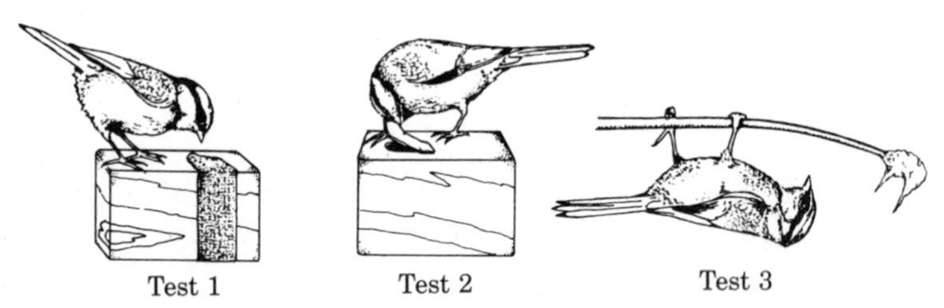

Table 1
Comparison of the Foraging Skills of Naive Hand-Reared Coal and Blue Titmice as Measured by the Time Taken to Secure Food from Three Artificial Dispensers

Species	Mean time (in seconds) to secure food		
	Test 1 (peck through cover)	*Test 2 (pull off cover)*	*Test 3 (hang upside down)*
Blue titmice	14	3.8	2.9
Coal titmice	22	5.0	6.1

Source: Partridge [571].

within a single species [268]. This difference was first noted when some American investigators attempted to duplicate Lindauer's results with swarms of the Italian honeybees rather than with swarms of the northern European race [363, 364]. The Italian bees did not exhibit the same preferences and indeed did not even discriminate between sheltered and unsheltered hives. They accepted smaller cavities and sites closer to the home hive from which they were emigrating than did northern bees.

All of these differences can be correlated with the reduced danger of winter mortality for the Italian bees. Because they live in warmer regions of Europe, they have not been under selection pressure to pass up sites that are not heavily protected [268]. There is also reduced selection pressure for large colony size, which appears to be adaptive primarily under cold climatic conditions. When the bees form a compact mass within the hive during the northern European winter, the outer layer serves to insulate the others from the cold. However, in freezing weather the outermost bees gradually die, an event that selects colonies that are large enough to provide numerous layers of insulation so that the queen and a sufficient worker force will survive the winter. During the mild winters in southern Europe, large colonies would not have a special survival advantage. Smaller colonies not only can occupy smaller cavities, they also can support themselves with food resources from smaller areas. Therefore, Italian bee swarms are not under pressure to disperse great distances from the home hive [268].

Application of the comparative approach to geographic races of the honeybee supports the hypothesis that aspects of habitat selection are shaped by the ecological pressures peculiar to a region. This in turn suggests that habitat selection by a swarm tends to promote the reproductive success of a colony.

Habitat Selection and Reproduction in Aphids

A more direct test of the predicted correlation between habitat choices and individual reproductive success is available in a study of poplar aphids, *Pemphigus balsamiferae* [777, 778]. In the spring in Utah, the eggs that female aphids laid in the bark of a cottonwood poplar tree the previous fall begin to hatch. From them emerges a new generation of tiny black females about 0.5 millimeters long. These unprepossessing animals make adaptive choices about where to live as they walk from the trunk to the places on limbs where leaves are just beginning to form. Each female, and there may be tens of thousands per tree, eventually selects a leaf, settles by its midrib almost always near the base, and in some way induces the formation of a hollow ball of tissue—the gall—in which she will live with the offspring she bears parthenogenetically (Figure 4). When her daughters are mature, the gall splits and they disperse to new plants in the summer. Tom Whitham found that females that settle on large leaves produce more numerous and heavier offspring than females on small leaves. If aphids can choose where to settle, they should pick large leaves. They do. In the trees examined by Whitham, there were 35 aphids for every 100 leaves.

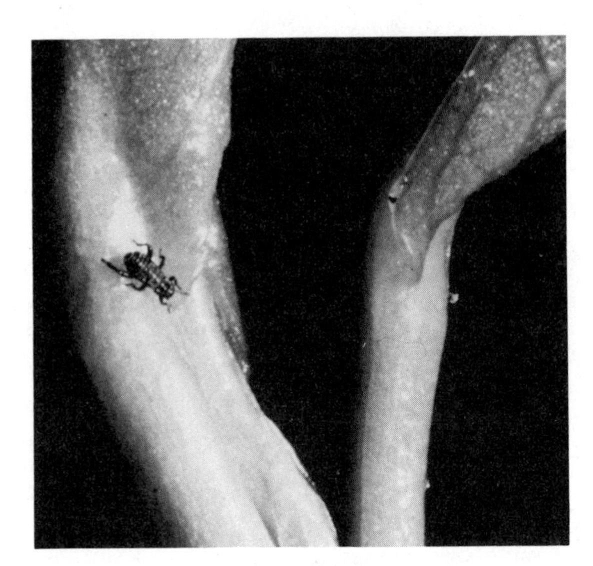

4 **Tiny stem mother aphid** searching for a site at which to form a gall on a poplar leaf. The female will select a relatively large leaf if one is available. Photograph by Thomas Whitham.

All the very large leaves on the poplars were occupied, although they made up only 2 percent of the total, whereas the smallest leaves (33 percent of the total) were avoided. Aphids have the ability to judge the potential productivity of leaves; they can find those that will provide the greatest flow of nutrients, which they will consume while nestled within their protective galls. They are so good at habitat selecting that the average aphid produces more than twice as many offspring as she would if she were to choose a leaf randomly.

The problem facing an aphid is that large leaves are in short supply and are quickly taken when the stem mothers emerge. An aphid that encounters a leaf with a female already on it has a choice: to settle with the original aphid or to hunt for an unoccupied leaf. A latecomer will have to form its gall farther out the central rib of an occupied leaf and will not get as much food as the individual nearer the base. Whitham showed that when aphids double up or triple up they choose much larger than average leaves. Single aphids settle on leaves with an average area of 16 square centimeters; two galls are on leaves with an average area of 20 square centimeters. The second colonist on a 20-cm^2 leaf does just as well as a single aphid on a leaf of 16 square centimeters (about 75 to 80 offspring on the average) and substantially better than a lone colonist on a 13-cm^2 leaf (an average of 60 offspring). In other words, a female *P. balsamiferae* somehow can determine the relative abundance of leaves of different sizes and the number of already established colonists on a leaf. She uses this information to do what selection theory predicts she should: select the leaf from among those available that is likely to make the greatest possible contribution to her reproductive success [778].

Homing Once a poplar aphid commits herself to settle on a leaf and begins to form the gall, she no longer has the option to move about and find a new living place or to return to her original home. Eventually she is sealed within her gall and before expiring becomes a factory for producing little aphids. But other animals retain their mobility after having made a habitat choice. Some species use this capacity to leave a specific homesite and then return to it later. The departure may be a brief one, as when a white-crowned sparrow leaves its nest to forage for food for its brood or when a limpet exits (slowly) from its resting spot grooved on a rock in the tide zone to graze on algae some distance away. On the other hand, homing journeys may be positively monumental, as in the ability of many migratory birds to travel thousands of miles from a wintering grounds and to return to precisely the same location where they bred the previous season. Thus, Adelie penguins taken 1200 miles from their rookery returned the next mating season and occupied the very sites from which they had been removed ten months earlier [209]. A considerably more speedy return from an experimental displacement was made by a shearwater (a long-winged sea bird) that was taken from its burrow in Wales and released in Boston, Massachusetts. This heroic individual was back on its nest 12.5 days later, having navigated 3000 miles of ocean in this time [769].

Table 2 summarizes some features of animals with homing ability. Hom-

Table 2
Convergent Evolution of Homing in Unrelated Animals

Species	Reference	Distance Traveled	Homesite Characteristics	Reason for Leaving Home
INSECTS				
Beetle larva	[4]	Several centimeters	Burrow with food	To forage
Digger wasp	[705]	Tens of meters	Nest burrow	To forage
Honeybee	[744]	Several kilometers	Hive with brood	To forage
CRUSTACEANS				
Isopod	[430]	Several meters	Nest burrow	To forage
MOLLUSKS				
Limpets	[141]	Several centimeters	Safe resting spot	To forage
VERTEBRATES				
Coho salmon	[640]	Hundreds of kilometers	Spawning site	To migrate
Black-headed gull	[705]	Several kilometers	Nest with young	To forage
	[705]	Hundreds of kilometers	Nesting site	To migrate
Laysan albatross	[769]	Thousands of kilometers	Nest with young	Experimental displacement
Wild dog	[734]	Tens of kilometers	Den with young	To forage

ing individuals usually possess a safe home base with dependent young. The cost of replacing such a site and its contents is extremely high. In this case, selection favors individuals with the capacity to relocate a home base that they are forced to leave because of the scarcity of food in its immediate vicinity. Additional benefits are derived from homing when animals make investments in the modification of the home environment. Many species build elaborate nests, burrows, or other structures, an energy investment that increases the value of the home site. This in turn raises the benefit-to-cost ratio of homing. Even the humble limpet gradually etches out a depression in its resting place (Figure 5), a depression into which its shell fits perfectly; this snug fit makes accidental displacement of the limpet by a violent wave less likely [141] and makes it all the more advantageous for the limpet to be able to find its way back to the spot after a bout of feeding.

Mechanisms of Homing

Homing ability is not provided free of charge. The animal must expend time and energy and take risks on a homing journey, and it must develop the physiological mechanisms that make the process possible. Some mechanisms are considerably more complex than others. A limpet leaves a chemical trail behind in the course of its meanderings, a trail that it can retrace to find a resting spot [141]. Far more mysterious is the capacity of

5 **Limpets and their homes.** Members of this species create slight depressions in rocks to which they return daily. Photograph by David R. Lindberg.

a migratory animal or a displaced bird to return home after a trip of hundreds of miles over totally unfamiliar terrain. If you or I were dropped off in a strange spot even a few miles from home, we would probably have no idea which way to go. Nor would we be any better off if we were given a compass by the persons studying our homing abilities. We would have to be told which direction to go; only then would the compass be helpful. Navigation to a destination requires both a compass sense and a map sense. You have to know where you are relative to your goal and you have to be able to orient your movements accordingly. How do some animals do this?

A number of key conclusions have emerged from studies on the appealing mystery of animal orientation and navigation. First, the map sense of animals with the ability to traverse unfamiliar regions remains unknown. No one really understands how navigating creatures know where their goal is. But better progress has been made in analyzing the compass sense of some species. There are clear demonstrations that the mechanisms vary from species to species in ways that can sometimes be related to ecological differences among them. Moreover, most homing species possess more than one compass system, often relying on a primary cue with one or more backup mechanisms available if the major cue is not available [211, 373].

Our focus will be on a comparative analysis of the compass sense of some insects and vertebrates. We begin by examining the navigational ability of honeybees and homing pigeons. Although these creatures are obviously not close relatives, they exhibit convergent evolution in their physiological and behavioral attributes vis-à-vis homing. Both are skilled navigators, as demonstrated by a honeybee's ability to make a beeline back to its hive after a meandering outward journey in search of food and a homing pigeon's ability to make a pigeonline back to its loft after having been released in a distant and strange location. Both animals have participated in ingenious experiments that have revealed a good deal about the basis of their homing ability.

Honeybees and homing pigeons are active during the daytime and, as one might suspect, both are able to use the sun's position in the sky as a directional guide [372, 744]. Even we can do this to some extent, knowing that the sun rises in the east and sets in the west—provided we know approximately what time of day it is. Every hour the sun moves 15° on its circular arc through the sky. One has to adjust for the sun's movement if one is to use its position as a compass. A bee leaving its hive notes the position of the sun in the sky relative to the hive and flies off on a foraging trip. It might spend 15–30 minutes on its trip and move into unfamiliar terrain in a search for food. If it were to try then to return home, orienting as if the sun were where it was at the start of its travels, the bee would not return precisely to its hive because the sun's position would have changed with the passage of time.

Honeybees rarely get lost, in part because they are skillful learners of visual landmarks but also because even in unfamiliar areas they use their biological clock (see Chapter 6) to compensate for the sun's movement [428]. This can be demonstrated by training some marked bees to fly to a sugar-water feeder some distance from the nest (say, 300 meters due east of the hive). One then can trap the workers inside the hive and move everything—lock, stock, and barrel—to a new location. At this site a new feeder is set up at a different orientation from the nest (say, 300 meters southeast). One observer stands at this feeder and another observer watches an empty feeder 300 meters due east of the hive. After two or three hours have passed, the hive is unplugged and the workers are free to go in search of food. They do not have familiar visual landmarks to guide them and yet the marked individuals remember that food is found 300 meters due east and they fly to the spot where the food source "should have been." They do not go to the new sugar feeder, a finding that shows that they are not tracking feeders by olfactory cues or some other basis. Instead, they have compensated for the 30° or 45° shift in the position of the sun that has taken place during the hours of their confinement. They are still able to fly due east to the place where a rich food source would have been (had the hive not been moved).

Pigeons too can be tricked into demonstrating how important a clock sense is if they are to orient accurately by the sun. The birds' compass orientation can be disturbed if they are induced to reset their biological clock [751]. This can be done by placing a pigeon in a room with artificial lighting only and then shifting the light and dark periods in the room out of phase with sunrise and sunset in the real world. For example, if sunrise is at 6 a.m. and sunset at 6 p.m., one might set the lights to go on at midnight and off at noon. A pigeon exposed to this routine for a number of days would experience a clock shift of six hours out of phase with the natural day. If taken from the room and released at 6 a.m. at a spot some distance from the loft, the bird will behave as if the time were six hours later (noon) and orient improperly. For example, let us say that the pigeon is released at a place 50 miles due west from its loft. Its map sense somehow tells it this and it attempts to orient itself to fly east. As Charles Walcott points out: "To fly east at 6:00 a.m., you fly roughly toward the sun, but because your clock tells you it is really noon, you know that the sun is in the south and that to fly east, you must fly 90 degrees to the left of the sun. And this is exactly what the birds do, although they presumably do not go through the reasoning process I have described." Figure 6 illustrates this story.

Backup Orientation Mechanisms in Honeybees and Pigeons

If a sun compass were the only mechanism available to bees and pigeons, their homing abilities should be severely affected by cloudy weather. But both species can forage and navigate successfully on totally overcast days [193, 798]. In fact, some pigeons can home accurately at night or when they have had frosted contact lenses placed over the eyes, devices that reduce their vision to a murky blur! Thus, these species have more than one compass mechanism, one of which may be a sensitivity to the

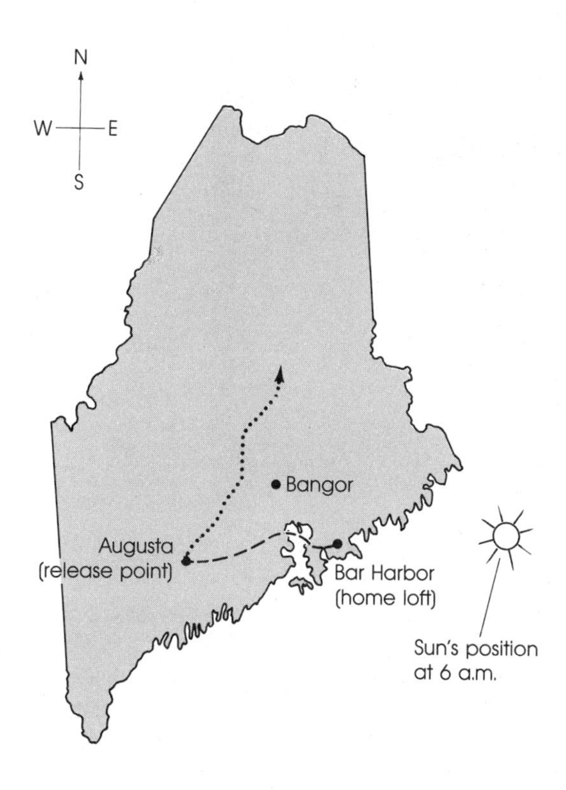

6 **Effect of shifting the biological clock** of homing pigeons: results of a hypothetical experiment. The birds were released at 6 a.m. near Augusta, Maine. The lower track was followed by a bird kept in natural light; the upper dotted flight path was taken by a bird whose clock had been shifted by six hours (see text). The clock-shifted bird flew more or less at a 90° angle to the left of the sun, "believing" that it was noon and that the sun therefore was in the south. Note that the clock-shifted bird exhibits good compass orientation, flying steadily in a northerly direction; the control pigeon demonstrates both compass orientation (flying a steady easterly route) and navigation ability (crossing unfamiliar regions to reach a specific goal—the home loft).

weak lines of magnetic force created by the earth's magnetic field. These lines run roughly north–south and, if detectable, might help orient a homing animal. Both pigeons and honeybees have magnetic compounds concentrated in certain tissues of their bodies; these compounds may be part of a magnetism detector [267]. Honeybee workers will orient their recruitment dances to the eight points of the magnetic compass if deprived of all celestial cues for some days. Normally these dances contain information about the angle between the hive, food, and the sun (Chapter 14). But whether bees use their magnetic sense to orient their dances on cloudy days is not certain, and there is even some evidence to the contrary [193].

The value of a magnetic sense as a backup compass has been unequivocally demonstrated with homing pigeons and some other birds [218], such as the garden warbler (Chapter 6). Charles Walcott altered the magnetic field about some pigeons, either by strapping a magnet to the birds or by outfitting his subjects with helmets consisting of a battery-powered Helmholtz coil (Figure 7). The altered magnetic field about the pigeons disoriented them when they were released far from home—but only on overcast days. If the sun was shining, the birds attended to the cues provided by sun position and ignored the signals detected by their magnetism sensors.

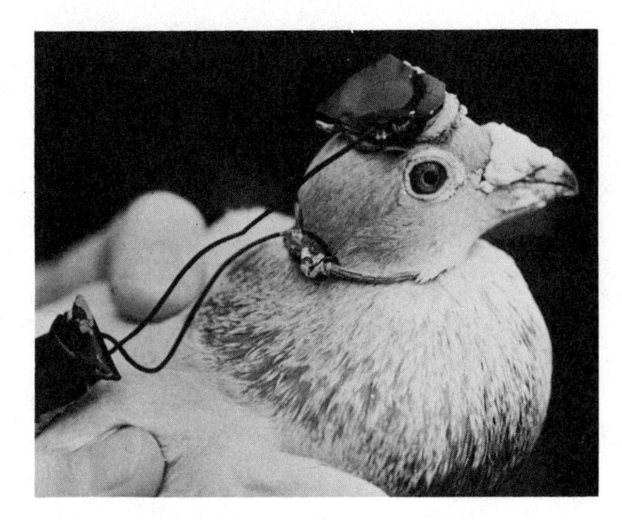

7 **Homing pigeon with Helmholtz coil** on its head. The coil can generate magnetism that interferes with the pigeon's ability to perceive the earth's magnetic field. Photograph by Charles Walcott.

But when reliable information provided by the sun was missing, the birds used the next best thing, information about lines of magnetic force, as an aid to getting home. In addition, homing pigeons can detect infrasound waves of very low frequencies, minute changes in barometric pressure, polarized light, and even perhaps area-specific odors, all of which may contribute to their navigational skill, although much remains to be learned about these and other perceptual abilities in relation to homing [372, 754].

The Orientation Mechanisms of Ants

There are many similarities in the orientation physiology of honeybees and pigeons that can be traced to the shared ecological problems they face. Both species must travel out from and back to a homesite under a variety of daytime conditions ranging from full sunlight to complete overcast. Having a battery of compass mechanisms enables the traveling bird or bee to negotiate an economically direct route home, using the most prominent cue available under the climatic conditions of the moment. We can extend this principle to the African ants *Cataglyphis bicolor* and *Paltothyreus tarsatus*. The first is a solitary, long-range forager in the dry flat deserts of North Africa; the second lives in dense tropical forests, where it feeds upon abundant termites. Which of these ants do you imagine employs a homing mechanism similar to the pigeon and honeybee?

I suspect (and hope) that most readers predicted convergence between the desert ant and the bird and honeybee. The ant *C. bicolor* lives in an open sunny environment where it can easily view the sky and derive compass information from the position of the sun. Rudiger Wehner has shown that the ant relies on a particular sun-related cue, the pattern of polarized light in the sky [763]. When sunlight strikes the earth's atmosphere, the wavelengths of light are scattered by molecules in the area. The light tends to become polarized (i.e., vibrate in a specific plane) in a manner dependent upon the angle of the incoming sunlight relative to the atmosphere and the observer on the ground. We cannot perceive polarized light in the sky, but if we could we would see a polarization pattern that changes as the sun's position in the sky changes over the day. There is a band of maximum polarization that stretches across the sky at right angles to incoming sunlight. Thus, the band shifts from the west after the sun has risen to the north as the sun sweeps through its southern arc (for a northern hemisphere observer) to the east as the sun begins its descent in the west at the end of the day (Figure 8).

A foraging *C. bicolor* employs in its compound eye a number of units that specifically monitor the pattern of polarized light [334]. It may travel 100 meters from its nest in a wandering course, find something to eat (a bit of cheese in Wehner's experiments), and then dash off directly for home with its prize—unless followed by a human holding a special filter above it that prevents the ant from seeing polarized light. Likewise, ants that have had the upper part of their eyes painted are unable to home successfully.

The reliance of *C. bicolor* on a form of sun compass orientation is in marked contrast to the orientation system of *P. tarsatus* [334]. When foraging ants of this species were displaced a mere 20 meters from their home on sunny clear days, they could not find their way back. To eliminate the possibility that the ants require a chemically marked trail to relocate their nest, Bert Hölldobler trained foragers to come to a food source (termites)

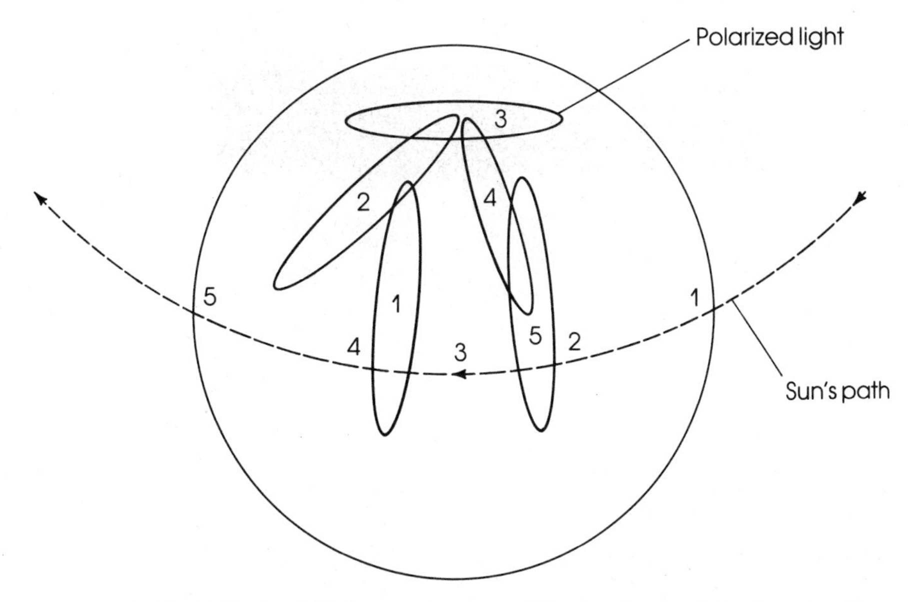

8 **Polarized light patterns** at different times of the day. As the sun crosses the sky, the region of maximum polarization shifts, providing a compass guide to some navigating animals.

placed about two meters from the nest entrance. After an ant had made several trips, he scraped away the upper layer of soil between the food and the entrance when an ant was away from the nest. It homed directly nonetheless.

Homing persisted even when the ant was surrounded by a visual barrier that was held in front of it wherever it faced on a trip back to the nest. Although it could not see exactly where it was going the ant still could orient properly. But this ability disappeared when the barrier received a lid that blocked the ant's view of the forest canopy above it. Under these conditions the forager wandered randomly.

In the laboratory Hölldobler extended his tests by training captive foragers to hunt for food in a small arena that could be covered by photographs of forest canopy that were 1.5 × 0.75 meters in size [334]. If the photograph was picked up and turned 180° in the midst of an ant's foraging journey, the insect "homed" in exactly the opposite direction from its actual nest entrance (Figure 9). These tests indicate that in this species food gatherers secure a visual image of their surroundings, including the canopy above them, when they exit from the nest. They use this information to orient their movements. The great ecological differences between *C. bicolor* and *P. tarsatus* are correlated with the different cues they use to orient themselves. Polarized light patterns offer a reliable compass aid for a long-distance hunter in a sunny desert; canopy orientation is superior for a short-range forager in a frequently overcast jungle environment.

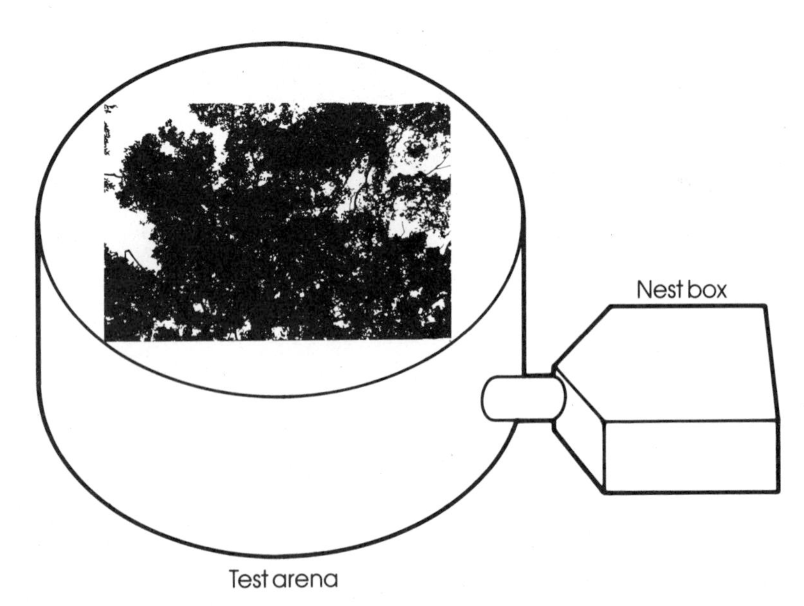

9 **Canopy patterns** provide orientation guides for the forest-dwelling ant *Paltothyreus tarsatus*. If a photograph of a canopy is experimentally shifted 180° above an ant foraging in an enclosure, the ant cannot find its way back to its nest box.

The Use of the Stars versus the Sun

Just as insects vary in their orientation systems, so too do bird species have different kinds of navigational abilities. Many relatively sedentary species almost surely rely primarily on visual landmarks to orient their movements in their permanent home ranges. But even migratory species do not always possess a dominant sun compass mechanism. This is because the majority of migrants travel at night, perhaps because they are safer then from hawks and because some daylight hours must be reserved for foraging. The garden warbler and indigo bunting, for example, engage in long-distance, nocturnal migration. Both species, like the pigeon, can sense and use the earth's magnetic field as a compass guide. This has been demonstrated by their ability to orient in the correct direction in completely enclosed cages during the migratory period (Chapter 6). Experimental deflection of the magnetic field about them alters the orientation of these birds [218, 292].

The magnetic sense of garden warblers, indigo buntings, and other nocturnal migrants is probably a backup system used on overcast nights when the pattern of stars in the sky is obscured [210]. A number of experiments have shown that birds held in cages in a planetarium will orient themselves during migratory periods according to the celestial cues provided by the stars. One can project on a planetarium ceiling a star pattern rotated 90° from the actual pattern, and the birds will shift their nighttime hoppings accordingly. This experiment shows that when celestial information is available birds will attend to it even if it is not in harmony with magnetic field information.

Stephen Emlen's study of indigo buntings is an especially instructive case history of how a songbird uses star position as a compass [210]. He demonstrated that if birds in a planetarium could see several major stars that are located in the northern sector of the sky—among them the North Star and the stars of the Big Dipper—they could orient appropriately for the season even if the other stars of the sky were artificially displaced and rearranged. The spatial relationship between the North Star and several major constellations clustered about it does not change during the night as the earth rotates (Figure 10). This provides a reliable fixed compass cue. Young buntings become imprinted on this portion of the night sky and will orient their migratory jumps to the south if given a chance to see the stars early in life. If they are held in experimental conditions in which they are exposed to diffuse light only, their migratory movements are random.

Migration

Indigo buntings, Adelie penguins, and many other animals employ their navigational abilities to leave one home and migrate to another, only to return in a few months to the same area [35, 554]. Tiny ruby-throated hummingbirds weighing ¼ ounce fly nonstop across 500 miles of the Gulf of Mexico twice a year; the Arctic tern travels a 22,000-mile round-trip course each year (the equivalent of seven trips across the continental United States); a migrating hoopoe (with the exquisite

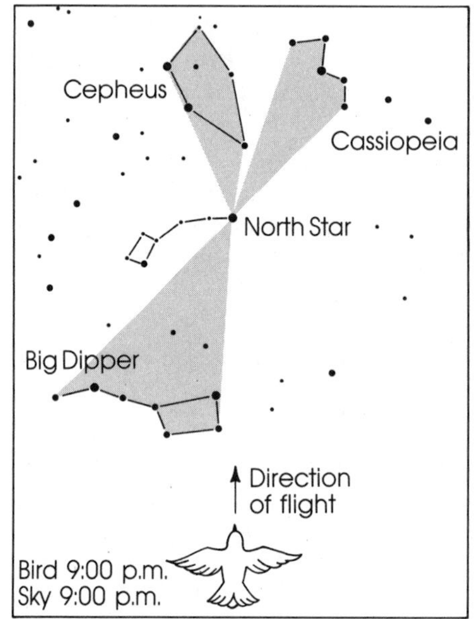

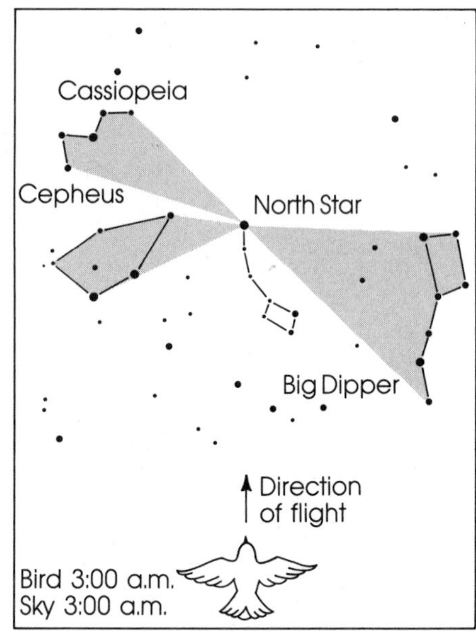

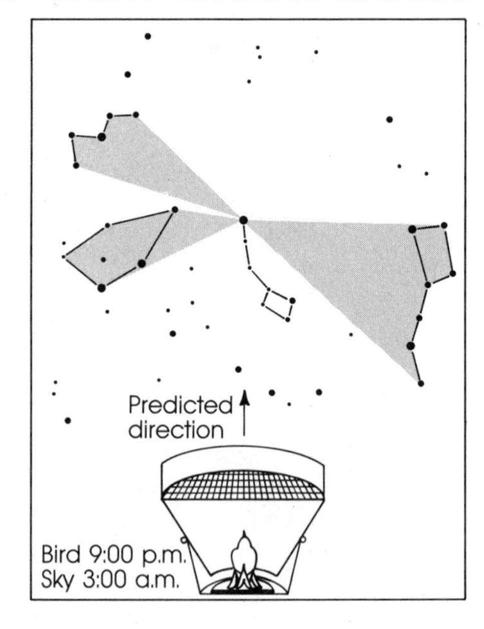

10 **Indigo buntings use the North Star as a compass.** The positions of most stars in the sky change over the course of a night, but not the North Star. Birds held in a cage in a planetarium in the spring were still able to orient in a northerly direction even when star positions were altered, provided the relationship between the North Star and the major constellations about it were not changed.

scientific name *Upupa epops*) was watched by an ornithologist named Lawrence Swan as it hopped up a Himalayan pass at 20,000 feet [689]!

Birds are not, however, the only migratory animals. Among the mammals, wildebeest, caribou, bison, seals, and whales make formidable journeys each year. Migratory reptiles include leather-back turtles, which move vast distances (up to 3000 miles of oceanic travel) between a breeding area in French Guiana and points near North America and Africa [586]. One population of green sea turtles nests on Ascension Island, a tiny speck of land (five miles wide) in the center of the Atlantic Ocean between Africa and Brazil (Figure 11). The adult female turtles visit the island only to deposit their eggs in beach sands. They then swim 1000 miles or so to warm, shallow water off Brazil, where they feed on marine vegetation for several years before returning, usually to the same beach, to lay another clutch of eggs [128].

There are many migratory invertebrates as well. Atlantic lobsters living off the continental shelf of the United States apparently move back and forth from deep water to coastal areas, a one-way trip frequently exceeding 50 miles [142]. Caribbean spiny lobsters have been observed marching along in a conga line of dozens of individuals, with each crustacean holding onto the one in front of it. This formation results in drag reduction and conserves the energy of migrating individuals [69]. Where they come from and where they go remain mysteries. Among the insects, monarch butterflies are justly famous because of the huge aggregations of migrants that form in trees at various locations along coastal California (Figure 12) and in central Mexico [101, 728]. Monarchs journey hundreds of miles to and from the winter areas; one marked individual was recaptured in Mexico, having traveled 1800 miles from southern Canada.

The Costs of Migration

If the proximate basis of navigation during migration is mysterious, the ultimate significance of this behavior is only slightly less puzzling. The evolutionary costs of migration are obviously major. Not only must the migratory individual invest in the development of the complex physiological systems (whatever they might be) that make navigation possible, but it may spend weeks or months each year on its energetically demanding journeys. Moreover, one can hardly overestimate the risks taken by many migrants. We mentioned earlier that one species of hawk, Elean-

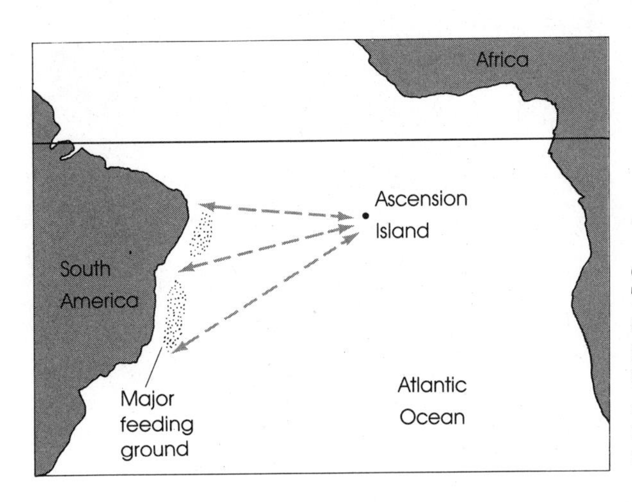

11 **Migratory movements** of some green sea turtles. The turtles journey between shallow-water feeding grounds off Brazil to the tiny mid-Atlantic island of Ascension, where the females lay their eggs in the sands of certain beaches.

ora's falcon, makes a living off exhausted songbirds that have crossed the Mediterranean Sea in the fall. Many other predators also have a field day with migrants. Lion prides compete for territories on the migratory route of the wildebeest and the winners grow fat while the antelope are passing through.

Many migrating individuals take action to reduce the costs of their trip. Thus, it is common for migrants to travel in groups, perhaps to dilute the risk of being captured by a predator (see Chapter 10). In Europe, a host of songbirds travel to central Africa by way of Spain and Gilbraltar in order to cross the Mediterranean at its narrowest point [769]. This lengthens the journey but perhaps lowers the chance that the bird will drown or arrive tired and vulnerable on the hunting grounds of a falcon. In light of this hypothesis, however, it is paradoxical that some North American songbirds, including several tiny warblers, migrate to South America from eastern Canada by way of the Atlantic Ocean (Figure 13) [789]. Radar watchers along the New England coast regularly see clouds of small migrant birds

12 **Cluster of wintering monarch butterflies** in Stinson Beach, California. Migrating monarchs select relatively warm and protected sites in which to spend the winter. Photograph by E. S. Ross.

on their screens flying out over the ocean during the fall migration. At first glance, this seems positively suicidal because to reach South America by a transoceanic route requires a journey of more than 3000 kilometers. One would think it would be far safer to travel along the coast of the United States and down through Mexico and Central America rather than inviting death by drowning. But there is no question that migratory songbirds do appear regularly on islands in the Atlantic and Caribbean, so that some, perhaps most, do survive the oceanic crossing.

There may be special advantages for the blindly courageous blackpoll warbler that attempts this trip. First, the sea route from Nova Scotia to Venezuela is about half as long as a land-based trek. Timothy and Janet Williams estimate that a blackpoll warbler flying under good weather

13 **Transatlantic route** of migrating blackpoll warblers. The small songbirds take advantage of prevailing wind patterns to fly across the Atlantic to their South American wintering grounds. Courtesy of Janet Williams.

conditions can go from Maine to South America in 80–90 hours of continuous flight [789]. Second, there are very few predators lying in wait in midocean or on the Greater Antilles chain of islands that it hopes (unconsciously) to reach. Furthermore, the oceanic route exploits some typical wind patterns: easterly winds in the north Atlantic that blow strongly after the passage of west-to-east moving cold fronts and westerly breezes in the southern part of the trip, which blow the birds to their island landfall.

The capacity of pigeons to sense barometric pressure changes and very low frequency sounds is probably possessed by songbird migrants as well. If so, these birds could be meteorologists waiting to begin their over-the-ocean journey until they detected the right combination of barometric indicators and acoustical cues (storm systems provide such cues, which human meteorologists can track and use as weather predictors). In any case, many songbirds do launch their migratory flights just after the passage of a cold front and therefore take advantage of tail winds to speed their travel.

For all the navigational and meteorological skills of these migrant birds, they still pay a steep price for their attempts to migrate, especially those that are forced into the water when unexpected head winds or storms develop over the Atlantic. The general point is that although migrants may reduce the costs of their travels, they cannot eliminate them. What ecological conditions can possibly elevate the benefits of migration enough to outweigh the dramatic costs of this behavior?

The Benefits of Migration

Migratory behavior has no single function but has evolved in response to several different ecological pressures. First, individuals may find it advantageous to move to an area when that region is rich in resources and then to leave it when the place becomes less suitable. Especially in temperate zones, there is great seasonal variation in productivity. Birds typically follow a migratory pattern in which they move from a warm wintering area with acceptable resources to temperate regions that have much more abundant food supplies, thanks to a spring and summer burst of insect reproduction. Presumably it pays individuals to expend great amounts of energy and time to reach these places because of the possibility of rearing large numbers of young. Competition for food and nest sites may be severe in tropical and semitropical regions. If so, selection will favor either sedentary birds that invest in adaptations that make them superior competitors or mobile species that invest in movements to areas that are "empty" before they arrive [106].

The other major migratory pattern involves movement from a feeding area to a place whose sole value is the protection it provides for breeding or for the birth of offspring. The Adelie and many other penguin species migrate solely for reproductive reasons. When waddling onto the Antarctic mainland, they move into an area totally bereft of food and far from the nearest feeding site (Figure 14). The windy, rock-strewn slopes chosen by the Adelie penguin offer the best of a poor situation for rearing young in the Antarctic region [209]. Likewise, marine mammals, fishes, and reptiles may abandon rich feeding grounds and swim without access to food hundreds or even thousands of miles to sites suitable only for reproduction [538]. Seals that mate and bear their young on land, as well as marine turtles that lay their eggs in sand, tend to choose areas that are isolated and well sheltered. The green sea turtle's choice of Ascension Island (among other places) may be highly adaptive despite the great costs involved in moving to such a small place so far from its feeding ground off Brazil because, owing to the island's isolation, predators have been absent (prior to human settlement) and the island has a number of U-shaped cove beaches well protected from Atlantic storms [128].

14 **200,000-pair Adelie penguin rookery** at Cape Crozier, Antarctica, after an unusual snowstorm during the breeding season. The adult birds build their nests on pebble-strewn ridges on the slope. They forage for food out at sea, returning to a main landing beach visible at the far left. Because the sea ice has broken up, they can swim directly to the rookery after foraging nearby, a fact that contributes to the efficient feeding of their voracious chicks (the all-dark birds). This photograph illustrates three requirements for a rookery site: open water late in the seaon, good landing beaches, and bare ground for nest sites. Photograph by David H. Thompson.

Whales and fishes are well adapted to bearing young and eggs in the water; however, some areas are very much better than others. In the case of salmon, the freshwater stream in which the fish was born has proved itself suitable for reproduction. As a result, individuals will, when mature, endure a long and difficult journey, risking death from exhaustion or predation, in order to mate and spawn in the same place. Whales exploit the extremely food-rich cold waters of northern and southern oceans, converting the tons of food they catch into tons of blubber. Then in the fall in the northern Atlantic and Pacific and in the spring in Antarctic seas, the animals move away from the feeding grounds to places where the females give birth and where mating occurs. The choice of tropical or temperate waters for calving grounds rather than frigid Arctic or Antarctic waters benefits the newborn whales. Furthermore, bearing an infant whale in a sheltered shallow cove or bay reduces the possibility of separation during storms and may facilitate defense of the young animal against attack.

Territoriality

Our final topic is TERRITORIALITY, the defense of living space against intruders. Just as there is diversity among animals with respect to habitat preferences, homing mechanisms, and migratory abilities, so too there are species that are extraordinarily aggressive in competing for a living area while others ignore or even tolerate their fellows in an undefended HOME RANGE. Many of the species mentioned in this chapter are territorial for at least a portion of the annual cycle. Even the tiny poplar aphid female can be ferocious in defense of a potential gall site at the base of a poplar leaf [778].

Territoriality by this aphid illustrates the costs and benefits of the trait. Fighting is an expensive occupation. Two evenly matched females may spend two days in a kicking and shoving match for a leaf (Figure 15). This is time that cannot be spent getting a gall established; the longer the delay, the greater the chance that the aphid will fall victim to a predator before she can become safely encapsulated in her gall. The combatants also probably run some risk of injury in the battle and the energy drain may weaken an aphid and speed its demise (although these factors have not been demonstrated for *Pemphigus*). It would seem that an intruder should leave occupied leaves and move to one that could be claimed without the costs of fighting.

15 **Territorial aphids.** Two *Pemphigus* females are engaged in a kicking contest to determine the owner of a preferred site on a poplar leaf. Courtesy of Thomas Whitham.

But because some leaves are much more valuable than others, the expense of territorial defense of a resource-rich leaf may be more than repaid by greatly improved reproductive success. As we noted earlier, very large poplar leaves are a limited resource on trees with many aphids; there may be 10–20 females for every premium leaf. The owner of the basal position on a 15-cm^2 leaf will on average have nearly twice the number of progeny as a female forced to settle on a more distal portion of the leaf (Table 3). Her reproductive success is also about double that of a sole owner of a 10-cm^2 leaf. Given these huge differences in expected fitness, we can predict that aphids will compete for basal sites on large leaves; and they do [779]. By removing some established stem mothers, Tom Whitham tested whether there was a population of females prevented from occupying preferred sites. In a tree with a high density of aphids, leaves from which stem mothers had been removed were nearly twice as likely to be occupied by an aphid as control leaves that had not been taken at the start of the removal experiment. In other words, some females are able to monopolize unusually productive sites; but if they are removed, "floater" females that had been prevented from settling on desirable leaves quickly occupy the now-open territories [777].

An evolutionary approach to territoriality suggests that other animals will also defend scarce, localized, and unusually valuable resources. Materials that can greatly elevate an individual's reproductive success should often be worth defending, particularly if the area that must be kept free from intruders is reasonably small and so will not require great time and energy investments to guard from intruders. (A territorial *Pemphigus* female excludes rivals from a segment that is only 3 millimeters long on the midrib of a poplar leaf.)

Comparative Tests of Territorial Function

There are a number of different ways to test whether animals engage in territorial behavior in an economically advantageous fashion. Sometimes the resources gained through monopolization of an area and the

Table 3
The Effect of Leaf Size and the Position of the Female's Gall on Reproductive Success of Female Poplar Aphids

Number of Galls per Leaf	Mean Leaf Size (cm)		Mean Number of Progeny Produced by		
			Basal Female	*Second Female*	*Third Female*
1	10.2		80	—	—
2	12.3		95	74	—
3	14.6		138	75	29

Source: Whitham [778].

costs of repelling intruders from a territory can be measured in the same units (e.g., calories gained and lost). In these cases one can employ an optimization model to generate quantitative predictions based on the assumption that animals should maximize their net gain from territorial behavior (for an example of this approach, see Chapter 9 on sunbird territoriality). Here, however, we shall use the more qualitative comparative method to test whether territoriality is correlated with ecological factors that raise the benefits and reduce the costs of defending valuable resources. There are, for example, territorial and nonterritorial limpets whose ecology we can compare. *Patella longicosta* aggressively repels all other limpets from an area of about 50 square centimeters by using its spiny shell to wedge intruders away. Its territory is covered with a species of nutritious algae that grows only when tended by a territorial individual. The resident limpet removes all competing species of algae and grazes on the richly productive garden that it creates and defends [87].

In contrast, another member of genus *Patella* living in the same South African tidal zones is not territorial and interacts little with other members of its species. This animal, *P. granularis,* is a wanderer, rarely remaining for prolonged stays in any one place, a trait that is correlated with its highly generalized browsing behavior. Thus, it does not create a valuable garden or a valuable resting spot; and, as expected, it does not defend its (temporary) living space [87].

Among ants, as well as among limpets, there are territorial and nonterritorial species. For example, the workers of the harvester ant *Pogonomyrmex rugosus* construct major highways from a colony's nest to areas with dense stands of seed-producing plants (Figure 16). If a colony builds a trail that comes close to another colony's route, workers from the two nests will battle, with many deaths on both sides, until one or the other moves its trail away from the area. In contrast, *Pogonomyrmex maricopa,* which lives in the same habitat but collects only seeds that are spread diffusely throughout its home range, is nonterritorial. Attacking other ants would not result in a large gain in food because no single small area has large quantities of its preferred seeds [161, 330, 331].

Other ants unrelated to *P. rugosus* have independently evolved territorial behavior that is focused on defendable concentrated resources [333]. Among these territorial species is *Myrmecocystus mimicus,* the honeypot ant, so called because it has a worker caste whose members are fed by other workers until their abdomens become enormously swollen (Figure 17). The honeypots are living food storage containers for their colony. This ant does not construct or defend a system of trails to a long-term food supply because it feeds on ephemeral sources of food, such as termite colonies, that it encounters here and there in its large foraging area. When workers from one colony find a rich patch of food and also encounter the nest of another colony nearby, some scouts from the foraging group rush back to their nest and lead other workers to the colony of their opponents. There they engage the enemy in a "territorial tournament" in which the members of each

16 **Ant highways** leading away from a nest of a harvester ant. These ants defend rich foraging areas. Photograph by Bert Hölldobler.

group employ ritualized threat displays (Figure 18). While large numbers of each colony are involved in the tournament, other foragers are gathering up as many of the vulnerable termites as possible. When the food supply is depleted, the threatening invaders withdraw and the interaction ends—unless the two sides are very unevenly matched. If the invaded colony is unable to muster an equivalent number of counterdisplayers, it may be raided and the queen destroyed. The honeypot ants and young workers are captured and taken back to the winner's colony, where they will become "slaves" for the winning queen [335].

Thus, the honeypot ants defend space only when they encounter a rich prize in a small area. In contrast, African weaver ants defend a large *permanent* territory, which may cover over 1800 square meters and contain a dozen or more trees in which the ants live [332, 335]. The territory is used as a foraging preserve, with workers ranging over the trees and ground in search of insect prey that have entered the area. In addition to the renewable resources provided by their territories, the trees provide leafy

17 **Honeypot ants.** Replete workers have distended abdomens that contain food reserves for the colony. They hang without moving from the roof of underground chambers in their nest. Photograph by R. R. Snelling.

nest sites. The workers build these nests by holding larvae in their mandibles and tying separate leaves together with silk from the immature ants. The leaf nests provide rearing places for the brood and retreats for the workers. Although there is only one queen in a colony, the total population of workers may be close to one-half million. They are distributed throughout the entire territory in the widely dispersed nests. The decentralized nature of the colony means that some workers are always near the boundaries of the territory. When intruders are detected at any point, the ants release an alarm-recruitment scent that attracts additional nearby workers. Other individuals lay an odor trail back to their nests over which still other recruits travel. Invaders are treated roughly (Figure 19). The large number of workers and their dispersed distribution, coupled with an efficient recruitment response, makes it possible for a weaver ant colony to monopolize an unusually large stable territory without exceptional expenditures of energy [335].

Territories and Reproductive Success in Songbirds

Our comparative analyses support the hypothesis that territoriality evolves when a useful and scarce resource, such as food, occurs in defensible clusters (Table 4). The presumption is that exclusive access to unusually large amounts of food enables the territory owner to survive and

18 **Territorial contest** at a nest of honeypot ants in the "tournament" phase of the conflict. At this stage the workers from the two opposing colonies give ritualized aggressive displays. Photograph by Bert Hölldobler.

reproduce more than nonterritorial individuals (or animals with inferior territories). It would be helpful, however, to test this assumption as Whitham did with his poplar aphids. Fortunately, there are other species in which the reproductive advantages of territorial possession have been demonstrated.

If suitable breeding sites really are in short supply, then one should be able to find nonterritorial animals in a population of a territorial species. These individuals should not be breeding. In her study of a population of the rufous-collared sparrow, Susan Smith captured many individuals in a mist-net (a fine-meshed, black, nylon net that can be strung between poles in areas traveled by birds) and gave each bird a unique color band combination [673]. Then she monitored her study site, plotting the location of known individuals and their breeding activity, if any. As is true for most songbird studies, she was able to define a set of exclusive territories owned by a breeding male and his mate. But there were other birds as well in the area, a population of "floaters" whose members, however, did not move

19 **Three weaver ants** about to dismember an intruder (center) from another colony. Weaver ants defend unusually large territories. Photograph by Bert Hölldobler.

randomly through the study site but instead had well-defined home ranges (undefended living spaces) within one or more territories of other birds. Male "floaters" were chased by male owners and female floaters received the same hostile treatment from female owners. But the nonterritorial birds were able to remain inconspicuous for most of the time and so "lurked" within territories permanently. They could not breed, however, because a singing male floater would not be tolerated by a resident nor would a nesting female intruder be permitted by the female territory holder. Instead, members of the underworld population waited for an opportunity to acquire a territory when a territorial bird of the appropriate sex disappeared. Immediately upon the demise of a male resident, a male floater in the territorial area would claim the site and the female in it. Female floaters did the same when the resident female died or moved away. Subordinates from outside the territory had little chance to secure the area in competition with an established floater, who, if able to make the transition, would then be able to begin reproducing [673].

Table 4
Ecological Factors That Influence the Adaptive Value of Territoriality versus Home Range Behavior

Selection Favors Territoriality	Ecological Variable	Selection Favors Home Range Behavior
Clumped	← Dispersion of valuable resource →	Scattered
Considerable	← Variation of quality of resources →	Little
Constant	← Reliability of resources →	Unpredictable
Moderate	← Competition for resources →	Very high

Another study that establishes a connection between territoriality and fitness is John Krebs's research on an European songbird, the great tit. In the spring, males of this species attempt to establish territories in woodland habitat [388]. This is where singing males are first heard in each new year and where fights, chases, and threat displays can first be observed. If the birds are color-banded, one can soon map the territories of the resident males. Krebs' study area became completely filled at a time when there were still other birds searching for breeding sites. These individuals settled in hedgerows adjacent to the woodland. They too established territories and attempted to breed. Given the bird's apparent preference for woodland, Krebs' predicted that great tits holding territories in this habitat would outreproduce birds with hedgerow territories. This proved to be the case, as only 2 of 9 nests (22 percent) succeeded in producing fledglings in the hedgerow location whereas 54 of 59 woodland nests (92 percent) yielded fledglings.

To test whether the hedgerow birds really were excluded from superior breeding territories that they would have preferred to occupy, Krebs "removed" (a euphemism for "shot and killed") six pairs from their woodland real estate. Within a few days, most of the territories were claimed by new defenders—most of whom were banded birds from the hedgerow area, which they had abandoned as soon as superior habitat became available (Figure 20).

These studies and others show that territorial disputes are centered on attempts to monopolize superior breeding places. The individuals that are unable to displace their rivals either adopt a nonterritorial waiting role or occupy second-rank territories where they have a poorer chance of reproducing than the birds that have been able to claim quality sites.

A Territorial Mammal

A mammal whose males compete for territories that contain food resources attractive to females is the impala [361]. Here, too, there is evidence that areas with exceptional food supplies are fewer in number than the male impala that try to defend them and that male reproductive success is correlated with acquisition of a territory. Prior to the onset of

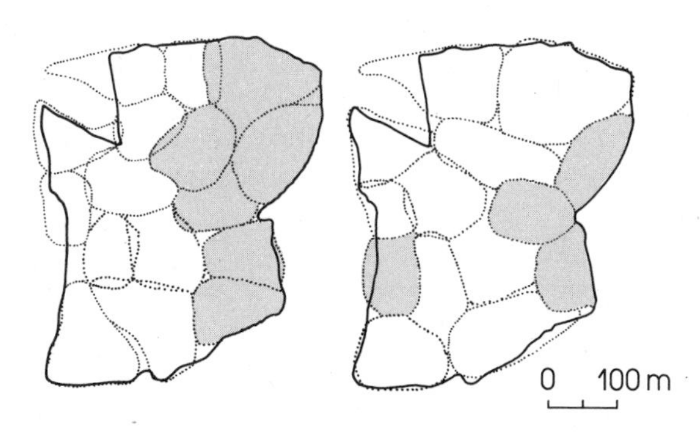

20 **Territorial replacement experiment.** On the left, the territorial boundaries of pairs of great titmice are shown prior to the removal of six pairs (*gray areas*). Three days later the territorial configuration was nearly the same after the arrival of four replacement pairs (*right, gray areas*).

reproductive activity by females, the males have engaged in considerable ritualized fighting with other members of the bachelor herds in which they live for much of the year. The primary activity is the clashing of their elegant horns; one male will interlock his horns with a rival's and the two will then push violently, attempting to propel the opponent backward. Through these interactions, a dominance hierarchy is established among the males, that is, some bucks can cause others to move aside merely by approaching them. The most dominant individuals—those that can force almost all other males to back down in horn wrestling combat—are those that move onto rich feeding grounds in the periods following the seasonal rains. Resident males attempt to monopolize these limited sites and will chase or battle with intruders in these areas.

A territory may measure 1–12 hectares [528]; it offers grazing material for the bands of females that come through the superior foraging areas. The duration of the stay of a band on a male's territory is a function of the quantity and (quality?) of the food his site contains. The longer the females stay, the greater the number of individuals that will cycle into estrus and become available for mating. Only territorial males have many opportunities to copulate, and some of these individuals succeed much more often than others. Males excluded from the superior grazing sites have few or no contacts with females. Their reproductive chances come when and if they are able to displace a territory owner and gain a chance to encounter feeding females, some of whom will be sexually receptive.

Why Do Nonterritorial Individuals Accept Their Status?

The ground rule for a territorial species is that individuals that fail to maintain a territory fail to reproduce. This raises an evolutionary puzzle; why do subordinate individuals so often accept their role in life without seriously challenging a dominant territory owner if they cannot reproduce without a territory of their own? One of the major findings of the early students of behavior was that opposing animals frequently resolve their disputes without violent fighting and often without even making physical contact. The use of threat displays, such as the territorial song of a titmouse or the lowered-head shake of an impala, is often sufficient to send a rival quickly on his way. In the well-studied red deer, males often roar at one another at substantial distances, with most intruders leaving before ever having used their antlers in anger against a territory owner (Figure 21) [136]. Even more serious clashes appear to have an element of comic opera or Marquis of Queensbury rules about them. After pushing and shoving harmlessly with locked horns, one buck impala will generally run away from the other. Likewise, after one or two body slams, a subordinate elephant seal generally lumbers off as fast as it can lumber, inchworming its blubbery body across the beach to the water while the victor bellows in noisy, but generally harmless, pursuit.

Observations of this sort have been interpreted in various ways. A com-

21 **Convergent evolution in threat displays.** In many unrelated species, aggressive individuals give signals that emphasize their size and endurance. (*Top left*) Roaring red deer. Photograph by Timothy Clutton-Brock. (*Top right*) Two elephant seals exchanging roars and pushes. Photograph by Burney Le-Boeuf. (*Bottom*) A charging bird of paradise. Photograph by Crawford Greenewalt.

mon view is that these largely symbolic interactions determine who will breed and who will not in ways that regulate population growth. Another hypothesis is that "gentlemanly" fighting helps males resolve disputes without leading to deaths that would dangerously decrease the population. Still another idea is that ritualized interactions ensure that genetically superior individuals will breed and will not be damaged by their fights with others. All three interpretations are clearly based on group selection theory because they imply that the nonterritorial (nonbreeding) animals and gentlemanly losers are an expendable population reserve that is sacrificed for the welfare of the species as whole. However, such self-sacrificing individuals, if they ever existed, would sooner or later be replaced by mutant types that behaved in ways that increased their fitness (see Chapter 1).

Thus, our working hypothesis is that the failure of subordinate individuals to breed must in some way advance their genetic interests. The solution to this paradox comes from studies of the sort that I have already described, studies that show that nonterritorial individuals often refrain from reproducing only *temporarily*. This leads to the prediction that subordinates will accept their status if in the long run this helps them reproduce more than persistence in attempts to supplant a dominant, territorial opponent. An allied prediction is, therefore, that the subordinate fraction of the population will tend to consist of young and inexperienced animals that can assess from brief threat interactions with other older, larger, or more powerful individuals that they could not displace these rivals no matter how hard they tried.

In many species, threat displays have evolved convergent properties that indicate something about the size and actual fighting capacity of the displayer [567]. Consider the full forward display of the green heron; in this display the bird erects its crest, back, and breast feathers to an extreme, while directly facing an opponent [511]. At the same time the bird flips its tail up and down, gapes open its bill to expose the deep red lining of its mouth, and with yellow-orange eyes bulging, rasps out a "raaah" call as it lunges forward at the enemy (Figure 22). Andrew Meyerriecks writes, "Of all the threat displays of the green heron, the full forward is the most effective because the opponent is almost always intimidated, retreating immediately." And small wonder.

The green heron's display is representative of a large group in which animals raise their feathers or hair, or expand expandable body parts (like anole dewlaps) while producing loud and deep sounds as they face (or stand parallel) to a rival (Figure 21). These displays help each opponent assess its rival's potential strength before ever engaging in direct physical interactions. If an individual can determine that it is not able to produce threat growls or roars of greater intensity or lower frequency than its opponent, almost surely it is smaller than its competitor because body size is strongly correlated with these aspects of acoustical production [136, 163, 524]. Likewise, while in a head-to-head position (or side-by-side), two individuals can

22 **Threat display of the green heron.** The contrast between a perched heron and an aggressive territory defender is striking. Note that the aggressive bird has fluffed its plumage to make itself look as large as possible.

probably determine visually which is the larger. Body size and strength go hand-in-hand, and therefore the larger of two contestants has an enormous advantage in fighting animals ranging from aphids to elephant seals [8, 405, 778]. An attempt by a smaller or weaker animal to supplant a stronger one is far more likely to lead to exhaustion, injury, or death than to success for the less powerful combatant. A relatively weaker individual would therefore often benefit from conserving his time and energy for later attempts to acquire a superior territory when the current resident has become older, less powerful, or dead.

This hypothesis matches the pattern seen in great tits, rufous-collared sparrows, and impala, species in which subordinate animals are generally younger and often smaller than dominant territorial owners [361, 388, 673]. Moreover, in all these and many other species, some subordinate individuals do eventually achieve territorial status and reap the reproductive benefits of their ascension to dominance. There is, therefore, every reason to believe that subordinate individuals that flee at the very sight of a rival are behaving in their own long-term genetic interests and not acting to further the welfare of their opponents or the species at large.

Summary

1 Animals that disperse from their birthplaces are generally able to select new sites in which to live on the basis of simple proximate cues. The choices they make should elevate their fitness and, in some cases (such as poplar aphids), the prediction has been successfully tested.

2 Having found a suitable living place, an animal will sometimes return to it if voluntarily or accidentally displaced. Homing has evolved convergently in many unrelated species whose members possess an unusually valuable homesite that they must leave on a daily basis to find food or on a seasonal basis to migrate to a wintering site. The variation in the kinds of mechanisms that underlie this ability can be traced to ecological variation in the sensory cues likely to be available to a navigating animal in its environment.

3 Migratory movement between two living places, often a summer breeding area and an overwintering area, is widespread in animal groups and is often associated with homing abilities. The capacity to migrate rests upon a neural foundation and requires an expenditure of time and energy resources coupled with a heightened risk of mortality. These costs, which are often substantial, are weighed against the potential benefits, which include the exploitation of seasonally fluctuating resources in separate regions and the utilization of a safe breeding site distant from feeding areas.

4 Despite the importance of finding a living place, not all animals defend their homesites. Territoriality, like homing and migration, carries with it a baggage of costs—the time and energy needed to acquire an exclusive preserve and to defend it against intruders and the additional risk of

predation associated with these activities. The benefit of territorial possession is the exclusive use of the resources contained within the territory.

5 A comparative survey of limpets, ants, and other animals supports the prediction that territoral behavior is generally focused on relatively small (i.e., economically defendable) areas that contain valuable resources that are in short supply elsewhere. This combination of factors should make the benefit-to-cost ratio of territoriality favorable for its evolution.

6 The prediction that possession of a territory in a territorial species should raise the fitness of the owner has also been tested and found to be true in some cases. Despite the correlation between possession of a territory and reproductive success, in many animals nonterritorial subordinates give way to residents after a brief exchange of threats with little or no actual physical fighting. This may often promote the long-term reproductive chances of the "loser" who lives to claim a territory on another day rather than exhausting itself in a futile effort to wrest a territory from a physically stronger rival.

Suggested Reading

The articles by Tom Whitham [777, 779] and Linda Partridge [570, 571], as well as Martin Lindauer's book, *Communication Among Social Bees* [428], offer modern analyses of habitat selection.

Stephen Emlen and William Keeton provide useful reviews of the underlying mechanisms of migration and homing [211, 373]. This is such a fast-developing field, however, that readers should consult the more recent literature as well [e.g., 636]. R. R. Baker has written a monumental review of migration in animals [35].

The classic exposition on the cost–benefit approach to territoriality is by Jerram Brown and Gordon Orians [108]. John Krebs's article [388] on territoriality in a song bird offers a clear example on how to test ideas about the adaptive value of territoriality.

Suggested Films

Adelie Penguins of the Antarctic. Color, 23 minutes. A film that shows the breeding cycle of the penguin with reference to migration and homing abilities.

Animal Landlord. Color, 25 minutes. An outstanding film on the ecology of territorial behavior in the impala antelope.

Bird Brain, The Mystery of Bird Navigation. Color, 27 minutes. An excellent film on current research into the proximate bases of bird migration.

The Year of the Wildebeest. Color, 54 minutes. The annual migration of the African wildebeest is followed in this often spectacular film.

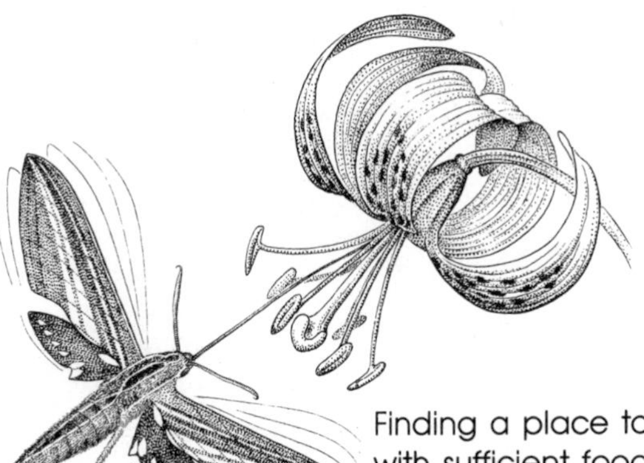

Finding a place to live almost always involves finding a place with sufficient food to sustain the individual and often its mate and offspring as well. But collecting food is usually not a simple task. Nutritious items are often in short supply, and they may be more than mildly reluctant to be consumed, if this reduces *their* fitness. Therefore, in order to survive and reproduce, animals must be adept at locating and capturing food. This requires decisions about what food items to feed upon, where to search or wait for meals, whether to defend a food patch, how to overcome a prey's defenses and so on. Behavioral ecologists have begun to analyze these decisions in some species. A useful working hypothesis has been that animals should feed in ways that maximize the energy gained during the time spent collecting food. This hypothesis can be used in certain cases to generate testable predictions about the kinds of food an animal should eat or the feeding tactics it should employ. This chapter provides some case histories of the approach and its utility (and limitations) for an understanding of adaptive feeding behavior. We then examine another area of current interest and controversy in feeding ecology, the hypothesis that the dietary differences among some animals are evolved responses to competition for food in the past. The chapter's final section employs the comparative method to test ideas on how predators overcome the specialized defenses of certain prey.

CHAPTER 9

The Ecology of Feeding Behavior

What is the Ultimate Goal of Feeding Behavior?

A familiar sight on many beaches around the world is shell-dropping by crows and gulls. A crow walking in a rocky intertidal zone may pick up a clam, snail, or whelk; fly up with the prey; and then drop it onto the rocks below. If the mollusk's shell shatters, the bird plucks out and eats the soft body of its victim. The adaptive significance of the bird's behavior seems straightforward. It cannot use its beak to crack open the extremely hard shell of certain mollusks. Therefore, it breaks its prey by dropping them on rocks. This seems adaptive. Case closed.

But with additional thought, one can devise adaptationist hypotheses about shell dropping that are much more refined and ambitious. When a hungry crow goes walking across a beach, it has many choices to make—which whelk to pick up, how high to fly up before dropping the prey, and how many times to keep trying if the whelk does not break on trial 1, 2, or 3. One can argue that crows should choose whelks that will yield the greatest possible energetic return for the time and energy invested in opening them. The logic behind this hypothesis is that the point of feeding behavior is to ingest sufficient energy to maintain one's body and to reproduce. Animals that consistently invest more energy in foraging than they gain in collected food obviously are destined for a short life. When they perish, so will their genes.

Selection should eliminate not only totally incompetent foragers but also those that forage less efficiently than their genetically different rivals. In Chapter 7 we briefly outlined the logic of an optimization approach. As applied to crow feeding behavior, we shall assume that a crow that maximizes its net energy gain when searching for food will ultimately leave more descendants on average than a less efficient individual. The less productive crow will either accumulate lower energy reserves or be forced to spend more time foraging. In either case its fitness is likely to be reduced through its relatively early demise or inability to produce large numbers of surviving offspring. Thus, as a working hypothesis we can assume that a shell-dropping crow should be an energy maximizer.

Reto Zach tested whether northwestern crows made optimal decisions that maximized the foraging efficiency of individuals [814]. (1) He observed that crows only dropped large whelks about 3.8–4.4 centimeters long. (2) They flew up about five meters to drop their selections. (3) They kept trying until the whelk broke, even if many flights were required. If their ultimate (unconscious) goal were energy maximization, (1) large whelks should be more likely to shatter after a drop of five meters than smaller ones, (2) drops of less than five meters should yield a much reduced breakage rate whereas drops of much more than five meters should not improve the chances of opening a mollusk a great deal, and (3) the chance that a whelk will break should be independent of the number of times it is dropped.

Zach tested each of these predictions in the following manner. He erected a 15-meter pole on a rocky beach and outfitted the pole with a platform whose height could be adjusted and from which whelks of various sizes could be dropped. He collected samples of small, medium, and large whelks and dropped them from different heights (Figure 1). He found that large whelks required significantly fewer five-meter drops before they broke than either medium-sized or small whelks. Second, the probability that a large whelk would break improved sharply as the height of the drop increased up to about five meters. After this height, the improvement in breakage rate was very small. Third, the chance that a large whelk will break is always about 25–30 percent on any drop. A crow that abandoned a "recalcitrant" whelk would do no better with a new one and would have to invest time and energy to find the replacement.

Zach went one step further by estimating the energy expended by his shell-dropping northwestern crows (using physiological data collected for

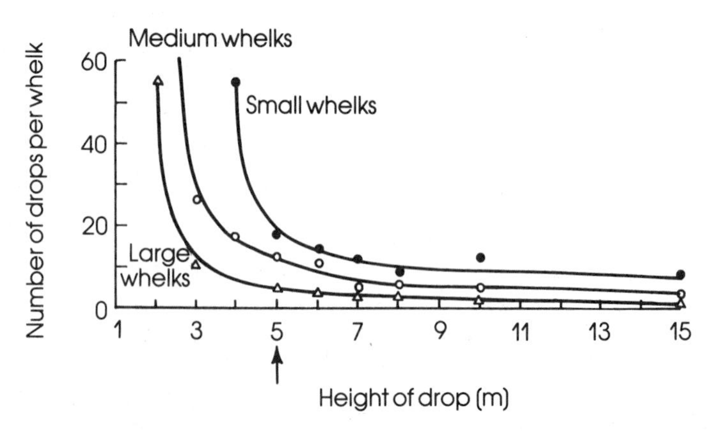

1 **Average number of drops** needed to break whelks of different sizes dropped from various heights. Large whelks break most easily. The probability of breakage does not improve much at heights greater than five meters. Crows select large whelks and fly up to drop them from a height of five meters.

another similar-sized crow species). He calculated the average number of calories required to open a large whelk (0.5 kilocalories). This figure was compared with the energy present in the body of a whelk (2.0 kilocalories); net gain = 1.5 kilocalories. In contrast, medium-sized whelks, which require many more flights, would yield a net *loss* of 0.3 kilocalories; and trying to open small whelks would have been an even poorer investment. Thus, the crows' rejection of all but large whelks is clearly adaptive, and their selection of dropping height and persistence in the face of failure enabled them to reap the greatest possible energy return for their foraging activities [814].

Minimizing Foraging Costs

The northwestern crow is remarkably good at keeping its caloric expenses down and thereby maximizing the energy gained from a large whelk. Is the bird an exception or are there other creatures like it? An optimality approach has been applied by David Vleck to the pocket gopher, *Thomomys bottae,* a common desert rodent that forages in a manner totally different from that of crows [741]. It is a burrower that digs tunnels close to the surface of the soil and feeds on the plants it encounters along the way. Burrowing requires energy and involves some decisions, including what to do with loose excavated soil. The gopher cannot store it in the tunnel because, once loosened, soil occupies more space than when it was compacted. The gopher solves this problem by building lateral exits up to the surface from the main horizontal burrow and pushing the loose dirt out of these exits. But this solution requires other decisions, particularly how frequently to build the ascending lateral offshoots.

In a laboratory setting it is possible to measure the energy expended by a pocket gopher as it pushes soil back horizontally and then up an inclined tunnel. In addition, the caloric cost for digging into hard-packed soil can be quantified. Knowing that the main horizontal tunnel lies about 25 centimeters beneath the surface, one can then calculate the cost of the various activities of a burrow-building gopher. At any point, the gopher has two options for dealing with the loose soil that has accumulated: (A) it can push the soil back a horizontal segment and up a lateral exit or (B) it can build a new lateral exit and then push the soil right up to the surface (Figure 2). Option B requires energy to build the new offshoot, but at some point this cost will be less than the energy spent pushing dirt back a long distance along the horizontal burrow to an already constructed exit. Given the soil conditions in his study area, Vleck calculated that the optimal distance between lateral shafts should be 1.22 meters. This distance minimized the costs of constructing a feeding tunnel. He then measured the actual distance between laterals in the burrows constructed by his gophers in the field. The observed mean distance was 1.33 meters, strong support for the argument that pocket gophers really do dig burrows in ways that reduce their energetic expenses to the minimum. This in turn should help the animal maximize its net energy gain while foraging.

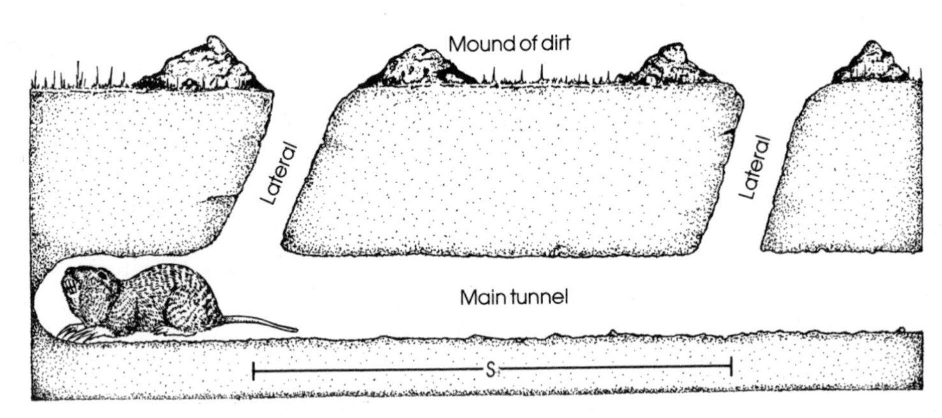

2 **Optimal burrowing behavior** of pocket gophers. To dig the least costly foraging tunnel, pocket gophers must balance the costs of digging a new lateral tunnel against the costs of pushing dirt back the main tunnel to the closest lateral exit. The key decision is how long to make the distance (S) between laterals.

Optimal Foraging over a Day

When a pocket gopher extends its burrow or a northwestern crow decides to eat a whelk, they are able to carry out the action in a highly efficient manner. But we can expand our horizons and ask, Do animals maximize their caloric gains over entire days or weeks or seasons? A bird whose daily foraging behavior has been studied is the golden-winged sunbird (Figure 3) [257–259]. It is particularly well suited for examination of the daily economics of foraging because during some times of the year the birds feed primarily on the nectar of conspicuous flowers. Individual birds may be territorial at patches of these flowers. One can count the number of flowers on which the territorial resident depends for its daily ration of nectar. One can also measure the rate of nectar production per flower. This is done by covering a number of flowers (to prevent animals from removing the nectar). After a period of time, a glass micropipette is inserted into the flower and the nectar taken up by capillary action. The sample can then be sealed and taken to the lab to determine its sugar (and thus caloric) content. Nectar production in a territory per day can be estimated by multiplying the number of flowers by the quantity of nectar produced per flower per day. If a bird can monopolize these flowers, then it alone can enjoy the nectar produced by them.

The *costs* of foraging for a territorial sunbird include the energy spent flying from flower to flower and the calories lost when chasing intruders from the area. Because physiological studies have shown what it costs for a bird of the sunbird's size to perch, forage, and pursue, one can quantify the calories spent while feeding at a territory and compare these with the caloric gains from having a private foraging preserve. This information can be used to test predictions about competing hypotheses on the ultimate goal of a sunbird's daily foraging behavior. Are sunbirds trying to maximize the net energy gained per day (net gain = gross caloric intake − total daily caloric expenditures) or are they trying to minimize the calories they spend to secure the energy they need to survive the day?

The actual caloric costs and gains vary from individual to individual and depend on the size and quality of the territory and the number of intruders the area attracts. But, the average difference between calories gained and calories expended by a sunbird is estimated to be close to zero (−190

3 **Golden-winged sunbird** feeding at its nectar-producing food plants. This species is capable of both territorial and home range behavior.

calories). This result clearly supports the cost-minimization hypothesis (Table 1) [590]. Because these foraging studies were done in the nonbreeding season, the birds did not have to forage for their brood or build up energy reserves for reproduction. They merely needed to maintain themselves, and therefore it is not too surprising that their goal is to meet their daily energy requirements with a minimum of metabolic expenses each day.

The primary way to minimize calories spent per day is to spend as much time perching as possible. This is reflected in the decisions the birds make about the kinds of flowering patches they will defend. As Table 2 shows, a bird that owns a territory with flowers producing nectar at the rate of two microliters per bloom per day potentially saves four hours of foraging time if alternative sites are generating nectar at only half this rate. In caloric terms, the added time spent perching instead of foraging represents 2400 calories saved, from which must be subtracted the calories expended to defend the rich patch against intruders. Each hour of defense flight burns up 2000 more calories than would be spent if the bird had abandoned territoriality and were foraging nonaggressively elsewhere. An hour's defense would be worthwhile if the bird were holding a two-microliter area, while forcing other sunbirds to forage in one-microliter patches (net caloric gain = 2400 − 2000 = 400 calories gained). But territoriality becomes disadvantageous if the bird spends an hour chasing intruders to protect flowers producing at the three-microliter rate if other undefended patches were producing at the two-microliter rate (net loss = 780 − 2000 = −1220 calories).

If sunbirds have evolved the ability to minimize their energy costs, they should be sensitive to the relative rates of nectar production in various patches and the time required to deal with intruders. In fact, when nectar production is uniformly high, the birds are not territorial and will feed without aggression. But even during a single day, if nectar productivity in different patches begins to diverge, some individuals will begin to take up territorial residence in the rich areas. In addition, as the density of birds

Table 1
Comparison of Predicted and Observed Values for Various Aspects of Sunbird Territorial Behavior: A Test of Two Hypotheses

Hypothesis	Net Energy Gain (calories)	Flowers in Territory	Hours Spent		
			Foraging	*Defending*	*Sitting*
Energy maximization[a]	28,000–48,000	6300–9600	5.7–8.1	1.9–4.3	0
Cost minimization[a]	0	1540–1600	1.7–2.6	0.2–0.4	7.2–7.9
Observed[b]	−160	1600	2.4	0.28	7.3

Source: Pyke [590].
[a] A range of possible potential values is given because some parameters in the mathematical model must be estimated.
[b] Mean values are given.

Table 2
Energetic Costs and Benefits of Territoriality by Sunbirds under Different Conditions

Activities	Expenses/Hour for Different Activities (calories)
Resting on perch	400
Foraging for nectar	1000
Chasing intruders from territory	3000

Nectar Production (microliters·blossom^{-1} day^{-1})	Time Required to Collect Sufficient Food for One Day (hours)
1	8
2	4
3	2.7
4	2

Production Rate (microliters/blossom)			
Territory	*Alternative Undefended Site*	Time Spent Resting instead of Foraging (hours)	Calories Saved[a]
2	1	8 − 4 = 4	2400
3	2	4 − 2.7 = 1.3	780
4	4	2 − 2 = 0	0

Source: Gill and Wolf [257].
[a] For the bird that owns a particular territory, the calories saved for each hour spent resting instead of foraging = 1000 − 400 = 600 calories. Total calories saved = 600 × number of hours spent resting instead of foraging.

increases, so does the cost of repelling intruders. In response to increased caloric expenditures, birds tend first to contract their territories, reducing the area from which they repel conspecifics. If the rate of intrusion continues to rise, the birds will abandon territoriality altogether. Thus, sunbirds (and other species as well) have the capacity to switch from foraging in a home range to defending a territory, the choice depending upon variable ecological factors that influence the economic return of their behavior [257].

Optimal Foraging over Prolonged Periods

Behavioral ecologists have used sunbirds to test the hypothesis that animals should be territorial when this enables them to meet their daily caloric requirements most efficiently. The same hypothesis has been applied to the pied wagtail, a European songbird (Figure 4), which like the sunbirds defends feeding territories during the nonbreeding season [162].

4 **Pied wagtail at its nest.** The winter feeding territories of these birds are not often used as breeding territories in the summer. Photograph by Eric Hosking.

The wagtail, however, eats insects not nectar. In southern Britain its winter territories consist of about 600-meter stretches of river bank. Territory owners consume the aquatic insects that are constantly being washed up by the river. Wagtails, like sunbirds, have the ability to adjust their territories. When insect renewal rates fall within an area, the territory owner often temporarily joins flocks of nonterritorial birds that forage widely over the countryside. But when the food supply is high, a territorial bird will remain at its site and vigorously defend it against outsiders.

But there are two puzzling and unpredicted aspects of their behavior. First, residents sometimes tolerate another bird on their territory for a time [164]. Second, even on the best of days during Nicholas Davies's study in the winter of 1974, the food-collecting rate for a territorial wagtail never exceeded 20 items per minute, whereas flocking birds collected 21–29 prey each minute during the same period [162]. Examination of the feces of the birds (one of the less glamorous activities of a behavioral ecologist) showed that the size of the insects consumed by territorial and nonterritorial birds was the same.

The solution to the first puzzle comes from the discovery that owners tolerate a satellite only on days of high food abundance. This reduces the cost of having a satellite share one's territory. Moreover, on these days intruders are especially likely to invade the territory. The satellite assumes 20–50 percent of the defense of the area against these outsiders, and as a result the resident actually increases its feeding rate by associating with a satellite under these conditions (Figure 5).

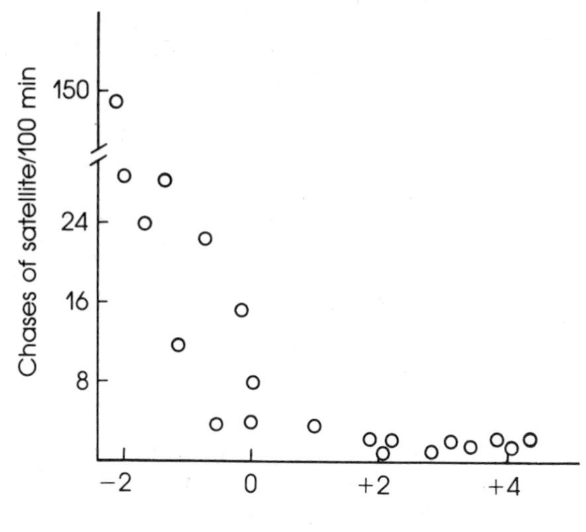

5 **Treatment of satellite** by territorial wagtail varies, depending on the economic value of having an assistant to repel other intruders. When satellites enable the owner to eat more food, they are tolerated. When satellites represent a drain on the territory's resources, owners chase them away.

The second paradox is less easily resolved. How can we account for the territorial birds' persistent return to areas that yield less prey than nondefended regions visited by flocking birds? Why bother to be territorial when the hunting is better elsewhere (see Chapter 8)? Davies suggests that when snow covers the meadows, the river edge may be the only place with a predictable and constant supply of food. To maintain its body weight, a wagtail must feed 90 percent of the time and collect about 7500 food items per day. A single day of starvation can kill it. Thus, territorial ownership may provide an element of insurance against exceptionally bad weather. Animals may consider not only the relative productivity of different areas, but also the reliability of production, in reaching a decision about whether to defend a foraging site or abandon it [162].

Constraints on Foraging Efficiency

A key ingredient of the optimality approach is the recognition that each strategy has its own mix of advantages and disadvantages. One foraging decision may maximize the average daily intake of food; another may minimize the risk of starvation on certain kinds of days. The optimal (fitness-maximizing) solution may be a compromise of some sort, like that exhibited by the territorial wagtails that adjust the time invested in protecting their "insurance policies" on any given day in relation to the productivity of their stretch of riverbank that day.

The importance of considering trade-offs between alternative strategies becomes especially clear when we examine the foraging behavior of species that face special dangers, such as predation, or other survival risks while collecting food. One example is the intertidal snail *Acanthina punctulata,* which feeds by drilling through the shell of its molluscan victims with a rasping organ and chemical secretions. This takes a long time; the thicker the shell, the longer the time it takes to drill through it. If a snail's ultimate goal were to maximize energy intake, it should select only prey with relatively thin vulnerable shells. It tends to make these choices during low tide. But as the tide turns, the snail becomes more and more likely to choose an "inferior" prey that will require a longer drilling time, an action that reduces the predator's net energy gain per unit time spent preparing food. By accepting an inferior prey as the tide comes in, however, the snail can move to a safe retreat in which it wedges itself against dangerous pounding waves. If it is not secure, a large wave might sweep it from its feeding area and batter it to pieces on the rocks. Therefore, the feeding behavior of *Acanthina* appears to represent a compromise between maximizing energy gain and minimizing the risk of death from wave action [508].

The Risk of Predation and Foraging Behavior

Food-collecting animals are often themselves collected by predators. This threat to survival, like the risk posed by a severe snowfall or a mighty wave, can be expected to modify the foraging behavior of a prey species. This expectation has been examined in a study of gray-eyed juncos [124], a small seed-eating bird that forages on the ground in scrubby areas in which lurk sharp-shinned and Cooper's hawks. Both hawks are major consumers of small birds. If a junco tried to eat seeds as fast as possible, it would not look up from the ground from time to time as it foraged for food. But juncos interrupt their visual search for seeds as many as 15 times *per minute*; they look up quickly and then resume hunting for seeds. The risk of predation favors animals that sacrifice short-term food intake in favor of long-term protection against surprise attack.

Juncos also form flocks (despite the competition for seeds that results) in order to regain some of the lost foraging time that must be devoted to surveillance for predators. The larger the flock, the more eyes there are to scan for predators and the less time any one individual spends in this activity (Figure 6). The savings in time is offset to some extent by increased aggression among flock members. The larger the flock, the greater the chance an individual has of becoming involved in a clash and this, coupled with increased competition for food, imposes an upper limit on flock size for any one set of conditions. Both the number of birds in a flock and the scanning time per individual are, however, adjustable. When Tom Caraco flew a tame hawk over his study site a number of times, the juncos devoted more of their time to scanning, reducing feeding rate in response to an increased perception of danger from predators. Although this diminished the daily intake of food, it also increased the chance that a junco would live to forage and reproduce on other days [124].

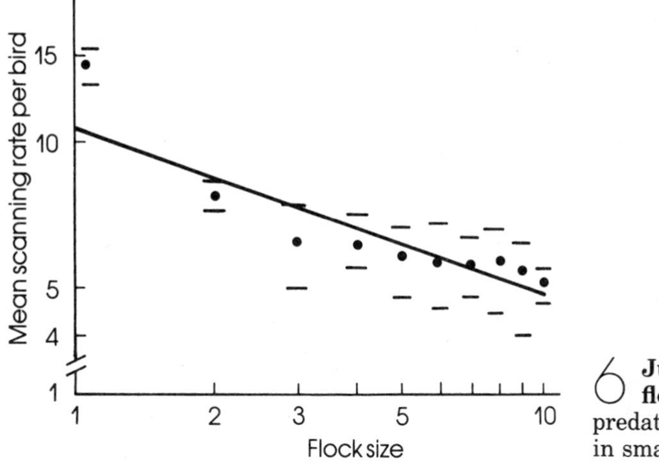

6 **Juncos in large flocks** look up less for predators than individuals in small flocks.

Nutritional Constraints on Foraging

There is still another reason why the simple prediction that animals should maximize their caloric intake over a time period will often fail. Individuals almost always require more than energy alone from their diets if they are to reproduce successfully. Northwestern crows do not eat whelks all day long. In fact, after consuming one or two they generally leave the beach and go off to hunt a different food elsewhere. Perhaps the advantages of having a varied diet may affect their selection of food. This has been documented for a lycosid jumping spider, whose diet is significantly more heterogeneous than would be predicted if energy maximization were the only function of its feeding behavior. Matthew Greenstone analyzed some of the principal food items in the spider's diet for the presence of nine essential amino acids (protein building blocks) that the animal cannot synthesize but must acquire from its foods [281]. He demonstrated great variation in the concentration of particular amino acids from prey to prey. It is probable that the need for an amino acid balance acts as a dietary constraint for the lycosid. This favors animals that vary their food intake rather than concentrating on the single food item that returns the most calories per unit time spent foraging. A prediction that follows from this argument is that lycosid spiders that are given a varied diet will outreproduce those offered a single prey species ad libitum; and they do.

The same kind of constraint operates on the food preferences of moose. All other things being equal, one would expect a moose to select a diet that maximized its net energy intake, because so much food is needed just to maintain its body. Gary Belovsky tested this prediction by analyzing the energy content of plant foods available to moose in Isle Royale National Park in Michigan [60]. If a moose were to maximize energy intake in the

7 **Tassel-eared squirrels** forage preferentially for pine needles low in toxins. Photograph by R. C. Farentinos.

summer, it should feed exclusively on the leaves of certain deciduous trees and assorted weeds. In fact, although moose do consume prodigious quantities of these high-calorie plants in the park, they also eat a great many aquatic plants, which are relatively low in energy per unit weight. But aquatic plants contain much higher sodium concentrations than terrestrial ones. Belovsky hypothesized that this critical element was a limiting resource for the moose and that the need to consume some minimum amount per day constrained the moose's food selection. He constructed a mathematical model based on this assumption and from it predicted what a moose should eat if it were trying (unconsciously) to consume as many calories as possible, subject to the constraint of ingesting a certain amount of sodium. The predicted diet closely matched the observed one. Furthermore, as one would expect, when moose feed upon aquatic plants they select those that are highest in sodium content. This selectiveness reduces the amount of time they must spend eating low-calorie foods and enables them to return sooner to the forest, where energy-rich plants are found.

Another kind of nutritional constraint on foraging behavior is imposed by the defensive properties of certain potential foods. Tassel-eared squirrels live in ponderosa pine forests (Figure 7). The pines supply their winter food, which consists largely of the cortical tissue from the twigs that support clusters of pine needles. The needles and bark are discarded and the squirrel then gnaws off the layers of cambium and phloem around the twig. Given the great number of trees within the home range of any squirrel, one might think that individuals would choose pines that were relatively lightly cropped because it would be easier to collect twigs there. But instead, the squirrels avoid many trees in order to feed at certain heavily defoliated trees. They can even discriminate between the twigs collected from these preferred trees and those of other ponderosas. The squirrels invest increased travel and search effort for a lower return per foraging time in order to collect food from a particular tree because not all trees have equally nutritious twigs. The preferred trees have much lower concentrations of a monoterpene (α-pinene) than the ignored trees. Terpenes are toxic substances for most animals, and the squirrels avoid them—and so probably avoid poisoning—even though this requires the added expenditures involved in getting food from a heavily cropped tree [231].

The conclusion that emerges from these and similar studies is that optimal foraging decisions are often influenced by considerations other than the number of calories that can be ingested during a period of feeding. Although some animals, like the crow, may be energy maximizers in the short run, others cannot do this because of the nature of their environments. Tests of optimality predictions have been helpful in showing when energy maximization is *not* the ultimate goal of animal foraging behavior. Modified optimality models will almost certainly become more precisely predictive in the future and so contribute to a further understanding of the effects of multiple constraints on foraging efficiency.

Has Competition Shaped the Diets of Animal Species?

We now turn from an analysis of the nutritional returns from an animal's diet to deal with the question, Why do species differ in the things they eat? There is practically no organic item that does not serve as food for at least one species, and what is the *pièce de résistance* for one species is unlikely to be so highly regarded by another. Because humans are omnivorous, we can readily comprehend how crows might eat whelks or sunbirds drink nectar, but consider the butterflies that travel with columns of army ants in order to feed on the droppings of the antbirds that also accompany the army ants [600]. There is a scorpion that specializes in finding, killing, and eating a species of burrowing cockroach that hides deep under the surface of sand dunes [109]. A marine amphipod that is only found on the tips of spines of the giant red sea urchin feeds on bits of organic debris swept to it on underwater currents. It captures this food with specially modified hairy antennae, which form a little net when the animal holds its antennae in an upright V. When the animal defecates, it saves its feces and uses them to construct a soft flexible rod on which it perches. In the summer months, algal diatoms grow in great profusion on the dung rods, and the amphipods graze (contentedly?) upon them [496]. Leaf-cutter ants are yet another animal with a remarkable diet. These animals search for flowers and leaves, which they cut into pieces and transport to an underground nest (Figure 8). But they do not eat the plants themselves; instead, they use them as garden material to grow certain fungi that convert molecules of leaf sugar (cellulose) into other sugars that ants can digest [593, 761].

The list of gastronomic novelties could be expanded almost indefinitely,

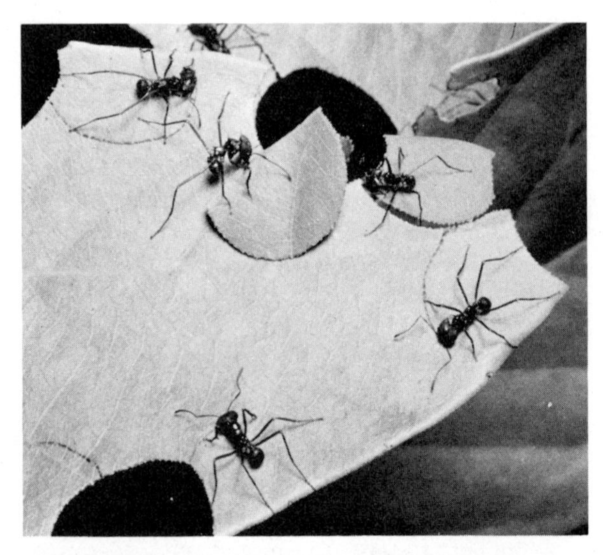

8 **Leaf-cutter ants.** A group of leaf cutters of the genus *Atta* are clipping leaves in Peru (*top*). Photograph by E. S. Ross. A leaf-cutter queen is surrounded by workers tending the spongelike fungal garden in the foreground (*bottom*). A large soldier appears in the background above the queen. Photograph by Neal A. Weber.

but the point is that there appears to have been an astounding radiation in diets, with each animal species diverging from its ancestors to consume things or combinations of foods not eaten by other animals. This point is further reinforced by an examination of some closely related clusters of species, each member of which has its own specialty. The classic example involves Darwin's finches [81, 402]. These superficially drab sparrowlike birds that occur only on the Galapagos islands off Ecuador feed on everything from tiny insects to huge, thick-hulled seeds (Figure 9). There are ground-feeding seed eaters with powerful crushing beaks, cactus consumers with beaks adapted for probing and slicing, tree-bud and leaf eaters with cutting beaks, and insect eaters with delicate forcepslike bills. When more than one of these species exploits the same category of foods, as in the case of the seed-eating ground finches, their beaks are specialized for one component of the total resource (e.g., large, medium, and small bills capable of crushing large, medium, and small seeds, respectively).

The traditional explanation for this and other cases of dietary divergence is that competition among members of different species conferred a reproductive advantage on those individuals that happened to avoid or reduce their overlap in food preferences. By exploiting food supplies not tapped by another species, an individual should have access to more energy and therefore have more opportunities to produce young than those members of its species struggling to secure the same resources required by another species. Because the diets of even closely related species (like the Darwin's finches) are different, many species can coexist in the same region, so the argument goes, each supported by its own food supplies. A very large number of cases of interspecific differences in foods taken or foraging tech-

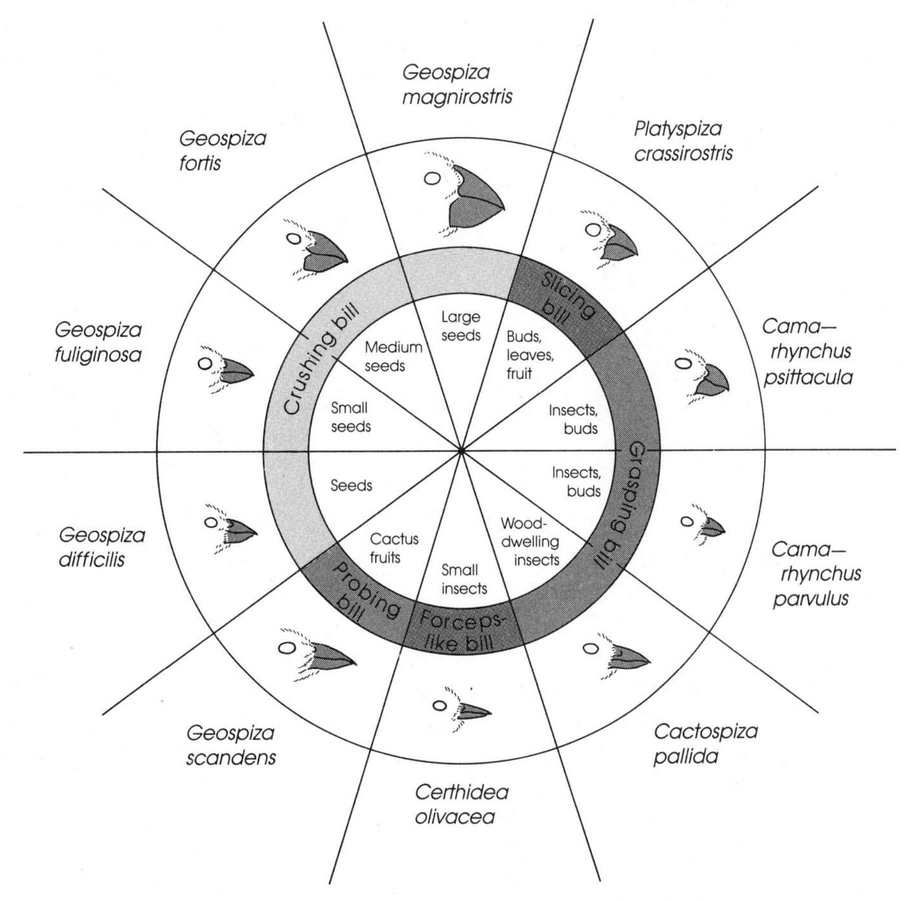

9 **Adaptive radiation** in bill shape and diet has taken place during the evolution of Darwin's finches.

niques employed have been interpreted as adaptations that reduce competition among species and permit the coexistence of animals that would otherwise displace one another [e.g., 81, 521].

More recently, however, a number of ecologists have challenged the traditional view [639, 784]. A key criticism can be phrased as a question: What is the probability that by chance alone two species would happen to share identical or very similar diets? Given that different species have separate ancestry and distinct genetic makeups from the moment of their reproductive isolation from one another, the probability must be low indeed. The different diets of animals could therefore be an incidental effect of the accidents and other pressures associated with their separate histories and have little or nothing to do with a reduction of ecological overlap in food demands.

Rodents, Seeds, and Competition

To illustrate the point that competition need not be responsible for different food preferences or foraging behavior among species, let us examine the case of kangaroo rats and pocket mice. There are several species of each of these seed-eating rodents that live in the same areas in the southwestern United States. They differ, however, in size, with the smallest kangaroo rats weighing at least three times as much as the largest pocket mouse. If the example of the Darwin's finches were to be our guide, we might expect that the larger kangaroo rats would take larger seeds and the smaller pocket mice would gather smaller seeds, thus apportioning food resources between them and avoiding direct feeding competition.

The first study on the diets of these animals in areas of geographic overlap reported that there was segregation between species in seed preferences [105]. This was taken to support the competition–avoidance hypothesis. But later research documented that in some areas at least the two groups of species were consuming exactly the same kinds of seeds [667]. These studies offered two other hypotheses to account for the coexistence of kangaroo rats and pocket mice. First, the two species appear to have different microhabitat preferences: the pocket mice staying close to cover, the kangaroo rats living in more open, sparsely vegetated locations in the same general area [614]. The spatial segregation of the species would mean that they would not come into contact frequently and therefore would not often compete directly for food. Second, the two species have different modes of foraging. Kangaroo rats have large powerful hindlegs (they really do look something like miniature kangaroos; Figure 10). The hindlegs propel them rapidly through their nocturnal world. Pocket mice have much shorter legs and travel at much slower speeds. Thus, kangaroo rats can afford to hunt for the rare clusters of wind-blown seeds in the desert, traveling economically in search of these patches, whereas pocket mice are gleaners that pick up each seed they encounter [603]. Once again the species might avoid competition because, although they eat the same seeds, they forage for them in different ways and so should utilize different sources.

But are the differences in body size, locomotion, microhabitat selection, and foraging tactics the product of selection for avoidance of food competition? One might argue that it is at least possible that the two kinds of animals first evolved different antipredator tactics: the kangaroo rats relying on rapid escape dashes and the pocket mice hiding in dense cover to avoid detection by their enemies. These divergent antipredator adaptations would impose constraints on the foraging behavior of the two kinds of rodents with the associated, but incidental, result that they do not collect food in an identical manner.

How then are we to resolve the question of competition's role in the evolution of foraging differences? Testable predictions are always helpful, and some have been forthcoming. For example, if two species really are in current competition with one another, then the removal of species A should result in an expansion of the foraging range and numbers of species B (and vice versa). A direct way to test this hypothesis is by experimentally removing one species and observing whether its suspected competitor responds in the predicted manner.

A test of COMPETITIVE RELEASE has been done with kangaroo rats and pocket mice in the following manner [527]. Eight square plots (50 meters × 50 meters) were fenced off; four were assigned to be controls and four were experimental exclosures from which all the resident kangaroo rats were removed. In both kinds of plots, holes were placed in the fence at intervals; but in the controls the holes were big enough for kangaroo rats to pass through, whereas in the experimental plots the passages were large enough only for the smaller pocket mice and other small rodents. For three years the populations within the exclosures were monitored (Figure 11). The removal of the large kangaroo rats did result eventually in a small, but significant, increase in the number of smaller seed-eating rodents of which the pocket mice were a component. In contrast, there was no evidence of competitive release for the small *omnivorous* rodents that should not be primary competitors with kangaroo rats.

10 **Merriam's kangaroo rat** is a seed-eating rodent that uses its large hindlegs to travel long distances in search of rich pockets of seeds. Photograph by George Gamboa.

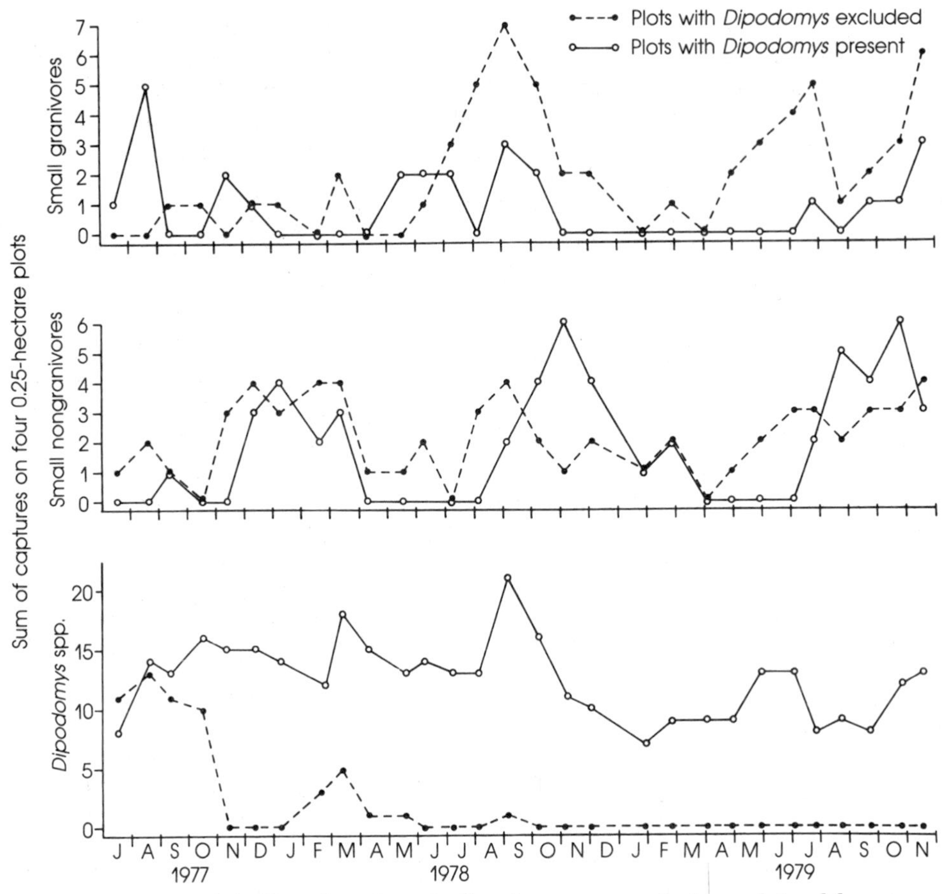

11 **Results of excluding kangaroo rats** from plots of desert. The total number of small grain-eating rodents captured in kangaroo rat exclosures (dashed line) increased over time relative to total number captured from plots from which the seed-eating *Dipodomys* were not removed (solid line). Removal of the kangaroo rats had no effect on the numbers of small omnivorous rodents.

This experiment indicates that some species may be in direct competition for a limited food resource. But many issues in the competition debate continue to be explored—a classic example of a topic that was once thought to be quite well understood but that now seems much less clear. In a way it is exciting that such a fundamental question—why do different species eat different foods—can still generate difficult puzzles for ecologists.

How to Capture and Consume Difficult Foods

Whether or not competition has been responsible for the radiation of animal diets, there is no question that the radiation has been impressive. Nothing is safe, not even prey that are extraordinarily well hidden or elusive, dangerous or poisonous. For example, the larvae of one bruchid beetle species eat only the seeds of a particular tropical tree. The seeds of this tree are laced with large amounts of a substance (L-canavanine) that kills most insects (because it gets incorporated in their proteins in place of the correct amino acid, with the result that defective enzymes are produced). The bruchid has a special enzyme that converts canavanine to ammonia, which it may then use as a source of nitrogen for the construction of molecules it needs [613].

The ability of the bruchid to eat poisonous foods is based on enzyme systems within its digestive tract. Other animals use behavioral means to detoxify certain foods. The Great Basin kangaroo rat has specially shaped incisors that shave off the salt-impregnated layer from the leaves of saltbush (Figure 12). It can then safely eat the nutritious interior tissues of the leaves [374].

Some mice and skunks are able to feed on tenebrionid beetles that possess abdominal glands that discharge irritating and poisonous chemicals in the face of an assailant. The grasshopper mouse counters this defensive maneuver by quickly grabbing the bettle before it fires and stuffing the tip of its abdomen into the sand. It then proceeds to eat its way through the insect from the head down [666]. Skunks use a different tactic, rolling the beetle vigorously on the ground with their forepaws until the glands are completely emptied, after which the then defenseless tenebrionid can be eaten with impunity [205].

There are some living things that actually gain an advantage by being eaten (partially), but even so may make it difficult for their potential consumers. Many plants participate in a symbiotic relationship with a pollinator species, providing nectar and pollen food for an animal that will in turn transport pollen to another plant. For this to be beneficial to the donor, the other plant should be a member of the same species. Flowers have evolved a host of characteristics that make it difficult for all but one (or a few) major pollinators to remove their nectar and pollen [506]. It is

12 **Great Basin kangaroo rat** feeds on saltbush at night. Photograph by G. J. Kenagy.

particularly important to foil ants and small flies, which are poor pollinators because of their limited mobility. Therefore, many flowers guard the approaches to the nectar with hairs, scales, or latex-containing tubes that burst if gouged by an ant's claws. Nectaries may be hidden deep in tubes, where they can be reached only by a long-tongued pollinator. The food rewards may be concealed in a capsule made of the flower petals. The capsules of Scotch broom and alfalfa will spring open, spraying pollen everywhere, but only if a sufficiently heavy insect alights upon the flower. Scotch broom, therefore, is profitable only for bumblebees that weigh enough to trip open the flowers (Figure 13). B. J. D. Meeuse writes that as the bees go "from one flower to another at breakneck speed, they cause explosions all over the place, the air is full of tiny pollen clouds, and the whole scene is strongly reminiscent of a Civil War picture or an old western, full of gunsmoke" [506].

The variation among flowering species in appearance and in the location of nectar and pollen encourages the pollinators that can reach the food rewards to be FLOWER-CONSTANT (i.e., to visit a series of individuals of the same species). In order to locate food-containing plants and to remove pollen and nectar efficiently, it often pays a pollinator to specialize. Bumblebees

13 **Coadaptation of plants and their pollinators.** (A) A bumblebee weighs enough to trip open the locked flower of Scotch broom. (B) The hawkmoth does not have to land to extract nectar from the delicately suspended Turk's cap lily flowers. Pollen adheres to its proboscis as it feeds. (C) To remove the nectar from a monkshood blossom, a bumblebee must learn through trial and error how to enter the flower and find the nectar.

and their food plants are good examples of this [315, 316]. Although any one bee could potentially forage at a number of different species of plants, individuals typically restrict themselves to a single type of flower at a time (Figure 14). In this way they can search for one kind of visual cue associated with a productive plant and acquire the special skills needed for the extraction of resources from its flowers. Bumblebees that have not had experience with monkshood flowers, for example, will often try to enter from the top; and even if they find the entrance, they may probe for nectar fruitlessly among the anthers. With experience, a bumblebee learns that it must enter the flower from the bottom and find the long, thin tube hidden within that contains a drop of nectar at the top, which can be reached with the bee's long tongue (Figure 13). Only after it has learned this can a bumblebee forage efficiently at monkshood. Thus, it is profitable for the worker to specialize on one species rather than cope with the different obstacles to food rewards in a dozen or so food plants on the same foraging trip.

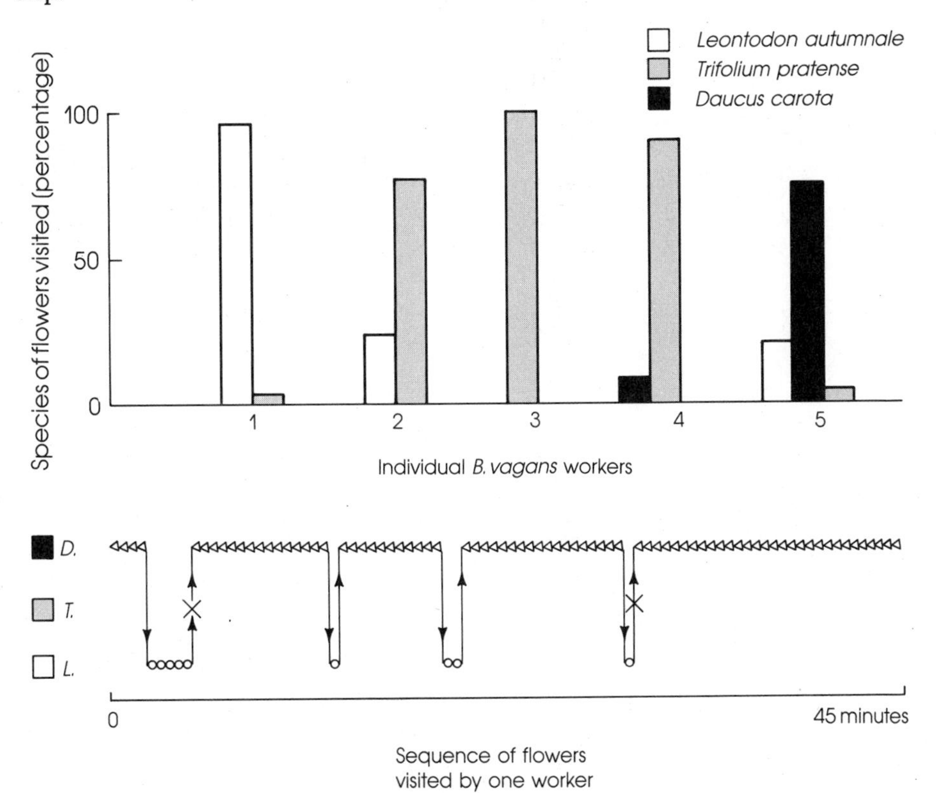

14 **Flower constancy** by individual bumblebee workers. (*Top*) Strong flower preferences are shown by five different workers of *B. vagans.* (*Bottom*) The sequence of visits to three flower species by one worker illustrates the strong tendency of worker bees to visit a series of flowers of one species before changing to another species.

How To Use Lures To Capture Prey

A bumblebee must visit flowering plants in order to extract their nectar. Other animals use lures to entice their food to come to them. These animals, which range from spiders to giant turtles, are almost without exception ambush specialists that lie in wait for mobile, elusive prey, which are then lured in for the kill. The very large majority of these predators use lures that consist of body parts or released scents that resemble a food item and so entice foraging prey to their doom (Table 3). For example, the alligator turtle's tongue has a projection that closely resembles a worm and this body part twitches about in a wormlike manner that may literally draw a hunting minnow into the mouth of the turtle. An angler fish has a bizarre projection from the front of its head, a projection that consists of a thin rod from which is suspended a bit of tissue with an uncanny resemblance to a small fish. This bait is waved about by the anglerfish, who explosively engulfs any small predatory fish that comes to investigate [577]. Another kind of tactic is employed by species of fishes that have light organs on various portions of their bodies; these light organs may, for reasons that are not entirely clear, attract other foraging (edible) creatures to them (Figure 15).

Still other predators employ mimetic sexual lures, rather than feeding lures, to secure their prey. One group that does this are female *Photuris* fireflies. James Lloyd has described these insects as the ultimate *femmes fatales* [435, 437]. They, like flashlight fishes, possess luminescent organs. A *Photuris* female can use the light on the tip of her abdomen to signal her sexual receptivity to a male of her own species *and* to mimic the receptivity of other fireflies in the genus *Photinus*. A searching male *Photinus* flies

Table 3
Animals That Use Lures to Attract Prey

Species	Type of Lure	Victim	
Alligator turtle	Food lure: "worm"	Fish	[781]
Antennarius angler fish	Food lure: "minnow"	Fish	[577]
Bothrops snake	Food lure: tail tip resembles an insect larva	Rodents	[279]
Zelus reduviid bug	Food lure: legs covered with "sugary droplets"	Insects	[200]
Ptilochraceus reduviid bug	Food lure: hairs on thorax carry attractive secretion	Ants	[793]
Hydrozoan siphonophore	Food lure: enlarged tips of stinging tentacles resemble edible zooplankton	Copepods	[588]
Deep sea angler fish	Food lure: luminescent organ	Crustaceans?	[781]
Flashlight fish	Food lure: luminescent organ	Fish?	[497]
Photuris firefly	Sexual lure: luminescent organ	*Photinus* males	[435]
Bola spider	Sexual lure: sex pheromone	Male moths	[196]

15 **Lure-using anglerfish.** An *Antennarius* anglerfish that resembles a lump of coral reef with a fishlike appendage at the top of its head (*top*). The anglerfish makes its lure move, the better to attract its victims (*bottom*). Photographs by David Grobecker.

through the summer night, producing a complex flash pattern (Figure 21, Chapter 13). If he sees a female's answering flash with the correct timing, he will quickly approach and attempt to mate. If he has been fooled by a *Photuris,* he will not live to regret it (Figure 16).

The bola spider's mimicry is equally devious. The spider releases a scent that mimics the sex pheromone of certain female moths. Males of these moths locate mates by tracking the scent plume. But a moth that flies upwind to a bola spider is likely to be caught when the spider swings its bola, a sticky globule on the end of a silken thread. If the blob hits the moth, the spider reels it in and eats it [196].

The lure-using animals are superb examples of species that use their victim's programmed responses to key stimuli to the victim's disadvantage, thereby attracting food that would otherwise be too mobile, elusive, or well-protected to be easily captured.

Tool-Using Animals

Some species use portions of their bodies as special aids to secure elusive prey; others employ inanimate objects as extensions of their body to manipulate or subdue difficult-to-capture prey. There is a tendency to think that because humans are intelligent and use tools, other tool-using creatures must also be exemplars of intelligence and adaptability. This is not true. For example, ant-lions (larval Neuroptera) and worm-lions (fly larvae) have convergently evolved trap-building and tool-using behavior (Figure 17) [713]. These animals construct pits in sandy areas frequented

16 ***Photuris* femme fatale** consumes a male *Photinus* firefly that has responded to her light flashes, which mimic the signal given by a sexually receptive female of his species. Photograph by James E. Lloyd.

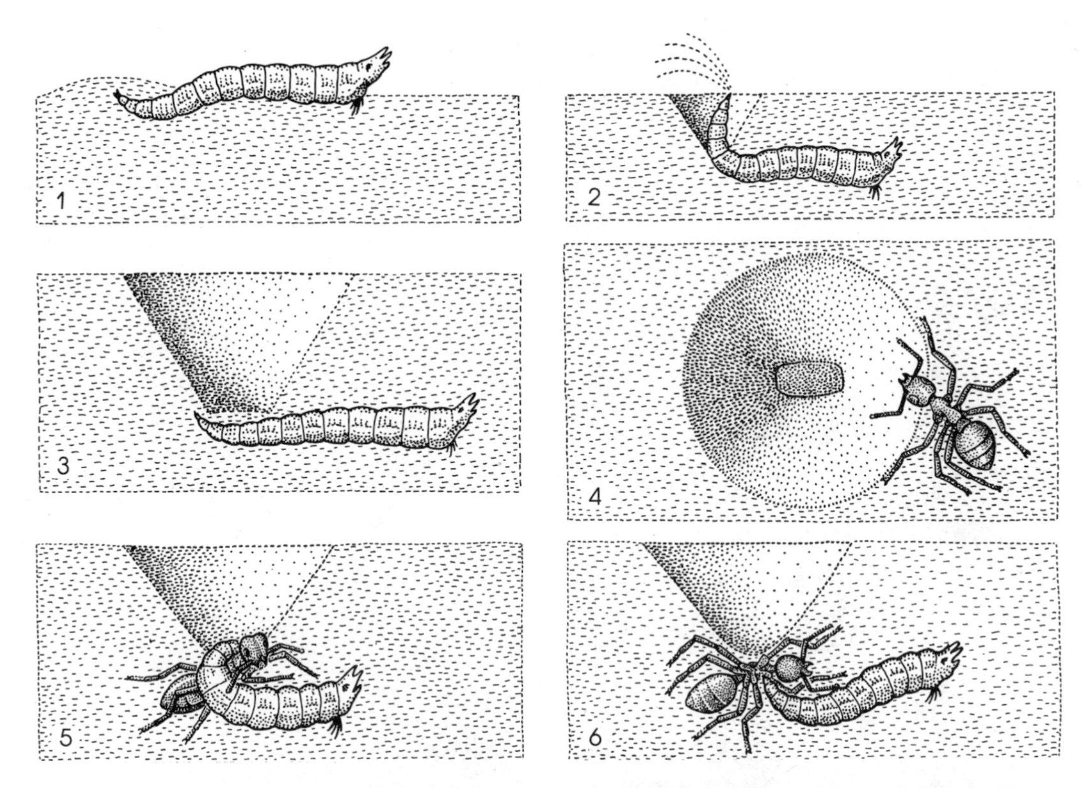

17 **Trap-building activities of the worm-lion,** a fly larva, *Vermileo comstocki.* (1, 2) The larva makes a pit. (3) The predator lies in wait. (4) An ant approaches the trap. (5) The ant slides into the pit and is seized. (6) The now-paralyzed ant is hauled under the surface of the sand and its body fluids are sucked out by the worm-lion.

by ants. When an ant topples over the lip of the trap and begins to slide toward the bottom where the predator awaits her, the ant-lion or worm-lion will speed the descent of its victim by tossing sand grains at it. The same movement is used to build traps and to clean them of debris. This mechanically performed tool-using behavior appears to increase the effectiveness of the trap.

The other animals that make good use of tools in feeding are the archer fish [455], the woodpecker finch (Figure 18) [402], the Egyptian vulture [734], the sea otter [298], the chimpanzee [734], and us. The archer fish maneuvers close to the surface of water in mangrove swamps and spits out water droplets in an accurate stream capable of striking insects and spiders on overhanging vegetation as much as four feet away. This knocks the prey into the water, where they can be easily picked up. Although the fish can leap some distance into the air in pursuit of these food items, water spitting may be energetically more economical for this purpose; and, as presently

18 **Tool-using animals.** A woodpecker finch of the Galapagos Islands is about to place a spine tool into a tunnel made by a wood-boring insect (*top left*). Photograph by Irenäus Eibl-Eibesfeldt. An Egyptian vulture is about to hurl a rock at an ostrich egg (*top right*). Photograph by Hugo and Jane van Lawick-Goodall. A sea otter is about to smash a clam, which it is holding in its paws, against another clam that is resting on the otter's chest (*bottom left*). Photograph by Karl W. Kenyon. A chimpanzee inserts a grass stem into a termite gallery (*bottom right*). Photograph by Leanne Nash.

evolved, the use of water droplets permits the fish to capture prey out of reach of its most acrobatic leaps.

The woodpecker finch works with a thorn or twig probe to tease wood-dwelling larvae out of chambers and cracks in wood where the bird's beak will not fit. The Egyptian vulture picks up rocks and drops them on ostrich eggs, sometimes breaking the thick shell. It then consumes the spilled contents. The sea otter collects mussels and other hard-shelled mollusks and places them on a rock anvil lying on the otter's chest. It cracks open the shellfish either by whacking them against the anvil or by striking them with another rock. The chimpanzee employs the most varied tool-using behavior of a nonhuman animal. It may place large sticks and branches into ant nests and eat the ants that cling to the stick. Or it may make a delicate probe, stripping a twig or vine of its leaves, and then insert it into an ant or termite nest, eating those insects unfortunate enough to grasp the foreign object with their feet or mandibles. Modifying an object so that it will be a better tool has been observed only in chimpanzees, humans, and woodpecker finches (these birds sometimes will shorten a too-long probe).

What ecological pressures do tool-using species share that might lead to the evolution of the behavior? In each case, by using a tool an individual gains access to food that is in one way or another well protected and relatively inaccessible to conspecifics without tools. Ants can scramble out of pits; insects perched above the water or in tiny tunnels in wood or in hard-packed termite and ant nests are difficult prey to capture for fish, finches, and chimpanzees, respectively, and hard-shelled mollusks and ostrich eggs are not easily opened by otters and Egyptian vultures. Manipulation of a tool means more calories for less energy expended for all these creatures.

If one wishes to understand something about the evolution of tool-using by humans, the fact that chimpanzees employ tools is important, but primarily because it indicates a potential for this behavior in the primate evolutionary line. It is more interesting that the other vertebrates that use tools for a substantial amount of food collecting—woodpecker finches, sea otters, and human beings—have all invaded niches that are unique for their phylogenetic groups [3]. If one accepts competition theory, one could hypothesize that the use of tools may have originally compensated in part for the absence of the biological equipment necessary for the direct exploitation of wood-dwelling larvae, hard-shelled mollusks, and moderately large mammals. This could have enabled the original tool-users to gain an advantage over others in their species in the competition with other species for similar foods.

Social Prey Capture

A very few animals use tools to augment their capacity to capture difficult foods. Many more species forage socially with the apparent goal of overcoming special aspects of prey resistance. Some competing hypotheses to account for group feeding by gulls were listed in Chapter 7. One suggestion made there was that gulls might counteract the escape maneuvers of schooling fishes by plunging en masse into groups of their prey. When threatened by predators, many schooling fishes use two responses that make it difficult for a solo enemy to capture members of the school

[569]. The first tactic is the "fountain effect" in which a pursued group splits in half and turns back to circle around the attacker, reforming as a single school behind the predator. The second is "flash expansion" in which the members of a school that is under attack appears to explode outward away from the onrushing predator. One counter to these tactics employed by some predators is to forage in coordinated groups. This has evolved independently in giant bluefin tuna (a fish), dolphins (marine mammals), and white pelicans (birds). The hunters form a bowed line and move in unison, driving the prey before them. The schools cannot divide and move behind any one member of the line because of its neighbors. When a sufficient number of prey have been assembled in the center of the living net, the hunters sweep in on them. At this time "flash expansion" of the school is likely to lead an individual prey closer to the mouth of an enemy rather than away to safety (Figure 19).

The formation of driving lines of predators requires sophisticated communication and cooperation among these aquatic predators. The same abilities have evolved in certain terrestrial social hunters. For example, groups of lionesses sometimes form a line parallel to a herd of prey. After they have crept close to their prey, some other lionesses may circle out and run conspicuously toward the herd at an angle that sends the game animals dashing toward the lionesses waiting in ambush. When the predators leap from their hiding places, panic ensues (understandably) and some prey may in running from one predator be captured by another [635].

Cooperative hunting by terrestrial carnivores has evolved independently in three families: cats (Felidae), dogs (Canidae), and hyenas (Hyaenidae). In each case, the social cats, dogs, and hyenas take prey that weigh up to 6–12 times as much as any one adult hunter [397, 505, 635]. The correlation between cooperative social hunting and the capture of unusually large prey suggests that for these unrelated animals, a primary function of group

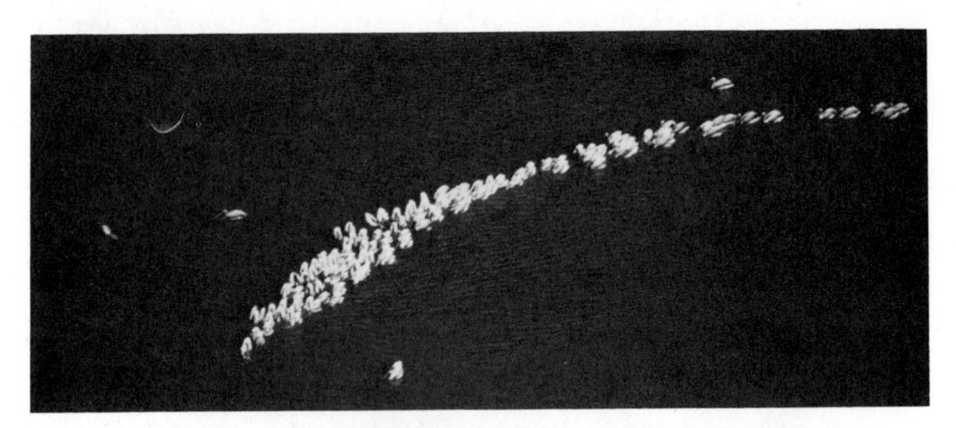

19 **Social foraging by white pelicans.** The birds cooperate to foil the defensive maneuvers of their prey—schooling fish. Photograph by Ralph and Betty Ann Schreiber.

foraging is overcoming the defenses of large herbivores. Because of their size, these creatures are difficult for most predators to subdue. They therefore offer a relatively untapped protein source for a predator that can kill them; but they are wary, powerful, and potentially dangerous. A giraffe can crush the skull of a lion with one kick of its massive leg, and a moose can do the same to a wolf. But through cooperation, the social predators confound the defenses of their large prey. When one lion or hunting dog closes on an adult zebra or wildebeest, others often quickly join, overwhelming the prey and ending any chance of effective defense (Figure 20). Moreover, by living and foraging in groups, the social carnivores are able to defend territories, protecting their cubs and kills from others that would be dangerous rivals.

The fact that these four unrelated carnivores share so many traits in common (Table 4) is testimony to the strength of convergent selection for adaptations that lead to successful exploitation of big game. Territoriality, boundary marking, and intolerance toward outsiders helps the clan, pride, or pack preserve the area where the integrated group can enjoy the benefits of communal hunting without interference from competitors.

Diversity in Social Feeding by Ants

One other way in which to test the hypothesis that group foraging is associated with the collection of large and difficult-to-capture prey is through a comparative study of ants. The entire spectrum of foraging group sizes is represented among these insects, from species that gather

20 **Cooperative behavior of wild dogs.** A pack pulls down a wildebeest in a communal kill. Through cooperation with others, individual wild dogs capture prey that a solitary animal could not kill. Photograph by Norman Myers.

Table 4
Behavioral Convergence in Four Species of Social Carnivores

	Social Carnivores				
Behavior	*Lion*	*Wild Dog*	*Wolf*	*Hyena*	**Solitary Carnivores**
Cooperative social hunting	X	X	X	X	No
Capture of big game	X	X	X	X	No
Sharing of kills	X	X	X	X	With offspring only
Group protection of kills	X	X	X	X	No
Territoriality	X[a]	No[b]	?	X[a]	Rare
Scent marking of hunting area	X	X	X	X	Some species
Killing of competitors including own species	X	?	X	X	Rare
Intragroup tolerance	X	X	X	X	—

Sources: Kruuk [397], Mech [505], Schaller [635], and van Lawick and van Lawick-Goodall [733].
[a] Pronounced only when lions form prides and hyenas form clans during times of food abundance.
[b] Defense of breeding area but not total hunting range.

food as isolated individuals to others that form huge groups that cooperate in the collection of food. We have already mentioned the North African ant *Cataglyphis bicolor,* which wanders about on its own gathering seeds and carrion [763]. It is typical of solitary foraging ants that they focus their attention on small, immobile items that one ant can easily collect and transport to the nest. Other species, like the *Leptothorax* ants, sometimes cooperate to a small degree, forming groups of two [329, 517]. A worker that has found a relatively large, but immobile, food source (a chunk of dead insect or a colony of aphids to be milked, for example) will return to the nest and regurgitate a food droplet to her fellow colony mates. Then she turns around, raises her abdomen in the air, extrudes the stinger, and releases a chemical signal. Another worker may approach her and touch her body with her antennae. The leader starts off with the follower close behind, running in tandem, all the way to the food site (Figure 21).

The species that are most similar ecologically to the social carnivores and the cooperative fish hunters are the ants that practice MASS RECRUITMENT to a food source (Figure 22) [793]. Army ants, having found a large

21 **Tandem running in ants.** The recruited ant places her head near the tip of the abdomen of the recruiter and follows her to food or some other item of interest. Photograph by Michael Möglich.

mobile living insect capable of defending itself against one or two attackers, will lay an odor trail from the prey to the main body of workers streaming through the forest. This draws large numbers of individuals to the site, where they cooperate in subduing and transporting the prey. Similarly, the famous fire ants that have become established in the southern United States are able to prey upon large insects that happen to pass close to a colony because of mass recruitment. An ant that discovers a potential victim rushes back to her nest, laying an ephemeral odor trail that is gone in less than two minutes. The pheromone arouses other workers, who hurry out along it; if they are excited by the discovery, they too run back, applying more trail pheromone. This can lead to an exponential increase of individuals at the food source. Because the trail fades so quickly, it does not provide the precision possible with tandem running methods or with more durable chemical trails. But because their prey may have moved a short distance from the point of initial discovery, the very imprecision of the fire ants' recruitment system may be advantageous. The ability to attract a swarm of workers in a very short period of time to the general area where a prey was found increases the probability that the prey will not escape and can be quickly killed, sectioned, and brought back to the colony in pieces.

22 **Cooperation in prey capture by ants.** A group of weaver ants are returning to their home with a large ant they have killed. Photograph by Bert Hölldobler.

A somewhat similar system is practiced by another ant, *Leptogenys ocillifera* [475]. This species has permanent "highways" in areas with a high density of earthworms and termites. Workers search near the main trails for freshly disturbed soil. Upon finding such a site, the discoverer hurries back along the main trail to the colony, laying an alerting trail. Many workers leave the colony and follow the trail back to the discovery site, where they cooperate in either digging up and dismembering a worm or collecting and killing termites.

Although many species of ants use chemical signals to recruit fellow workers, they do so in a variety of ways that reflect the utility of different patterns of recruitment for foods that vary in their mobility, size, dispersion, and distance from the nest. This divergence in ant behavior supports the argument that group cooperation is favored in foraging when the victim has attributes that make it difficult to capture or subdue.

Summary

1. Optimality theory is the use of evolutionary theory to make specific predictions about what an optimal solution to an ecological problem should be. (Optimal solutions are those that raise an individual's personal fitness more than any alternative would.) The prediction can then be tested against reality.
2. The optimality hypothesis that animals should maximize energy gained per unit of foraging time in order to maximize their reproductive success has been used extensively in the study of feeding behavior. Although some animals do behave in ways that match predictions derived from this hypothesis, many others compromise energy maximization to deal with such things as the need for a balanced diet or the need to avoid predators while foraging.
3. Typically each species has its own diet and unique set of foraging tactics. There is some experimental evidence that suggests that competition among species may currently shape the dietary preferences of some animals, although some questions remain about the generality of this conclusion.
4. Many species consume foods that appear to be difficult to locate or capture. Several unrelated ambush predators have convergently evolved food (or sex) lures that entice certain elusive and mobile prey within easy capturing range. A small group of predators employs tools to extract energy from some well-armored or well-protected food items. A third category of animals forages cooperatively (1) to overcome the defenses of schooling fish or (2) to capture and subdue large, mobile, and potentially dangerous prey.

Suggested Reading

A general review of the literature on predatory behavior is provided in Eberhard Curio's *The Ethology of Predation* [154].

Reto Zach's article [814] on optimal foraging in the northwestern crow is a model of clarity. Equally interesting are the articles on the sunbirds by Frank Gill and Larry Wolf [257–259] and by Graham Pyke [590]. John Krebs has written a general review [391] of the optimal foraging approach to feeding behavior.

For a delightful book on the relation between flowers and their pollinators, see M. J. D. Meeuse's *The Story of Pollination* [506]. Bernd Heinrich's book, *Bumblebee Economics* [316], is a good companion to this chapter as it deals with optimality theory, competition theory, and plant–pollinator interactions as applied to bumblebees. The book is thoroughly readable.

Suggested Films

Army Ants, A Study in Social Behavior. Color, 19 minutes. A film that shows a social predator at work.

Food Handling in Kangaroo Rats. Color, 10 minutes. Shows the Great Basin kangaroo rat shaving the salt-laden tissues from saltbush leaves.

The Galapagos Finches. Color, 22 minutes. A look at the various finches and their ecological divergence.

The Hyena Story. Color, 52 minutes. A film on the feeding behavior of a social carnivore.

Insect Parasitism. Color, 18 minutes. An outstanding film on the relation between the alder woodwasp and its parasites. The film shows how four parasitic species exploit the identical food resource in four different ways.

Life on a Thread. Color, 20 minutes. Web building and prey catching by *Areneus dimidiatus,* an orb-web spider.

The Mussel Specialist. Color, 25 minutes. A film on the oystercatcher, a bird that feeds largely on one food, mussels.

Predatory Behavior of the Grasshopper Mouse. Color, 10 minutes. Contains a segment showing a grasshopper mouse consuming an *Eleodes* stink beetle.

The Social Cat. Color, 25 minutes. An excellent film on lion behavior, including hunting.

The Tool Users (05905). Color, 14 minutes. A sketch of several tool-using animals.

Tool-Using Species (05983). Color, 23 minutes. Tool-using by wild chimpanzees.

Because predators are so adept at overcoming the defenses of their prey, they have exerted a profound evolutionary influence on the animals that they eat. When you consider that a dead animal has the greatest difficulty reproducing, selection should often favor individuals whose behavior helps them delay the inevitable moment of reckoning. Nevertheless, identifying behavioral traits that actually have this effect is sometimes challenging. We have presented many hypotheses in earlier chapters on the possible antipredator functions of certain behavior patterns ranging from eggshell removal by black-headed gulls to the escape dive of bat-pursued moths and lacewings. The proposed effect of these actions was either to make it hard for a predator to detect a prey or to make it difficult for a predator to capture a prey that it had spotted. This chapter categorizes antipredator adaptations into these two groups: antidetection and anticapture tactics. Under these categories we shall consider examples of convergent evolution in such things as cryptic behavior, techniques to deflect attack, chemical repellents, and the spectrum of antipredator effects of group living. Our goal is to employ experimental and comparative approaches to test hypotheses about the possible survival value of an animal's behavior.

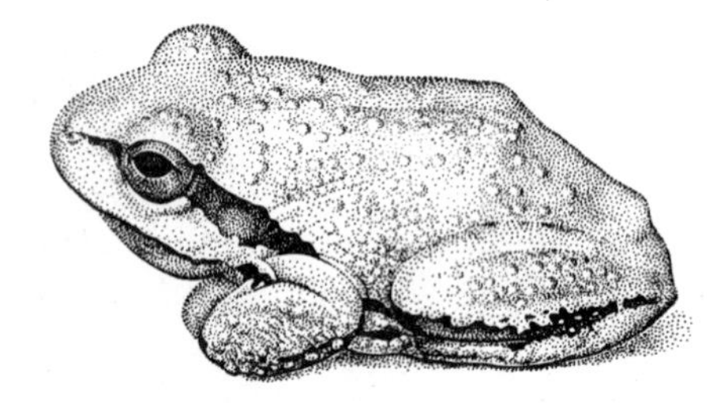

CHAPTER 10

The Ecology of Antipredator Behavior

Making Prey Location More Difficult

Any person who spends time observing animals in the field will occasionally come across species whose resemblance to their background is so detailed that it takes one's breath away (Figure 1). I could have included in Figure 1 a picture of a bark-mimicking insect that my wife found on an Australian eucalyptus tree. Its mimicry of eucalyptus bark is so perfect, however, that the creature cannot be seen in the photograph I took of it. If we have great difficulty seeing these animals, it is a reasonable deduction that the enemies of cryptic prey probably have a hard time also. Camouflage should discourage visually hunting predators from searching for certain prey because predators that inspected every leaf or every inch of tree trunk for leaf- or bark-mimicking insects would probably spend far more energy than they gained. Let us review some of the evidence that some animals are able to hide successfully from their enemies.

Cryptic coloration (camouflage) requires appropriate background selection by a prey if it is to be concealed. Therefore, we can predict first that mobile, cryptically colored prey will choose the substrate for which their color pattern and shape are best suited. Second, visually hunting predators should have greater difficulty finding prey species on the "correct" substrate than on other backgrounds. A test of the first prediction has been done with certain moths whose wings have contrasting stripes that to humans resemble the grooves in the bark of particular kinds of trees. If the moths are attempting (unconsciously) to resemble bark, they should orient themselves so that their stripes match the orientation of the bark's pattern. They do. Moths whose stripes run parallel to the long axis of their body orient themselves with their head pointing up or down on tree trunks so that the wing stripes blend in with vertical grooves in the bark of the trunk. Other species with horizontal stripes orient their bodies parallel to the ground when resting on a trunk so that once again, the animal matches its background [459, 630, 631].

Background matching has also been recorded for a number of lizards. Two closely related Australian species, *Amphibolurus decresii* and *A. vad-*

1 **Cryptic coloration and behavior.** A leaf-mimicking grasshopper (*upper left*). A moth larva that resembles the twig of a eucalyptus tree (*upper right*). A geometrid caterpillar has hooked dried grass fragments onto its back for concealment (*lower left*). A young ringed plover is crouched for safety, blending into its background (*lower right*). Photographs by the author.

nappa, are light pinkish yellow and dark reddish brown, respectively. In the field they live in the same general area but consistently are found on rocks that are appropriate for their body hues. When taken to an enclosure with light quartzite rocks and dark reddish limestone resting sites, the lizards chose different resting sites. The yellowish *A. decresii* perched on the limestone only one-third of the time whereas *A. vadnappa* greatly preferred the red limestone. Because a small lizard-eating hawk is abundant in the area where these reptiles live, their ability to pick the "right" color background is likely to be adaptive [255].

A test of the second prediction was done many years ago by H. B. D. Kettlewell in his classic study of industrial melanism in moths [377]. Lepidopterists knew that a completely black form of the moth *Biston betularia* had become much more numerous in Britain after 1850, largely replacing the typical salt-and-pepper form of the moth in many areas near big cities. Kettlewell's explanation for the spread of the melanic form was that they were less conspicuous to bird predators in woodlands blackened with soot from urban factories. The salt-and-pepper form remained abundant in non-polluted forests because in this environment whitish wings blended in nicely against lichen-covered tree trunks.

Kettlewell tested his hypothesis in several ways. In one experiment he placed samples of the two forms of moths on the same tree trunk (Figure 2) and observed the reaction of insectivorous birds from a blind. On sooty tree trunks, the whitish forms were found and eaten much more quickly than melanic individuals; the reverse was true when the experiment was conducted with a lichen-covered tree as the background.

2 **Typical and melanic forms** of the salt-and-pepper moth (*Biston betularia*) on a lichen-covered tree trunk (*left*) and a soot-blackened tree trunk (*right*). The conspicuousness of the two forms varies considerably, depending on their choice of a background. Photographs by H. B. D. Kettlewell.

In their ingenious modern test of the importance of background selection and resting orientation, Alexandra Pietrewicz and Alan Kamil employed operant conditioning techniques (Chapter 4) to test the prey detection abilities of North American blue jays [576]. Captive jays were trained to peck at a key when they saw a moth in a slide projected on a screen in their view. If a food-deprived bird correctly detected the image of a moth within a given time interval and pecked at the food key ten times, it received a part of a mealworm as a reward. But if it made a mistake and pecked when there really was no moth in the picture, the jay received no food and had to wait a relatively long time before it had another opportunity to scan for a moth and give a rewarded response. The experimenters used a great many different kinds of slides, some with and others without a moth in the photograph. The species of moths were varied, as was their orientation on backgrounds of different sorts. By projecting a series of these slides to their jays, they were able to measure the number of times a bird overlooked prey that were in the picture. Pietrewicz and Kamil found that when a moth species was on a background appropriate for its color pattern and in the orientation that it normally selected, it was less likely to be detected by a jay. For example, *Catocala relicta* rests head up with its whitish forewings over its body on white birch and other light-barked trees (Figure 3). In choice experiments it will choose birch bark over darker backgrounds as resting places [632]. As predicted, the jays failed to see the moth 10–20 percent more often when examining photos of *C. relicta* pinned to birch bark rather than to other backgrounds. Moreover, if the moth was oriented vertically on a birch trunk, the birds overlooked it more often than if its long axis was horizontally oriented on the tree trunk. These results show the adaptive value of the moth's ability to complement its camouflaged appearance through its selection of a daytime resting spot.

Removing Evidence of One's Presence

Some cryptically colored prey advance their concealment with actions other than matching their color pattern to the right background. When a moth larva feeds on its food plant, it creates a cue—chewed leaf—that a visually hunting predator might use to approximate the location of a prey. A small bird's hunting time per caterpillar could be enormously reduced if it could productively restrict its search to the vicinity of damaged leaves rather than having to hunt through an entire tree for its victims. But Bernd Heinrich discovered that a number of common caterpillars of the Minnesota northwoods either (1) removed themselves from their feeding sites during the daytime or (2) eliminated evidence of their foraging [317]. In the first category are the caterpillars of *Sphecodina abbotti*. When dawn comes, these animals abandon the grape leaves they have been consuming at night to go to resting sites some distance away on the vine stem, where they are beautifully camouflaged. A bird that looked by damaged leaves for this prey would not find any. In the second category are the larvae of another *Catocala* moth, *C. cerogama,* which after eating most of a leaf neatly snips through the petiole, dropping what is left to the forest floor

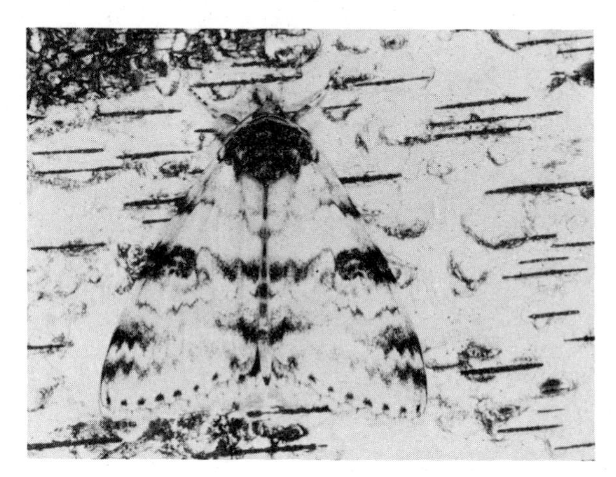

3 **Match between cryptic coloration and resting position.** The moth, *Catocala relicta,* orients its body so that its wing pattern matches the dark lines in birch bark. Photograph by H. J. Vermes, courtesy of Theodore Sargent.

where it can offer no clue about the location of the caterpillar in the tree.

In order to test the hypothesis that these edible caterpillars are trying to hide from their predators, Heinrich contrasted the behavior of these cryptic moth larvae with those species that are protected by spines, hairs, and toxins. For protected species, an investment in time and energy to remove the evidence of their feeding would not be repaid by a reduction in predation. They rely on warning coloration and behavior to repel their predators directly (see below). They remain on the leaves on which they feed [317].

Heinrich has also conducted feeding experiments with captive birds to demonstrate that they can learn, as predicted, to inspect artificially or naturally damaged leaves in preference to intact ones when food is associated with the damaged leaves [318]. Wild-caught chickadees were permitted to enter an enclosure containing ten fresh-cut birch or chokecherry tree tops. Two of the ten trees had damaged leaves and only these trees contained food rewards in the form of impaled mealworms or small edible caterpillars. At first the birds spent some time hopping about in the undamaged trees. But after a few trials, they began to go directly to those trees with damaged leaves (Figure 4). The position of trees was changed before each test to remove the possibility that the chickadees had formed an association between food and a particular location in the enclosure. Therefore, it seems likely that birds can use cues made by feeding caterpillars. If so, selection will favor individuals of edible species that remove the evidence of their activities.

Making Capture More Difficult

Staying out of sight or detection range of a predator is an extremely useful thing to do, but it imposes considerable constraints on what an animal can do. Cryptic moth caterpillars give up foraging for many hours during the day and remain absolutely motionless on the underside of a leaf or on a twig in order not to give their location away to a predator. A pied wagtail that needs to gather thousands

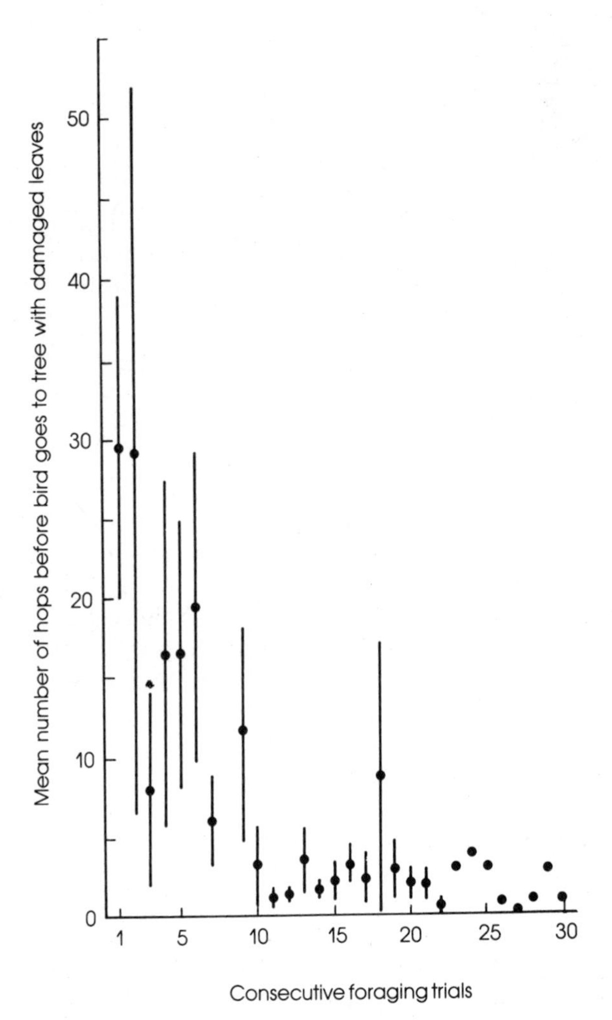

4 **Chickadees learn quickly to visit damaged leaves** in search of prey. The mean number of hops before a captive bird went to a tree with damaged leaves in an enclosure declined rapidly as the chickadees learned to associate this cue with the presence of food rewards.

of food items on a winter's day or a large African antelope living on an open grassy plain cannot easily hide from its enemies all day long. Even camouflaged caterpillars sometimes get discovered and may gain by being able to respond when detected. We noted earlier that some caterpillars can sense approaching wasp enemies and will regurgitate food fluid, thrash about, and sometimes tumble off their feeding perch upon hearing the vibrations created by buzzing wasp wings [692].

The noctuid caterpillar's sensory mechanisms permit it to be vigilant, detecting predators at sufficient distance to take evasive action. Vigilance is one of the commonest of all antipredator tactics, and for good reason. If a caterpillar were to begin to take action to avoid capture only after a predator had come in contact with it, the chances of its early demise would be greatly increased [692]. By detecting an enemy at distance before an attack occurs, the prey can prepare for escape or, better yet, can move away far enough to discourage contact with the predator altogether. The response of African antelope and zebras to lions they can see is usually remarkably blasé because they know precisely how close lions have to be before they are likely to charge; lions usually do not waste their time chasing animals that have spotted them at a distance and have kept out of attack range [635].

When there are hunting lions about, one species of African antelope, Thomson's gazelle, may run a little ways, but not as fast as possible, and then begin bounding high in the air with stiff-legged leaps. This behavior ("stotting") has been assigned many different possible functions. One suggestion has been that the behavior announces to a lion or other predator that it has been seen and therefore should save its time and energy by not attacking, a decision that would also save the gazelle time and energy as well. Another possibility, however, is that by leaping high in the air an alarmed gazelle is able to scan better for predators hiding in ambush. Because in some places lions usually hunt gazelles in groups, an antelope that dashed directly away from the first lion it spotted might race directly into the jaws of an ambush set by its hunters. Thus, it may pay to sacrifice some speed initially in order to determine what the safest escape route might be. Stotting, therefore, might be a vigilant counteradaptation to the ambushing techniques of socially hunting lions [578]. But as often is true in behavioral ecology, working hypotheses are easier to propose than to test, and this idea, appealing though it is, remains to be critically tested.

Misdirecting a Predator's Attack

Although an animal may be vigilant, it may nonetheless be surprised by a hunter on some occasions. If this happens, it may still be able to protect itself to some degree. One tactic is to try to cover vital body parts, because the individual that keeps from being instantly immobilized may eventually escape. Predators typically direct their assault against an animal's head because damage to the brain quickly renders a prey totally helpless. A convergent response to this tactic is to conceal the head when threatened by tucking it under harder, less vulnerable, body parts. Turtles are masters of this behavior, but so too are animals as different as the familiar crustacean pill bugs and mammals like the armadillo, the hedgehog, and the Australian spiny echidna. Wingless females of a cockroach found in the Phillipines do the same thing, rolling up into a hard-shelled ball when a predator bothers them. This also protects the female's soft-bodied young offspring, which ride about on their mother's underside and become encapsulated during an attack [616].

An alternative way to protect one's head is to induce a predator to strike at some other, expendable, portion of the body. The eyes on an animal's head often provide visual cues that predators use to orient their initial attack [100]. Some fish and insects take advantage of this by having a false head with eye spots on a region of the body that can be removed without mortally wounding the animal [781]. The hairstreak butterflies offer many examples of this phenomenon (Figure 5). To test the hypothesis that false heads really do deflect attacks, Robert Robbins predicted that hairstreaks with more elaborate false heads would show more symmetrical wing damage (the sort inflicted by bird predators) than hairstreaks with less complete false heads on the tips of their hindwings [606]. There are numerous species of butterflies with pseudoheads that may or may not include tails that seem to mimic antennae, eyespots, enlarged hindwing tips, and contrasting lines on the wings that point to the false head. Robbins ranked a set of Colombian species on a scale of one to four in terms of the perfection of false head mimicry. He then collected a large sample of these hairstreaks, which he inspected for beak-inflicted damage in the area of the false head. As expected, there was a significant correlation between the degree of completeness of false head mimicry and the percentage of specimens with wing damage.

The behavior of these butterflies supports the attack deflection hypothesis because perched hairstreaks usually hold their wings together and constantly move the hind wings up and down over a small angle. This keeps the eyespots and pseudoantennae in motion. It is well known that movement is a cue that directs attack in many predators, and therefore wing movement presumably enhances the effectiveness of the false head wing patterns, although there is no direct evidence on this point.

The use of a moving body part as a deflection lure is employed by any number of lizards and snakes, which twitch their tails when threatened by

5 **False head wing pattern** in the gray hairstreak viewed with its wings closed and wings open. The bright orange eyespots and black and white false antennae on the wings are much more conspicuous than the butterfly's actual head. Photographs by the author.

a predator, distracting their enemies from their heads (Figure 6). In addition, many lizards have evolved a tail with a fragile connection to the rest of the body, an anatomical anomaly that facilitates the removal of this appendage. As a final touch, the tail may thrash wildly on the ground after breaking off, keeping the predator's attention while the lizard escapes [278]. The detached leg of a daddy longlegs harvestman does exactly the same thing, thanks to a special neural mechanism in each leg [515].

How To Make a Predator Hesitate

Some deflection lures work by occupying a predator while the prey, or what is left of it, escapes. Another way to buy time for escape is to startle an enemy sufficiently to reduce the efficiency of its attack. Again the Lepidoptera provide many good examples. Some adult moths will, when disturbed by a bird's peck, begin flapping their wings and rocking from side to side. The thought is that a bird might be taken aback by this unexpected transformation of a moth into a whirling dervish. While the bird hesitates, the moth is actually warming up its flight muscles and may succeed in suddenly taking off and escaping before the bird can react [74].

Other moth species hold their cryptic forewings over brightly colored hindwings. In this group are the *Catocola* underwing moths. By releasing various species to caged birds, Theodore Sargent has shown that in flight the brightly patterned hindwings draw the attack of pursuing birds (supporting the deflection hypothesis). When they are resting, with their hindwings covered with cryptic forewings, and are grabbed by a jay, the moths will struggle, exposing the hindwings, which typically have orange, red, or yellow bands on a dark background. But in many species, a small proportion

6 **Distraction display by a desert gecko.** This lizard raises and curls its tail as an expendable lure when threatened by a potential predator (*left*). Photograph by Justin Congdon. The same species—an individual has lived to regenerate its tail after losing it to a predator (*right*). Photograph by the author.

of the moths have all-black hindwings. These rare individuals are more likely to escape than those with the more common color pattern. Sargent suggests that birds come to expect a color pattern flash when they pick up an underwing moth; when instead the underwings are entirely black, this anomalous result startles some birds, causing them to gape involuntarily and allowing the moth to escape [632].

Still other moths have huge eye spots on their hidden hindwings. Figure 7 shows one of the more spectacular representatives of this group after it has been poked in the thorax. The front wings have been flipped up, exposing two large staring "eyes." The sudden change in the moth's appearance might persuade some predators to hesitate in attacking, as has been shown experimentally [75] (Table 1). This might provide the moth with a few seconds to prepare for flight, or the bird might simply decide to go elsewhere for a meal. It has been suggested that moth "eyes" mimic those of owls and that insectivorous birds, major enemies of the moth, may have an innate fear of these stimuli because owls are their predators [100].

Producing cues that mimic enemies of their predators is common among threatened snakes, birds, mammals, and even some insects, which generate deep growls, loud hisses, or sharp chirps when under attack. For example, the harmless hog-nosed snake remains immobile until very closely approached, at which time it suddenly jerks about and hisses vehemently. If my own response is at all representative of the snake's major enemies, the reptile may be able to slip off while the startled predator retreats. (In my

7 **Large moth with small eyespots** on the upper wings. The small dots may serve to deflect attacking birds from vital areas; the large eyespots on the hindwings may be quickly exposed to frighten away a predator. Photograph by Thomas Eisner.

Table 1
Number of Times a Presentation of a Butterfly, *Nymphalis io,* Elicited a Strong Escape Response from Yellow Buntings[a]

Bird	Normal Butterfly (with large eyespots)	Altered Butterfly (with eyespots blotted out)
1	56	16
2	11	5
3	8	4
4	18	1
5	18	3
6	17	2
Total	128	31

Source: Blest [75].
[a] Equal numbers of each type of presentation were made over a period of 4 days.

two encounters with this species, its unexpected vocalization prompted me to leap back with far more agility than I customarily exhibit.) As E. S. Morton points out, large animals can easily produce deep or loud sounds [524]. Prey species that growl or hiss suddenly when in danger may trigger a startle response in a predator because the sounds may be associated with the close approach of something large and potentially dangerous to it. A startled hunter may hesitate just long enough to permit its prey to escape.

The startling effect of insect stridulations has been well documented in several cases. For example, crows often pick up and drop large passalid beetles a couple of times before crushing and consuming them. But they are much slower to make a second contact with an intact squeaking beetle than one whose acoustical apparatus has been glued together to render the beetle silent [111]. Likewise, both mice and jumping spiders take much longer to subdue a stridulating mutillid wasp than one that has been silenced. Even when the wasp's stinger has been removed, it takes mice five times longer to kill a noisy mutillid than a silent one [480].

Whether the stridulations startle these predators because of their similarity to the sounds made by large biting or stinging insects is not known. A more easily interpreted case involves imitation of the visual cues that signal "snake" by the large, two- to six-inch larvae of several moths. These caterpillars have evolved the capacity to respond to being touched by forming a triangular head complete with snake eyes (Figure 8). Moreover, the "snake" will strike accurately at the object that touched it [100, 781]. One assumes that this secures the attention of a would-be predator. These mimics may be especially designed to repel lizards because a transformed larva closely resembles the head of a certain vine snake that is fond of lizard meals.

Fighting Back

The most direct way to deal with predators is to deter them with the threat or application of physical resistance or chemical warfare. These are costly and risky maneuvers, but the options of a

8 **Mexican vine snake, *Oxybelis aeneus*** (*left*), and its possible mimic, a snake caterpillar (*right*). The anterior end of the larva has dropped off the vine and the animal has changed its body shape to create a triangular snake head complete with realistic eyes. Photographs by James D. Jenkins and Lincoln P. Brower.

cornered prey are limited. We have already noted that large herbivores are dangerous because of their hooves and horns. But even small animals have evolved very potent repelling devices, as we shall see in our examination of the defenses of black widow spiders.

Although you might think that the black widow would be protected primarily by her poisonous bite, other animals are far less intimidated by this species than are humans. It takes time for the spider to maneuver itself into position to use its fangs, and a small mammal, like a deer mouse, can exploit this to immobilize the spider with its own quicker bite. Deer mice find black widows completely edible, and in the laboratory they have been shown to capture this prey safely. But about half the interactions in an experimental arena ended with the mouse repelled and the spider alive [738]. In a large majority of these cases the cause of the spider's survival was traced to the rapid secretion and deployment of a strand of silk, which the spider produced as soon as it was jostled (Figure 9). The silk strand was adorned with droplets of a profoundly adhesive substance. With its hindlegs, the spider applied the strand to its attacker's face. When this happened, the mouse typically recoiled and attempted, sometimes by rolling frantically on the ground, to clean itself of this material. In nature, the spider would have easily escaped during this time by scrambling from its web to an enclosed retreat in a crevice of some sort.

Rich Vetter tested the importance of the sticky silk strand by blocking the spinnerets of some spiders with wax. When attacked by mice, these individuals were three times as likely to die as intact spiders able to defend themselves with their adhesive chemical. Although the glue is not toxic (consumption of the viscous substance did not affect the mice), it could potentially mat a mammal's hair and interfere with its protective and thermoregulatory properties. Mice that give top priority to the removal of adhesive fluids will avoid damage to their fur.

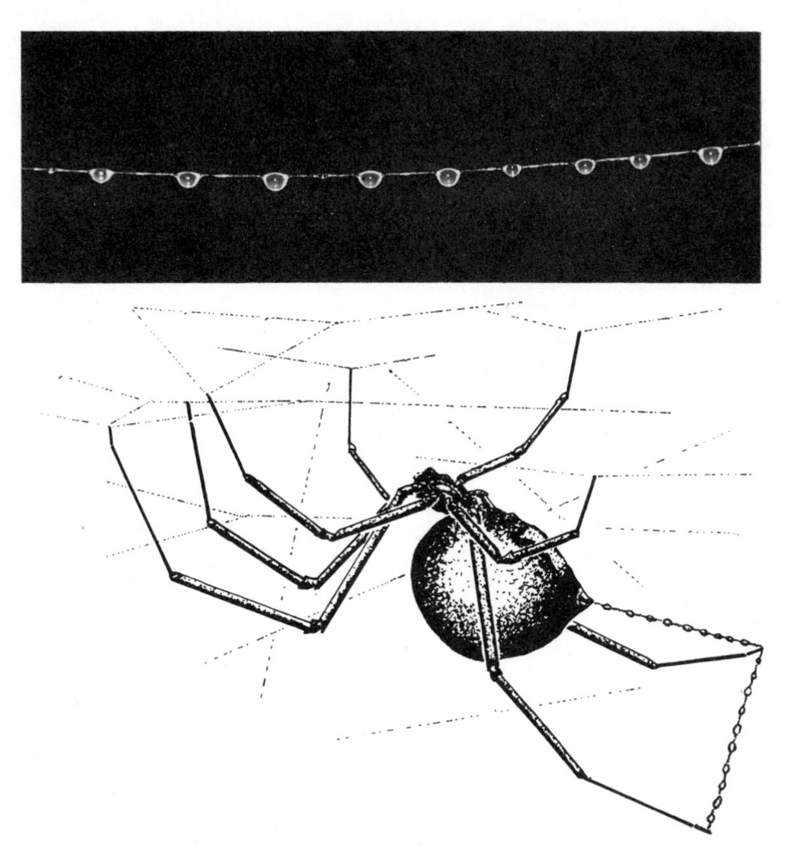

9 **Black widow spider silk.** The viscous defensive silk strand (*above*) is displayed by an endangered black widow spider (*below*) who will apply the silk to a predatory enemy. Courtesy of Richard Vetter.

A Comparative Analysis of Predator Repellents

Although black widows do not use toxic materials to repel some of their enemies, many other species do. Like the black widow, however, other prey species match the nature of the repellent to the kind of enemy they must confront [205]. For example, ants that live in acacias are well adapted to drive off herbivorous invertebrates that would consume the plant on which they depend. The ants are active and agile, quick to approach and to bite or sting insect invaders. Humans are relatively little bothered by these species. In contrast, an African ant (*Pachysima*) that lives in the hollow stem of a particular tree behaves differently. It is slow and clumsy and makes no effort to catch invertebrates. Its chief enemies are probably large browsing herbivores that trim the tree, an action that prevents tree growth and thereby blocks expansion of the ant colony. Unlike acacia ants, *Pachysima* move doggedly to a bare patch of skin on a human (and presumably other mammals), where they insert their stingers deeply. No pain is felt for a moment, but there then follows a most unpleasant and deep pain that lasts several days [360]. Daniel Janzen reports, "While hundreds of stings of *Pseudomyrmex* [an acacia ant] can be tolerated if there is a compelling reason to invade the tree, 1–5 *Pachysima* stings were enough to drive me away from a tree, leaving me reluctant to return."

Along these lines, a comparison of the potency of the stings of honeybees of the genus *Apis* is also instructive [647]. The temperate zone honeybee (*A. mellifera*) has colonies with many thousands of workers. If the colony is threatened by a major vertebrate enemy, like a raccoon or mouse, the workers will readily plunge their stingers into the robber's skin. Because the stinger is barbed, it cannot be extracted by the bee after it enters flexible skin. As a result, the sting and its associated poison sac are pulled from the worker's abdomen as she tries to free herself from the predator. Although the worker dies (see Chapter 14, The Evolution of Eusocial Insects), the colony's assailant carries with it the maximum dose of toxin, which will continue to be pumped through the stinger unless it is removed.

There are three other species of *Apis* that live in southeast Asia, where they are subject to attack by both invertebrate enemies (mainly weaver ants) and vertebrate predators ranging from birds and tree shrews to humans. The bees vary greatly in size, aggressiveness, and the potency of their stings [647]. The smallest species, *A. florea,* builds a comb on a limb in low vegetation (Figure 10). Colonies are small; this fact and the small size of the workers means that predators that gain access to the nest have no difficulty overwhelming its defenses. The stings hardly bother a human honey collector. But this does not mean that the bee is defenseless. They deal with ants by making it almost impossible for these enemies to reach the colony. The workers coat the branch on which the nest is supported with rings of a chemical substance so sticky that ants cannot walk to the colony without becoming entrapped. Other insects subject to ant predation have independently evolved similar sorts of ant guards with which they protect their nests [365, 723] or eggs [321] (Figure 11).

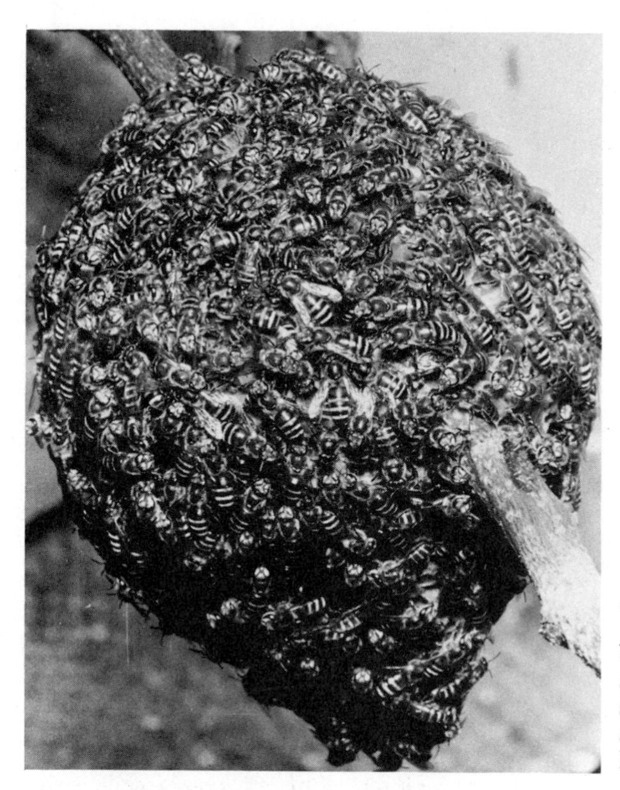

10 **Exposed comb** of *Apis florea*. These bees place sticky substances around the limb on either side of the nest to repel ants. Photograph by Martin Lindauer.

The response of *A. florea* to vertebrate predators is first to attack in waves. However, if overmatched, the entire colony abandons its nest readily and absconds to a new location rather than persist in futile resistance. The primary line of defense against such enemies may be the relative inconspicuousness of the small colony, which is not easy to see in dense vegetation.

In contrast, the nests of *A. cerana* and *A. dorsata* are much easier to locate. *Apis cerana* is a medium-sized bee that forms large colonies within cavities in hard wood. The workers stream in and out of a narrow entrance that is monitored by a group of guard bees. In this respect they are rather like the honeybee *A. mellifera.* These bees have little trouble with ants because they are larger and because all attackers must pass through the restricted entrance, which can be easily protected. Even when threatened by vertebrate predators, the workers are slow to leave the nest to attack because of the ease with which small enemies can be turned back at the entrance. Larger ones cannot pass through the entrance into the nest. In order to gain access to a colony of *A. cerana,* the predator must be powerful enough to rip open a tree trunk.

The stings of *A. cerana* are more potent than those of *A. florea,* but they pale by comparison with those of *A. dorsata.* This is a giant bee that forms

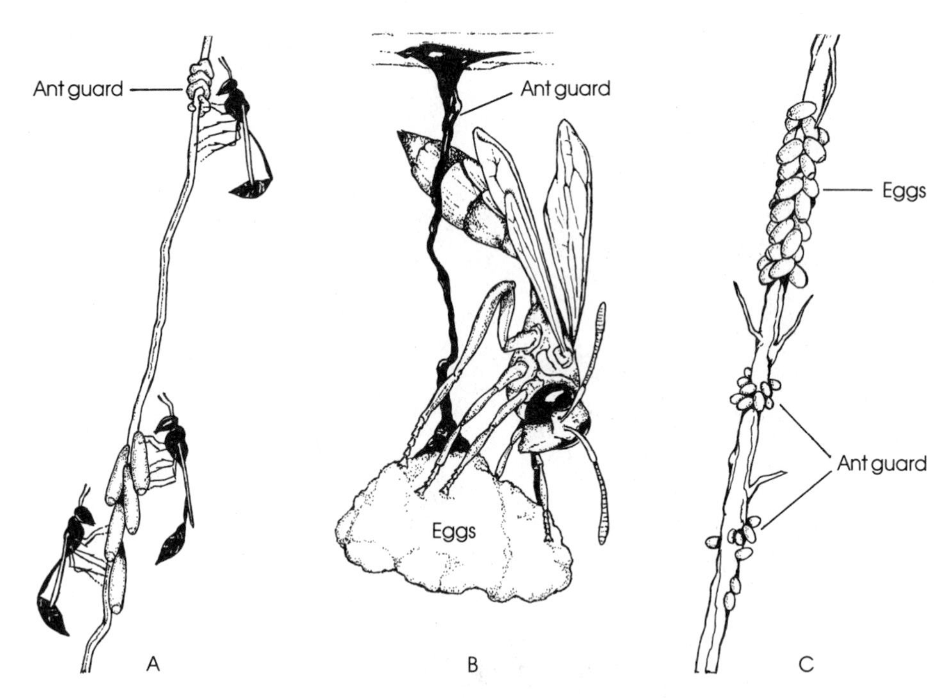

11 **Ant guards** have evolved independently (A) in some stenogastrine wasps, (B) in polistine wasps, and (C) in a neuropteran owl-fly. The *Polistes* wasp applies a repellent to the pedicel of its nest. The other two insects build sticky or repellent rings on the access route to the nest or egg mass.

large, highly visible colonies on the branches of trees many meters above the ground. Ants pose no risk to them because of the size advantage of the bees, but vertebrate predators can easily detect the conspicuous nests. Because the nests are suspended at great height, however, larger vertebrates are unlikely to be able to reach them easily. Any that try are greeted by workers of unusual ferocity. After a binocular case was swung by one colony for three minutes and then retrieved, researchers counted 97 stingers in it. The large size of these bees enables them to store large quantities of toxin, and few animals can tolerate their intensely painful stings. Thus, each bee species has different kinds of major predators that try to collect the bee's brood and wax [647]. This variation corresponds with the relative stinging ability of each species and their other defensive stratagems (Table 2).

Chemical Repellents of Vertebrates

Insects and spiders are not the only creatures that have the ability to use chemical deterrents to attack, as anyone who has had the misfortune to irritate a skunk knows all too well. Among other vertebrates

Table 2
Comparative Defenses of Three Species of *Apis* Honeybees

Factor	*A. florea*	*A. cerana*	*A. dorsata*
Attributes of bees	Small exposed nest Weak stings Ant guards	Nest hidden in cavity Moderate stings Defense of entrance	Large exposed nest Powerful stings Violent group attack
Detection by predator	Difficult, except for ants	Moderately easy	Easy
Approach by predator	Easy, except for ants	Difficult	Difficult, except for good climbers and fliers
Consumption by predator	Easy	Moderately difficult	Difficult

Source: Seeley, Seeley, and Akratanakul [647].

that possess chemical defenses are certain salamanders. Some species employ adhesive repellents that are functionally similar to those used by the black widow. When *Plethodon cinereus* is grasped by a garter snake, it goes into a paroxysm of writhing and thrashing coupled with the release of secretions from its tail and body. The salamander tries to wrap its tail about the head of the snake and keep from being swallowed. Because the secretions are sticky, its looped tail tends to adhere to the body of the snake, a situation that makes it more difficult and time-consuming for the predator to ingest the prey. The goal of the salamander may be to frustrate attempts to be eaten long enough to induce the snake to release it, somewhat the worse for wear, but still alive [28].

Another consequence of salamander glue is seen in Figure 12. Small snakes can become stuck to themselves if they attack certain salamanders. The salamander *Ensatina eschscholtzi* tries to strike an attacker in the head with its secretion-laden tail; if the salamander succeeds, the snake will often come away with its mouth glued shut, a most desirable outcome from the salamander's perspective [28].

Still other salamanders have evolved toxins rather than adhesives, although these must be released as skin secretions rather than injected into an enemy in the manner of honeybees [96]. The familiar red eft (*Notophthalmus viridescens*) that wanders over the forest floor in moist woodlands can cover its body with a deadly tetrodotoxin exudate [584]. Minute amounts of eft skin injected into a mouse will kill it promptly. Birds appear to be sensitive to these poisons, too, and quickly learn to avoid efts [348].

The red eft is a classic illustration of warning coloration and behavior. Its body is brilliantly colored (orange to crimson), and the animal moves slowly and conspicuously in full view of its potential predators. Reds, oranges, yellows, and blacks are common colors for toxic species. It has been suggested that these animals rely on warning coloration and behavior, the

12 **Result of salamander secretions.** A small garter snake became glued to itself after a struggle with a large salamander (*Batrachoseps attenuatus*). Photograph by Stevan Arnold.

better to remind experienced predators of past unpleasant experiences with other noxious, warningly colored species. This reasonable idea has been tested by seeing whether young chickens can learn faster to avoid a prey that blends into its background or that stands out from it. After having been fed for a few days on food crumbs dyed blue or green, some chicks were offered a mix of the crumbs on either a blue or a green arena floor. These crumbs, however, were made equally noxious by addition of quinine in the dye. The rapidity with which the birds came to avoid the adulterated baits was dependent on the contrast between the item and the background on which it rested [262]. An aversion to green was more quickly formed on a blue background, and fewer blue baits than green ones were taken when the substrate was green. If it is generally true that animals can learn more quickly to avoid things that are conspicuous, the adaptive value of warning coloration would be clearer.

Batesian Mimicry

After young chicks have had experience with quinine-dipped food crumbs of a certain color, they will avoid them on sight alone. The ability of predators to learn to reject certain color patterns (or auditory or olfactory cues) without making contact with the prey has been exploited by a host of deceptive species [781]. These animals, although perfectly edible, have evolved a close resemblance to a protected species. This is Batesian mimicry, named after its discoverer Henry Bates, a nineteenth-century English naturalist and explorer. It is probably safe to say that every organism with a powerful predator-repellent has one or more edible species that mimic it and so deceive some predators that avoid the model (see Figure 23, Chapter 4).

There are mimetic salamanders that look like red efts to humans (Figure 13). Inexperienced captive birds that are offered *Pseudotriton ruber* will often eat them, one after another, with no ill effect. But if they are then given a red eft as a meal, the bird will taste and reject the eft and thereafter have nothing to do with *Pseudotriton*—which is also red-bodied [348].

Studies of this sort are generally done in the laboratory under carefully controlled conditions. To create a test that more closely approximated conditions in the wild, the Brodies distributed a large number of escape-proof trays containing salamanders in a pine woods where wild birds could feed from the trays [95]. Four moist leaves were placed in each tray with its salamander occupant, which in some cases was a toxic red eft, in others a striped cryptic form of the edible *Plethodon cinereus,* and in still others a red-bodied form of this salamander. (It is not uncommon for animal species to occur in more than one color form, as the human species itself illustrates.) Is the red-bodied morph of *P. cinereus* a Batesian mimic? If it is, one would

13 **Red eft *Notophthalmus viridescens* and its Batesian mimic *Plethodon cinereus.*** Predators that have experienced the toxic eft leave the edible mimic alone. Photographs by E. D. Brodie, Jr.

expect that more red forms than striped ones would survive feeding trials with foraging birds.

Bird predators evidently found the caged salamanders, a conclusion based on the presence of bird droppings in some trays. In cases in which leaves had been displaced from or thrown about the tray (almost certainly by foraging birds), the percentage of surviving red *P. cinereus* was three times as great as the survival percentage of the striped form [95]. Apparently when birds inspected a tray and found a red salamander in it, they often left it alone whether it was a toxic red eft or an edible *P. cinereus*. (Red efts were almost never removed from their trays during the experiment.)

Associating with a Protected Species

Batesian mimics derive protection by looking like a dangerous, poisonous, or noxious species. Other animals use the defenses of well-protected species in a different way. This can be as simple as living in association with a repellent species as is true for those oropendulas (a blackbird) that nest by colonies of social wasps [669]. More active symbioses involve hermit crabs that collect sea anemones and transfer them to their backs, where the stinging tentacles of the anemones drive back the octopi that consume undefended crabs [615].

The sea anemones do not appear to suffer from the loan of their defensive apparatus to the crabs, but the same cannot be said of some remarkable cases of recycled antipredator defenses. Some marine slugs (nudibranchs) eat the stinging tentacles of various jellyfishes and corals. The tentacles are covered with cells that fire poisonous hairs and barbs into most of their enemies, but the nudibranchs are immune. In fact, they store any untriggered cells in special sacs and use them against their own predators [818]. Similarly, hedgehogs are able to eat toads with no ill effect despite the poisons in toad skin (which have felled many a naive dog). Hedgehogs often partially skin their victims. One might think that this was done to reduce contact with the toxins in the skin. But after polishing off a toad, a hedgehog may munch on the deadly parotid glands in the skin and then froth at the mouth, licking the combination of saliva and toad poison onto its spines (Figure 14). For good measure, it may also wipe its body with the skin of its prey [94].

Hedgehogs and oropendulas do not have to manufacture their own toxins, but they do pay a price for their use of another species' defenses. Hedgehogs require the enzyme systems that enable them to cope with toad poisons, and oropendulas may have to search longer to find nesting sites with suitable wasps present. Although it seems likely that these animals do derive protection from their associations with other species, there are no direct tests of this hypothesis. One edible species for which there is quantitative evidence is the lycaenid butterfly *Glaucopsyche lygdamus* [575]. The caterpillars of this butterfly are generally attended by a retinue of ants of the genus *Formica*. The worker ants are fed from "honeydew" glands on the larva's back. At least two hypotheses have been offered to account for

14 **Hedgehog anoints itself** with foam containing toad toxins, the toad having been recently eaten by the hedgehog. Photograph by E. D. Brodie, Jr.

the ant–butterfly association: (1) the butterfly feeds the ants to keep them from killing it and (2) the butterfly feeds the ants to enlist their aid in repelling enemies.

The lycaenid does have enemies, particularly small parasitoid wasps and flies, which lay their eggs on caterpillars. A parasitoid's offspring will eventually consume the caterpillar before it reaches reproductive age. To examine whether being with ants confers protection from these enemies, Naomi Pierce and Paul Mead prevented ants from tending some larvae while permitting other caterpillars to retain their keepers. In order to create a group of untended larvae, they placed a sticky ring of Tanglefoot about the base of a plant on which one or more caterpillars were feeding. Ants are unable to cross this viscous barrier. After establishing the two groups of caterpillars, each larva was inspected daily in order to collect it when it had reached full size and was ready to pupate. The caterpillars were then held until they pupated or until they died from the emergence of parasitoids from their bodies. The experiment was done in two locations, and in both places many more braconid wasps and tachinid flies appeared from the untended ants than from lycaenids that had ant guards with them (Table 3). This experimental evidence, in conjunction with direct observations of ants driving off or killing parasites (Figure 15), shows that *Formica* ants are helpful protectors of these butterfly larvae [575].

Cooperative Defense Against Predators

Some animals gain protection by associating with members of another species; many animals enhance their survival chances by living with others of their own species. Group defense is sometimes an effective way to deal with predators that no one individual could repel, as we have already seen in the case of ground-nesting gulls (Chapter 7) and honeybees. Figure 16 illustrates some addi-

15 **Ants guard butterfly larvae.** (*Top*) An ant attends a lycaenid butterfly larva. (*Bottom*) An ant attacks a braconid wasp parasitoid that could destroy the larva. Photographs by Naomi Pierce and Paul Mead.

Table 3
The Effect on Parasite Success of Experimentally Preventing Ants from Guarding Lycaenid Butterfly Larvae

	Larvae without Ants		**Larvae with Ant Guards**	
Site	*Percentage Parasitized*	*Sample Size*	*Percentage Parasitized*	*Sample Size*
Gold Basin	42[a] (36%)[b]	38	18 (7%)	57
Naked Hills	48 (11%)	27	23 (0%)	39

Source: Pierce and Mead [575].
[a] The percentage of larvae that yielded either a tachinid fly or braconid wasp.
[b] The figure in parentheses is the percentage of larvae parasitized by the braconid wasp.

16 **Group defense** is used by musk oxen in the Canadian tundra and by sandwich terns threatened by a black-headed gull flying overhead. Photographs by Ted Grant (Information Canada Photothéque) and by J. Veen.

tional examples of cooperation in dealing with enemies from the vertebrates, but it is among the social insects that examples of group defense abound [793].

In many termites and ants, a specialized soldier caste has evolved; members of this caste are usually larger than other workers and are armed with formidable mandibles. Their sole function is to protect the colony against dangerous invaders. Soldiers communicate with one another via alarm signals, particularly chemical scents, that attract large numbers to a point of combat. The reaction of soldiers to intruders varies among species: they may chop their opponents in half with shearing mandibles, or stab them with piercing jaws, or snap them away with the peculiar "finger-snap" mandibles possessed by some termite species. Alternatively, a colony's defenders may douse their enemies with acidic sprays or—and this is a favorite of mine—entangle them with a sticky secretion. Nasutitermitine soldiers have become highly modified to project streams of a viscous substance accurately at intruders (often ants) at a nest entrance or along a foraging column of workers. This termite manufactures and stores resinous materials in liquid form in a huge frontal gland operated by powerful muscles. When alarmed, the blind soldier somehow manages to point its "nose" toward the enemy and, contracting the head muscles, fires a liquid thread that becomes more sticky on exposure to the air and entangles the foe. The defensive spray also contains alarm substances that draw still more soldiers to the area, where they too may engage the invaders [793] (Figure 17).

Several species of ants practice a variant on this theme. The ant's abdomen consists largely of a gland filled with an entangling glue. In effect, the ant is a living grenade, for it rushes at an enemy, simultaneously constricting its abdominal muscles so violently that its body wall bursts and the gland explodes. The released glue entraps the intruders and prevents them from harming the colony [404, 476].

Not only can groups of social insects carry out effective repelling attacks, but they can be so intimidating that some predators will flee before even making contact with the colony. It would be the unusual mammalian predator that would approach the mass of warningly colored wasps shown in Figure 18. Many species of tropical social wasps engage in communal defensive displays. One species builds a comb against a tree trunk or limb and then covers it with a thin sheet of papery material. Its response to a potential predator has been vividly described by Howard Evans and Mary Jane Eberhard.

"When undisturbed most of the wasps remain inside the nest, in the ample space between envelope and comb. When disturbed they produce a warning signal by somehow vibrating the nest envelope in unison from within, producing a rhythmic drumming sound audible several yards away. If further aroused hundreds of individuals rush out onto its outer surface, where they continue the coordinated thumping, at the same time raising and lowering their wings in synchrony with the sound so that the nest is suddenly a rhythmically throbbing mass of aroused wasps . . . it is little wonder that the natives of regions inhabited by these wasps consider it an 'outstanding act of valor' to approach their nests" [226].

Sociality and Alarm Signals

Cooperative warning display and attack are just two of a host of possible advantages social animals may gain in their attempts to deal with predators. One of the entertaining aspects of studying animals that

17 **A group of nasute termite workers** is spraying silk strands over a fruit fly intruder. The sticky silk entangles the intruder. Photograph by E. Ernst.

live together is developing alternative interpretations of a group action and then trying to devise a way to discriminate among the many possibilities. To illustrate this point, let us analyze the tendency of white-tailed deer to raise their tail when disturbed by a potential predator (Figure 19). This behavior exposes the white underside of the tail as a deer begins to bound off through the forest. For a sample of possible explanations for this behavior, consider the following list of hypotheses (326), which is far from exhaustive.

Hypothesis 1: Tail-flagging is an alarm signal that warns an animal's offspring of danger.

Hypothesis 2: Tail-flagging is an alarm signal that coordinates the group's escape dash, helping the alarm-giver better foil the predator (e.g., by being a member of a tightly bunched herd that the enemy cannot disperse to single out a victim).

18 **Tropical social wasps** cluster on the outside of their nest. These insects provide examples of warning coloration and behavior. Photograph by W. D. Hamilton.

Hypothesis 3: Tail-flagging is a signal, but not one of alarm, because the intended receiver is the predator, not other deer. By providing the signal, the deer announces to an enemy that it has been detected and therefore further pursuit will be unproductive. If alerted deer are difficult to capture, it could be in a predator's interest to ignore tail-flagging prey and this would help the signaling deer as well.

Hypothesis 4: The behavior has no evolved antipredator function but is merely an incidental effect of the physiological changes that occur when an alarmed animal prepares itself for escape.

All four ideas have at least some plausibility and therefore our problem is how to discriminate among them. This requires that we use a hypothesis to develop predictions that are incompatible with other explanations. For example, both hypotheses 1 and 2, but not 3 or 4, generate the prediction that deer in groups should tail-flag more often than solitary deer. One would predict from hypothesis 1, but not hypothesis 2, that adults associating with their offspring will tail-flag more than adults without nearby juvenile descendants.

19 **Alarmed white-tailed deer** lift their tails high in the air and expose the conspicuous white underside, perhaps as a signal to a potential predator that it has been seen. Photograph by Jon E. Cates.

Some data have been collected on these points [326]. In a Michigan population, tail-flagging occurs 90–95 percent of the time when deer flee from humans, whether the prey are in groups or not. Moreover, herds of unrelated males were just as likely to exhibit the response as groups composed of does and their yearling offspring. These results throw cold water on the social alarm signal hypotheses and encourage us to examine hypotheses 3 and 4. A prediction from the third hypothesis is that solitary deer should raise their tails more often when encountering an ambush predator (e.g., mountain lion) than a long-distance running predator (e.g., wolves). Ambush predators typically have low endurance and rely heavily on surprise to make their kills. If they are detected by a prey at a distance, they rarely chase after it. Long-distance predators, however, should be much less likely to be deterred by an "I have seen you" signal because of their ability to wear down a victim in a long chase. No information is available on this prediction, and therefore the issue remains unresolved.

Sociality and Improved Vigilance

Even though the function, if any, of tail-flagging in deer remains obscure, in theory the ability to warn others to raise one's fitness is a potential benefit of group living. (See Chapter 14 for a more completely tested case of alarm signaling in a social mammal.) As well as this possible benefit and the advantages derived from group defense, we can identify three other antipredator advantages that may be gained by some kinds of social animals in certain environments: (1) improved vigilance, (2) the selfish herd effect, and (3) the dilution effect.

Earlier the improved vigilance hypothesis was discussed in the context of gray-eyed junco feeding behavior [124]. By joining a flock, an individual junco can detect the escape flight of its companions and so join them in fleeing from a predator that it had not personally detected. In an experimental test of this hypothesis, G. V. N. Powell released an artificial hawk model that traveled along a line over a cage in which lived (a) a single starling or (b) a group of ten of these despised, but adaptable, birds. The average reaction time of the solo bird was significantly slower than that of a bird in a group, despite the fact that single birds spent about four times as much time looking around for danger as a bird in a flock [584] (Table 4).

R. E. Kenward took this test one step closer to reality by releasing a hungry, trained goshawk at a standard distance from flocks of pigeons [375]. As predicted, the larger the flock, the sooner the prey took flight and the lower the probability that the goshawk would make a kill (Figure 20). Although this experiment provides strong support for the vigilance hypothesis, it should be noted that it does not prove that flocking pigeons are safer in big groups than in small ones. If being in a large flock attracts disproportionately more predators, the benefits gained by improved vigilance (or dilution; see later) could be negated by a greatly increased frequency of attacks, raising the risk that an individual would eventually be killed.

The Selfish Herd Hypothesis

There is no indication that starlings or wood pigeons attempt to warn each other of danger. Instead, the birds in a group appear to be

Table 4
Effect of Group Size on Caged Starling's Response to an Artificial Hawk Model

	Group size	
Behavior	*One bird (61)*[a]	*Ten birds (117)*[a]
Number of times per minute each bird stops foraging and looks around	23.4	11.4
Percentage of foraging time spent in surveillance	47	12
Number of times hawk model released	24	33
Mean time (seconds) from appearance of model over cage to flight reaction by bird	4.1	3.2

Source: Powell [584].
[a] Number of observations of group.

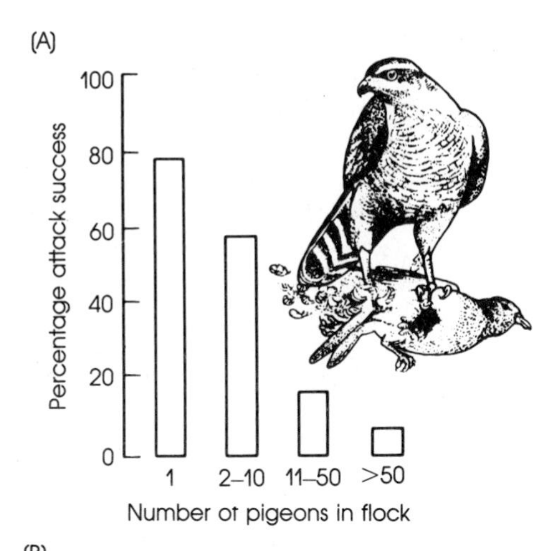

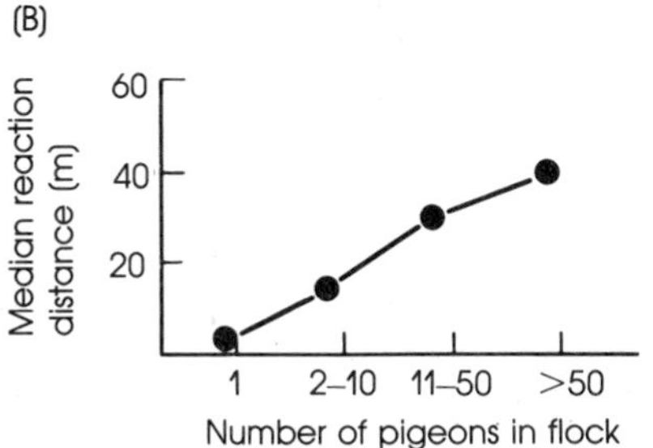

20 **Flocking behavior** helps protect individual wood pigeons from goshawk predators. The larger the flock, the sooner the birds are likely to detect an approaching goshawk and the lower the predator's chances of making a kill.

taking advantage of the cues provided by their fellows. Thus, what looks like the coordinated flight response of an alarmed flock of birds may be the result of individuals that are trying (unconsciously) to avoid being left behind to face a predator alone. The birds might even be competing for the safest position in the group. To emphasize the possibility that what looks like a cooperative group may actually be composed of reproductively competitive (= selfish) individuals that are always trying to keep others between themselves and potential danger, W. D. Hamilton coined the phrase *the selfish herd* [302].

When Adelie penguins leave their breeding groups to go out to sea to feed, they often gather in groups on rocky ledges by the ocean and then jump into the water together to swim out to the foraging grounds. The value of this becomes clearer when one realizes that by every jumping-off point there is likely to be a leopard seal lurking in the water. The seal can only capture and kill a certain small number of Adelies in a short time. By swimming out in a group through the danger zone, many penguins will escape while the seal is engaged in dispatching one or two unfortunate ones (the dilution effect; see later). If you had to run a leopard seal gauntlet, you would probably do your best to be neither the first nor the last one into the water. Adelie penguins appear to have a similar goal. By going out in the middle of a wave of paddling penguins, an individual can use his companions as a living shield to protect him against the predator. If true, penguin flocks would qualify as selfish herds.

In a previous edition of this book I suggested that a penguin might go so far as to try to lure another individual into leaping first by lunging toward the water as if he were going in. At that time I was unaware of a book written in 1914; the author, M. G. Levick, claimed that a mass of penguins would sometimes actually try to push a fellow penguin into the water, presumably to test for the presence of a leopard seal [416]. Levick wrote, "When they succeeded in pushing one of their number over, all would crane their necks over the edge."

Selfish herds need not always be composed of murderers, but if the thesis is correct, we should at least be able to detect competition for the central positions within some groups. Central locations should be the safest (because an animal there will be surrounded by others that will act as a buffer against predation), and they might be the most productive places to be (because the central animals will spend less time looking for predators and more time foraging) [124, 357, 431]. There is evidence that in ground-nesting gulls the central nesting territories are claimed by older, dominant birds that force younger, subordinate individuals to the outer edges of the area, where they are first to face incoming predators [707].

Similar support for this prediction comes from research on bluegill sunfish [289]. These animals breed in colonies, with males defending nesting territories against rivals. Females visit a colony to lay their eggs. There is intense competition among males for central territories, and larger (generally older) males are more likely to win this competition. Females prefer to lay their eggs in these territories, which are far safer from predators than peripheral nests. It is the peripheral male that is first to confront a foraging bullhead or cannibalistic bluegill, and as a result his eggs are twice as likely to be eaten by a predator as those of a male on a central nest (see Table 2, Chapter 14).

If peripheral individuals in a selfish herd of bluegills or black-headed gulls are at special risk, why do they accept their inferior position? The argument developed earlier on the evolution of dominance and subordinance applies here as well. If a younger or weaker individual has no reasonable probability of forcing a more powerful rival to yield his superior position, then the subordinate animal has two options: to nest on the outskirts of a group and incidentally shield the central animals or to nest as a solitary individual. In the bluegill, solitary nesters (which sometimes occur in this species) experience slightly higher egg predation than peripheral, colonial males. Thus, given its options, a subordinate fish may benefit slightly by nesting in a group even though more dominant males will use him as a living shield against their enemies [289].

The Dilution Effect

Perhaps the simplest antipredator advantage of living in a group is to overwhelm the consumption capacity of the local predators.

Imagine that a region contains ten predators, each of which eats one prey per day. If 100 prey live in this area, each prey runs a 10 percent chance of being killed per day. But if 1000 prey can cluster together, without altering the distribution pattern of their killers, the risk of dying per day will drop to 1 percent. We can express the dilution hypothesis in terms of Adelie penguins and leopard seals. If a seal can kill one penguin every three minutes, then a penguin in a group of 100 passing through the predator's foraging zone has ten times the survival chance of a penguin in a group of 10 (provided each takes about three minutes to get through the danger zone).

An amphibian that also illustrates the dilution effect is *Bufo boreas* [29]. The major enemy of this toad is a garter snake that hunts the margins of ponds for toads that are undergoing metamorphosis from the aquatic tadpole to the terrestrial adult. When a *Bufo* is in the midst of this transformation, it cannot swim as well as a tadpole because of its reduced tail and emerging legs. But it cannot hop as well as a full-fledged adult either. Therefore, it is highly vulnerable, and the snakes take full advantage of this situation. In the premetamorphosis and immediate postmetamorphosis stages, the toads come together to form large, dense aggregations, clumping to such an extent that mounds of individuals form in some places. Because metamorphosis is synchronized, there is as large a cohort as possible to cluster together during the most dangerous time in their life cycle. A snake that finds a cluster will become satiated before it eats all the toads, a situation that improves an individual prey's chances of reaching full adulthood.

As adults, male toads and frogs often call conspicuously. Some predators, among them a bat and a possum, are known to zero in on calling males of many tropical frogs (see Figure 25, Chapter 13). In response, some species, like *Physalaemeus pustulosus,* form reasonably dense groups whose members chorus together more or less synchronously. The predation rate per male is negatively correlated with the number of individuals in a chorus (Figure 21) [628]. Any protection the calling frogs derive from grouping together almost surely comes from the dilution effect because the animals do not counterattack their predators nor do they have the capacity to escape through improved vigilance (calling frogs are generally so intent on attracting a mate that their level of vigilance is extremely low).

The Monarch Butterfly

Having reviewed the spectrum of antipredator behavior, we turn now to a more detailed look at a single species' repertoire.

The monarch butterfly is a fascinating animal whose behavioral ecology reveals the deep imprint of the effects of predation pressure on all aspects of its life cycle. During one phase of this cycle, the dilution effect appears to play an important role in the survival of individuals. Earlier we noted that overwintering aggregations of monarchs form at traditional sites in coastal California and mountainous central Mexico (Figure 12, Chapter 8). One previously stated hypothesis for the selection of these locations is their temperate, humid climate, which helps the torpid

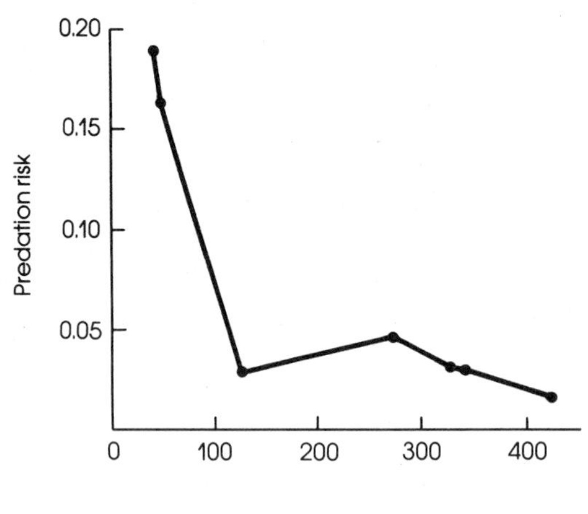

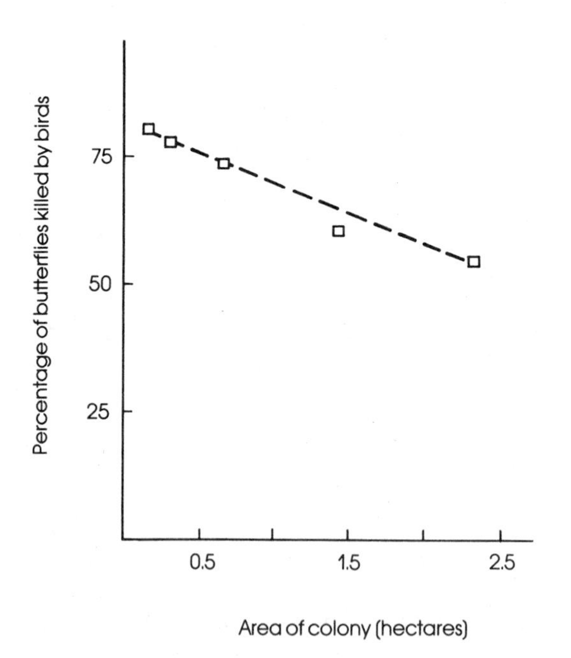

21 **Dilution effect.** Individuals in large groups of the frog *Physalaemus pustulosus* and the butterfly *Danaus plexippus* are less likely to be eaten than members of smaller groups.

butterflies avoid freezing and dehydration during the winter. But if thermoregulatory and metabolic considerations were the only factors involved in habitat selection by the butterfly, we would expect to find them far more broadly distributed than they are. Instead, they are clustered in almost incredible densities at certain restricted locations (tens of thousands may rest upon a single tree), a situation that swamps the feeding capacity of their enemies. Because the monarchs are often cold and incapable of flight in their winter retreats, they cannot escape from their predators. In Mexico, orioles and grosbeaks kill hundreds of thousands of monarchs over the course of the winter. Lincoln Brower and his associates have shown that the risk of predation for any one monarch is inversely proportional to the number of butterflies clustered within an aggregation site (Figure 21). Because the birds seem to feed on the edges of a colony, selection favors individuals that are densely clustered within a large colony [122].

It is more than mildly surprising to find birds preying so heavily on monarch butterflies because the butterflies exhibit classic warning coloration (orange and black stripes). Moreover, it has been proved that some monarchs contain toxic cardiac glycosides [99]. A blue jay that eats a poisonous monarch becomes ill and vomits 15–30 minutes after its meal (Figure 22). These birds find internal illness an intensely unpleasant learning experience. In the laboratory, a single encounter that leads a jay to vomit is sufficient to educate the predator to avoid the butterfly thereafter. If the caged bird does attack again, a quick peck provides a refresher course; the taste of the monarch's wings apparently reminds the predator of the disastrous consequences of eating a substance with this taste. In this context it is significant that the highest concentrations of the toxins occur in the wings of the butterfly rather than in the abdomen or thorax [103].

22 **Blue jay eating a monarch butterfly** and then vomiting. After this experience the jay will avoid monarchs on sight. Photographs by Lincoln P. Brower.

Under natural conditions, a butterfly grasped by the wing and then released can often fly off with no difficulty, and beak-marked specimens are not uncommon.

But why is it advantageous to carry toxic chemicals in one's tissues if one must be eaten to enable the toxins to take full effect? This is a general problem that applies to the many other organisms that are warningly colored and cause delayed vomiting or internal injury *only* when consumed. A vomited monarch does not fly off into the sunset. It has been widely argued, however, that by being consumed a noxious animal may still derive indirect genetic benefits by educating a predator to avoid a specific color pattern, which nearby *relatives* share. These individuals not only share the dead animal's appearance, they share some genes in common, which they may better propagate because the educated predator will avoid them (see Chapter 14). This rather elaborate hypothesis would be unnecessary if it could be shown that the emetic effects of the toxins were incidental to their immediate disgusting properties, which could lead a predator to release a captured monarch while it was still alive. Experiments done with captive, hand-reared birds have revealed that noxious prey are often quickly dropped on a first encounter before severe damage has been done to them [783]. If the same is true for interactions between monarchs and some of their predators, selection may have favored toxic individuals simply because they taste bad and not because they cause emesis when swallowed. Vomiting and the learning it promotes may be protective adaptations of the bird, which rids itself of ingested toxins and avoids a second mistake.

Monarchs and Their Food Plants

Monarchs do not manufacture their own chemical deterrents but simply incorporate some poisons that may be present in their food plants [99, 104]. That this is possible is testimony to the selection herbivores have exerted on plants. Plants are generally not eager to sacrifice portions of themselves to an animal (except to a helpful pollinator or seed disperser). The plant whose leaves are consumed by caterpillars or by a mammalian browser loses an investment in photosynthetic equipment whose function was to collect energy that would ultimately contribute to the making and dispersal of gametes. To prevent the loss of leaves, plants have evolved a range of adaptations almost identical to those employed by insects against their enemies [201]. The plants may be covered with spines, thorns, hairs, barbed hooks, and other mechanical repellents. Or they may possess chemical deterrents, including bitter caffeine or quinine, nauseating and perhaps disorienting hallucinogens (such as the active ingredients in marijuana or peyote), insect hormones that prevent metamorphosis, sticky resins that entrap attackers, and hydrogen cyanide. However, competition among herbivores has favored mutant individuals that possess a special capacity to overcome a plant's defense system and thereby gain access to an abundant source of untapped food.

23 **Female monarch butterfly** laying an egg on a milkweed plant. Her choice of the species of milkweed will determine the palatability of her offspring to bird predators. Photograph by the author.

Monarch females lay their eggs primarily on milkweeds (Figure 23). Some members of this plant family have evolved powerful toxins (cardiac glycosides) that interfere with many basic cellular activities [99]. Monarch larvae, however, can consume these plants with relative immunity. Moreover, they have evolved the capacity to store the poisons in their tissues and to retain those stored poisons when they undergo metamorphosis to adulthood. Proof that the butterfly acquires toxins from its foods came when L. P. Brower and his associates "persuaded" some monarchs (through artificial selection) to eat and survive on a cabbage diet. When the adults reared on this harmless food plant were fed to jays, they were eaten with no ill effect. In contrast, jays that consumed butterflies reared on the milkweed *Asclepias curassavica* always vomited a short time later [104]. Moreover, chemical analysis showed that the glycosides in the milkweeds and in the butterflies reared on them were identical. Thus, the monarch that feeds on a toxic food plant benefits from the evolved defenses of the plant. But it probably pays a price as well. The larva must devote some of its metabolic energy to gather the toxins and to store them safely. If it did not have these expenses, it presumably could grow faster or larger, reaching maturity sooner or as a stronger individual. Larger size, greater strength, and earlier maturation, in turn, could affect fitness favorably.

In fact, monarch females in nature oviposit on a variety of food plants, ranging from the very poisonous to the completely nontoxic [188]. Any choice of oviposition material carries with it trade-offs (Figure 24), but an advantage for those females that lay eggs on edible food plants is the production of offspring that may be able to exploit those conspecifics that have been reared on toxic food plants. "Automimics" is the label Brower

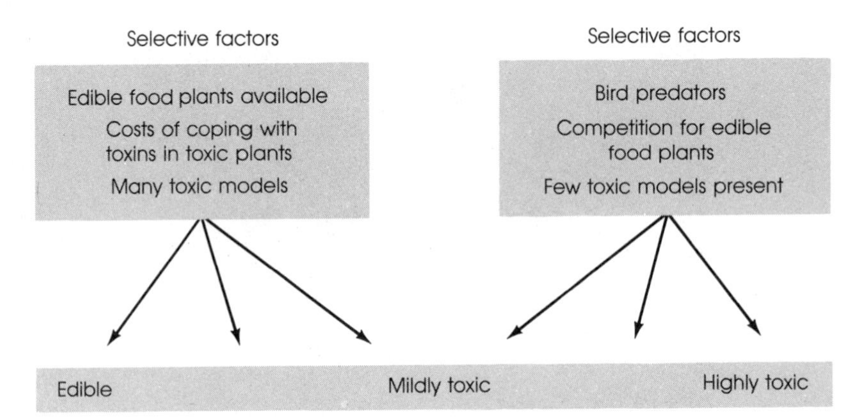

24 **Conflicting selection pressures** affect the optimal choice of food plant for egg laying by monarch butterfly females. The arrows indicate the kind of oviposition site that selection should favor given certain environmental factors. Because environmental pressures vary, the best oviposition site can range from completely edible milkweeds to highly toxic ones.

has given to those edible individuals that gain some protection by resembling unpalatable fellow members of their own species. Predators with experience with *Asclepias*-reared monarchs are not likely to have much to do with a slightly noxious *Gonolobus*-reared individual, if they avoid the monarch pattern on sight.

But this brings us back full circle to the Mexican monarch eaters. These birds apparently have learned that there is variation among monarchs in palatability. They overcome "automimicry" by throwing away some (toxic) monarchs after pecking at them while accepting others after tasting them. Moreover, they may avoid whatever toxins are present in some mildly poisonous monarchs by eating just the abdomen and the muscles of the thorax, which contain the lowest concentrations of poison [237].

Monarchs and Their Mimics

Because monarchs are a protected species, other animals have evolved a mimetic resemblance to them. The famous Batesian mimic of the monarch is the viceroy, a butterfly belonging to a completely different family. Jane Brower has shown that scrub jays will gobble down viceroys if offered them alone [97]. However, birds that are first given a series of monarchs will avoid the viceroy as religiously as they would reject another monarch. The viceroy's resemblance to its model is close, differing only in its smaller size and the presence of two black lines on the hindwings. The mimic's behavior, although not dramatically similar to that of the monarch, is not dissimilar, and the two are often together in the same area.

The monarch also has a number of Müllerian mimics, among them the queen butterfly. The queen, like the monarch, is an unpalatable butterfly, often feeding on the same food plants. Mimicry of this sort does not involve deception of a predator but, instead, is intended to make it easier for a bird to learn to avoid members of both species. A bird that attacks species A may learn to avoid both A and B if both share the same color pattern. Müllerian mimicry evolves because it reduces the chances that an individual will be one of the unlucky ones killed by predators as they learn to avoid a particular color pattern [100].*

The existence of close mimics, both Batesian and Müllerian, poses a potential reproductive problem of species recognition for the monarch and may have influenced various aspects of the monarch's courtship. It is possible, but not proved, that olfactory cues may be somewhat more significant factors in species recognition in the queen and monarch than for butterflies that do not belong to mimetic complexes [102].

Thus, the antipredator behavior of the monarch is linked with disparate components of its ecological web (food plants, bird predators, herbivorous competitors, Batesian and Müllerian mimics) and with many aspects of its behavior (Table 5). Whatever the monarch does has ecological consequences for the other members of its community and for its own key biological activities. This is true for all animal species.

Summary

1 Antipredator behavior is a dominant feature of the repertoire of most animals. *Antidetection* adaptations help individuals avoid interactions with their predators; *anticapture* adaptations help individuals reduce the probability of being captured, once detected.

2 The most prevalent form of antidetection behavior is cryptic behavior in which an animal enhances the effectiveness of its camouflaged color pattern by selecting the appropriate resting background or by removing cues that predators might use to locate it.

3 The spectrum of anticapture responses include (a) vigilant behavior, (b) misdirecting a predator's attack to an expendable body part, (c) responses that make a predator hesitate during an attack, and (d) employing mechanical or chemical means to repel an attack by a predator. Examples of convergent evolution are numerous within these categories.

*Let us say that species A and species B each has a population of 1000 in an area and that each has a distinct color pattern. If the birds kill 100 insects while learning to avoid one color pattern, the risk to each individual of being a learning experience is 1 in 10. If, however, they share a common color pattern (AB), then the birds will be taking 100 victims from a combined population of 2000 animals that share the same color pattern. This reduces the risk of a given individual to 1 in 20. Often Müllerian mimicry complexes in the tropics involve dozens of species that share a highly conspicuous color pattern (red and black stripes, orange and black stripes, etc.) as well as highly conspicuous warning behavior (slow fluttering flight and clumping in communal roosts).

Table 5
Ecological and Evolutionary Interactions among Monarch Butterflies, Birds, and Plants

Evolution of toxic food plants in response to herbivore pressure ↓	
Evolution of resistance to food plant poisons by monarchs (escape from competition for food for offspring) ↓	Oviposition behavior (female lays eggs on toxic plant species) Larval feeding behavior
Evolution of toxic monarchs (escape from predation) ↓	Conspicuous flight and clustering on fall migration
Evolution of automimicry by some monarchs (escape from costs of consuming toxic foods as larvae) ↓	Expansion of food plant choice by females
Evolution of Batesian and Müllerian mimicry by other butterfly species (exploitation of bird predators' learning abilities)	May affect courtship behavior of monarch with selection favoring a heavy reliance on chemical cues

4 Individuals often use other animals to improve their chances of survival. This may involve exploitation of the warning coloration and behavior of a toxic species by an edible one (Batesian mimicry), a strategy that may deceive some predators. Alternatively, the members of a less well protected species may join or use a portion of a member of another species for their own defense. A third possibility is for members of the same species to form groups for more effective defense.

5 The antipredator functions of group living are diverse and include (a) group attack on a predator, (b) the ability to warn offspring and other relatives of impending attack, (c) increased vigilance at lower individual cost, (d) the use of fellow group members as a physical shield against predation, and (e) the union of forces to swamp the feeding capacity of the local predators.

6 The monarch butterfly possesses an array of antipredator responses ranging from clustering together in dense groups on the overwintering grounds to the incorporation of toxins by the larvae for use against the monarch's enemies. The pressure of predation has not only influenced the evolution of winter habitat selection and larval feeding behavior but also the choice of oviposition sites by females and perhaps the courtship behavior of the butterfly as well.

Suggested Reading

Malcolm Edmunds' *Defense in Animals* [199] updates Hugh Cott's classic *Adaptive Coloration in Animals* [143] as a good general book on antipredator adaptations.

Wolfgang Wickler's *Mimicry in Plants and Animals* contains a useful review of many amazing examples of mimetic defenses of animals [781].

Thomas Eisner's article on chemical defenses of arthropods [206] and Lincoln Brower's article on the monarch butterfly documents the remarkable antipredator adaptations of these creatures [99].

Bernd Heinrich's articles on how predation has affected the evolution of caterpillar feeding behavior are good examples of how to test hypotheses on potential antipredator behavior [317, 318].

Suggested Films

Baboon Social Organization. Color, 17 minutes. Outlines how the savannah-dwelling baboons cooperate to repel predators.

The Monarch Butterfly Story. Color, 11 minutes. A sketch of the life cycle of this insect.

Patterns for Survival. Color, 27 minutes. A film cataloging some remarkable color pattern and behavioral adaptations of insects that help them foil their enemies.

Polar Ecology, Predator and Prey. Color, 22 minutes. The interactions between rodents and predators in the Arctic tundra.

Strategy for Survival. Color, 30 minutes. An attractive film on the feeding, antipredator, and migratory adaptations of the monarch butterfly.

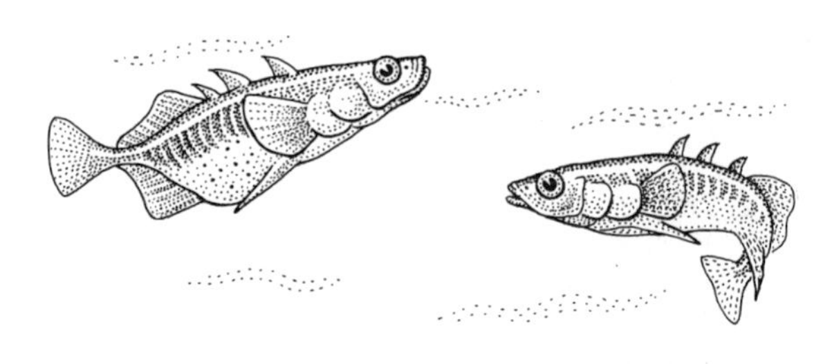

However skillful an animal is in finding good living space, foraging efficiently, or repelling predators, these abilities only count in the long run if the individual succeeds in passing on its genes. If an animal fails to do this, it will have no lasting influence on the evolution of its species. Therefore, reproductive behavior is the central focus of natural selection. Typically animals propagate their genes by making copies of them and donating the copies to an offspring. But there are a bewildering variety of ways to do this, and our goal in this chapter is to show how one can make some sense of this variety. We shall begin by describing the unusual reproductive behavior of a damselfly to illustrate three major topics: the evolutionary puzzle of sexual reproduction, the role of sexual selection in shaping reproductive behavior, and the evolution of diversity in mating systems. This chapter discusses the first of these topics, explaining why modern biologists have trouble understanding why sexual reproduction ever evolved. Although this puzzle has not yet been completely resolved, it is widely accepted that this mode of reproduction creates a special category of individual selection called sexual selection. An understanding of sexual selection can help us understand why male and female reproductive strategies are so often strikingly different. The chapter attempts to explain why males are usually more eager to copulate than females, more willing to fight for mates, and more likely to show intraspecific variation in the tactics used to acquire mates. After dealing with these issues, the chapter analyzes certain postcopulatory behavior patterns of males to show that sexual competition among males does not end with the race to secure mates but influences all aspects of male reproductive behavior.

CHAPTER 11

The Ecology of Reproductive Behavior: Sexual Selection and Male Competition

Reproduction in a Damselfly

Almost everyone living in the eastern United States has seen the damselfly *Calopteryx maculata* at one time or another (Figure 1). This striking insect, a relative of the dragonflies, has a brilliant iridescent green abdomen and large, jet black wings; it is usually found perched by the edges of a stream. Most observers, charmed by the jewellike beauty of the insect, are unaware that the life of this creature, like that of most sexually reproducing species, is marked by sexual violence and competition. The damselfly reproduces sexually, by definition, because the species is composed of males and females that produce gametes by meiosis. During the production of sex cells, sections of chromosomes are exchanged and the genome of the individual is reduced by one-half so that each gamete has only one copy of each gene, instead of two (Chapter 1). Thus, when a male and a female gamete unite, the resulting offspring has two copies of each gene, just like its parents.

The damselflies that are most often seen are males. Their maleness does not reside in their brilliant coloration or in their penis but in the relatively small size of their gametes. Males of any species are identified by their "sperm," which are smaller than the eggs of the females. A male of *Calopteryx* produces vast numbers of tiny sperm, each of which carries half the individual's genotype and almost nothing else. A female of this species produces many fewer, larger eggs, which contain a nutritional supplement in addition to a set of genes from the female.

Adult males of *C. maculata* spend much of each day trying to find females that will accept their sperm [746]. Some males are territorial, claiming a stretch of stream bank one to three meters long from which they repel all other males. Intruders are chased and may be struck in flight. Physical contact is rare, however, because most invaders dart from the territory as fast as they can fly as soon as they are detected. Occasionally, however, a nonterritorial male will persist in reentering the territory of a rival, and prolonged pursuit flights and weaving aerial clashes take place. These encounters may lead eventually to the overthrow of the resident and his replacement by the intruder.

Males are willing to fight for possession of a bit of stream bank because females come to water to lay their eggs in barely submerged rootlets and other underwater vegetation. When a female comes fluttering down to the stream, the resident male will dart toward her and then turn and land close to a suitable oviposition site, spreading his black wings and raising the white tip of his abdomen in a conspicuous display. The female may or may not land briefly at the spot, and she may or may not fly up to land in nearby vegetation with the male close behind. If she does perch, the male will hover in front of her with wings whirring in a courtship display before he attempts to land on her back. He moves forward to grasp the area in the front of her thorax with specialized claspers at the tip of his abdomen. The male then pulls the end of his abdomen close to the point where his abdomen joins the thorax, forming a loop. This maneuver involves transfer of sperm from the sperm-producing organs at the tip of his abdomen to a "penis" at the base of the abdomen [746].

After this rather bizarre transfer has taken place, a receptive female will swing her abdomen under the male's body and place her genitalia over the penis (Figure 1). After several minutes the pair uncouples, and the male flies back to a perch in his territory. The female typically follows after a pause, returning to the oviposition site advertised by her mate. While the male surveys his domain, the female perches on floating plants on the surface of the water and probes underwater with her abdomen for good places in which to insert her eggs.

This sketch of the behavior of a damselfly is sufficient to raise a number of key evolutionary questions around which this chapter is organized. First, why do the damselflies reproduce sexually instead of asexually? Why should a female accept sperm from a male if all she gets from him is a sample of his genes?

Second, why do female damselflies make large eggs and help their offspring get a good start in life by ovipositing in certain areas whereas males offer no nutritional or other developmental benefits for their progeny?

Third, why do males of this species fight for access to females whereas females sometimes turn down opportunities to copulate?

These questions could be asked of the large majority of animal species, and it is only recently that some tentative answers have been developed.

The Puzzle of Sexual Reproduction

The fundamental question for any sexually reproducing species is, Why bother with sex? Look how complicated it is for the damselfly. How much simpler it is for animals that reproduce asexually. If the point of life is to inject one's genes into the next generation, what more direct way to do so than to run off photocopies of one's genotype and place them in offspring? Imagine a mutant female damselfly able to reproduce parthenogenetically. All of her offspring will be capable of reproducing on their own (and so can be considered egg-producing females), whereas the sons of a sexual female must mate with

1 **Sexually reproducing insect,** the damselfly, *Calopteryx maculata*. (*Above*) A territorial male on his streamside perch. (*Below*) A female copulating with a territory owner. The male grasps the female with his terminal claspers while she twists her abdomen under his to make contact with his sperm-transferring organ. Photographs by the author.

sexual, egg-producing females in order to propagate their mother's genes. If we assume that the sons and daughters of a sexually reproducing female require the same amount of maternal resources per gamete, then for every son produced, a sexual female loses one daughter. If one half of the progeny of the sexual female are sons and the other half daughters (which is the typical case), then parthenogens should have twice as many surviving daughters as sexual females [490]. (We assume that a parthenogenetically produced daughter has the same survival chances as a sexually produced one.) Each of the parthenogenetic female's daughters should also have twice

Table 1
Advantage of Parthenogenesis If Sexual and Parthenogenetic Females Produce the Same Average Number of Offspring

Generation	Sexual Individuals		Parthenogenetic Females	Proportion of Females That Are Parthenogenetic
	Males	*Females*		
1	49	49	1	1/50 = 0.02
2	49	49	2	2/51 = 0.04
3	49	49	4	4/53 = 0.08
4[a]	45	45	7	7/52 = 0.13
5[a]	43	43	13	13/56 = 0.23
⋮	⋮	⋮	⋮	⋮
n	0	0	100	1.0

[a] To keep the total population of this species about 100, across the board cuts of 10 percent and 5 percent were made in generations 4 and 5, respectively.

as many female offspring as the genetically distinct sexual females in the population. The proportion of parthenogenetic females will therefore rapidly increase (if our assumptions are correct) and the end result will be the extinction of the sexually reproducing females and males (Table 1).

Now although most species are similar to the damselfly in that the male's contribution to his offspring consists of little more than a set of genes, there are animals in which the two "sexes" produce gametes of equal size. (In this case, there are no males or females, but there could be sexual reproduction if the gametes were produced meiotically and fused to produce new individuals.) In many other species, males offer parental care or nutritional benefits to their offspring. In this case, when a female accepted from a male genes with which to fertilize her eggs, she would receive more than just the genes. The other benefits could more than double her production of offspring and so enable her to outreproduce a parthenogenetic competitor. But as G. C. Williams points out, this situation would still be vulnerable to invasion by a mutant parthenogen that would accept the large gamete or other material goods from a donor "male" but would not incorporate the genes of the donor in her offspring [788]. Such an individual would double her production of offspring and also double the propagation of her genes by avoiding meiosis and the reduction in chromosome number in each gamete that this entails. Thus, a mutant allele whose female carriers accepted assistance from male partners while rejecting their genes should spread throughout the population. The competing allele for sexual reproduction may or may not appear in an offspring of sexually reproducing females, depending on meiotic chance.

Thus, whether or not males offer females more than their genes, there would seem to be great advantages for females that reproduce asexually [787]. Yet, as we are all very much aware, sexual reproduction is still with us. Why is this so?

Individual Selection and Sexual Reproduction

It is well known that sexual reproduction results in the continual formation of novel genotypes [182]. In the making of gametes during meiosis, the genetic information in the original reproductive cell is thoroughly scrambled because of recombination. As a result, the egg or sperm that is produced may carry any one of a huge number of possible combinations of genes. The genotypes that are produced by the union of eggs and sperm are literally guaranteed to be unique (except in identical twins). The fact that almost every individual has a different combination of genes in a sexually reproducing species ensures a substantial amount of phenotypic diversity.

This process produces variant individuals that may, in times of rapid environmental change, survive and reproduce while most others die. This may have the effect of perpetuating populations that would otherwise have become extinct had the parental generation been reproducing asexually and thus not introducing any new variation into subsequent generations. And many biologists have suggested that sexual reproduction has evolved because it ensures the future survival of species. This argument, however, supposes that individuals are gratuitously providing a certain number of variant offspring for the sole purpose of establishing a genetic bank for the long-term advantage of the population. If such foresighted individuals did exist, logic suggests that they would soon be eliminated by other genetically different types that acted ("selfishly") to maximize their immediate reproductive success (Chapter 1). But if sexual reproduction confers a fitness advantage to individuals, it would not be necessary to resort to a group selectionist argument for its occurrence in animal species. If under some circumstances, sexually reproducing offspring survive better or reproduce more than asexually produced ones, then we would be able to understand why sex is maintained in populations despite the apparent benefits of parthenogenetic reproduction.

The circumstances that might favor individuals that reproduced sexually, despite the costs of this technique, are contained in unpredictable and variable environments [787]. Most environments are not uniform. They change from place to place, from season to season, and from year to year. Because this is so, there is often no guarantee that the environment would remain the same for an individual's offspring even if they were to live and reproduce themselves in exactly the same spot where the parent lived and reproduced. Moreover, it is sometimes advantageous for offspring to disperse to find relatively unexploited areas. Because of spatial variation in the environment, dispersing animals are unlikely to find a place with conditions identical to those a parent experienced. Therefore, an individual that reproduced asexually might very well have progeny beautifully suited for conditions that no longer existed (even in the place where the parent had lived) or that were impossible to find.

Williams argues that sexual reproduction is analogous to the strategy one employs in a lottery [787]. You do not usually win a prize by buying a large number of tickets with exactly the same number. For species that live in variable habitats, the vacant slots in the environment that potentially can be exploited by offspring represent the different prizes in a lottery. Winning tickets are those offspring that find and are capable of filling these variable slots. By reproducing sexually, an individual places different samples of half his genotype in a variety of different progeny, who will have a chance to distribute themselves over a range of habitats rather than all being highly dependent on the same set of ecological conditions. Moreover, sexually reproducing individuals may occasionally hit the jackpot by producing a genotypically unusual individual that happens to be capable of exploiting a particular habitat in a highly successful way.

Sexual reproduction is a gamble, but not a wild one. Bets are hedged because the generation of variable offspring is controlled. Progeny are different from their parents, but almost never radically different because developmental homeostasis (Chapter 3) guarantees that they will possess the key adaptations of their parents and their species. But by being somewhat variable, some may be especially well suited to take advantage of the inevitable and unpredictable variation in some environments over space and time. In addition, genetically variable offspring may compete less with each other for the same resources than they would if they were genetically identical [489]. Finally, by creating genotypically unique offspring, parents of some species may make it difficult for certain parasites or pathogens to decimate their entire brood [198]. Some of their progeny may have unusual recombinant disease-resistance systems that permit them to survive in the face of a diversity of dangerous bacteria or infectious viral agents (another unpredictable component of some environments) [304, 359].

If Williams's lottery analogy is correct, we should find sexual reproduction linked to relatively unpredictable ecological situations. One way to test this prediction is to look at species that employ both modes of reproduction. In a species of freshwater *Hydra* that lives in the northern United States, individuals reproduce asexually as newly hatched polyps in the spring and throughout the summer [167]. When fall arrives, the hydras begin a sexual phase, producing eggs and sperm. When an egg is fertilized, an embryo is formed; this embryo detaches itself from the parent and attaches to the substrate. It then spends the winter in a quiescent state before hatching out in the spring and beginning the phase of growth and asexual budding. Thus, an individual hydra produces variable offspring at just that time when the conditions its progeny will face are least predictable (because after a winter has passed the aquatic environment may be very different from its state the previous year).

Likewise, female aphids of many species reproduce parthenogenetically (Figure 2) when they colonize a host plant early in the spring after emerging from an overwintering site [219]. Because there are likely to be few other aphids on the plant at this time, a female's offspring will be able to live with her, consuming the same materials she feeds upon. The female can take full advantage of the power of parthenogenesis. But eventually some

2 **Parthenogenetic insect,** the rose aphid. She is shown here giving birth to a daughter. Photograph by E. S. Ross.

female descendants of the original founder female abandon the advantages associated with not having to produce sons. The sexual generation creates offspring that have novel genotypes and that will overwinter in a dormant state, as do the embryos of *Hydra hymanae,* before emerging the next spring to face altered conditions and an unpredictable search for new host plants (Figure 3). Thus, the alternation of sexual and asexual behavior in hydras, aphids, and other insects matches Williams's model fairly well. Nevertheless, a variety of other factors may contribute to the evolution and maintenance of asexual reproduction in some species [152], just as there are other ideas on the origins and advantages of sexual reproduction [57, 66, 490].

Male and Female Reproductive Strategies

Whatever the advantages of sex, there are many profound evolutionary consequences related to the adoption of this mode of reproducing. The heart of the matter is that male and female reproductive strategies are basically different. As noted, females by definition allocate more energy per gamete than males do. Females of *C. maculata* produce a goodly number of eggs during their lifetime,

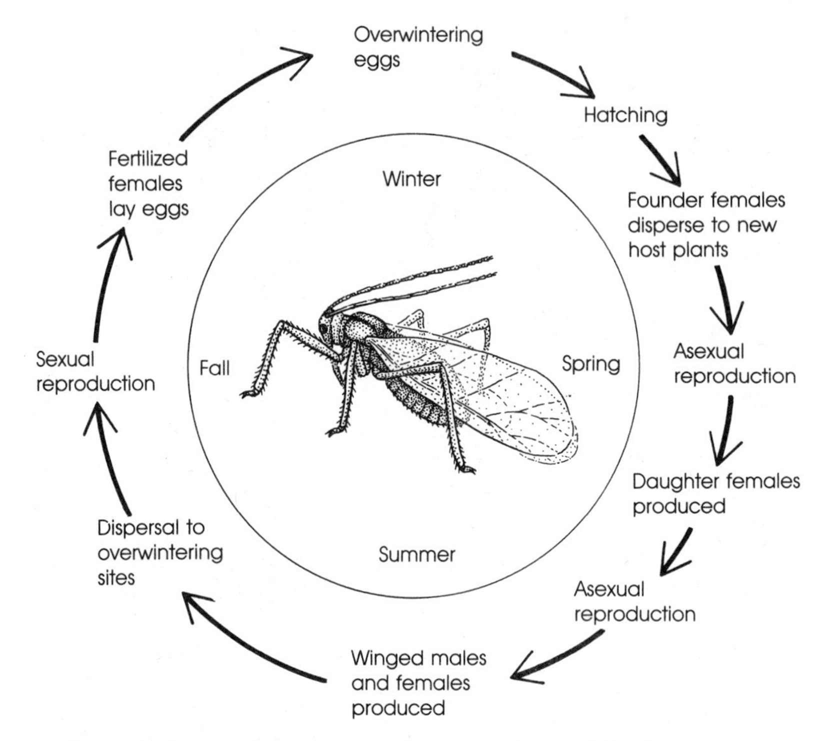

3 **Aphid life cycle pattern.** In a typical aphid life history, sexually produced offspring disperse widely and face somewhat unpredictable environmental conditions. The asexual offspring often remain in the same habitat with their mother.

but each egg is substantially larger than any sperm. The size differential between gametes can be much more impressive than that between male and female *Calopteryx*. For example, among birds it is not uncommon for a single egg to weigh 15–20 percent of the weight of the female, and some go as high as 30 percent [403]. With the same amount of resources, a male could produce trillions of sperm. Male birds often do make and expend millions of sperm in a single insemination attempt, but the male that used sperm equal to 5 percent of his body weight in a single season would be a sexual athlete of the first degree. Among mammals, the same correlation between gamete size and numbers produced holds true. A human female has only a few hundred cells that can develop into large eggs [155]. In contrast, a single male could theoretically fertilize all the women in the world because he can produce billions of minute sperm in his lifetime.

The fundamental difference between the two sexes may have had its origins in competing selection pressures favoring gametes that were good developers versus those that were good fertilizers [14, 568]. Imagine an ancestral sexual population in which there were no males and females, only individuals that produced gametes meiotically. As with almost any trait, there was bound to have been a certain amount of variation in the size of gametes produced in such a population. If so, selection might have favored both the relatively large ones (because these carried greater energy stores that promoted the survival of the zygote after fertilization had occurred) and the relatively small ones (because these were more mobile and better in the race to fertilize other gametes). If selection favored ever more extreme variants, the outcome would have been the evolution of distinct sexes, with males producing large numbers of highly mobile, lightweight gametes (sperm) that are adept at fertilizing eggs in competition with other sperm (Figure 4). On the other hand, females have evolved eggs whose large size and nutritional contents enhance their chances of development following fertilization.

There is an alternative hypothesis developed by Jeffrey Baylis that downplays the mobility advantages of small gametes and focuses on the production rate advantages of these gametes [51]. Simply by virtue of being small, sperm can almost surely be produced much faster than eggs. This means that males can afford to be gamete donors because they can replace what they give up quickly and relatively inexpensively. Males typically give their mates a number of sperm greatly in excess of the number of eggs to be fertilized. Humans are a good example. Although a female produces one or at most two eggs to be fertilized at any one time, males donate about 200 million sperm per ejaculate. This could help swamp a competitor's sperm that were also present in the female's reproductive tract and racing toward a mature egg. It is consistent with this hypothesis that the testes of male chimpanzees are much larger in proportion to body weight than those of men and gorillas. Chimp females in estrus regularly mate with several males, whereas gorilla and human females are much less likely to do so [307]. A second hypothesis is that females make it profitable for a male to offer them a large number of gametes by making fertilization contingent upon it [311]. From a large sample of sperm, all but the very best ones could be eliminated by physiological obstacles in the female's

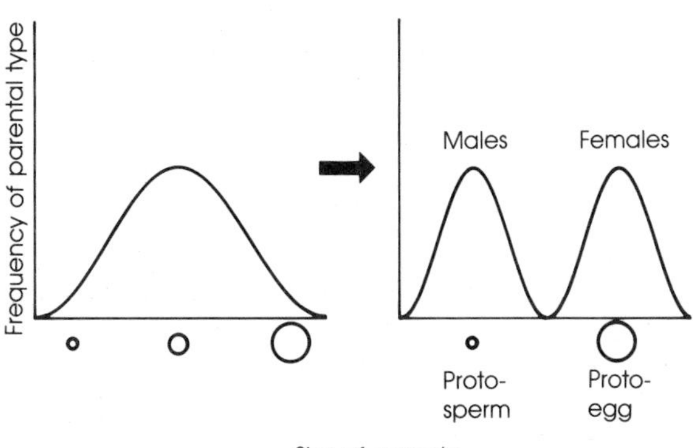

4 **Gamete evolution.** Selection against gametes of intermediate size could have led to the evolution of individuals (females) that produce large, developmentally advantaged gametes (eggs) and individuals (males) that produce small, mobile, easily replaced gametes (sperm).

reproductive tract. If mechanisms of sperm selection exist within the female, they could help her "choose" genetically superior combinations for her offspring and so help explain the maintenance of sexual reproduction despite the seeming advantages of parthenogenesis.

Parental Investment

Whatever its evolutionary basis, the male strategy of making a great many inexpensive gametes is in sharp contrast to the female strategy of producing many fewer, much larger gametes. The difference between males and females of a species in their allocation of resources per gamete can be expressed as a difference in their PARENTAL INVESTMENT per offspring [716]. Parental investment is whatever increases the probability that an existing offspring will survive to reproduce at the cost of the parent's ability to generate additional offspring. Males and females probably have access to equal amounts of energy to invest in gametes. Therefore, the female's "decision" to allocate relatively more energy per gamete means that she sacrifices opportunities to produce additional gametes in favor of increasing an egg's chances of developing after it is fertilized. The typical male makes a sperm that contains little or no resources that will help the zygote develop. Thus, the male's parental investment per offspring may be zero, but the female's parental investment per offspring will usually be substantial.

The concept of parental investment was invented by Robert Trivers, who noted that there is more than one way in which a parent can sacrifice for an existing offspring [716]. The female *Calopteryx* not only allocates considerable energy to each gamete, she also goes to the trouble of seeking out superior oviposition sites. While placing her eggs in plant tissues where her offspring will presumably have a better chance of surviving than if they were broadcast at random in the environment, the female is also vulnerable to attack by predators. The energy spent, time lost, and risks taken on behalf of one fertilized egg, at a cost to the female of future chances to reproduce, all constitute parental investments on her part (Figure 5).

One of the clearest examples of the gametic cost of helping an existing offspring comes from those tropical frogs whose females feed their tadpoles unfertilized eggs, thereby forfeiting some potential offspring to assist already living progeny [589]. Female animals often make these kinds of postfertilization investments to promote the welfare of their progeny, but males do so much less frequently. In many mammals, for example, the male impregnates a female and then departs. The female nourishes the embryo(s) that results, and then cares for and feeds her offspring with milk after they are born. These activities place a heavy energetic demand on a female, and they may lead to her death, either through increased physiological stress or through heightened exposure to predators. The maternal care given to an embryo or dependent offspring is often a major component of her parental investment.

5 **Female parental investment.** Baboon mothers feed and protect their offspring for many months after they give birth (*top left*). Photograph by Leanne Nash. Lizard females that are carrying a clutch of large eggs in their bodies are probably more vulnerable to predators (*top right*). Photograph by R. Shine. Crocodile mothers help their newly hatched offspring by carrying them in their mouths from the nest to the safety of the river (*bottom left*). Photograph by Jonathan Blair. Females of this cichlid fish secrete onto their bodies a nutritious substance that their young consume (*bottom right*). Photograph by A. van den Nieuwenhuizen.

Why Is Parental Care More Often Provided by Females?

The reason why females are more likely to offer parental care than males is not fully understood, but a number of hypotheses have been advanced. The first of these argues that the cost of parental care is greater for males than for females in terms of lost gametes. Because parental care requires energy, it diverts some amount of resources away from gamete production for both a male and a female. But the *number* of eggs that will not be manufactured is less than the number of sperm lost for the same amount of resources diverted to parental behavior. This means that the *potential* loss of offspring is greater for a male than for a female. Thus, the benefit:cost ratio of parental care is inherently less favorable for males than for females. The ecological conditions that promote parental care have to be extreme to make parental investment pay for males. For many species, ecological factors may favor female parenting but may not be severe enough to make parental care a favorable investment for males [794].

A second hypothesis is the "reliability of paternity" argument [290]. If a female lays a fertilized egg or gives birth to an offspring, this progeny will definitely have 50 percent of her genes. A male has no such assurance in many cases, especially if the species practices internal fertilization. Even though he may have copulated with a female, she may have mated with another individual whose sperm were used to fertilize her eggs. To the extent that a male runs the risk of caring for progeny other than his own, the benefit of parental care falls for a male, thus reducing the benefit-to-cost ratio of male parenting and making its evolution less likely than maternal care. This hypothesis, although intuitively appealing, cannot apply if the reliability of paternity is the same for parental and nonparental males within a population, as it must often be [771]. Imagine a species with a mix of males, mostly nonparental types, but with a few parental mutants. The fact that mutant males take care of some offspring will not in itself affect the "fidelity" of their mate(s). Thus, the reliability of paternity will be the same for the two kinds of males in the population, and its effect on the evolution of the parental behavior by males will cancel out. Paternal care can spread through the population, even if the reliability of paternity is low, *provided* the paternal male improves the survival rate of those eggs he does fertilize enough to compensate for any reduction in the number of mates he can find.

A hypothetical numerical example may help make the argument clearer. We shall set the proportion of eggs actually fertilized by a male when he copulates with a female at 0.40, a low figure. Because he is largely occupied with paternal care, the average paternal male copulates with only two females (each with an average of 10 eggs) whereas the nonpaternal types acquire five mates (giving them access to a total of 50 eggs). But the survival of protected offspring is 50 percent versus 10 percent for eggs that lack a paternal guardian. The reproductive success of the paternal male will be $0.40 \times 20 \times 0.50 = 4$ surviving offspring that bear his genes; the average fitness of the nonparental males is less ($0.40 \times 50 \times 0.10 = 2$). Thus, a low reliability of paternity is not in itself an absolute barrier to the evolution of parental care by males.

The third hypothesis for the prevalence of maternal care is perhaps the most straightforward of all [290]. Its simple point is that, because they carry the eggs, females are often in a position to do something helpful for the offspring when they emerge from the female's body. The father has no direct control over the release of offspring by his mate and may not be in the neighborhood when this happens. Even if he could direct helpful parental care and even if it would be in his genetic interest to do so, in many cases his physical separation from his progeny make paternal behavior impossible.

Tests that would enable us to discriminate conclusively among these hypotheses are not available at the moment. There is a correlation between male parental care and external fertilization in fishes [290]. But the correlation supports both the reliability of paternity argument and the proximity to offspring hypothesis. A male that fertilizes a clutch of eggs laid in a nest in his territory is likely to fertilize a high proportion of the eggs *and* he is close to his offspring and can help them. A male of another species that injects his sperm into a female that departs, not only is less likely to fertilize her eggs (because she might accept sperm from another male), but also is less likely to be around when the young are laid or born. Therefore, he does not have the same opportunity to exhibit paternal care as the male that fertilizes eggs externally.

Sexual Differences and Sexual Selection

Although the basis for the differences are not completely clear, no one doubts that females (1) produce larger gametes and (2) more often exhibit parental behavior than males. Thus, there is typically a large difference in parental investment per offspring by males and females of the same species. Robert Trivers proposed that sexual selection occurs because of this imbalance [716].

SEXUAL SELECTION is a component of individual selection that is created by the pressures males and females exert on conspecifics as they compete for mates and choose among potential partners. Charles Darwin was the first to define and discuss sexual selection. He recognized that males usually fight among themselves for opportunities to fertilize females, which appear to select among the males; but he was not entirely clear why this was the typical pattern. According to Trivers, it is the inequality in parental investment per offspring between the sexes that generates competition and mate choice. Whichever sex makes the greater parental investment per offspring will be a limited and valuable resource for the other. Thus, there will usually be competition among males for access to females, who will have the option of choosing among the many males that want to copulate with them. They should select individuals that will have the greatest positive effect on their fitness.

The test of Trivers's position comes from the exceptions to the typical case. There are some species in which females compete for males; and, as we shall see, in these cases males do (as predicted) provide unusual amounts of parental care or other limited resources to their mates. But far more often males compete in many ways to gain access to females and their eggs. To the extent that a male can prevent his rivals from mating with females and fertilizing their eggs, he should gain genetically. This generates the typical INTRASEXUAL SELECTION component of sexual selection [155, 794].

The damselfly, *C. maculata,* illustrates the impact of intrasexual selection on male behavior. The control of egg-laying sites by some males enables these individuals to monopolize the areas where receptive females are most likely to occur. As a result they have many opportunities to copulate and fertilize eggs with their abundant sperm. But the males that are excluded from prime oviposition and mate-contact points do not resign themselves to genetic failure. Instead, they move along the streamside looking for females that are about to go down to the water and may be willing to mate. Alternatively they try to "steal" ovipositing females from a territory by biting and pulling at the wings of an egg-laying female that has been left unguarded by her territorial mate. (He may be driving away another intruder or copulating with a new female that has arrived within his territory.) Sometimes this causes the harassed female to fly up to nearby vegetation and occasionally she will mate with the nonterritorial male [746].

The most dramatic product of intrasexual selection has its effect when one male succeeds in mating with a female who still has sperm stored from a previous mating. Females of *C. maculata* (like most female insects) have a special organ (the spermatheca) that holds the gametes received from a male until they are needed to fertilize mature eggs. The sperm reservoir creates a problem for a male that mates with a nonvirgin female. Whose sperm will she use? Jonathan Waage discovered that males of *C. maculata* possess two traits that increase the probability that their sperm will actually fertilize a mate's eggs [747]. The first tactic has a refreshing directness to it. The male uses his penis to scoop out any rival sperm in the female's spermatheca prior to releasing his own gametes into her. To this end, the penis has a shape that fits neatly into the spermatheca, and it possesses spiny lateral horns that hook into stored ejaculate and help remove it (Figure 6). By comparing the amount of sperm in samples of females captured before, in the middle of, and after copulation, Waage showed that males are able to remove 88–100% of the sperm their partners had stored.

The second option for improvement of egg fertilizations is restricted to territory owners, which stand watch over their mates (Figure 7) and are generally adept at detecting and repelling intruders. If a territory owner's mate is ovipositing, she will probably be using his sperm. If she is chased from the water and copulates again, he will lose some egg fertilization opportunities. Again Waage has shown that the male's guarding is effective in inducing females to remain longer in a male's territory than they would if not protected from assault [748]. In one population, females with a guarding male oviposited uninterruptedly for a mean of 14 minutes. When fe-

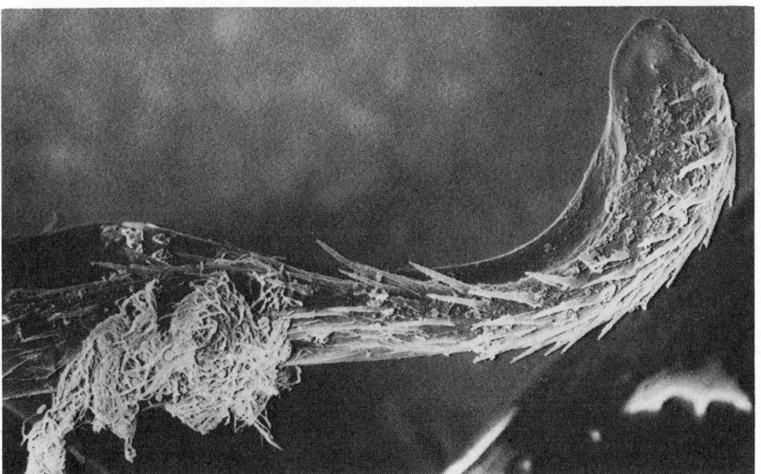

6 **Instrument of sperm removal** in *C. maculata*. Males of this species use their penis to remove rival ejaculate from their mate's sperm storage organ. The penis, with its backward-pointing spines and lateral horns (*top*). A close up of a lateral horn, showing a sperm mass caught in its spiny hairs (*bottom*). Photographs by Jonathan Waage.

males laid their eggs in areas without a guarding male, oviposition lasted for only two minutes (an average from many observations) before the female was chased from the water by a passing male intent on copulation. Thus, males of *Calopteryx* behave in ways that are easily interpreted to be adaptations that help individuals fertilize as many eggs as possible, within the constraints imposed by their rivals, who are also trying to do the same thing.

7 **Mate-guarding male of** *C. maculata* (wings spread) on the oviposition material in his territory, with three ovipositing females. The male will try to prevent other males from disturbing his females. Photograph by the author.

In contrast to the copulatory "eagerness" of males, females of this species regularly forego some chances to copulate. This is a widespread feature of female behavior in "typical" species in which females make a greater parental investment per offspring than their mates. The inequality of parental investments means that a reproductive "mistake" is far more serious for females than for males [550]. If a male impregnates a female that is unlikely to produce viable young, he has lost little that cannot be replaced in a short time. But if a female mates with a male of the wrong species or with an inferior male of her own species, she may lose an entire clutch of eggs. Thus, females gain more than males through a discriminating choice of partners. If either sex does reject potential mates, EPIGAMIC SELECTION is the result. This is the component of sexual selection in which the sex (usually the female) that is in demand exerts selection that favors individuals of the other sex that have the preferred attributes [155, 794].

Females of *C. maculata* behave as though they were selecting partners with characteristics likely to enhance their fitness. They are more likely to accept the courtship of territorial males than the copulatory attempts of subordinate intruders. As a result, they usually gain entry into superior oviposition sites controlled by their partners. In addition to the material benefits from mating with territory owners, females may also derive some genetic benefits. If the characteristics of territorial males contribute to reproductive success and are heritable, females that mate with such males may confer a genetic advantage on their sons, who may become superior competitors as a result.

Having illustrated the two components of sexual selection with reference to damselfly behavior, the remainder of this chapter will deal with the multiple evolutionary effects of intrasexual selection throughout the animal kingdom. Chapter 12 will deal with epigamic selection and female reproductive strategies.

Intrasexual Selection and Male Reproductive Competition

If male reproductive success is rarely limited by the number of gametes he produces but instead by how many eggs he fertilizes, sexual selection should favor males with traits that (1) help an individual have a chance to copulate with as many different females as possible, (2) increase the probability that his sperm will be used by his mates to fertilize their eggs, and (3) reduce the fitness of competitor males (Table 2).

Let us consider the first category of characteristics. A common feature of males of many species is a remarkably low threshold for sexual arousal [550]. This trait expresses itself in dramatic fashion in the frequency with which males engage in sexual behavior with biologically inappropriate stimuli (Figure 8). The ease with which males can become sexually aroused has been exploited by ethologists, human pornographers, some plants, and

Table 2
Results of Intrasexual Selection on the Evolution of Male Behavior

Adaptations that play a role in the competition to secure copulations

A. A low threshold for mating attempts

B. Dominance behavior (the monopolization of females): the exclusion of other males from
   1. The vicinity of a female that is or will become receptive
   2. A harem of females
   3. Positions of high status in multimale groups
   4. Areas with useful resources attractive to females
   5. Display sites attractive to females

C. Subordinate behavior (coping with dominant males)
   1. Submissive behavior and the (temporary) postponement of attempts to reproduce
   2. Courtship of females only when dominant males are absent or occupied
   3. Sneak copulations
   4. Rape

Adaptations that help a male ensure that his sperm will be used by females he has inseminated

A. Guarding behavior
   1. Temporary guarding of a recently inseminated female
   2. Prolonged protection of a female or harem against other males

B. Behavior that reduces the likelihood that an unguarded female will copulate again
   1. The insertion of mating plugs in females after copulation
   2. The use of biochemical or behavioral signals to activate mechanisms within the female that reduce her receptivity

Adaptations that lower the reproductive success of competitors

A. Sexual interference
   1. Interruption of another male's courtship
   2. Female mimicry by males that induces other males to waste time, energy, or sperm

B. Attempts to injure competitors
   1. Assault on other males, including homosexual rape
   2. Assault on mates or offspring of other males

8 **Mating mistakes made by males.** The drive to copulate is often so strong in male animals that they sometimes try to mate with inappropriate objects. A European toad clasps a human finger as it would a female of its species (*left*). Photograph by Tony Allan. An Australian beetle attempts to copulate with a beer bottle. Note the extended penis (*right*). Photograph by David Rentz and Darryl Gwynne.

some predatory species like the bola spider and firefly *femme fatales,* which use models or mimetic releasers of copulatory behavior for their own purposes (Chapter 9). The sexual enthusiasm typical of males may contribute to the prevalence of male homosexual behavior in humans and many other animals, which contrasts sharply with the relative rarity of female homosexuality [690].

Males also often attempt to force unwilling females to mate. For example, a study of the crabeater seal showed that males associate themselves with females that are resting on ice floes with an unweaned pup [663]. A male will at intervals make attempts to copulate despite the fact that females do not become receptive until after the pup is weaned and independent. Nor is the male gentle in his approaches; he frequently bites the female after being rebuffed (and is bitten in return). D. B. Siniff saw males and females covered in blood as a result of their sexual disputes. Although perhaps an extreme example, nothing could illustrate more clearly that sexual reproduction is not a gloriously cooperative enterprise designed to perpetuate the species. A male crabeater seal gains by inseminating a female as soon as possible because the longer he waits the more likely he is to be supplanted by a rival [663]. Moreover, the sooner he copulates, the sooner he can get on with the search for a second mate. The female, on the other hand, has an interest in seeing to it that her pup is not prematurely pushed into independence. It may also be to her advantage not to mate with the first male that comes along because a better one might appear later. Sexual conflict is the result.

Observations on the "hypersexuality" of males are consistent with the hypothesis that because male fitness is generally a function of the number of eggs fertilized, males are under selection pressure to take every opportunity (and then some) to copulate in order to maximize the transmission of their genes. A low threshold for sexual behavior should often help males avoid missing a potential mate (a benefit of the trait), but it also means that males are more likely than females to engage in misdirected or resisted copulatory attempts (a cost of the trait).

We can test this hypothesis by identifying the conditions under which males will *not* be sexually indiscriminate. If, for example, males have several simultaneous opportunities to mate, they should pick the female that will yield the greatest rate of egg gain. In some species, males live with their mates and therefore have the chance to copulate with these individuals repeatedly. They may also have occasional chances to copulate with unfamiliar females. We can predict that under these circumstances males should favor novel females over familiar ones. We have already discussed an example of a male sexual preference for unfamiliar mates in describing house mouse reproductive behavior (Chapter 6). A territorial male lives with his harem and becomes habituated to his pregnant females (for obvious adaptive reasons) but is quick to respond sexually to a new addition to his group. Moreover, although the harem master may find his pregnant mates sexually neutral, other males that happen to contact them do not treat them in this fashion [458, 642].

The "Coolidge Effect"* is exhibited by many mammalian males (but not in obligatorily monogamous species [155]). In rats, sheep, cattle, rhesus monkeys, and humans, males that have copulated to satiation with one female can be speedily rejuvenated by the opportunity to copulate with a new female [690]. Nor is this phenomenon restricted solely to mammals, as it occurs in a number of bees that can identify their recent mates and avoid them in order to search for new partners to inseminate [7, 46].

Male rejection of potential copulatory partners is also linked with ecological conditions that (1) make copulation costly for a male and (2) cause females to vary in their fecundity. Our expectation of copulatory eagerness was predicated on the assumption that males can quickly and cheaply replenish their supply of gametes and any other materials passed on to a female. Sometimes this assumption clearly does not apply. If the male's ejaculate is very large, it may represent a significant physiological expense and may limit the number of copulations a male can hope to have in a period of time. Under these circumstances, a male should exercise choice if there is variation among females [176, 625].

An example is provided by *Anabrus simplex,* the Mormon cricket, which despite its common name has no religious affiliation and is not a cricket.

*Gordon Bermant [690] tells the story of how the "Coolidge Effect" got its name: "One day the President and Mrs. Coolidge were visiting a government farm. Soon after their arrival they were taken off on separate tours. When Mrs. Coolidge passed the chicken pens she paused to ask the man in charge if the rooster copulates more than once each day. "Dozens of times" was the reply. "Please tell that to the President," Mrs. Coolidge requested. When the President passed the pens and was told about the rooster, he asked "Same hen every time?" "Oh no, Mr. President, a different one each time." The President nodded slowly, then said "Tell that to Mrs. Coolidge."

Males of this large flightless katydid transfer an enormous spermatophore to their mates (Figure 9) along with their sperm [293]. The spermatophore is evidently nutritious because the female consumes it after copulation and sperm transfer are complete. Donation of a spermatophore reduces a male's body weight by about 25 percent; therefore it is reasonable to assume that the transferred materials cannot be quickly and easily replaced. This limits the number of times a male can copulate during his lifetime and favors individuals that choose superior mates. When Mormon cricket populations reach a high density, bands of them march across the countryside eating farmers' crops and mating as they go. Under these circumstances, males have access to many potential mates. When they begin to stridulate from a perch, announcing their readiness to mate, females come running. (Note that this sex role reversal is in keeping with the prediction that when one sex provides a limited resource during copulation, the other sex will compete for chances to mate.) In order to copulate, a female must mount the male, who then inserts his genitalia and transfers sperm and the spermatophylax. But males refuse to transfer a spermatophore to some females. In Darryl Gwynne's study, the average weight of rejected females was significantly less than those that were "permitted" to copulate. By mating with heavier females, males transferred their sperm to more fecund individuals. A male that rejected a 3.2-gram female in favor of a 3.8-gram mate gained about 50 percent more eggs as a result of his choice [293].

The same preference for large, more fecund females has been demonstrated in the wood frog, *Rana sylvatica* [89, 347], and other animals [660] in which the cost of "copulation" for a male is time. A male frog must

9 **Material benefit of copulation.** A female Mormon cricket consuming the huge white spermatophylax she received from her mating partner. Photograph by Darryl Gwynne.

remain in amplexus with a mate until she reaches an oviposition spot and gradually lays her clutch of eggs while he fertilizes it externally. Because wood frogs breed in large aggregations that may form on just one night each year, a male has only a short time to secure his partner(s). But there are a great many females present to choose among. Males grasp and release relatively small females until they find a large one with whom they remain until the egg mass is fertilized.

Risk-Taking and Fighting by Males

Even though males of a few species are discriminating in their selection of mates, under most circumstances males are much less likely to turn down a partner than are females. It is also a general rule that males will take greater risks to secure copulations than females. The genetic gains for a successful male are potentially so great that males can "afford" to engage in costly actions if there is some probability of achieving multiple copulations as a result.

The risks that males take in the mating race are diverse and include things we have already discussed, like being attracted to a lethal pseudo-female when searching for the real thing [435]. Males of many species are also prone to predation while trying to attract a female to them [114, 628] (see Figure 25, Chapter 13). Reproductive competition among conspecific males favors individuals that employ conspicuous signals, such as loud calls or bright visual displays, to attract mates. These activities make it easy for a female to locate a male, but they also make it easy for predators, too. If a *Photinus* firefly is not lured to his death by a *Photuris* femme fatale, he may be hawked out of the air by a *Photuris* predator that uses his bright flashes as cues to his position (Figure 10).

It is significant that in species in which females do the calling, the signal is much more likely to be a subtle chemical scent (a pheromone) than a highly conspicuous visual or acoustical message [280]. Because minute amounts of pheromone are released by females, the signal probably cannot be easily detected and exploited by predators, although there are exceptions [740]. A female moth has the luxury of using a difficult-to-locate and relatively inconspicuous attractant because her males, under pressure to copulate frequently, have evolved large antennae and great sensitivity to the pheromonal stimulus [637, 638].

A higher exposure to predators is not the only consequence of the male motivation to find or attract mates. Males often create risks for each other by fighting for possession of or access to females (Figure 11). We earlier noted that dominant male cowbirds are both willing and capable of killing rivals that threaten to attract females from them (Chapter 4). In their studies of red deer, Timothy Clutton-Brock and his associates found that each year 6 percent of the male population sustained a serious injury in clashes with other males [137]. Injuries of this sort not only pose the risk of lethal infection, but they may increase the vulnerability of a deer to its predators. Death and injuries resulting from fighting among males occur regularly in invertebrates. Some territorial bees can sever the wings of intruders, and dragonfly males may drown after having been blasted into the water during an aerial clash with rivals [700].

The importance of fighting ability to male reproductive success in many species is seen in the direct correlation between male territorial ownership and mating success in many birds (e.g., the black-headed gull) and mammals (e.g., the impala). Winners of territorial contests also outreproduce losers in some species of fishes, lizards, and insects [699, 718, 757]. In species in which groups of individuals live together, the adult males of the group usually form a dominance hierarchy based on the ability of individuals to displace one another from desired resources, particularly females.

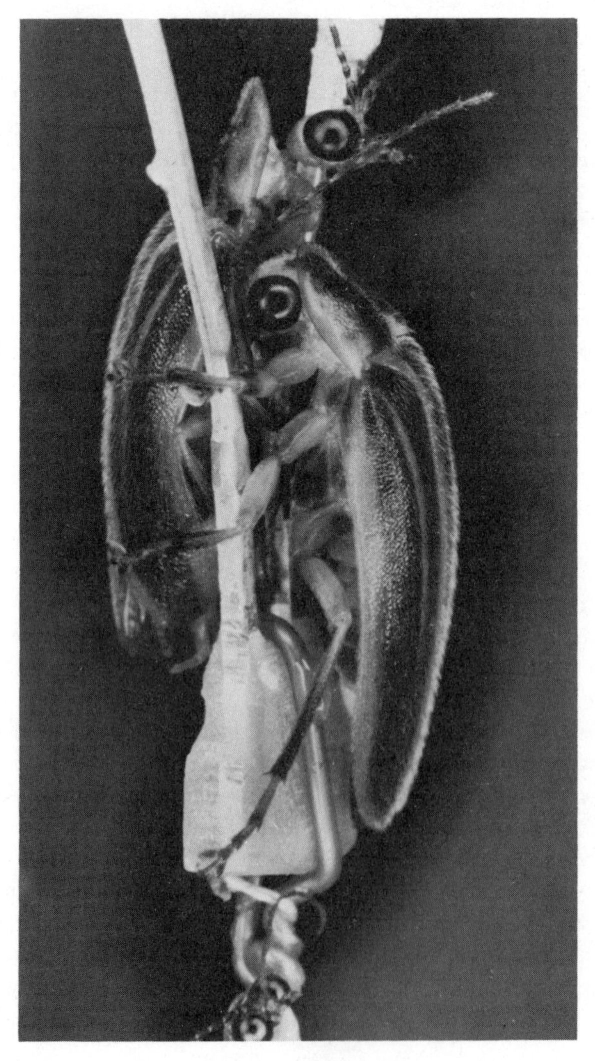

10 **Signaling for mates is risky** for males of *Photinus* fireflies. The light flashes of a flying male attract hunting female *Photuris* fireflies. One can experimentally draw in the predatory females with light-emitting diodes (the pale object at the bottom) dangled at the end of a fishing line. Here two female *Photuris* have arrived nearly simultaneously at the artificial "male"; one female has killed the other. Photograph by James E. Lloyd and Steve Wing.

Although doubt has been expressed from time to time about the relation between male dominance status and the monopolization of females, there are now many studies that have documented that higher ranking individuals outreproduce males of lower status [24, 346, 453].

One such study involves the southern elephant seal. T. S. McCann watched a breeding population of this species hauled out on a beach on South Georgia Island in the Atlantic Ocean [495]. He ranked a set of ten individually recognized males on the basis of the outcome of aggressive interactions among them. These impressively massive animals roar at one another and sometimes face each other chest-to-chest to slam their upper bodies together and bite one another about the neck. Losers can be readily identified when they drop their aggressive stance, often retreating with the winner in pursuit. (Most interactions, however, do not reach this stage but are resolved with a threat or two.) The top five males were easily ranked. The number 1 male had from 14 to 157 encounters with the other top nine males and won them all. The number 2 male elicited submission from all but the top male. The hierarchical status of the sixth- to tenth-ranked males was less clear, as males within this group sometimes won and sometimes lost when paired against each other. McCann recorded the number of copulations that each individual secured over the breeding season (Figure 12). These data clearly support the hypothesis that dominance (and thus fighting ability) lead to sexual success.

But when Glen Hausfater counted copulations in a troop of the baboon *Papio cyanocephalus,* he discovered only a weak correlation between dominance ranking and the total number of copulations achieved by a male [313]. The dominance hierarchy within the group was clear; and yet, during the estrous phase of female receptivity, males of both low and high status were likely to copulate with a female. The solution to this puzzle came from the discovery that a high-ranking male is very likely to guard an estrous female during the third day before the female's sexual skin shrinks in size. It is on this day that the female usually ovulates. In other words, a dominant male does not closely guard an estrous female early and late in her cycle (and at this time she may copulate with low-ranking baboons). But when a female is likely to be fertilized, the dominant individual mates with her and prevents other males from approaching her [313]. Thus, although dominance status is not closely tied to the number of copulations a male achieves, it is strongly correlated with the number of *effective* matings a male secures (Figure 12).

Species Differences in Fighting Intensity

Although combat over females by males is common, it is by no means universal; and even among the species where it occurs, there is considerable diversity in the intensity with which males compete with one another. Thus, males of some species fight to the death to mate; in others, they may threaten or wrestle with one another but do not kill opponents [801]. Interspecific variation of this sort gives us an opportunity to test the

◀ 11 **Fighting males** in competition for females. Two male gray kangaroos are "boxing" in an aggressive encounter (*top left*). Photograph by Gordon Sanson. Two mountain big horn sheep collide in a dominance dispute (*bottom*). Photograph by Valerius Geist. A male tassel-eared squirrel pulls a copulating rival from a female (*top right*). Photograph by R. C. Farentinos. A male tree frog tries to displace an opponent from the back of a female (*center right*). Photograph by Joseph Bagnara.

hypothesis that males should allocate time, energy, and risk-taking to fighting to the extent that it improves their fitness. A prediction from this hypothesis is that fighting to the death should only occur if alternative mating chances are very limited [774]. Thus, we do not expect extreme violence to occur regularly in potentially long-lived species because males of this sort may move up a dominance hierarchy or into a vacated territory if they avoid losing their lives in an early battle.

But in certain insects, particularly some parasitic wasps and fig wasps, death by dismemberment is a common fate for a loser male. In these animals, males typically emerge in numbers from a victimized host or a fig

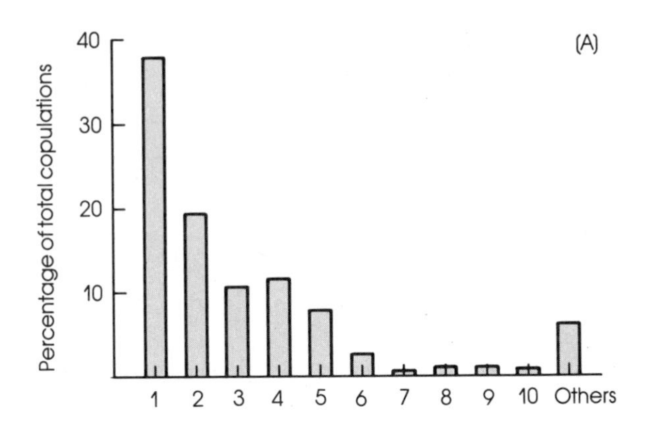

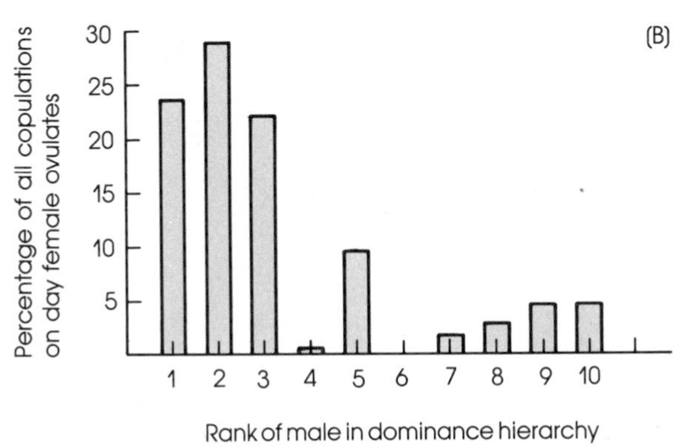

12 **Male dominance and copulatory success.** (A) There is a strong correlation between a male's rank and the percentage of all copulations he achieves in the southern elephant seal. (B) High-ranking male baboons monopolize females on the day their mates ovulate.

and fight for the right to fertilize the large number of receptive females that will emerge shortly thereafter [700]. Thus, the potential gain for a winner is high, a situation that enables males to make a large investment in combat and still gain (if they win). In these cases, alternative options for a male are also limited. A male parasitic wasp probably has little chance of finding a suitable host from which females were about to emerge; and even if he did, he would have to contend with the males that emerged from it. In the fig wasps, fighting males are wingless, a condition that greatly reduces their options. These males stay where they emerge and battle it out, slicing each other to pieces with their powerful jaws until only one survivor remains to claim the spoils of victory.

A more general prediction, which can apply to males that do not fight to the death, is that the greater the potential reproductive return for a winner, the greater the investment a male will make to be an effective fighter. In most animals, the larger the male the more likely he is to win a contest in which physical strength is important. This rule applies to creatures ranging from bees [8] to crayfish [682] to elephant seals [405]. Large size is expensive because it requires an investment of calories and time and because, once achieved, it must be maintained even though heavy body mass can interfere with other aspects of reproductive competition and foraging efficiency [651]. Male lions, for example, are larger than female lions and have evolved impressive manes with which to intimidate rivals. These traits, however, make them more conspicuous to their prey; therefore, whenever possible, males do little hunting, leaving it to the females, which are better designed for prey capture [635].

One way to measure the *degree* of investment of males in fighting ability is to compare the ratio of male and female body sizes (the sexual dimorphism of the species). Richard Alexander and his co-workers predicted that in species in which some males could secure relatively many mates, the degree of sexual dimorphism would be greater than in species whose males were limited to one to two females per breeding season [16]. When multiple copulations are possible, the intensity of competition should increase and lead to selection for attributes (such as large body size) that facilitate success in combat. This prediction is supported by the observation that male mammals are usually much larger than females when the potential for exceptional mating success is high (Figure 13). Thus, an elephant seal male may have as many as 100 mates in his harem, and males are about 60 percent longer than females and much, much heavier. Monogamous pinnipeds, in contrast, show little or no sexual dimorphism in body length (and weight) [16].

How To Cope with Dominant Males

In the competition for receptive females, territories, or high social status, there are of necessity both winners and losers. Historically, the losers have been thought to be superfluous animals not needed for the preservation of the species, that is, animals whose only possible function

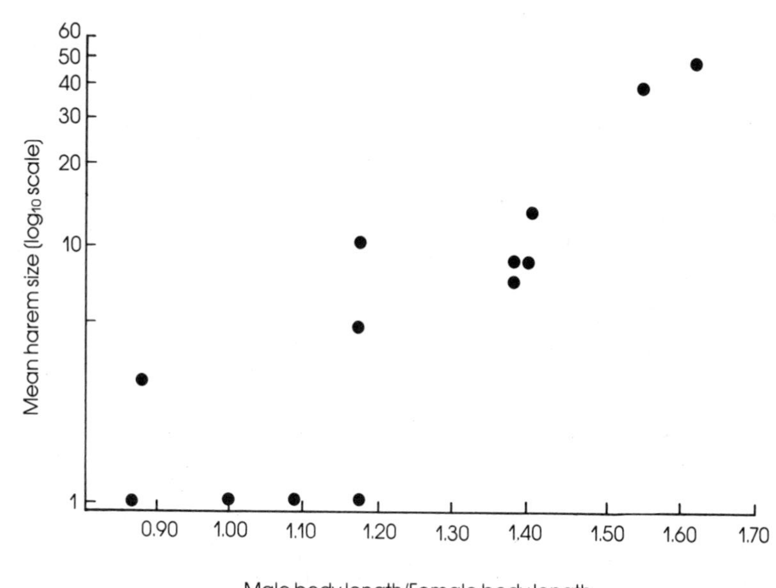

13 **Potential for polygyny and male body size** in species of pinnipeds (seals, sea lions, and elephant seals). When some males can acquire large harems, the males of that species tend to weigh much more than their females. In monogamous species, males and females weigh about the same.

was to provide a population reserve in the event that dominant individuals were unfortunately lost as a result of old age, predation, or disease. But application of individual selectionist thinking instead of group selectionism has resulted in a reevaluation of the adaptive significance of the behavior of subordinate animals [491, 567, 786].

These individuals typically adopt a role that appears to minimize the costs of aggressive interactions with dominant males. This helps the animal to survive long enough to reproduce eventually. As noted before, it can hardly pay a small or inexperienced male to get severely thrashed by a larger or more experienced opponent. Risky physical interactions among males are relatively rare in most cases. For example, only 4 percent of 4000 fights between southern elephant seals involved actual contact between the two males [495]. As a rule, a large dominant male need merely threaten a subordinate to have that individual retreat or engage in a submissive behavior that signals acceptance of his inferior status. Subordinate individuals assess their chances of winning, find them poor, and avoid combat until conditions are better.

Because body size is so important in determining who will win an actual fight, it is not surprising that assessment of an opponent is often based on cues that correlate reliably with body weight. The European toad, *Bufo*

bufo, is a species in which body size is an excellent predictor of fighting success [163]. A male that has grasped a female is almost never supplanted by a smaller opponent but may be displaced by a larger one (see Figure 11). When a male finds a pair, he may try to pull his rival from the female. As soon as contact is made, the defender croaks. Large males have deeper pitched croaks than small toads, whose vocal apparatus is simply incapable of generating low-frequency sounds. An attacker might use this information to determine whether it was profitable to continue his attack. Nicholas Davies and Timothy Halliday tested this idea by placing mating pairs of toads in tanks with an attacker for a 30-minute trial [163]. The paired males had been silenced by looping a rubber band under their arms and through their mouths. Whenever the second male touched the pair, a tape recorder supplied a 5-second call of either low or high pitch. Small defenders were much less frequently and less persistently attacked if the other male heard a deep-pitched call (Figure 14). Thus, deep croaks do deter potential attackers (although tactile cues also play a role in determining attack frequency and duration, as one can see from the higher overall attack rate on small toads). By using acoustical signals, males could in nature accurately determine their chances of successfully displacing a defending male. Smaller toads could avoid costly fights as a result.

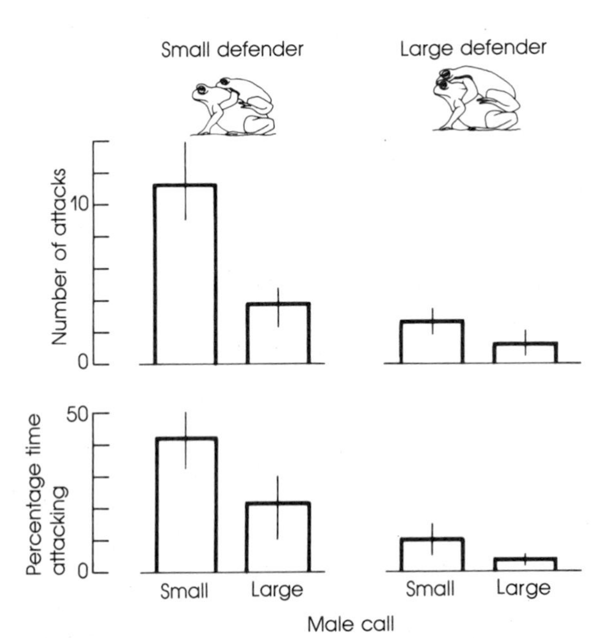

14 **Deep croak experiment.** Single male toads make fewer contacts and interact less with silenced mating rivals when a tape of a low-frequency call is played than when a higher-frequency call is played.

Sneak Copulations

Subordinate males may do more than merely try to avoid more powerful opponents. They may also try to acquire some mates along the way in the manner of nonterritorial males of *C. maculata,* which lurk about other males' territories waiting for the occasional chance to copulate. Two common tactics of subordinates are female mimicry and satellite behavior. In elephant seals, roe deer, and a variety of fishes, some males behave like females and so can sometimes enter another male's territory or harem without being attacked. Once there, the pseudofemale may on the sly fertilize females that would normally mate only with the territorial male [140, 243, 288, 729]. Satellites occur in many of the species of crickets, toads (Figure 15), and frogs in which some males employ loud acoustical signals to attract females; the silent satellites slip in close to a calling male and intercept females that were zeroing in on the caller [119, 574, 766].

There are many additional examples of the coexistence of two or more very different reproductive tactics within a species. For example, the dominant male in a group of chimpanzees may form an exclusive consort relationship with an estrous female, thereby preventing other males from copulating with her. Lower ranking males may attempt to induce a receptive female to leave the group (and the alpha male) to forage (and mate) with them for a time. Sometimes they also join the top male and other group members in repeatedly copulating with a female that the alpha male

15 **Satellite and calling males** of a spadefoot toad. The calling male has his throat pouch inflated. Two silent males (arrows) wait to intercept females attracted to the caller. Photograph by Brian Sullivan.

chooses not to monopolize [725]. In the bluehead wrasse, smaller males excluded from territorial ownership of an egg-laying site on the edge of a coral reef wait in groups by a territory for an egg-laden female to pass by [757]. The little fish may swarm about such a female, subjecting her to tactile stimulation, which sometimes triggers spawning and gives these individuals a chance to fertilize some eggs (Figure 16).

Alternative Behavioral Traits: Their Evolutionary Maintenance

The co-occurrence of alternative traits within a population raises a theoretical question: Why hasn't selection eliminated all but the most reproductively successful of the alternatives? The answer to this ques-

16 **Alternative mating tactics** in the bluehead wrasse. (*Top*) A large territorial male over a coral reef spawning site. (*Bottom*) Small, nonterritorial males swarming about an intercepted female in an attempt to induce her to release her eggs. Photographs by Steven Hoffmann.

tion in many cases is that the different categories of males do not differ genetically with respect to their reproductive behavior, and therefore selection cannot eliminate one variant in favor of the other(s). The dominant (or territorial) role and the subordinate (or nonterritorial) role within a species often represent different options that can be adopted by any male in the population, the choice being dependent upon the conditions the male happens to encounter [169]. A young male damselfly that finds all the prime territories occupied by older males may temporarily engage in mate stealing when possible while waiting for vacancies to occur. Likewise a bluehead wrasse that would be unable to displace a large territory owner may try to fertilize some eggs by the swarming tactic but would probably quickly enter an empty territory if the resident were removed. In the jargon of game theory, these options are part of one CONDITIONAL STRATEGY [43, 169].

The genetic basis for a conditional strategy leads to the development of neural (or hormonal) systems that enable a male to respond to changing conditions by changing its behavior. (Despite the connotations of the word "strategy," there is absolutely no requirement that the individual be consciously aware of what it is doing.) Conditional strategies are often advantageous because the flexibility they provide may help a male salvage some reproductive opportunities, even if excluded from the most profitable method of acquiring mates. In Richard Dawkins's terms, they help subordinate, younger, or weaker males make the best of a bad job [169]. Imagine a population consisting of some males with the conditional ability to adopt a subordinate role and other genetically different males that persist at all costs in trying to become a territorial or dominant male. If, as is often the case, the persistent male would exhaust itself in futile attacks on stronger males, the conditional strategy will be the EVOLUTIONARILY STABLE STRATEGY (ESS) [487]. An ESS is a behavioral program that cannot be replaced over evolutionary time by an alternative behavior (in this case, the alternative strategy of unconditional aggression), provided present circumstances persist.

One can best demonstrate that a behavioral variant is a conditional strategy by showing (1) that one option is reproductively more successful than the other(s) and (2) that individuals that are forced to practice the low-yield options will switch to the superior practice when conditions permit. This has been done nicely with scorpionflies of the genus *Panorpa* [699]. In some species of these insects, three different tactics for mating exist within the same population: (1) some males defend dead insects that attract receptive females, which feed upon the carrion; (2) others secrete salivary materials on leaves and wait for females to come to consume this nutritional gift; and (3) still others offer their mates nothing at all but pounce upon them to secure a forced copulation (Figure 17).

Randy Thornhill placed groups of ten male and ten female *Panorpa* in cages with two dead crickets. Some of the male *Panorpa* were large, others medium-sized, and still others were relatively small. They competed for the crickets and, not surprisingly, the larger males won. These individuals

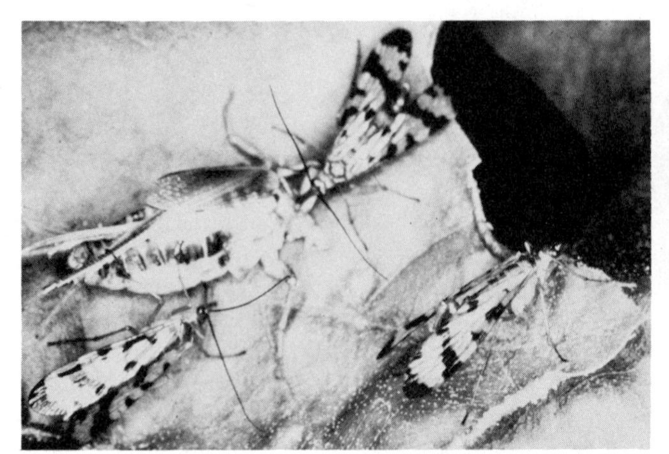

17 **Conditional strategy** of *Panorpa* scorpionflies. A male can either defend a dead insect (*left*) or produce a salivary mass to attract females (*right*). Photographs by Nancy Thornhill and Thomas Moore. A third option is to try to force females to mate without offering a meal of any sort.

monopolized the crickets and secured 60 percent of all the matings, averaging nearly six copulations per individual over the course of the experiment. The medium-sized males generally attempted to attract mates with salivary gifts but were much less successful (gaining about two copulations per male). The small males were unable to claim crickets and appeared incapable of generating sufficient saliva to be attractive to females. They employed the forced copulation route but were least successful of all (averaging only about one copulation per male). These results strongly suggest that the different tactics do not produce equal fitness gains.

In additional experiments of a basically similar nature, Thornhill removed the males that were able to defend carrion from an enclosure [699]. Other individuals that had previously been excluded from these items promptly abandoned their salivary mounds and moved to claim the dead crickets as these became available. Males that had not possessed a cricket or a salivary mound took the salivary secretions of other males when these were abandoned. As predicted from a conditional strategy hypothesis, male *Panorpa* are able to change their behavior, their choice of behavior depending on the nature of the competition they face. Moreover, they choose whichever option yields the highest possible rate of copulations at that moment.

The Coexistence of Two Distinct Strategies

Not all alternative tactics need to be the products of a single conditional strategy. As John Maynard Smith has shown with the algebra of game theory, it is possible for a MIXED ESS to persist indefinitely within a population. A mixed ESS is one composed of two pure strategies, each with its own genetically distinct basis, *or* one in which the population is made up of individuals that all play two roles with the same fixed probabilities [43]. For example, in most species males and females represent two different strategies with respect to gamete production. Typically, once a male, always a male, and the same is true for females; in which case, a mixed ESS results from the coexistence of two pure strategies. But in some animals, sex changes are possible, with sperm-producing males becoming egg-producing females and vice versa. Therefore, it is theoretically possible for such a species to possess a mixed ESS in which all individuals practiced the same mixed strategy (e.g., each individual might spend half its life as a male and half as a female, although I know of no such species in nature).

However achieved, a mixed ESS is most likely to evolve when the fitness gains associated with any one option are FREQUENCY DEPENDENT. Frequency-dependent selection occurs when a behavioral type enjoys high reproductive success when it is rare, but low fitness when it is the commonly practiced option. For example, males and females coexist in an evolutionarily stable fashion because if males come to outnumber females in a population, each female will enjoy higher fitness on average than each male. But if there are more females than males, the average fitness of a male will exceed that of a female.

Imagine a parent that invests more in the production of sons than daughters when the sex ratio in the population as whole is one male for every five females. Such a parent would propagate its genes much more effectively than one that allocated its parental resources primarily to create daughters. (Each son can expect to have five times greater reproductive success than a daughter.) But if the sex ratio becomes strongly male-biased, parents that invest in daughters will have greater genetic success than those that generate sons. When a son or daughter require the same parental investment, the only evolutionarily stable strategy (in an outbred species) is for a parent to invest equally in the two sexes. Only then will the average genetic return per unit investment in a son or daughter be the same. This produces the typical 1:1 sex ratio and the stable maintenance of the two distinct male and female reproductive strategies in sexually reproducing organisms [238].

We can analyze alternative mating tactics within a sex in much the same way. Bluegill sunfish males practice either a territorial, nest-defending role or a nonterritorial, sneaker–satellite role [288]. It has been shown that males must adopt one or the other of these alternatives at an early age and almost certainly cannot switch later. Thus, we seem to be dealing with two pure strategies [184]. Although territorial males appear to have more opportunities to fertilize eggs in any one breeding season, the nonterritorial types enjoy a major benefit. They begin to reproduce at a much earlier age (at two to three years instead of six or seven) because they do not have to be large and aggressive in order to defend a territory. Probably many males that commit themselves to the territorial strategy do not live long enough to be able to begin to practice it [288]. It is just possible that the average reproductive benefits of the two tactics are balanced when the frequency of the territorial types is high enough to support a certain number of sneaker males. (On the other hand, the two alternatives might be the two options of a conditional strategy; males that were, for example, malnourished in the first year of life and were unlikely to grow large might use this conditional cue to switch to the developmental path leading to a nonterritorial tactic.)

The maintenance of two distinct strategies within the same population has not been definitely established for many behavioral traits, but satellite and calling tactics in crickets is one such case. William Cade has shown through artificial selection experiments (Figure 18) that there is a genetic component to the behavioral differences between males that call frequently and those that rarely call [120]. By definition this indicates that the two characteristics are pure strategies because each has a distinct genetic basis. Here, too, an element of frequency-dependent selection seems certain to be involved because silent males presumably rely on callers to attract females that they can intercept. The lethal parasitoid flies that attack calling males

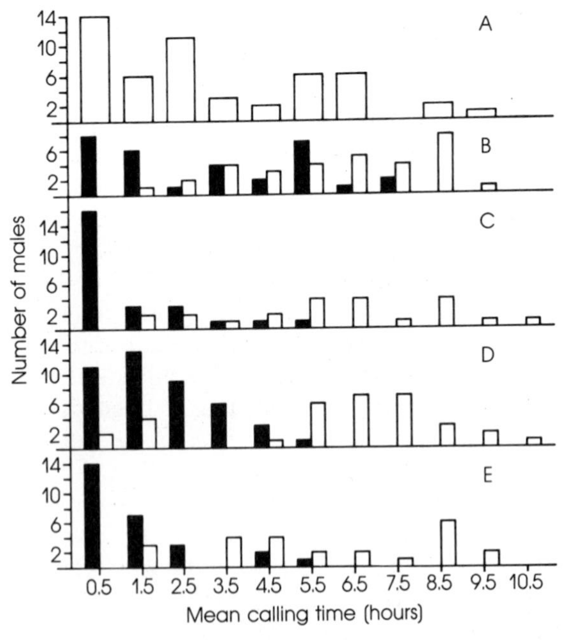

18 **Artificial selection experiment** on calling behavior in the cricket *Gryllus integer.* By selecting males at either end of the initial frequency distribution (A) of mean calling time per night, two distinct populations of males were produced by generation E. One group was largely silent; the other was composed of highly active callers.

may also play a role in maintaining genetic variation in the population because the density of these enemies varies a great deal from year to year or from place to place. As a result, the selective advantage of calling probably fluctuates, a situation that contributes to the maintenance of the competing alleles [118, 120].

Although Cade's work shows that there is a genetic difference between the two traits, he has not yet demonstrated that the two types have equal fitness when there is a certain ratio of callers to noncallers. Work with another species, the green tree frog, suggests that the mating success of calling males and silent satellite males might be equal in this species [574]. In the field, about one-fifth of all calling males have a satellite stationed near them. When a receptive female was experimentally released near a caller and a satellite, the probability that she would be intercepted by the satellite was 43 percent. The success of the silent option might be higher still over the long run because satellites do not pay the various costs associated with calling. Unfortunately, in this case there are no data on the possible genetic bases of the satellite and calling traits.

Evolutionarily stable strategy theory provides a way to make precise quantitative predictions about what evolution should produce, predictions that can then be tested against reality [43]. It therefore offers a promising new way to analyze evolutionary predictions about animal behavior. But its primary contribution to date has been to demolish the idea that there can be only one set of behavioral traits exhibited by a species. In the past, individuals that have behaved differently from the most common pattern often have been considered to be abnormal mutants of low fitness or have simply been ignored as irrelevant. This is no longer a tenable way to approach behavioral variation within a species. The recognition that the fitness gains from one behavioral option are affected by how many other individuals are practicing the alternative(s) is a real advance in behavioral ecology.

The Protection of Inseminated Females

Securing copulations is a major ultimate goal of male behavior in most species; but, as the damselfly *C. maculata* illustrates, copulation itself is no guarantee that a male will fertilize his mate's eggs. In species that practice internal fertilization, a male always runs the risk that his mate will copulate with another male and that, disaster of disasters, the rival's gametes will fertilize some or all of the female's eggs. (The sneaker males of the bluegill sunfish show that egg fertilization "thievery" can also occur in species with external, as well as internal, fertilization.)

In his extremely important review article, G. A. Parker showed that, for a host of insects, the last male to copulate with a female before she laid her eggs usually won the SPERM COMPETITION contest [566]. There is evidence for a "last male advantage" from studies of birds as well [819]. The mechanisms that achieve fertilization for the last male are varied; but whatever their mode of action, their evolutionary consequences are profound. Many aspects of male behavior appear to have evolved in response to the selective advantages of preventing inseminated females from receiving or using another male's sperm.

One tactic is to guard a mate physically until she lays her eggs. We have already noted that territorial males of *C. maculata* perch near a recent partner while she oviposits. A male guard repels all intruders who would otherwise chase the female from the water, mate with her, and unceremoniously remove the previous male's sperm in the process. Guarding behavior is widespread in damselflies, but not all species exhibit the same technique as *C. maculata*. A much more common method is for the male damselfly to retain his grip on the female's prothorax after she has accepted his sperm. He will remain *in tandem* with his mate while she oviposits (Figure 19). Once in a tandem position, the male's grip is unbreakable; and therefore another male cannot adopt the tandem position, which is a necessary prelude to copulation. CONTACT GUARDING by male damselflies is an extremely effective method of preventing a mate from copulating again before she has finished laying her clutch of eggs. Why then do males of *C. maculata* employ the less effective system of noncontact guarding?

If we compare *C. maculata* with the contact guarding species, two differences emerge [5]. (1) The tandem guarders are usually nonterritorial insects; and (2) receptive males vastly outnumber receptive females at oviposition areas. These factors mean that a male gives up very little by

19 **Tandem (contact) guarding** in a damselfly. The male remains in tandem with his mate while she oviposits. While in tandem the male cannot be separated from his mate. Photograph by the author.

remaining in tandem with his mate. During this guarding time, the male cannot lose his territory because he does not have one; nor is he likely to miss other mating chances because the probability of capturing two females in a short time is so low. But males of *C. maculata* that abandon their territories for even a short time may lose their sites to intruders that quickly adopt vacated territories as their own. Perhaps as importantly, a territorial male often has large numbers of females arriving on his territory over the course of an afternoon. If he were to remain in tandem with one mate for several hours until she finished egg laying, he would lose some opportunities to copulate with other females [5]. Therefore, it is plausible that males of this species sacrifice a certain amount of guarding certainty in order better to retain their territories and to be free to mate with newly arrived females. In order to move from "plausible" to "almost certain," precise measurements are needed on (A) the average number of egg fertilizations gained as a result of the freedom to mate multiply *minus* (B) the average number of egg fertilizations lost as a result of using the less effective noncontact guarding method. For noncontact guarding damselflies, (A) should exceed (B).

Forced Copulation and Mate Guarding in Birds

We can test the hypothesis that mate guarding carries with it both costs and benefits that vary in different environments by predicting that males will guard their mates only when the risk of sperm displacement is high. Many birds (e.g., the black-headed gulls and white-crowned sparrows) form monogamous pair bonds in which a male and a female remain together to rear their brood. In a number of these species, however, males also seek copulations outside the pair bond—an example of alternative reproductive tactics possessed by the same individual. Bank swallows provide an excellent example [54]. These colonial nesting birds live in high density (Figure 11, Chapter 7), and a female traveling to and from her nest may be mobbed by a group of males and mated. In this way some males may fertilize eggs of females that are paired with other males. But this activity requires that a male leave his own partner undefended. At times a male is reluctant to do this but instead flies right behind his mate, wherever she goes, to fend off attacks from other males. As predicted, close continual guarding of a mate occurs *only* during the period just before and during egg laying (when a female's eggs may be fertilized) (Figure 20). At this time the benefit of mate guarding increases dramatically (because if a male fails to protect his spouse, she is likely to copulate with other males whose sperm may fertilize her eggs). The benefits of close guarding during these critical days outweigh the costs (the time and energy expended in guarding flights and the loss of opportunities for matings outside the pair bond). But once this period is past, the benefit:cost ratio falls and males sometimes attempt to locate other females that appear to be in the egg-laying process [54].

A similar picture emerges from studies of some other birds. Males of

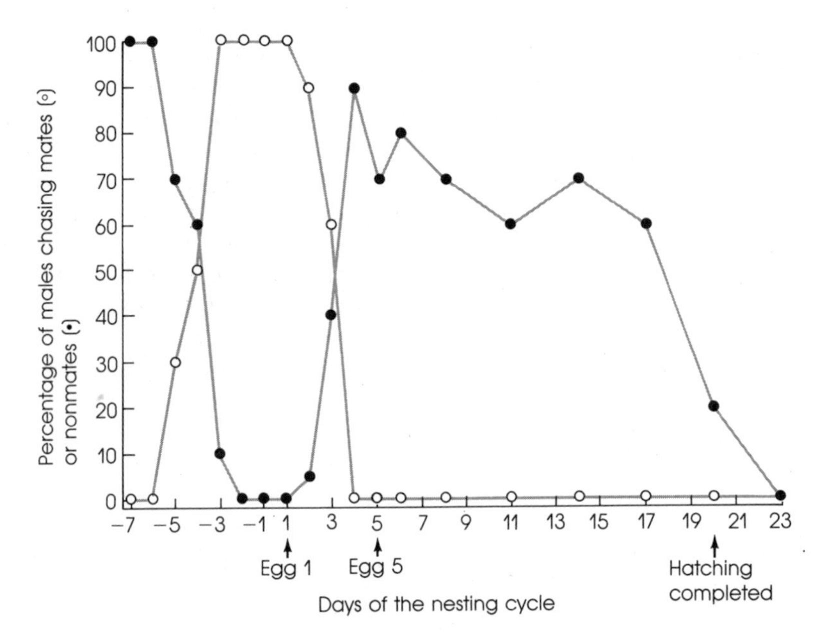

20 **Strategy of male behavior** in bank swallows. Males fly with their own mates on the several days shortly before the first egg is laid. These days are when their mate's clutch may be fertilized. At other times, males pursue the partners of other males and seek to copulate with them.

green-winged teal [502] and magpies [72] sometimes pursue and force copulation upon females to whom they are not bonded. The female's regular mate attempts to prevent this from happening by remaining close to his partner during the time she is fertile and by keeping her constantly in view. Male magpies become much less attentive when their mates have laid their eggs and are no longer vulnerable to forced copulation.

In ducks, but not magpies, a male responds to the "rape" of his mate by immediately copulating with her, perhaps to supplant or dilute rival sperm with his own [42, 502]. A similar function has been assigned to the unusually frequent copulation of males and their mates in many colonial birds. Repeated copulations with one's partner may represent a kind of "insurance mating policy" for males whose females live in close proximity to so many other males with whom they might occasionally copulate, either willingly or unwillingly [72]. Through regular copulation a male could either swamp the sperm of opponents or improve his chances of providing sperm at the moment when egg fertilization is most likely.

Some male insects also appear to use repeat matings to increase their chances of successful fertilizations. A prime case involves the giant waterbugs. When a female belostomatid water bug lays her eggs she does so on the back of her mating partner (Figure 21). He copulates with the female

21 **"Cuckolded" male water bug** with an offspring that it has cared for but that was sired by another male. The cuckolded male participated in an experiment in which his sperm duct was severed, which made it impossible for him to transfer sperm during copulation. The male's mate had previously copulated with another male carrying the gene for the stripe down the bug's back. She evidently retained enough sperm from this mate to fertilize the eggs she donated to the male in the photograph. One of the eggs has hatched, and the nymph clearly shows the genetic marker the adult male lacks. Photograph by Robert L. Smith.

before she begins to oviposit, and again after two or three eggs have been laid, and again in recurrent bouts until the female has deposited her entire clutch. Because females do store sperm from previous copulations, a male may avoid caring for the offspring of another male by repeatedly replenishing the supply of his gametes within his mate [670, 671].

Mating Plugs and Repellents

Even if a male leaves a mating partner unguarded, he can do things that reduce the likelihood that she will copulate again, at least for some time. The most direct method involves sealing the female's genital opening with a substance applied during or immediately following copulation. In some garter snake species, the male, after inseminating the female, secretes into the female's cloaca a substance that hardens to form a tough gelatinous plug that remains in place for several days [175]. Because garter snakes mate in aggregations, the first male to copulate with a female apparently finds it advantageous to invest in the production of a copulatory plug to prevent other males in the group from supplanting his sperm.

In insects, mating swarms are a common phenomenon, and mating plugs are correspondingly well represented [700]. A virgin queen honeybee flies to a mating station about which many dozens or hundreds of drones have gathered. She is pursued and captured in flight by a male who, upon completion of copulation, propulsively fires his genitalia into the female's genital opening. With these structures detached from his body, the drone dies, but presumably the implanted genitalia act as a physical barrier to other males, reducing (but not completely preventing) the possibility of multiple matings by the queen [512]. In the swarming ceratopogonid midges, the male also sacrifices himself to prevent other males from inseminating his mate; but in these species he is partly eaten by his mate while *in copula*. The male genitalia remain intact in the cannibalistic female, blocking further sexual activity on her part for some time [186].

The death of a drone honeybee or a ceratopogonid midge may seem a rather stiff price to pay to increase the chance that a mate will not copulate again, but one must consider the alternatives for males of these species. Both form swarms that are enormously male-biased, so that the odds against one male capturing two or more females are very great. As a result, a suicidal male need not improve his egg fertilization gains with one mate very much in order to make his action genetically profitable.

Males that can live to mate again with some reasonable probability do not make the ultimate sacrifice and, instead of using themselves as a mating plug, may simply apply a repellent odor to their mates. (Actually it is the smell of the mating plug that deters male garter snakes from a mated female [250].) In the tropical butterfly *Heliconius erato,* Larry Gilbert has found that males have specialized glands that produce a repellent material while the male is in the late pupal stage [256]. This prevents adult males from waiting on the pupal case until the butterfly emerges in order to copulate with it (which they do if the pupa is a female) (Figure 22). By warning off mate-locating males, the pupal male avoids the risk of being damaged by mistaken would-be suitors. After metamorphosis to the adult stage, the male transfers the same repellent material to the female during copulation. She stores it in an abdominal gland. By exuding the substance from "stink clubs" at the tip of her abdomen, the female can probably repel ardent males. Although Gilbert feels that the donor male is able to force monogamy on the female against her interests, the fact that she has apparently evolved a special storage organ for the male antiaphrodisiac makes this explanation less plausible to me. Instead, the female may use the male-donated pheromone selectively to repel males when this is to her advantage. But she does not release the odor when she wishes to copulate again to acquire additional sperm or another valuable spermatophore. Even though a male cannot prevent his mate from accepting a new mate eventually, his scent makes it easier for her to signal her lack of receptivity when this is to her benefit. This could help her discourage

22 **Two males** of *Heliconius charitonius* are waiting for a virgin female to emerge from the pupal case on which they are perched. Photograph by Larry Gilbert.

potential mates from persisting in fruitless sexual attempts. It is to the donor's advantage if his mate does not have to flee constantly from other courting males but instead can repel them quickly and use her time and energy to produce and lay eggs fertilized with his sperm.

(The cost of superfluous copulations is a general problem for females. One of the most ingenious solutions has been evolved by females of the ground beetle *Pterostichus lucublandus*. This is one of a group of beetles that expels a liquid spray that deters predators. The females, however, also spray unwanted suitors; a male covered with the repellent attempts to clean his body with his mouthparts and then collapses. He remains in a coma for several hours before reviving [381].)

Heliconius males may not be the only animals that cooperate with their mates by providing them with chemical substances or signals that help the female control her receptivity to the advantage of both parties. In the American anole, stimulation from the male's penis during copulation has this effect [145]; it is conceivable that one of the reasons why copulation lasts as long as it does in the anole is to ensure that the female's "I-have-been-mated" receptors are stimulated, thus shutting down her willingness to copulate for a week or more. When sperm transfer occurs in mallard ducks, the male performs a postcopulatory bridling display in front of his mate [295]. Although a female mallard will copulate repeatedly, the signal she receives from her mate may enable her to copulate selectively—only with her pair-bonded male and not with others. Likewise, crickets have a "postcopulatory triumphal song" [62]. (I detect a hint of male chauvinism in this title.) Rats produce ultrasonic cries after copulating [44], lions [635] and rhesus monkey males [276] bite their mates (Figure 23), and male house sparrows [688] peck at the napes of their partners. These and other behavior patterns that occur during or just after a successful mating may signal the female that she has received sperm. This will help her regulate when and with whom she will copulate again. She accepts the signals because it is to her benefit to do so, but this also incidentally enhances her mate's genetic success as well.

Sexual Interference

One final category of traits that might be favored by sexual selection could be classified as *spiteful* acts that, although costly to the actor, greatly reduce the fitness of another (unrelated) individual [301]. By preventing a competitor from reproducing, a male prevents rival genes from being transmitted to the next generation and thereby increases the relative proportion of his own genes in that generation. There are many examples of sexual interference that at first glance appear purely spiteful. The salamander *Plethodon jordani* provides an elegant example of a case in which a male interrupts the courtship of

23 **A male lion bites** the neck of his partner during copulation. This may provide a signal of sperm transfer that helps the female regulate her sexual receptivity. Photograph by George Schaller.

an opponent and so prevents him from mating [26]. Receptive females of this species are rare and widely dispersed. If a male happens to come upon a courting pair, he may move behind the courting male, straddling the tail of the other male and sliding his chin along it. This behavior mimics the response of a highly receptive female (Figure 24). The reaction of the duped male is to walk forward; the other male follows while the real female usually wanders off. After a period of the straddling walk, the leading male deposits a spermatophore on the ground. If there really were a female behind him, she would walk forward and press her cloaca over the spermatophore. The male interferer does not do this, of course, and is often chased when the duped male discovers him. But by this time, the duped male has usually lost a valuable opportunity to mate with a female.

Another equally striking method of reducing a competitor's fitness occurs in some acanthocephalan worms [1]. These marine invertebrates make copulatory plugs by applying a cementlike material to the female's genital opening following copulation. But a male will also approach another male and rape him. On these occasions no sperm are transferred, but the cement is, firmly sealing the genitals of the unfortunate rape victim so that he can no longer reproduce.

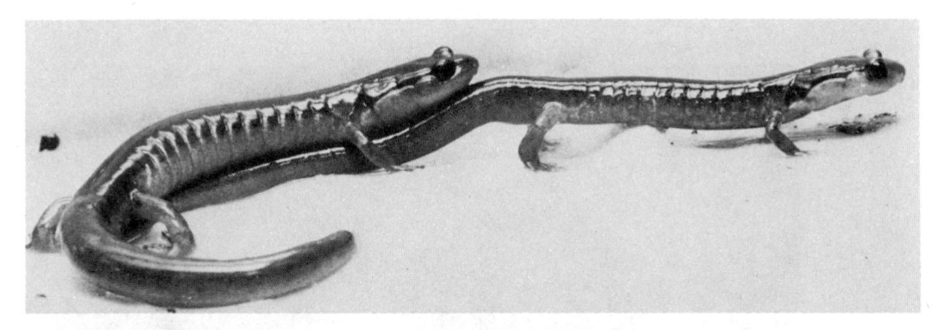

24 **Heterosexual and homosexual courtship** in the salamander *Plethodon jordani.* A female has straddled the male's tail to signal the male to move forward and deposit a spermatophore (*top*). A male has interrupted a courting pair and supplanted the female (*bottom*); he now mimics female behavior while the courted female moves away. Photographs by Stevan J. Arnold.

A different way in which to reduce the relative reproductive success of competitors would be to monopolize resources above and beyond those that the owner could actually consume or use. Possession of a "superterritory" would prevent other animals from using the materials in the area and so might lower their genetic success [737]. But this hypothesis has been subjected to scrutiny and found to have a defect that makes purely spiteful resource monopolization or sexual interference less plausible from an evolutionary perspective [722]. To interfere sexually or to hold a superterritory carries a cost in time, energy expenses, and risks taken. The actor derives a *relative* benefit in terms of a reduction in the fitness of a competitor or two. But all other unaffected individuals also enjoy these benefits and at no cost to themselves. A purely spiteful action therefore tends to reduce a very few individuals' reproductive success but raises that of many others besides the actor, a consequence that dilutes or eliminates its *net* benefit to the spiteful individual. Therefore, we expect that cases of apparent spite will carry with them some immediate positive reproductive benefits for the individual. In the salamander or acanthocephalan worm, it is possible that the "spiteful" male actually increases its own chances of ultimately copulating with some of the females that the other males lose. This is almost certainly true for the kind of infanticide that lions (Figure 25), langurs,

25 **Infanticide** by a nomadic lion that has found the cub of another male. Photograph by George Schaller.

and house mice practice (Chapter 1). Although superficially these actions appear only to damage the males whose offspring are killed, they also increase the speed with which the affected mothers come into estrus again, thereby increasing the rate of offspring production for the infanticidal male.

Summary

1 Asexually reproducing females need not produce sons and should enjoy a great advantage over sexual females as a result. The persistence of sexual reproduction in many species suggests that sexual females enjoy some fitness gains that more than compensate for the costs of producing sons. One possibility is that their genotypically variable offspring may be able to exploit various environments different from those the parent exploited as well as being able to combat parasites and pathogens that might decimate a genotypically uniform brood.

2 Sexual reproduction creates a social environment of conflict and competition among individuals, as each tries to maximize its genetic contribution to subsequent generations. Males make many small gametes and typically exhibit little parental care but instead try to fertilize as many eggs as possible. Females make many fewer, larger gametes and often invest additionally in existing offspring at the expense of producing other progeny (PARENTAL INVESTMENT).

3 Sexual selection occurs if genetically different individuals differ in their reproductive success (1) as a result of competition within one sex for mates or (2) because of their differential attractiveness to members of the other sex. Because females usually make a greater parental investment per offspring, they are a limited resource for which males compete. Intrasexual selection among males appears responsible for a host of male attributes, including a low threshold for copulatory attempts and a readiness to fight to monopolize mates.

4 Within a species, males often differ in the competitive tactics they employ to acquire copulations—the most familiar example being dominant and subordinate behavioral roles. Dominance and subordinance are options within a CONDITIONAL STRATEGY. A male can adopt either role, his choice depending on the social conditions he experiences. There may also be some cases in which two genetically different strategies can co-occur in the same species in an evolutionarily stable state.

5 A copulation may or may not yield egg fertilizations for a male. The risk that a male's sperm will be superseded by those of a rival has led to many postcopulatory traits, most commonly the guarding of a mate until she has laid her eggs.

Suggested Reading

An entertaining article by James Lloyd is recommended as an introduction to the behavioral ecology of reproduction [439].

G. C. Williams's *Sex and Evolution* and W. D. Hamilton's review of this book are helpful analyses of the evolution of sexual reproduction [303, 787]. See also J. F. Wittenberger [801] and Graham Bell [57] for a comparison of competing hypotheses.

Robert Trivers's article on parental investment has been extremely influential in the rebirth of interest in sexual selection [716]. General treatment of the effects of sexual selection are found in books by David Barash, Richard Dawkins, Michael Ghiselin, E. O. Wilson, and J. F. Wittenberger [43, 168, 253, 794, 801].

Richard Dawkins and William Cade have written understandable accounts of intraspecific variation in mating tactics [119, 169]. Jonathan Waage's articles on *Calopteryx maculata* provide an unusually clear and complete description of reproductive competition in an animal [746–748].

Suggested Films

Aggressive Behavior in Mature Male Bison. Color, 12 minutes. How male bison compete for dominance status.

Courtship Behavior of the Queen Butterfly. Color, 31 minutes. How males of this species secure female mates, with emphasis on chemical communication between the sexes.

In Search of a Mate. Color, 25 minutes.

The Pelicaniform Birds. Color, 16 minutes. A film that contains some good footage on the bizarre courtship displays of these birds.

Rhesus Monkeys of Santiago Island. Color, 33 minutes. The behavior of a polygynous mammal in which dominance plays a major role in determining the reproductive success of males.

Spiders: Aggression and Mating Behavior (05892). Color, 17 minutes.

Having described some ideas on the evolution of sexual reproduction and the basis for sexual selection, let us now examine whether female choice occurs and what its consequences for female reproductive success might be. Because females are typically in great demand as copulatory partners, they potentially have the opportunity to select among the males willing to mate with them. A male really only has two categories of inducements for a female: his genes and any material benefits he might transfer to her. If females do make adaptive choices, they should prefer those males that advance their reproductive success either because they offer superior genes or better material benefits. Evidence that helps test this prediction is reviewed in the first section of the chapter.

The concluding portion of the chapter deals with the behavioral ecology of mating systems. Male and female strategies combine to determine how many mating partners an individual will have in the course of a breeding season. This can be used as the primary criterion to define a mating system as monogamous, polygynous, or polyandrous. One of the real achievements of behavioral ecology has been to identify how sexual selection and key ecological variables interact to create the mating system of a species. We shall explore how competition among males in different environments determines the number of partners a male or female can expect to have during a breeding season.

CHAPTER 12

The Ecology of Reproductive Behavior: Female Choice and Mating Systems

Female Choice

A female's reproductive success rarely increases the more often she copulates. Her success will be determined in part by how many eggs she can produce and by their fitness after fertilization. If males vary in their effects on offspring quality, then females should discriminate in favor of those individuals that will contribute the most to them. If they are discriminating, females will exert epigamic (intersexual) selection on their mates.

That female choice can occur is supported by two observations. First, as a general rule, a female can prevent a male from inseminating her or fertilizing her eggs against her will. For example, a female damselfly has to twist her abdomen around to receive her partner's sperm. If she refuses to do this, there is nothing the male can do to force her to accept his gametes. If a female sunfish chooses not to release her eggs in his nest, the male that defends the site cannot fertilize her eggs. There are exceptions in which males force copulation on females, but even here a female may be able, through physiological mechanisms, to prevent the sperm she unwillingly received from fertilizing her eggs [699].

Second, it is a fundamental generalization that females almost never receive gametes from a male of another species [492]. Hybrid offspring usually have low viability; and therefore females that make the mistake of producing them are presumably selected against. The traditional view is that females choose males of their own species on the basis of the ability of these individuals to perform distinctive (and often complex) courtship signals. Males often do advertise their readiness to mate with species-specific signals such as the territorial song of the white-crowned sparrow, which cannot be confused with that of a song sparrow, a white-throated sparrow, or a black-chinned sparrow (Figure 1). Although there are reasons for thinking that courtship signals may have other major functions besides announcement of species membership (see later), this behavior is likely to contribute to species identification. Whether on the basis of male courtship or some other cue, females make the appropriate species discrimination. If

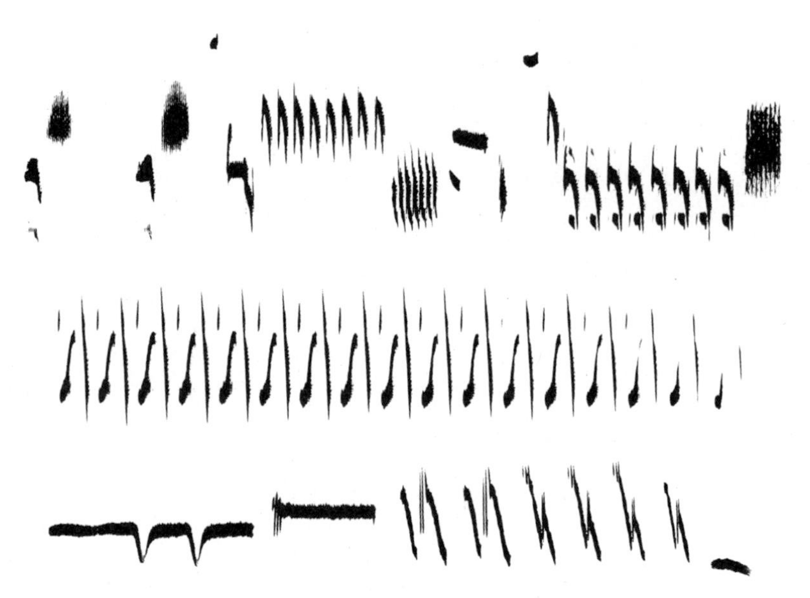

1 **Diversity among species of sparrows** in the sound patterns of male song: spectrograms of a song sparrow (*top*), swamp sparrow (*middle*), and a white-crowned sparrow (*bottom*). The differences among species may reduce the risk of hybridization. Courtesy of Peter Marler.

they can do this, they should be able to discriminate among males of their own species on the basis of various cues.

Nevertheless, it has proved difficult to demonstrate unequivocally that females actually do have preferences for certain kinds of conspecific males. Consider the case of *C. maculata.* As noted, females of this damselfly are more likely to copulate with a territorial male than a nonterritorial individual. Although this might seem firm evidence for female choice, there is the strong possibility that dominant males outreproduce the others because they often interfere with mating attempts by subordinate males [746]. Thus, what looks like (and may be) a female preference for territorial males that perform courtship displays may be in part or wholly the result of intrasexual competition among males. In order to demonstrate active female choice, it is not enough to show that some males mate more than others because this could be the result of male interactions rather than female preferences [700].

In order to search for evidence that females do select superior males of their own species, we must consider what constitutes a superior mate. One simple, but important, attribute is a male's ability to provide sufficient sperm to fertilize a female's eggs. Evidence that females can potentially discriminate in favor of males that provide large numbers of sperm comes from a variety of sources, including fruit flies [464] and a fish, the lemon tetra [530]. Males of this species externally fertilize a female's eggs when she releases them. A male tetra has a limited supply of sperm, and therefore the more frequently he spawns the less likely he is to fertilize all of his mate's eggs (Figure 2). When given a choice between a sperm-depleted male and one that has not yet spawned, females associate preferentially with the sperm-rich males, thereby improving the fertilization chances of their valuable eggs.

More generally, males can influence a female's reproductive success by offering her superior genes that raise the fitness of her offspring or by providing material benefits that are translated into more offspring or progeny of unusually high reproductive value (Table 1). The typical male strategy is to invest entirely in securing copulations. To provide nutritional gifts, parental care, or other useful resources diverts energy from gamete production and mate location and therefore is not expected to be a common component of male strategy (but see later). Without doubt, there are many species whose males give their partners only genes. How can a female determine (unconsciously) which male has superior genes?

Male Genetic Quality and Female Choice

One indication that females can make mating choices purely on the basis of male genetic quality comes from observations that females of some species actively avoid their brothers and fathers. Inbreeding often results in an increased number of congenital defects in offspring, so avoidance of close relatives is probably adaptive [597]. The proximate mechanisms of a female preference for nonrelatives varies among species. In a species of *Peromyscus,* deer mice females that are reared with their brothers or fathers delay their sexual maturation and so avoid being fertilized by a close relative [324]. Females of the great tit prefer males whose songs are similar, but *not* identical, to their father's song [500]. Inbreeding in olive

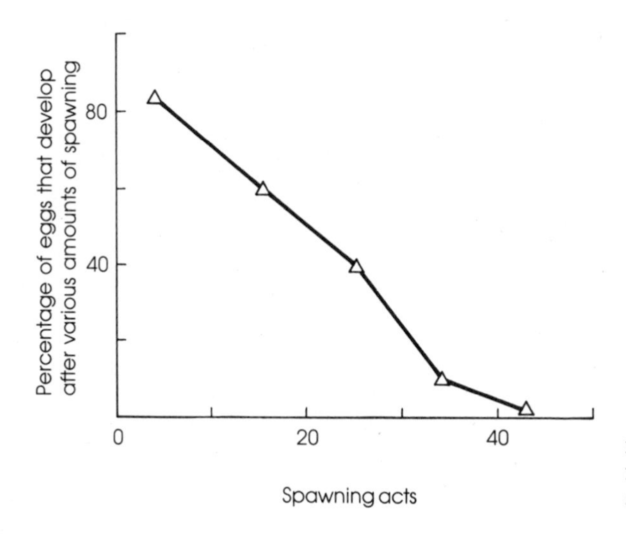

2 Percentage of eggs that develop declines in the lemon tetra as the number of spawning acts by a male increases.

Table 1
Attributes That Females May Use in Selecting Males from among Members of Their Own Species

Indicators of high genetic quality

A. The genetic complementarity of a male
   1. Male signals that he is a nonrelative
   2. Male signals ability to provide large sample of gametes

B. The capacity to dominate rival males
   1. Evidence of high social ranking
   2. Response to female's incitation of male–male aggression

C. The capacity for prolonged survival
   1. Behavioral or morphological cues associated with old age
   2. Demonstration of male's sound physiological condition

Indicators of capacity to provide useful resources or services

A. Material benefits offered by potential mate
   1. Nutritional nuptial gifts provided prior to or during copulation
   2. Access to valuable resources controlled by male

B. Paternal services offered by male
   1. Indicators of foraging skill
   2. Demonstration of readiness to defend mate and offspring

baboons is blocked because females choose males that have transferred into their troop from other bands rather than copulating with males that grew up with them in their natal group [561]. Males of the black-tailed prairie dog usually disperse from the group in which they were born and move away from their recent mates. Thus, males rarely have opportunities to mate with their sisters and daughters, who remain behind. But if these opportunities exist, a female prevents inbreeding by rarely coming into estrus when the father (or brother) is present. If she does become receptive, she will stay away from her close male relatives [339].

How to Pick a Dominant Male

If nonrelatives have only genes to offer, it is a natural deduction that females should prefer dominant males as mates. High male status, large body size, good foraging ability, superior survival skills, and good physical condition are all interrelated. Males rarely are able to dominate their rivals physically unless they are relatively strong, well-fed, powerful, and experienced animals. They rarely can achieve these attributes unless they are able to survive a long time, consume sufficient amounts of food, and avoid debilitating diseases and parasites [241]. A female gains by acquiring the genetic basis for these traits (assuming they are heritable) for insertion in her eggs. Her sons would be more likely to enjoy the mating success that comes from dominance; her daughters could benefit as well through improved disease resistance or foraging ability.

Unfortunately there has been no ironclad demonstration that female fitness is increased by the active selection of a dominant male in a species in which males make no contribution other than the genetic one. Nevertheless, there are aspects of female reproductive physiology and behavior in some animals that seem certain to help them fertilize their eggs with sperm from dominant males. The first of these traits is a tendency to be receptive only in the presence of a group of males. In species as different as sage grouse, Uganda kob (an antelope), certain frogs, and many insects, males form mating aggregations that females visit solely to acquire sperm for their eggs [84]. By going to groups rather than to solitary males, females improve their chances of copulating with a truly dominant individual. Even if males are accepted passively by females, their mates would have had to overcome competition from many other individuals to reach them.

But female choice in these situations often appears to be more active than passive, with individuals avoiding certain males and moving within the group to points that are most highly contested. In sage grouse, the owners of a few central territories in an aggregation of males secure the overwhelming majority of matings [785]. Males of the black swallowtail butterfly compete for conspicuous landmarks (hilltops) in certain open areas. They form loose groups of territory owners in these places. R. C. Lederhouse has shown that the territorial sites that males prefer and that they will fight for most intensely are also the very places to which females fly when they are ready to mate [407].

Rather than going to where the tested males are, a female may announce in a conspicuous manner a readiness to mate, thereby attracting to herself a group of males that will compete for the chance to copulate [144]. Sexual advertisement by female baboons and chimps, which live with or near groups of mature males, includes a combination of olfactory and visual signals. Significantly, gorilla females lack pronounced estrous swellings near their genitalia. They live in a harem with a single dominant male who has fought to gain control of the group; inciting aggression within the group does nothing to promote a female's chances of mating with a dominant gorilla [307].

Female incitation of male aggressive competition by vocal signals is likely for alligators [249] and for some western chipmunks [121]. During the mating season females of these animals produce conspicuous calls that attract males. A female chipmunk may call noisily for four hours a day for three to four days before coming into estrus. During this time, several males may gather in the area; and when the female finally does become receptive, the males fight for the chance to inseminate her.

Testing Male Dominance

One might think that once a female had begun to copulate her options were limited. But mate choice can be exercised before, during, and after a copulation. When a male elephant seal tries to mate with a female, she invariably screams loudly, a cry that to human observers conveys more protest than pleasure (Figure 3). If her partner happens to be a subordinate that has slipped into the group unnoticed by the harem master, the dominant male will "race" over on the double; and the subordinate generally flees before having begun or completed the copulation. Only a truly dominant male can subdue and mate with a noisy female without sexual interference from his rivals [144].

There are also cases in which a female requires a lengthy interaction with a partner, which would seem to test the male's dominance status. In the cricket *Teleogryllus commodus,* females accept from their males sperm encapsulated in a small spermatophore. After the transfer of the spermatophore (which is accomplished promptly), a female becomes "restless" and wanders about, forcing her mate to move after her, often chirping as he goes in an attempt to regain contact with her. If a male cannot or will not do this, he may well be a subordinate or satellite male unwilling to risk encounters with a dominant territorial male. In such cases, the female eats the spermatophore, which is attached externally to her body, before many or any of the sperm have migrated into her reproductive tract. Moreover, she remains receptive to another male whose sperm she is likely to use. But males that can keep in close contact with their mates, chirping as necessary, are presumably likely to be powerful territory owners that have intimidated nearby rivals and can move freely and noisily about their territory. In this case, the females do not remove the spermatophore prematurely; and after an hour or so, they lose their sexual receptivity [445].

Fundamentally similar phenomena occur in those rodents whose females

3 **Copulation in elephant seals.** The male has thrown a flipper over the female and is attempting to mate; the female is screaming at the top of her lungs. Photograph by Burney LeBoeuf.

will not ovulate unless they experience a complex and lengthy copulatory interaction [544]. Male cactus mice copulate with a female several times in succession over a period of some minutes. But only the first copulation is associated with sperm transfer. Although the male does not ejaculate on the subsequent pairings, these intromissions are essential if the female is to ovulate and become pregnant. One can speculate that the additional copulations may be required by the female to test the degree of control of her mate over other males. If he cannot copulate repeatedly without interruption, he is either a sneak subordinate or a male whose dominance is in question. Under these circumstances, the female will fail to produce progeny carrying his genes and instead waits until she can breed with a truly dominant male [177].

Even after having become *pregnant,* a female can still potentially exercise mate choice. Again small rodents provide possible examples of this phenomenon. In some cases females that have developing embryos within them may resorb the embryos if exposed to the odor of a strange male during their pregnancies. In nature these species have territorial males that would prevent a strange male from approaching their mates. Thus, the scent of a strange male indicates that the dominance of the harem master has been challenged. By resorbing their embryos, females make themselves sexually available sooner to a new harem master, if a takeover occurs. Through resorption, a female avoids the risk that the new male will kill her infants, and she also conserves energy for the production of a new brood that will carry the genes of the newly dominant male [401, 642].

The Function of Courtship

Having presented the possibility that females might test the dominance status of their mates by the "quality" of their copulations with them, it requires no great leap of the imagination to speculate that courtship might also serve the same function from the female's perspective. One of the enduring puzzles of animal behavior is why the males of some species engage in courtship displays of such extreme complexity and elaborateness. The bizarre nature of the courtship antics of birds-of-paradise, bowerbirds, and many grouse far exceeds any requirement for species identification [260, 406]. The same is true for many insects. I was once entertained by a small otitid fly whose males use seven different courtship displays; they tap their mates on the head, wave their wings in front of the female, and bob back and forth on the perch, and so on, stringing together one display after another in rapid-fire sequence like a miniature windup toy [6].

An equally strange species is the three-wattled bellbird. Males have evolved three pendulous outgrowths that dangle about the beak. The male produces a monotonous series of deep-ringing notes of stupendous volume. Each calling male has his own territory, with a prominent perch from which he announces his ownership of the area. Females appear to fly from one territory to another, alighting on a limb of the calling tree. The male approaches the female, and what happens next has been described by Barbara Snow [674]: "As the male opens his beak wide to utter the call he leans right over the [female] visitor. The latter leans as far away as possible, clinging precariously to the very end of the branch. The delivery of the call, with the visitor flinching at its tremendous noise, is an amazing sight and can only be described as an ordeal by sound which few visitors withstand" (Figure 4).

In order to account for the evolution of strange and extreme behavior of this sort, R. A. Fisher developed the "runaway selection" hypothesis [238]. According to this scenario, females in an ancestral population might have a slight preference for certain male characteristics, either by chance alone or because the preferred traits were indicative of the genetic superiority of the male. The sons of females that mate with preferred males would inherit the desirable characteristics and would enjoy higher fitness in part simply because they possessed the key cues females found attractive. Females that found these traits attractive would gain by producing the preferred sons. This starts the runaway process—a positive feedback loop favoring ever more exaggerated key courtship stimuli and females that find them attractive. The process will be stabilized only when natural selection against too costly or risky displays balances sexual selection in favor of traits that are appealing to females. Thus, if at one time female three-wattled bellbirds

4 **A male bellbird** advertises his genetic quality (?) by vocalizing at full volume in the ear of a female that has come to his calling perch in his territory.

preferred males that could give loud calls because such males could deter intruders from a greater distance, they now favor exaggeratedly noisy males because the mating preference has taken on a life of its own, a preference ultimately enabling the females to produce exceptionally attractive sons.

There are, however, other hypotheses to account for the evolution of complex courtship. One is the test-of-dominance (or survival skills) hypothesis. The ability of a male to engage in a protracted courtship could provide information about the ability of the male to interact sexually without interruptions from rivals. Moreover, the ability to produce a complex or physiologically costly courtship could also indicate that the male was relatively old and experienced, with good stamina and in prime physiological condition—correlates of dominance and survival ability. Thus, a female bellbird could be visiting the males in her neighborhood to see how frequently they were singing and how much energy they were putting into each call. Support for this hypothesis comes from a detailed study of the village indigobird, a totally unrelated species that exhibits an extraordinary degree of behavioral convergence with the bellbird [572]. Males, which have a stunningly elongated tail, claim prominent landmark trees from which they give their powerful songs. Females make the rounds, visiting many males and being treated to an elaborate courtship flight and vocalizations. Only after many visits will a female select a partner. An exhaustive analysis of the call sites' characteristics, habitat attributes of the areas, and the behavioral traits of males revealed that by far the best correlate of mating success was that proportion of time a territorial male spent giving his calls [572]. Calling endurance could plausibly be correlated with a number of attributes like dominance, foraging skill, and physiological soundness that would be useful to pass on to one's male and female progeny.

The extreme extension of the idea that females may test the genetic quality of their mates through analysis of courtship behavior is the "handicap principle." Amotz Zahavi has argued that females of some species prefer a male with traits that *reduce* his survival chances but that announce that he has superior genes because he has managed to survive despite his handicap [815]. Thus, males that have long plumes, elaborate horns, conspicuous coloration, and bizarre displays may be more attractive to females than individuals that lack these handicapping adornments (Figure 5).

The only direct demonstration that female animals actually prefer males with extreme ornaments comes from a study of a ploceid widowbird, a fairly close relative of the village indigobird. Male widowbirds possess an extraordinary tail, about one meter in length, which is prominently employed in the bird's courtship displays. Malte Andersson cut large portions from the tails of some males he captured and glued the stolen tail segments to other individuals, artificially lengthening their plumes. The males of both types continued to hold their territories successfully but about four times as many females settled on territories of the super long-tailed birds as on the areas defended by the artificially short-tailed birds [19].

5 **Handicaps** that may advertise the male's age and competence to prospective mates. Male widowbirds (*top left*) and sage grouse (*top right*) have evolved remarkable plumage that is used in even more remarkable displays to attract females. A male fiddler crab (*bottom left*) has a large display claw that cannot be used to gather food. An elk (*bottom right*) has a magnificent, but costly, rack of antlers. Photographs from top left to bottom right by Malte Andersson, Haven L. Wiley, the author, and the San Diego Zoo.

But is the widowbird's tail a handicap in Zahavi's terms? Although mathematical tests of Zahavi's idea have convinced some authors that a handicap is unlikely to evolve [56, 389], a modified version of the argument continues to be appealing. In one sense, all traits are handicaps because they have a survival-reducing cost. This is obvious in the case of a widowbird's tail, which is physiologically expensive to construct and maintain and which almost surely makes the male more vulnerable to predation. But a human brain and the wings of a swift and the proboscis of a butterfly also have physiological costs (as well as survival benefits) and may expose individuals to risk from certain predators (e.g., a brain virus) that they would not have to endure if they lacked the structure. Traits with a superior benefit-to-cost ratio will spread through populations despite their costs. The extravagant feathers of a bird-of-paradise or a widowbird have a cost (lowering the survival chances of an individual) and a benefit (announcement of the competence of a surviving male). An individual without plumes would avoid the costs of the trait but would not be able to advertise his ability to acquire resources and avoid predators nearly as well as the "handicapped" bird [816]. A female might prefer to mate with a plumed bird because she can be certain of his competence. A male without the plumes might be an older, competent male with great survival ability, but he might also be a younger individual with untested abilities. Young males would surely be more numerous than older ones and in the absence of the means to discriminate between them a female might do well to choose a male with unambiguously informative properties.

The handicap principle, like most of the other ideas in this section, remains controversial and essentially untested. Much more work needs to be done to develop testable predictions that discriminate among alternative hypotheses before we can claim with confidence that females choose males whose genes alone will enhance the fitness of their offspring.

Female Choice and the Material Benefits of Copulation

Luckily for persons who would like to think that females can make theoretically sound choices, there are species in which males offer their mates more than just their gametes (Figure 6). In these species females ought to choose males whose material offerings are better than average and there is some evidence that they do.

One of the best examples of female choice occurs in the black-tipped hangingfly *Hylobittacus apicalis*. Males of this insect belong to a substantial group that give their mates a nutritional gift in return for a copulation (Table 2). For *H. apicalis,* the gift is a dead insect that males capture, kill, and transfer to a prospective mate. She feeds upon the nuptial present while sperm transfer takes place. The size of the insect and its nutritional contents and quality can vary. Randy Thornhill has demonstrated that whether a female mates at all and how long she copulates is related to the kind of prey she receives [697]. Unpalatable ladybird beetles are rejected outright, along with their donors. Very small edible items are eaten, but the female leaves her partner as soon as the food is gone. Only if the prey is large does copulation last a full 20 minutes or so.

Length of copulation in the hangingfly is proportional to the number of sperm a female will accept. If a mating lasts less than five minutes, she will not have received any sperm at all. Thereafter, the amount of sperm accepted increases steadily until the 20-minute mark (Figure 7). Not only does the duration of copulation regulate how many sperm a female will take from a mate, it also determines whether or not she will become unreceptive and start laying a number of eggs. If the female has been given

6 **Male animals that provide material benefits** to their mates. (*Top left*) A male golden eagle, on the left, watches his mate feed their eaglet with a morsel of grouse that the male had brought to the nest. Photograph by Niall Rankin; courtesy of Eric Hosking. (*Top right*) A paternal frog carries tadpoles on his back. Photograph by Roy McDiarmid. (*Bottom left*) A male scorpionfly is carrying a nuptial gift upon which his mate will feed. (*Bottom right*) A male thynnine wasp is regurgitating liquid food to his wingless mate. Photograph by the author.

only a 12-minute meal, she will leave her mate and, to add injury to insult, she will seek out another male. She then feeds upon his gift and accepts his sperm, which she presumably uses to fertilize the eggs she lays during her nonreceptive phase. In this way females exercise active mate choice, creating selection pressures that favor males that provide large, nutritious, nuptial presents [698].

Table 2
Examples of Nutritional Benefits Provided by Male Insects to Their Mates

A. Food collected by the male and passed to the female during or before copulation
   1. Prey items: Hangingflies and empidid flies
   2. Nectar: Thynnine wasps regurgitate or secrete nectar that they have collected to their mates during copulation

B. Food that is controlled by the male and that he permits potential mates to consume
   1. Prey items: Scorpionfly males permit their mates to feed on insect carrion that they find and monopolize
   2. Nectar: Many megachilid bees defend territories at rich patches of food plant and permit only their mates to remove the pollen from the flowering plants

C. Nutritional resources synthesized by the male and passed to the female during copulation
   1. Spermatophores and seminal fluid: Many butterflies, grasshoppers, and other male insects package their sperm in protein-rich containers or fluids, which are transferred during mating and then digested by the female
   2. Glandular secretions: Some cockroaches and beetles secrete presumably nutritious fluids, which their mates consume
   3. Part or all of the male's body: The males of some cricketlike species let their mates eat their fleshy hindwings; in mantids and others, the female may completely cannibalize her partner

Source: Thornhill and Alcock [700].

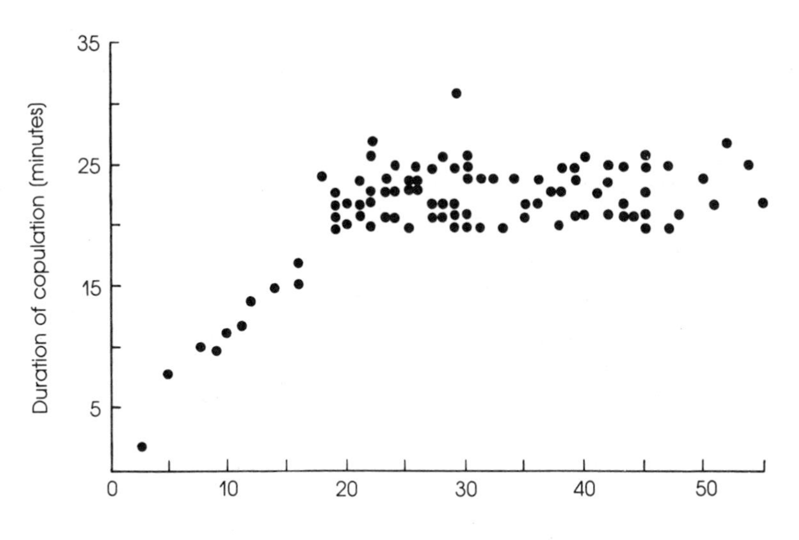

7 **Female choice** in *Hylobittacus apicalis*. The female controls the number of sperm she receives by regulating the length of copulation in relation to the size of the prey she receives from her mate.

It seems obvious that the mating preferences of female hangingflies raise their fitness. By tying egg fertilizations to being fed, a female receives food that she need not hunt for herself. Not only does this save her time and energy, it also reduces the risk that she will be captured in a spider web while foraging (a major danger for prey-collecting males).

How might prey offering by males have originated? Currently it is easy to see that a mutant male that tried to copulate without offering any food would not propagate many of his genes. But presumably at one time male ancestors of modern hangingflies were typical males in that they offered their mates genes alone. A mutant male in a population of this sort would seem to be at a disadvantage because time spent collecting nuptial gifts is time that cannot be spent tracking down mates. However, a gift-giver could gain some compensatory advantages. His mate would not have to go food collecting herself but could get right to the business of producing and laying eggs that would carry his genes. If the average increase in fecundity of gift-receiving females was great enough, the gift-giving mutant male could enjoy high fitness and the trait could spread.

Access to Monopolized Resources

A far more common way for females to get resources from their mates is to go to a territory controlled by a male and to exchange a copulation in return for access to the goods within the area. The damselfly *C. maculata* is a classic example of a species whose males do not collect a material benefit for a mate but station themselves by useful resources—in this case, oviposition sites. In order to lay her eggs in a superior spot, a female may have to mate with the site's owner before he permits her to exploit the oviposition material [746].

Bullfrogs are fundamentally similar to damselflies in that males compete for control of superior egg-laying locations in an aquatic environment. Male bullfrogs form choruses in areas where the water temperature and vegetational characteristics are best suited for egg development. Females come to the chorus and move toward a male. Amplexus usually does not occur until a female actually touches a male, an observation that suggests that a female can mate with the individual of her choice [347]. Large males are picked far more often than small ones, particularly by large females—perhaps for improved egg fertilization reasons but also because large males are able to defeat small ones and can control the very best egg deposition spots. Because females almost always lay their eggs in their male's territory, they benefit directly if the water temperature does not get too high and if the risk of leech predation is relatively low. Richard Howard has shown that the embryo mortality rate is lower in territories controlled by larger males than in sites monopolized by their smaller rivals (Figure 8).

Variation in Male Parental Care and Female Choice

In the damselfly, hangingfly, and bullfrog, males do not provide resources or protection *for their offspring*. The hangingfly transfers food that may be used by a mate to create additional eggs; the damselfly and

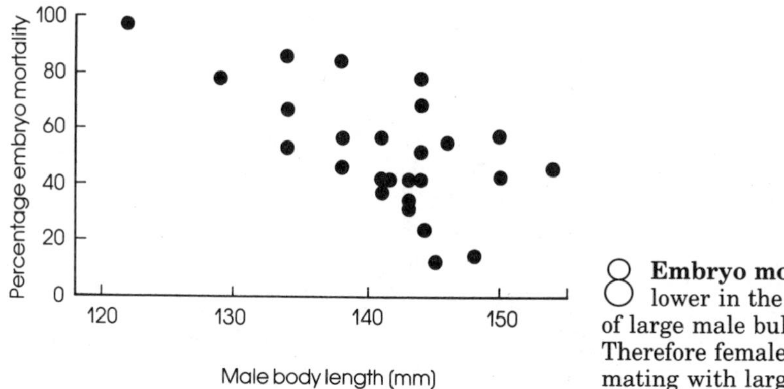

8 **Embryo mortality** is lower in the territories of large male bullfrogs. Therefore females gain by mating with large males.

bullfrog merely control areas that happen to be better-than-average for the development of a female's fertilized eggs. Therefore, these males cannot be said to make a large parental investment per offspring. Instead their activities are categorized as MATING EFFORT in which the male's allocation of time, energy, and risk taking is designed to assist him in gaining access to females and their eggs [14, 454]. But because this kind of mating effort yields material benefits for females, they can make choices that help them secure superior resources from their partners [700].

There are species, however, in which males actively provide for the welfare of existing offspring, and in these species variation in the potential paternal investment per offspring could be a relevant factor for female mate choice. The problem is, How can a female tell what a male will do when his young are available to be helped? It has been suggested for some birds that the nature of a male's courtship provides cues about his ability to be a good father. If a male collects food for his nestlings and fledglings, his foraging skill is important for the survival of his young. As argued earlier, male foraging ability may be indicated indirectly by his size and physiological condition, which in turn might be signaled by the intensity, persistence, or complexity of his courtship displays. Clive Catchpool believes that this is the explanation for the preference of female sedge warblers for males that employ a greater number of syllable types in their complex songs [131]. The males that sing songs with a greater repertoire size attract females sooner than their less vocally varied rivals (Figure 9). Catchpool notes that male sedge warblers defend such small territories that both parents must forage off the nesting territory a good deal, especially when providing for their progeny. Therefore, females cannot assess their reproductive chances by analyzing a territory's qualities directly and so must judge males on the basis of their songs. (In two other relatives of the sedge warbler in which males defend relatively large territories, females probably analyze territory quality directly. In these species the male's songs are much simpler than those of the sedge warbler [131].)

But why should song complexity be an indicator of male parentalism? Perhaps the ability to generate a varied song requires the male to be in

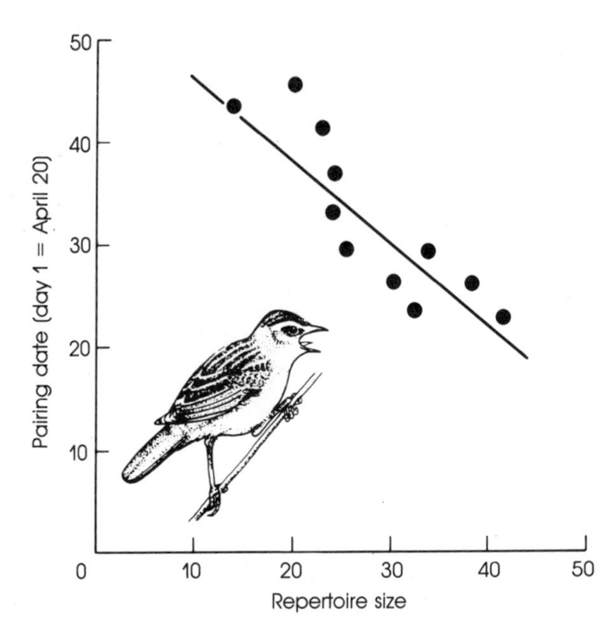

9 **Male sedge warblers** with larger song repertoires acquire mates before males with small repertoires.

good physical shape or to have experienced good conditions as a nestling or fledgling. If so, the male might be better able to forage actively than a weaker individual. In another songbird, the great tit, males with relatively large song repertoires are more likely to own territories that have produced surviving young in the past. Strangely, however, there is no evidence from this species that females prefer versatile songsters [501].

For some other birds, females might be able to assess the male's food-collecting ability in a satisfyingly direct way through the quality of his courtship feeding. It is common for male birds, like hangingflies, to feed their mates before copulating with them. Moreover, in the red-billed gull and some other species, the probability of copulation is much increased if the male feeds the female prior to his attempt [691]. A female could judge her would-be partner's parental skills by analyzing his courtship feedings (Figure 10). (Male common terns that provide their mates with more food than average do indeed feed their young at a higher rate [536].)

The story is complicated, however, by the absence of data showing that males that feed their mates inadequately are rejected during early courtship. Courtship feeding by male red-billed gulls occurs infrequently during the early phase of pair formation and reaches its peak in the period immediately before egg laying. If a female were to separate late in the breeding season, she would not have time to choose another mate and rear her brood successfully [691]. (But the possibility exists that inadequate courtship feeding could be the basis for rejection of a previous year's partner in some birds that can potentially retain the same mate over a period of years.)

10 **Courtship feeding** by a male red-billed gull. The female is behind the male and is reaching up to take regurgitated food from his mouth. Photograph by C. R. Tasker and J. A. Mills.

A final speculative case in which a male's capacity for parental care may be assessed by a female prior to permitting fertilization may occur in lions. Lionesses depend heavily upon the protection offered by adult males in the pride. If the males are ousted by strangers, the newcomers will often destroy any young cubs in the group [68]. Therefore, one could predict that a female should not produce a litter unless her mate were in peak condition and in complete control of the pride. A lioness may be able to test her mate's physical condition and ability to prevent intruders from entering the area by the male's copulatory behavior. An estrous female is receptive for two to three days, during which time she copulates about once every 15 minutes [635]. Despite the energetic and frequent matings, pregnancies are rare— a fact suggesting that females are sensitive to some aspect of the male's behavior and require special conditions to ovulate and be fertilized.

Even this abbreviated survey of examples of possible female choice shows the wide range of potential material and paternal benefits offered by males and the equally wide range of direct and indirect ways that females might assess the value of potential mates. Although our emphasis has been on female choosiness, I again note that the greater the parental investment or the more costly the benefits offered by a male, the more selective he is expected to be in committing his paternal care or resources to a female and the more active females are predicted to be in initiating courtship. This prediction is consistent with information gathered from humans (see Chapter 15), stickleback fish [419], doves [221], giant water bugs [670], Mormon crickets (Chapter 11), and butterflies [624]. In each of these animals, acceptance of a mate may commit a male to assist her in various ways, with the result that males exert epigamic selection on the opposite sex, as suggested by Trivers. As more is learned about the nature of male donations during or after copulation, a better appreciation of the reciprocal nature of mate selection in animals will become possible.

Mating Systems

Sexual selection determines the often conflicting strategies of males and females, strategies that together shape the mating system of a species. Defining the different kinds of mating systems is not easy, however, because there are several competing criteria. Traditionally, mating systems have been separated on the basis of (1) how many individuals a male or female of a species may copulate with, (2) whether a male and a female form a pair bond and cooperate in parental care, and (3) how long the pair bond is maintained [e.g., 403, 650, 651, 801]. The number of permutations of these variables is large and has resulted in some complicated lists and intimidating terms, such as polybrachygyny. I believe there are advantages to restricting a classification of mating systems to the first criterion only, namely, the number of mates acquired by an individual in one breeding season. There are two strategies for each sex: to mate with just one partner per breeding period (male and female MONOGAMY) or to copulate with several different individuals (male POLYGYNY and female POLYANDRY). It is entirely possible for all four options to be exhibited within a single species. Table 3 presents a classification of mating options that subdivides polygyny and polyandry into a number of functional types.

In order to account for the diversity in mating options, we shall try to relate male and female behavior to elements of the environment that influence the success of alternative tactics. Stephen Emlen and Lew Oring have added an extra dimension to sexual selection theory by pointing out that the disparity between the sexes in parental investment is only one determinant of the intensity of sexual selection. They argue convincingly that the mating strategies of males and females are strongly affected by a number of key ecological factors, such as the strength of predation, the distribution of food or other critical resources, the duration of the breeding season, and the requirements of the young. These variables place constraints on the ability of males to monopolize numbers of females and on the ability of females to choose among potential partners. The result is a wide variety of mating systems [217].

Monogamy

On the basis of sexual selection theory, we would predict that monogamy should rarely be practiced by males; and this is true [550]. The quintessential male strategy is to fertilize as many eggs as possible, and this is often correlated with mating with as many different females as possible. Thus, males usually try to be polygynous. But in some environments the potential for polygyny is much reduced because receptive females are scarce and widely distributed. If so, a male that encounters a potential mate can conceivably gain more from guarding that female than by searching for other females. The extreme case of the honeybee has already been presented. Drones, you will recall, practice monogamy in the most dramatic fashion possible by suicidally donating their genitalia to a mate [512]. In other species, males less dramatically sacrifice some mate-locating time to guard a prospective mate or to prevent an already inseminated female from mating with another male.

Table 3
Mating Systems of Males and Females

MONOGAMY: An individual male or female mates with only one partner per breeding season.

POLYGYNY: Individual males may mate with more than one female per breeding season.

A. Some males can economically monopolize access to several females
   1. FEMALE DEFENSE POLYGYNY: Females live in permanent groups, which males defend directly.
   2. RESOURCE DEFENSE POLYGYNY: Females do not live in permanent groups but are spatially concentrated at food, nesting sites, or other resources, which some males can control.

B. Males cannot economically monopolize access to females directly or indirectly
   1. LEK POLYGYNY: Males compete for high dominance ranking within a group, usually at a traditional display arena; the winner is often selected by many females.
   2. SCRAMBLE COMPETITION POLYGYNY: Females may be clustered spatially or temporally but cannot be defended because of high male density; males race to contact as many receptive females as possible without engaging in territorial defense of an area.

POLYANDRY: Individual females may mate with more than one male per breeding season.

A. Polyandry without sex role reversal
   1. SPERM REPLENISHMENT POLYANDRY: Females may mate more than once to secure additional sperm with which to fertilize a new clutch of eggs.
   2. "PROSTITUTION" POLYANDRY: Females may mate with more than one male in order to gain access to the resources that males control and offer only to their mates.

B. Polyandry with sex role reversal
   1. RESOURCE DEFENSE POLYANDRY: Females control resources attractive to more than one male; males may provide even greater parental care per offspring than females.
   2. LEK POLYANDRY: Females provide neither resources nor greater parental investment per offspring than male; they compete for dominance positions in female aggregations because paternal males select high ranking females as mates.

A relatively monogamous shrimp, *Hymenocera picta* (Figure 11), illustrates how monogamy of this sort might evolve [649, 782]. Females of this species require sperm to fertilize each clutch of eggs, which they produce at three-week intervals. Receptivity is limited to a very brief period, three to five hours after moulting, which occurs every three weeks or so. As a result of the restricted receptivity of females and their rare and scattered distribution, it would be difficult, risky, and energetically expensive for males to search for a series of receptive females. Instead, when a male finds a female, he stays with her, repels any other males if they appear, and mates with his partner when she eventually moults. (The shrimp is not strictly monogamous, because if the male happens to encounter a moulting female other than his mate he will copulate with her. But these chances are probably very rare under natural conditions.)

11 **Monogamous shrimp** *Hymenocera picta.* A male remains with a female when he finds her. He will mate with his partner when she becomes receptive. Photograph by Uta Seibt and Wolfgang Wickler.

By far the most familiar examples of monogamy are found in the birds, where about 90 percent of all species practice pair bonding during a breeding season [403]. Although mate guarding plays a role in the male's sacrifice of opportunities to acquire multiple mates (Chapter 11), it is not the sole factor that can make monogamy adaptive for males. As in the case of hangingflies, male birds have the opportunity to provide material benefits for a mate in the form of nuptial feeding and parental care. These benefits may increase the survival chances and fecundity of a mate, enhancing a helpful mate's fitness in the process. Moreover, ecological circumstances that produce synchrony in female breeding can make it advantageous for a male to forego the polygyny route and help one mate reproduce as much as possible. If a black-headed gull were to desert one female after having copulated with her, his chance of finding other willing mates would be slight. Most females come into breeding condition at about the same time. They require a prolonged courtship before mating. By the time copulation is achieved, the number of unpaired females available for a polygynist is small. Moreover, an abandoned mate would have little chance of producing surviving young because she would not have help in incubation or in the care of the young after they hatched. If the opportunities for productive polygyny are few, males can gain more by providing to a mate useful services that raise her reproductive success. Under these circumstances a female selects an unbonded male as a partner, thereby securing for herself as much of a male's help as possible rather than having to share the assistance of a male with his other mate(s). Monogamy is the outcome (unless an already paired male can deceive a female into thinking he has no partner [2]).

In contrast to birds, monogamy in mammals is rare, probably because female mammals nurture their embryos internally and feed infants with their milk [550]. This makes male assistance relatively less valuable to the female and to the male. Under these circumstances, attempts by a male to fertilize as many females as possible may be more profitable than providing a little parental care to his existing offspring. A few carnivorous mammals, like foxes and wolves, are monogamous. In these species males potentially can help their mates a great deal by bringing food to them and their offspring (animal prey are usually scarce and protein-rich), by helping defend a feeding preserve, and by protecting infants against their enemies, including infanticidal male intruders. The variety of useful tasks a male carnivore can do for his mate may elevate the potential benefit of parental care for the male and the value of monogamy for the female [217].

Female carnivores can help enforce monogamy on their mates by repelling intruder females in solitary species or by inhibiting the development of sexual receptivity in female pack members in group-living species. For example, the dominant female wolf's aggression prevents all other females in the pack from coming into heat [505]. (This is to their advantage if their progeny would be killed by the alpha female or would starve if the alpha female's offspring monopolized the food available for pups.)

Female Defense Polygyny

By far the most common male mating system is polygyny, for reasons already discussed. In many species some males are able to exercise their potential for polygyny because females or the resources they require are clumped spatially, a situation that enables some powerful males to monopolize mates directly or indirectly. The argument developed by Stephen Emlen and Lew Oring is that if females or the things they use are concentrated in a small area, one male can *economically* defend that area and lay claim to the copulatory rewards that ownership provides [217]. Female defense polygyny can evolve whenever any of several ecological pressures cause females to gather in groups. For example, gorilla females aggregate apparently for protection from predators. A male that can travel with and defend a cluster of females gains access to many potential mates without having to defend a large area [307].

Lionesses form groups, not for protection against predators, but because of the opportunities for cooperative hunting and defense of their kills against competitors. This creates a ready-made harem for the male or males that can defend them against rival males [635]. The same is true for males of the tropical bat, *Phyllostomus hastatus,* whose females also live in stable groups [498]. The members of a group, although unrelated to one another, always reassemble to roost together in the same spot after a night of

12 **Female defense polygyny** in the bat *Phyllostomus hastatus.* The large male at the top monopolizes a roosting cluster of females. Photograph by Gary McCracken and Jack Bradbury.

foraging. Radio-tracking studies indicate that the females of a cluster tend to forage in the same general area. Occasionally groups of females leave the roost in a cave simultaneously and gather in a presumably rich feeding area previously exploited separately by one individual. Thus, clustering by females may promote cooperative food sharing. In any event, the existence of such groups is tailor-made for female defense polygyny. One male can easily monitor the daytime activities of about 20 females (Figure 12). Gary McCracken and Jack Bradbury have shown, through analysis of key proteins in harem masters and the offspring of harem females, that a territorial male probably fathers 60–90 percent of these offspring. Thus, the distribution of females enables some males to be fabulously successful—fathering more than 50 bat pups during a multiyear tenure as harem master [498].

Female defense polygyny is not restricted solely to mammals. For example, females of various wasps nest in small clusters, perhaps because they can use the brood cell walls of other females to economize on their own nest construction. Whatever the reason, a male that can control a cluster of brood cells gains a harem of females when the new generation of wasps emerges (Figure 13). These females are receptive when they exit from their cell, a fact that the resident male exploits for his reproductive advantage [195, 668].

13 **Female defense polygyny** in a eumenid wasp that defends a cluster of brood cells from which a number of receptive virgins may emerge. Photograph by the author.

Resource Defense Polygyny

If receptive females do not live together, a male may still be able to monopolize many mates if he controls a limited resource that they use. This indirect route to polygyny is only economically feasible if the resources are clumped so that a male need not defend a huge territory. Good oviposition sites for bullfrogs and black-winged damselflies are restricted to small locations that one male can readily defend. Females that come to these places to lay their eggs can be controlled by the male that owns the resource they wish to use.

Sometimes the feeding requirements of females are specific enough and the food distributed patchily enough to permit some males to defend valuable centers of food. These individuals can mate with a disproportionate number of female foragers. An impala male that controls a rich grazing pasture containing the plants favored by females practices resource-defense polygyny [361]. The unrelated vicuña of South America has evolved a convergent system, with males competing for the best foraging areas (which are in short supply) and mating with the females that join them there [240]. Desert woodrats living on the Danzante Island in the Gulf of California primarily consume portions of ironwood trees, which occur on the island in restricted locations. Some males can monopolize an area with two or three ironwoods, each with its resident female, who will mate with the polygynous male [735].

We noted that most birds are monogamous but that some are polygynous; among these polygynous species are some examples of resource defense polygyny. An Asian honeyguide has a specialized food requirement—beeswax. This food is available in large amounts in small spaces: the exposed hanging combs of the giant honeybee *Apis dorsata,* whose defensive behavior was described in Chapter 10. The honeyguides are immune to the stings of the bee and can feed with relative impunity on the comb. The wax produced by one large colony of *A. dorsata* can support several of these birds. Males that establish year-round territories at a honeybee nest permit a number of females, but no additional adult males, to feed on the beeswax they control (Figure 14). In return, the females sometimes copulate with the territory owner [150].

The honeyguide is an example of a bird whose males can sometimes provide such large amounts of a valuable material that females gain more by mating with a polygynist male than by choosing an unbonded individual. Resource defense polygyny in birds should be associated with territorial defense of unusually productive areas because a female that mates with a polygynous male loses some or all of the parental care this male might offer. If she were not compensated by increased availability of food (or a safer nest site), selection would presumably favor females that avoided already-mated males.

14 **Resource defense polygyny** in a bird, the orange-rumped honeyguide. Males defend combs of the giant honeybee and mate with the females attracted to this essential resource.

Female Reproductive Success and Resource Defense Polygyny

Tests of this prediction come from several sources. Polygyny is common in the familiar red-winged blackbird. Although it is possible that females select males in part on the basis of cues related to genetic quality [760, 809], there is good evidence that the number of females a male has living with him is linked to the productivity of his territory [412, 551]. Moreover, females do make choices that help them achieve as much reproductive success as their rivals. Michael Monahan studied a population of redwings nesting in a hayfield; some males had small harems and others large ones [518]. Females collected 80 percent of their food for their nestlings within the breeding territory. Therefore, its productivity was essential for their reproductive success. The number of fledglings produced per female in harems of different sizes did not vary significantly (Figure 15). This shows that females can in some way assess their reproductive chances in different territories, weighing the number of nesting females already present against the foraging value or safety of a site.

Similar results have been achieved for two other birds unrelated to redwings. For the lark bunting, *Calamospiza melanocorys,* Wanda Pleszcynska was able to predict which males in her various study plots would have zero, one, or two mates simply by using a light meter to measure how

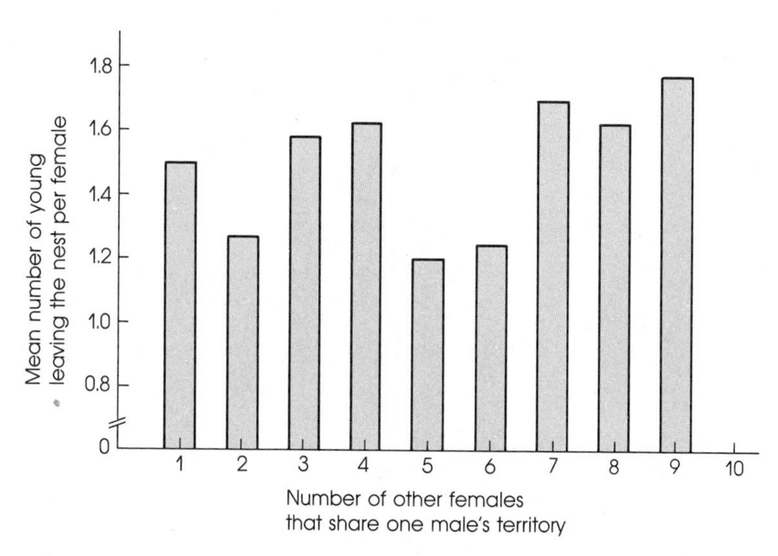

15 **Polygyny and female reproductive success,** as measured by the number of young leaving the nest for each female red-winged blackbird. Females that mate monogamously do not do better than females that have joined males with other mates. Males with harems own superior territories, the rich resources of which enable their families to enjoy the same reproductive success as monogamous females.

much light penetrated the vegetational cover in the grassland habitat of this species [579, 580]. There is considerable variation in the density of cover in lark bunting territories. This affects how well the nest is shaded, and degree of shadiness appears to be the key factor influencing fledging success in the species. As one would expect, the territories with the densest plant cover sometimes attracted two females; the males that owned these locations practiced polygyny. Moreover, second females that joined already-mated males produced as many young on average as females that joined an unmated male at the same time in other, less heavily vegetated territories. The same ability to pick a mated partner whose territory quality compensates for the presence of another female has been documented for the indigo bunting, *Passerina cyanea* (Table 4). Work with these birds supports the argument that females will join a harem when this decision confers as much reproductive success as would monogamous alternatives [126, 127].

But one study appears to contradict this hypothesis. In the yellow-bellied marmot, a large rodent of mountain meadows in the west, there is considerable variation in the number of females that live with one male in his territory [187]. A comparison of the number of young produced per female in harems of different sizes reveals that the average annual reproductive success of a female falls as the size of her group increases (Figure 16). Two (untested) hypotheses have been proposed to account for this result. The first argues that females in large groups may sacrifice some annual reproductive success in order to gain better protection from predators that will enable them to live longer and maximize their *lifetime* reproductive success [208]. The larger the group, the more eyes to detect hunting coyotes and golden eagles and the longer a female should live. Unlike lark buntings, whose females probably have a life expectancy of a year or two and who therefore should attempt to maximize their annual output of offspring, a female marmot can potentially live many years. Her fitness is *not* deter-

Table 4
Reproductive Success of Male and Female Indigo Buntings Engaged in Different Mating Systems

	Number of Pairs in Which Age of Male Was		Reproductive Success[a]	
Pair Bond	*≥ Two Years*	*One Year*	*Males*	*Each Female*
Polygyny	12	2	2.6	1.3
Monogamy				
Season-long	34	6	1.6	1.6
Short-term[b]	22	30	0.4	0.4

Source: Carey and Nolan [127].
[a] Mean number of offspring fledged per individual.
[b] Pair bond lasted less than three-quarters of the breeding season.

mined by how well she does in just one year but by the sum of each year's production of young.

The second hypothesis states that there is competition among females for access to the very few males with good territories [801]. If an unmated female can force herself into a small harem, she may do better there than if she were to settle for an unmated male on a poor territory. But her presence depresses the reproductive success of the already-established females who should try to keep her out but may not be able to. Long-term studies of the survival rate and reproductive success of marked individuals in harems of different sizes are needed to discriminate between the predator-avoidance and reproductive-competition hypotheses. It is clear, however, that a marmot male's reproductive success is increased by the acquisition of a harem; the foraging area he controls provides the basis for attracting females to him.

Lek Polygyny

There are many territorial species whose males neither defend useful resources for mates or clusters of females that happen to be living together. Instead, these males defend purely "symbolic" territories, often at a traditional display site or arena (the *lek*), which contain no ready-made harem nor any resource of utility to potential mates. Females do visit these arenas but just to select a mate; when copulation is completed, the relationship between a male and a female ends.

The white-bearded manakin is representative of this group [424, 675]. Each male's territory contains nothing of value to a female, consisting only of a sapling or two in the forest and a bare patch of ground underneath the perch site, which the bird clears of leaves and debris to make a display court (Figure 17). There may be as many as 70 courts in a display ground 10 by 20 yards in area. From a small branch a few feet above the ground, the male performs a vigorous display, which begins with a series of rapid jumps from perch to perch and which is accompanied by explosive sounds

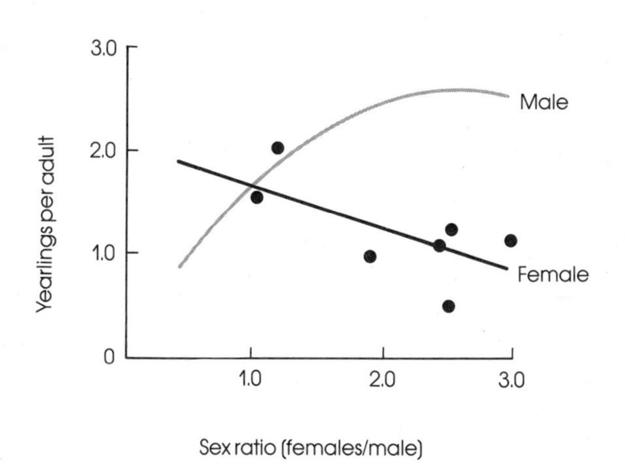

16 **Reproductive success of female marmots** is lower in large harems than in small harems.

produced when the bird snaps his clublike wing feathers together in flight. Following the introductory phase of the display, the male pauses with its body tensed and then begins to jump to the ground with a snap and immediately back to a perch with a buzz and then back to the ground "so fast that he seems to be bouncing and exploding like a firecracker" [675]. When groups of males are displaying together, the resultant pandemonium is spectacular; and the noise may carry 100 yards or more in the jungle.

When a female arrives at the lek, she observes the males and then may move to a perch near a male's court. The male flies to her and performs a visual display. If the female is receptive, she follows him to his court; a number of mutual displays ensues, typically terminating with the "grunt

17 **Small arena of the white-bearded manakin,** showing five males at their individual territories. Each male creates a court for his acrobatic displays by clearing the area below his perch of all leaves. The male in the right foreground is about to begin a courtship display consisting of a series of jumps to the ground and back to his perch. Two females are watching the displaying males from the periphery of the display area.

jump" in which the male flies to the court floor and then immediately jumps back to the sapling above the female. He then slides down the trunk of the sapling while beating his wings and slips onto the back of the perched female. After mating, the female leaves to begin nesting, and the male remains to await new arrivals. Alan Lill's superb studies of individually marked manakins revealed that females select males on a highly nonrandom basis. In one lek of ten males, a single male secured nearly 75 percent of the 438 copulations recorded in this arena; his next nearest competitor mated 56 times (13 percent) whereas six of the males together copulated a total of just 10 times [425]. Mate selection does not appear to depend upon the way in which males display or upon size and plumage differences among males, but upon which perch site a male owns. Male and female manakins apparently share the perception of what constitutes the most desirable territory, and the male that is able to claim this court and repel other males from it enjoys enormous reproductive success. When Lill removed the most successful reproducers from their perches, lower-ranking males quickly moved to these sites and almost immediately began to enjoy disproportionate reproductive success.

The willingness of apparently subordinate males to remain at the lek and spend many hours in energetic displays probably stems from the reluctance of females to copulate with solitary males. It is to the female's advantage to have an opportunity to compare the competitive ability of males directly; female preferences may result in the formation of leks. By remaining at a lek for a number of years (white-bearded manakins may live ten years or more), a male may eventually advance to the preferred territory and enjoy reproductive success.

In order to figure out why male manakins behave in such a bizarre fashion, we should consider the ecological practicality of their other options. Female defense polygyny is clearly out of the question because female manakins do not live in permanent groups but instead nest in isolation from one another. Monogamy and resource defense polygyny, however, cannot be so easily dismissed. After all, most male birds do try to control a territory whose resources will attract one or more females. But perhaps male manakins refuse to enter into resource-defending contests because the tropical fruit-bearing trees that provide most of the bird's diet come into fruit irregularly and unpredictably. A male that tried to hold a territory about one tree might have a long wait before it became attractive to females. A male that tried to defend an area large enough to always contain at least one fruiting tree would have to expend vast amounts of energy to do so. When a tree does begin to produce fruit, it often becomes covered with food, drawing in a host of fruit-eating animals. This would make defense of the resource economically unattractive for a would-be territorial male manakin. Thus, the environment of this species makes it difficult for males to monopolize females directly or indirectly, and they are forced to resort to competition for certain display positions within a lek because females prefer to mate with the owner of a particular display site (in order to secure "good genes"?).

But why do females not prefer a monogamous male to help with incubation duties and feeding of the young? Perhaps such males would be chosen, but males refuse to cooperate because the benefit in terms of improved offspring production would be outweighed by the cost of reduced mating opportunities. (In tropical rain forests, manakins may breed nearly year round, unlike the highly seasonal and synchronized breeding of monogamous temperate zone birds.) It is even possible that male assistance is actually detrimental on balance to female manakins. She and her offspring can feed on abundant fruits that can be gathered in a short time each day. Thus, she does not gain much by having a male help her incubate or feed the young. She might lose overall if the male's activities about the nest increase the chance of a predator finding her brood. There is considerable evidence that nest predation is the major problem for tropical forest birds [676]. Whether male help would reduce the male's reproductive success or is actively rejected by the female, the end product is the same. Males are "liberated" from resource control and parental duties and instead attempt to demonstrate their social dominance to selective females [426].

Lek formation occurs in a number of other birds and mammals that are subject to ecological conditions similar to those operating on manakins. In fact, one lek-forming mammal, the bizarre West African bat *Hypsignathus monstrosus,* is a remarkably close ecological equivalent of the white-bearded manakin [83] (Figure 18). The bat feeds at tropical fruit trees, which are widely scattered and bear fruit unpredictably. Males are unable to monopolize food for their mates and offspring. Moreover, because females bear and nurse their progeny in relative freedom from predators, there is little males can provide in the way of useful parental care. The males gather in the evening during the prolonged breeding season along river edges at traditional display sites. Each individual defends a territory that is ten meters in diameter and that is centered on a perch high in a tree, from which he hangs while producing a loud cry that sounds like "a glass being rapped hard on a porcelain sink." A receptive female flies to the lek and approaches several males in turn (each of whom responds with special vocalizations and wing-flapping displays) before finally settling on one individual with whom she copulates. As in the case of the manakins, position in the lek is important in determining male reproductive success, which is highly biased toward a few individuals. At one lek in 1974, 6 percent of the males secured 79 percent of observed copulations [83].

Thus, lek polygyny in both birds and mammals is correlated with ecological factors that (1) make resource defense and parental investment by the male disadvantageous to the male (or his mate) and (2) make it difficult for a male to monopolize a harem of females. These conditions may also apply to the origins of leks in insects [700], frogs [767], and other animals [801].

Scramble Competition Polygyny

I do not wish to leave the impression that all polygynous males are territorial. Many are not and instead engage in a race with competitor males for mates. Those that succeed in copulating with more than one female have employed nonterritorial, scramble competition to secure several mates. The ecological correlates of this mating system are very similar

18 **Hammer-headed bat,** *Hypsignathus monstrosus*. Males of this West African mammal form leks in much the same manner as the white-bearded manakins. Photograph by A. R. Devez.

to lek polygyny. One of the puzzles of behavioral ecology is why dispersed females and dispersed resources should in some cases facilitate the evolution of lek polygyny and in other cases lead to scramble competition polygyny.

There is an important, distinctive, ecological correlate of the most dramatic kind of scramble competition polygyny, the EXPLOSIVE BREEDING ASSEMBLAGE [217]. This mating system is associated with a short breeding season—unlike lek polygyny, which usually occurs in species with prolonged breeding periods. We have already mentioned the case of the wood frog, in which aggregations of receptive females may form in one pond on one night of the year only [89]. Similarly restricted breeding periods occur in other frogs and toads as well [768]. Because males experience intense pressure to be available to mate at these times, large concentrations of rival males form in small areas—a condition that does not favor the evolution of territoriality because of the high costs of repelling a large number of intruders. Male wood frogs avoid territorial expenses and instead rush about trying to encounter as many highly fecund females in as short a time as possible. Females usually accept the first male to reach them because of the advantage of synchronous breeding and egg deposition [89].

Even if the breeding season is not particularly truncated, other factors may produce large numbers of males in small areas, and the costs of trying to be territorial or dominant may rise to prohibitive levels (Figure 19). In some female defense or resource defense species, males cease to be territorial when male densities pass a certain point; and a scramble for females ensues [227]. Male dungflies at low density will defend a portion of a dung pat that attracts egg-laden females. At high density, the same males abandon all attempts to repel invaders and instead try to find incoming females before their opponents [79].

A current view of the evolution of diversity in male mating strategies is summarized in Figure 20.

19 **Scramble competition** may play a role in the acquisition of mates in the horseshoe crab. The male first to find a female gets directly behind her to fertilize her eggs. Other late-arriving males are clustered about the pair. Photograph by James Lloyd.

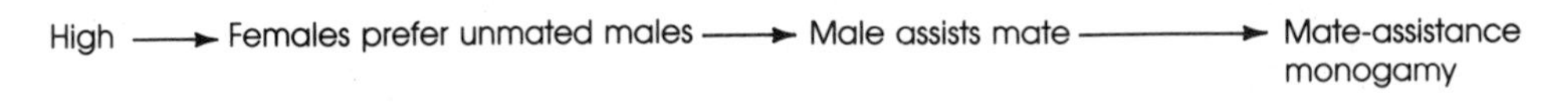

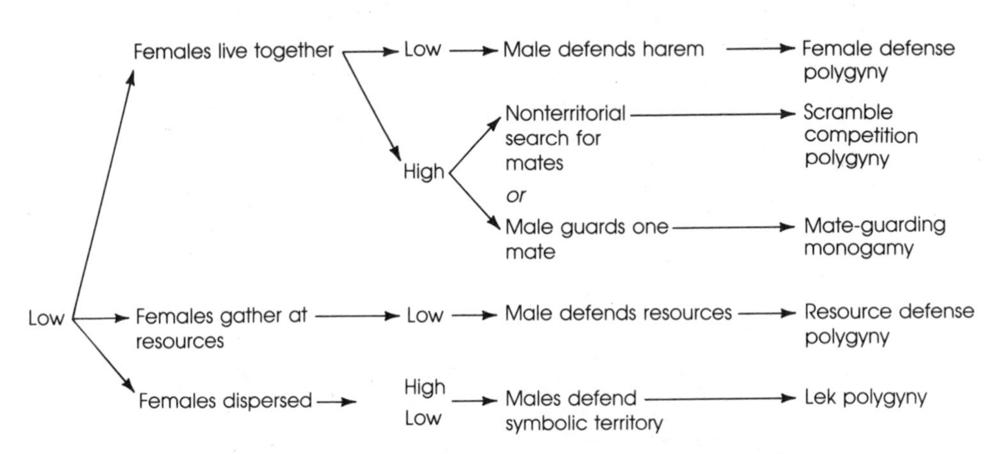

20 **Summary of mating system theory.** Different mating systems are thought to evolve in response to ecological factors that determine male potential to monopolize mates or to assist females in producing and rearing offspring.

Polyandry

As a general rule, females are far less likely to mate with a number of males than males are to mate with several females. Females often have a limited number of eggs that can potentially be fertilized by the sperm received from a single male. After having copulated with a superior male, a female may only suffer the time and energy costs of additional mating attempts with no compensatory advantages. In addition, polyandry is especially rare in species in which the male and female form a pair bond with shared parental duties. In these cases, although a female might benefit from having a bevy of males as her assistants, the risk to a male of providing services for another male's offspring is so high that males almost never enter into a polyandrous pair-bond arrangement [550]. Because polyandry is generally disadvantageous to either males or females, it is the rarest of the mating systems.

Nevertheless, it does occur. Females that lay eggs at intervals during a protracted breeding season, like anoles (Chapter 6) and fruit flies [541], may mate a number of times, often with different males, in order to fertilize the different eggs or clutches of eggs. These "supplementary" copulations may not only restore depleted sperm supplies but may also enable the female to increase the genotypic variability of her offspring. That sperm-replenishment polyandry can be adaptive is shown in the maintenance of high fertility in female *Drosophila* fruit flies that have access to more than one mate at appropriate intervals [463, 591] (Figure 21).

Polyandry may also result if some males control valuable resources that females can use only by copulating with the resource owners. Females of some polygynous hummingbirds may mate with a male that owns a rich nectar-producing flower patch each time they come to the patch [802]. An exactly analogous situation exists in a highly territorial megachilid bee whose males guard clumps of flowers that provide pollen and nectar to females. A female bee may visit several territories in the space of 15 to 30 minutes and may copulate with each resident male in turn (Figure 22) [7]. This kind of mating system has been labeled, perhaps uncharitably, "prostitution polyandry" [802]. It seems to apply to animals such as the black-winged damselfly whose females usually have to mate with the resident male if they are to enter his territory to exploit the oviposition material he controls. In cases of this sort, the territorial male may treat uncooperative females harshly, like any other intruder in which he has no reproductive stake.

Similarly, in species in which males offer food presents as inducements to mate, females might be expected to copulate with a number of males in order to receive the presents. We have discussed a classic example: the black-tipped hangingfly, whose females go through life copulating with one male after another and each time receive a nuptial gift from their partners. Acceptance of several males is the rule in those insects (including grass-

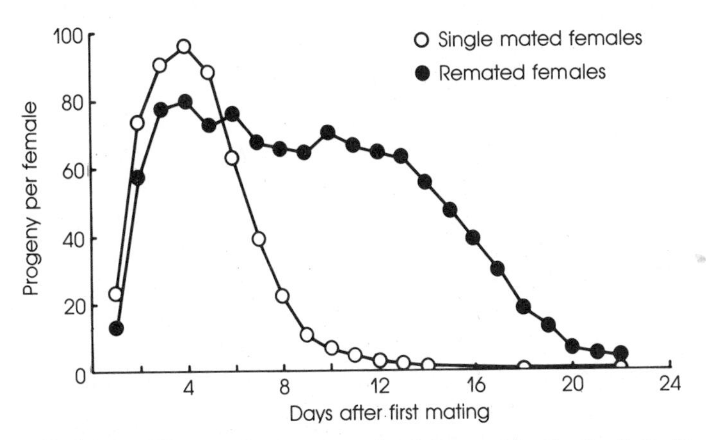

21 **Fertility and sperm replenishment** in the fruit fly *Drosophila melanogaster.* Females need to mate at five- to six-day intervals to maintain high fertility.

22 **"Prostitution" polyandry.** Females of this megachilid bee mate "on demand" with territory-owning males to gain access to the flower resources males control. Photograph by the author.

hoppers, katydids, and butterflies) in which the male transfers nutritionally valuable materials to the female along with his sperm [293, 626, 700].

Sex Role Reversal Polyandry

In the cases we have described, females for the most part behave like typical females in that they are selective in their choice of mates and more likely to engage in parental behavior than their males. Polyandry is most striking, however, in species in which there is a partial or complete sex role reversal, with males assuming all or most of the parental duties and females even competing, like typical males, for access to mates.

How can it be to a male's advantage to become a sole parent? This behavior is reasonably common in those fishes [76, 604] in which nearly continuous nest guarding helps combat intense egg predation [488]. A male's evolutionary options are to guard the eggs he fertilizes or to search for more mates. The benefits of guarding rise (1) as the predation rate on eggs increases, (2) as the number of receptive females a male can find away from the nest falls, and (3) as the difference between the effectiveness of male versus female nest protection increases. After laying her eggs, a female fish is likely to require a period of time to feed and to restore her energy reserves. A male with stored fats should be able to remain on guard more consistently and to defend the eggs more aggressively than a depleted female. This might favor males that assumed the guarding role immediately after the eggs were laid [488]. If the male were able to attract other females to deposit their clutches in his nest, he could elevate his fitness sharply at little additional cost (presumably it takes just about as much energy to defend a nest with 50 eggs as one with 500). This is what fishes like the stickleback and bluegill sunfish do.

The intensity of predation on large protein-rich eggs in aquatic environments may also have contributed to female polyandry and male parental care in giant water bugs [671]. The male, by accepting the eggs on his back, may in some environments provide much more effective defense of the brood than an energy-depleted female could. The male has enough space on his back to accept eggs from several females. Once burdened with some eggs, additional ones can be tended without much extra expense.

Male polygyny coupled with female polyandry also occurs in a few birds whose social systems closely resemble that of the stickleback and giant water bug. For example, tropical tinamous are ground-nesting partridgelike birds whose eggs are extremely attractive to many predators. Males perform the task of incubation without assistance from their mates. A male often receives clutches of eggs from several females in a short period and incubates the eggs of all his mates simultaneously. After producing one clutch, the female sets about making another one. She may donate this second clutch to her first mate if his clutch has been destroyed, or she may seek another male to incubate these eggs [550].

In all the examples discussed so far, the males behave more or less like typical polygynous males except that they provide parental care. Their parental behavior is not incompatible with acquiring multiple mates, and this seems to be their major goal. Indeed, having eggs to defend on a territory (or one's back) might even be an effective way for a male to attract additional mates.

There are other species in which the extent of sex role reversal is far more dramatic. Perhaps the best-studied species of this sort is the spotted sandpiper, a common shorebird that has been the subject of an intensive long-term research project led by Lew Oring (483, 552, 553). The reproductive strategy of females has been to adopt a malelike role. They are larger and more aggressive than males. They arrive on the breeding grounds first, whereas in most birds it is the males that precede the females. Once at the breeding site, females establish large territories, which they defend against other females with much fighting (Figure 23). A female's territory may

23 **Resource defense polyandry** is practiced by the spotted sandpiper. Here two females are engaged in a territorial dispute at the joint boundary of their territories. Photograph by Steven Maxson, courtesy of Lew Oring.

then attract a number of males (often two to four), which set up their own smaller territories within one female's domain. A male is smaller, less aggressive, and subordinate to the female with whom he mates and from which he accepts a clutch of eggs. While the female continues to defend her territory and to lay clutches of eggs for other males, her previous mates do all the work of incubation and parental care.

Perhaps this system evolved from one in which males took on incubation duties to permit their mates to produce a second clutch as quickly as possible. Predation is high on spotted sandpiper nests, which like those of the tinamous are located on the ground. If a male lost his clutch, his mate could provide him with another one in time to bring off a brood. Females do sometimes lay second clutches for their first mate, but they sometimes copulate with a second male in their territory and lay their eggs in his nest. The first male has no other option than to be parental, tending whatever eggs his mate chooses to give him.

Female spotted sandpipers, having been largely freed from the task of incubation, are able to invest in the defense of a resource—good nesting habitat—that attracts males to them. Another related case of resource defense polyandry is exhibited by the jacana, a tropical marshland bird (Figure 24). Like the spotted sandpiper, sex role reversal is pronounced. Females are larger than males and fight for large territories in the patchy

24 **The polyandrous jacana,** *Jacana spinosa*. Females of this marsh-nesting species are larger than males and claim territories in which several males brood clutches of eggs produced by the territory owner. Male jacana in photograph responds to a potential predator approaching his chicks. Photograph by Terry Mace.

marshland habitat where they breed. The female that can claim a superior patch of marsh is able to attract several males to her territory. She mates with each individual and provides each with a clutch of eggs, which he cares for by himself [367, 368].

One more polyandrous bird is the northern phalarope, another marsh-nesting species in which males provide all the parental care for the brood. In this species females do not claim territories (perhaps the quality of nesting habitat is constantly and unpredictably changing because it is dependent upon variable rainfall patterns). In some cases large numbers of females have been seen aggregating at a prime breeding area, where they display and compete with one another for access to males (lek polyandry). Thus, when resource monopolization does not occur, the more selective of the sexes (in this unusual case, the male) may choose a mate on the basis of dominance ranking alone [217].

The behavior of males and females in cases of sex role reversal is powerful support for Trivers's hypothesis that the relative parental investment per offspring by males and females is a prime mover in the evolution of mating systems [716]. Typically, males compete to copulate with females, but as males provide an increasingly large parental investment per offspring, females begin to compete more and more strongly among themselves for opportunities to mate, and males become more and more selective. A recognition of the significance of the relative parental investments of the sexes and the ecological determinants of male success in monopolizing mates has provided a coherent and successful way to solve some of the mysteries of mating systems.

Summary

1 Epigamic sexual selection is usually generated by females that prefer males with certain characteristics. Female mate choice theoretically could be based on male traits that indicate the possession of superior genes *or* that signal the male's ability to provide superior material benefits such as food, a nesting location, or parental care.
2 Evidence that females exercise active mate choice is limited, especially with respect to the "good genes" criterion. But females of some species do avoid mating with close relatives, and there is considerable suggestive evidence that females prefer to mate with dominant males. Dominant mates are often long-lived, competent survivors and therefore may possess "good genes." Male courtship performance may also provide indirect indicators of male longevity and physiological condition.
3 There is better evidence that when males offer more than genes alone, females discriminate among them on the basis of what they have to offer. Females may prefer males that provide larger nuptial gifts, or access to superior resources, or superior parental care.
4 Mating systems can be defined purely on the number of mates an individual has during a breeding season. Sexual selection generally favors males that try to mate with more than one female, leading to polygyny. Monogamy by both males and females can occur through a combination of female preferences for unmated helpful partners and various ecological factors that reduce the ability of a male to monopolize more than one female.
5 If females or resources are clumped, female defense polygyny or resource defense polygyny results because some males can economically defend the productive locations for mating. If, however, these forms of territoriality are too costly, males may either attempt to demonstrate social dominance through "symbolic" lek territoriality or engage in nonterritorial scramble competition for access to mates.
6 Polyandry is the rarest mating system because a male rarely gains if his

mate copulates with another male and because female fitness is not often promoted by frequent copulation. Nevertheless, in some species females mate with more than one male to secure additional sperm or added material benefits. In its most extreme form, polyandry is associated with sex role reversal in which males provide most or all of the parental care and females defend territories or compete for dominance in order to attract several males to them.

Suggested Reading

J. Wittenberger's *Animal Social Behavior* [801] provides an excellent review of most topics in this chapter and has the virtue of outlining competing hypotheses clearly.

Randy Thornhill's articles on hangingflies are highly recommended for persons interested in demonstrations of female choice [696–699].

Stephen Emlen and Lew Oring provide an updated analysis of the role of sexual selection in the evolution of mating systems [217], building on the classic article by Gordon Orians [550].

Alan Lill's meticulous studies of manakin behavior are excellent models for investigations into lek polygyny [424–426].

Suggested Films

Gelada. Color, 18 minutes. Illustrates the behavior of a baboon with female defense polygyny.

The Guanaco of Patagonia, A Study of Behavior and Ecology. Color, about 30 minutes. Shows the annual cycle of the guanaco, a South American mammal that engages in resource defense polygyny.

Life in a Weaverbird Colony. Color, about 20 minutes. An example of a colonial, polygynous bird species.

The Northern Elephant Seal. Color, 14 minutes. On the remarkable harem-defending elephant seals of California.

Reproductive Behavior of the Black Grouse. Color, 24 minutes. A film on a bird whose mating system is lek polygyny.

Reproductive Behavior of the Polyandrous Spotted Sandpiper. Color, about 20 minutes. On a bird with partial sex role reversal.

The Uganda Kob, Territoriality and Mating Behavior. Color, 20 minutes. A film about the African antelope that practices lek polygyny.

The final component of an evolutionary approach to behavior lies in answering the fourth basic behavioral question: What is the history of a behavior pattern? Ultimately, why an animal behaves the way it does is linked with the ancestry of the population to which it belongs. In this chapter we shall focus directly on the evolutionary process and attempt to trace the history of selected behaviors. If one could go back in time to follow the evolution of a species, it would be relatively easy to establish how behavior patterns have changed in response to changes in natural selection. In the absence of a time machine, however, determining what extinct animals were doing thousands of years ago is a difficult task—but not an impossible one. Sometimes fossil evidence can be used in conjunction with modern comparative material to determine the probable behavior of an extinct species. The more widely useful procedure, however, is to compare the behavior of a group of living species in order to reconstruct a plausible evolutionary sequence leading to a modern behavior of particular interest. Here we shall describe the comparative procedures one applies to trace the history of an unusual feeding pattern and some remarkable communication signals. Because of the focus on behavioral displays, this is a convenient place to discuss a number of general evolutionary aspects of communication. Therefore, the chapter will conclude with an examination of the historical and ecological factors responsible for diversity in the ways animals convey messages to one another.

CHAPTER 13

The Evolutionary History of Behavior

Does Behavior Ever Fossilize? A direct way to approach the problem of the history of a behavior is to try to learn from fossilized remains how extinct animals behaved. One can make surprisingly many inferences about the activities of extinct species from their fossils. For example, the tracks left by dinosaurs in soft mud have been used to estimate how fast these animals moved—about eight meters per second in some cases, a speed close to the maximum attainable by a world-class human sprinter [232]. In some sites the tracks of certain species are all oriented in the same direction, an observation suggesting that these dinosaurs were social, traveling in a herd [794]. The discovery of a group of young dinosaurs clustered in an apparent nest suggests that their species must have engaged in parental behavior because the dinosaurs were too large to have been recently hatched. They had to have been tended by an adult in order to have survived as long as they did without leaving their birthplace [344].

Sadly there are no more dinosaurs, unless one wishes to count birds as living representatives of the group [235], and therefore we cannot use their fossils to determine stages in the behavioral evolution of a living species. But one can sometimes go back from a modern animal into the past in search for antecedents of an interesting current behavior. For example, living birds are remarkable for their ability to fly. Anyone interested in flight would surely want to know what the earliest known bird was capable of doing. The extinct *Archeopteryx* is the most ancient feathered, winged animal. Its fossilized remains occur in rock strata dated at 150 million years. There has been considerable discussion about whether this bird ever flew, in part because its breast bone (sternum) lacks a prominent keel for the attachment of the large breast muscles that power flight in modern birds. Some researchers have argued that it was really a small dinosaur that used its feathers for insulation and perhaps employed its wings as insect traps, but not for flying [555]. But other ornithologists have recently pointed out that the central support (or *rachis*) of the feathers of *Archeopteryx* (which have been beautifully preserved in some cases) is not located in the center of the feather but is set off to one side (Figure 1). This

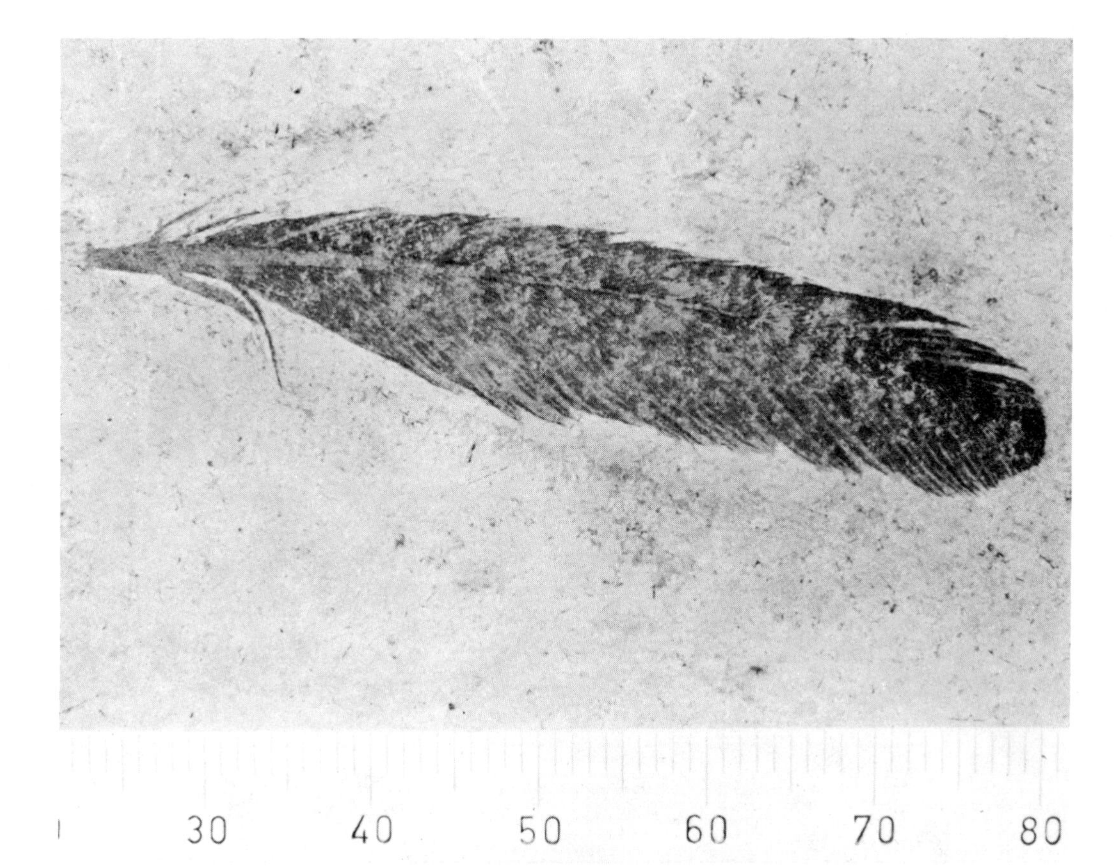

1 **Fossil wing feather** of *Archeopteryx*. The feather is asymmetric, as are the wing feathers of the modern flying birds. Courtesy of Dr. John Ostrom.

asymmetric arrangement is typical of all modern flying birds, whereas in a host of unrelated flightless species the rachis is placed directly in the center of the wing feathers. The asymmetry of the feathers of flying birds is related to their aerodynamic function, and therefore one can infer with confidence that *Archeopteryx* flew, although whether by gliding or flapping remains uncertain [236].

In contrast, there is no doubt that some extinct birds of the family Plotopteridae did not fly [547]. These creatures, which have no living relatives, are more recent in the geologic record than *Archeopteryx*. Their wing bones were massive and dense, totally unlike the delicate hollow bones of modern flying birds but remarkably similar to the flightless penguins and flightless great auk (which became extinct in recent times). It seems certain that these birds must have used their wings, like penguins and great auks, as paddles for underwater diving. Presumably they, like the penguins, evolved from species that could fly and secondarily lost this ability as a result of selection for powerful underwater locomotion.

The Evolution of Bipedalism in Man

Another unusual mode of locomotion is bipedal striding by human beings. Although a few other primates occasionally walk upright for short distances, we are the only living primate habitually to walk in this manner [727]. Our skeleton is greatly modified, particularly with respect to our feet and pelvis, in ways that facilitate our distinctive locomotory behavior. But how did our hominid ancestors walk? The general feeling has been that members of the genus *Australopithecus,* from which we are believed to have descended [752], probably were bipedal but were less efficient striders than modern man (Figure 2). More recently, however, fossil footprints have been discovered in the rock formed from a rain of volcanic ash dated between 3.5 and 4.0 million years old. The makers of these footprints were almost certainly australopithecines, and their footprints are almost identical to those of their modern living descendants (Figure 3). The tracks have a prominent heel mark, modern placement of the big toe in relation to the rest of the foot, and an alignment of prints that is the same as those made by living people [314, 776]. Thus, these creatures, whose skulls were much different from our own, apparently walked bipedally as efficiently as you or I.

2 **Artist's reconstruction of** *Australopithecus.* This painting depicts *Australopithecus* when he was generally thought to have walked bipedally in a shuffling fashion, with knees bent. Courtesy of Zdenêk Burian.

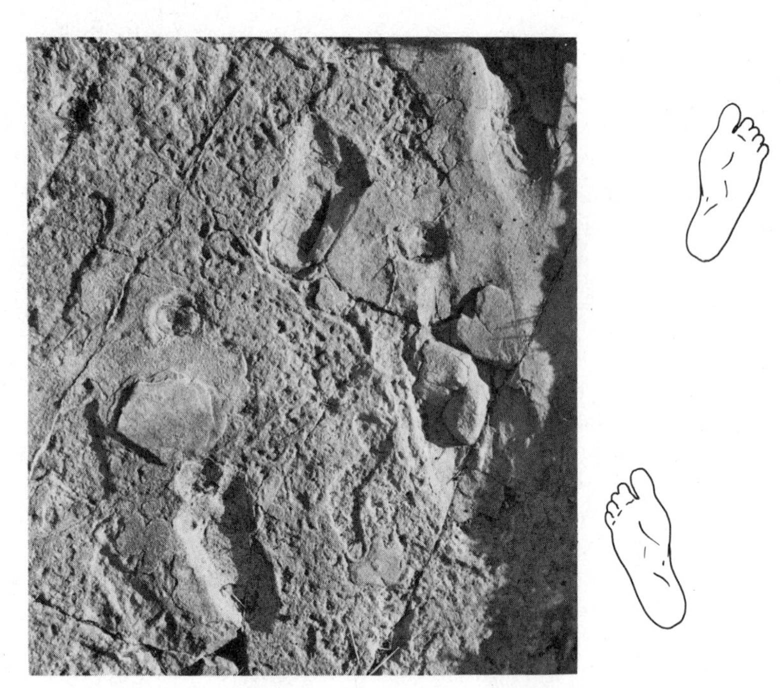

3 **Fossil footprints** of *Australopithecus* indicate that this extinct hominid walked upright in a completely modern way. Photograph courtesy of Mary Leakey and R. I. M. Campbell.

The Reconstruction of Historical Pathways without Fossils

We have seen that it is possible to make inferences about the behavior of extinct animals by comparing their fossilized remains with those of modern animals, whose behavior is known. But the examples presented here also show that fossil evidence does not automatically reveal the historical pathway leading to an existing behavior. Judging from *Archeopteryx* and *Australopithecus,* investigators have concluded that some unusual behavioral traits may be more ancient than once suspected and that the intermediate stages must be missing from the fossil record—perhaps because the evolutionary transition from one stage to another was very rapid. Therefore, we cannot often rely on the geologic evidence to trace the evolutionary steps leading to a behavior of interest. But this does not mean that it is impossible to make educated guesses about the history of a trait.

Consider the example of blood-sucking by a moth. One night, while in the course of his superbly exotic occupation as an observer of moths on Malayan water buffaloes, Hans Bänziger captured a specimen of *Calpe eustrigata* (Figure 4). Suspecting that it might be a species that consumes droplets of mammalian blood excreted by mosquitoes gorging themselves on a host, he made a slight cut on his finger and offered it to the caged moth [38]: "The moth climbed onto my finger and did in fact plunge its proboscis into the blood, but it appeared to imbibe none. Instead it stuck its straight, lancelike proboscis into the wound and, without any regard for the donor, penetrated the flesh. The pain I felt caused me to utter a cry of—joy! I had discovered a moth which pierces to obtain blood."

This is the only moth of some 200,000 species known to have adopted this manner of feeding. When one thinks of butterflies and moths feeding, one envisions them sucking the nectar of flowers with a long and delicate proboscis. Some species, however, exploit another supplier of concentrated sugar solutions, the juices from overripe fruits. Still others consume the exudate from surfaces of fruits partly eaten by birds, bats, or rodents. In certain populations of these species in the past, it is probable that individuals that happened to have a heavier than usual concentration of small rasping hairs at the tip of the proboscis enjoyed a selective advantage. They could scrape at damaged surfaces and extract the fruit juices that flowed from the scraped site. This, in turn, is an adaptation that could become further modified in some populations for piercing the skin of soft fruits. A few moths are capable of stabbing into blackberries and raspberries, not relying on other animals to wound the fruits. The kind of proboscis that can do this may, through selection for variants with more powerful components and cutting rasps, become capable of probing through thick-skinned fruits to the juices below.

But many moths and butterflies, particularly the males (for reasons that are unknown), also seek out sources of sodium salts. A common sight in many parts of the world is a cluster of butterflies by the margin of a mudpuddle, probing the soil with their probosci. It is unlikely that they are drinking water [185], and for one species at least it has been definitely established that mudpuddling males are attracted to sites with high concentrations of sodium [25]. It is not surprising, therefore, that many Lepidoptera also feed on the exquisite ambrosia that is found in liquefying carcasses, in moist piles of animal dung, or in pools of urine (also a rich source of sodium ions). Still other moths turn directly to living animals, consuming salty sweat and other secretions as well as blood from wounds

4 **Blood-sucking moth,** *Calpe eustrigata,* is feasting on the blood of a Malayan tapir. Photograph by Hans Bänziger.

and mosquito bites. The most striking examples of this phenomenon are those moths that are especially attracted to the eye secretions of mammals (Figure 5).

Calpe eustrigata probably evolved from a fruit piercer, inasmuch as other members of the genus are fruit specialists and *C. eustrigata* itself feeds in this manner at times. If individuals in a fruit piercing population were also attracted to animals to consume eye secretions or blood excreted by mosquitoes, they might initially have supplemented their diet with some additional fluids extracted by rasping the skin. Later mutants might have employed their powerful proboscis to cut straight through the skin of tapirs and other animals to extract blood directly from their victims [39, 40].

In developing a scenario for the evolution of blood-sucking behavior by a moth, Hans Bänziger and Anthony Downes used certain assumptions and rules of logic. The first assumption is that all behavioral traits, like other adaptations, have a genetic component and are shaped by natural selection. The second assumption is that those species that share common ancestry (and thus elements of a genetic program) are likely to behave in a similar manner. Behavior patterns may often be retained in new species as they evolve from an ancestral population in which the traits originated. It is an empirical observation that species whose close relationship has been established by traditional taxonomic techniques (the examination of anatomical and external characteristics) are likely to show behavioral similarities. Most moths do, in fact, feed upon the nectar of flowers.

5 **Tear-drinking moths,** *Lobocraspis griseifusa,* are clustered about the eye of a banteng (wild cow) in northern Thailand. Photograph by Hans Bänziger.

If assumptions one and two are correct, it follows that behaviors shared by many closely related species are probably all derived from an ancestral behavior pattern employed by a distant ancestor. If several related species share a behavioral characteristic, it is unlikely (but not impossible) that each has evolved the trait independently. The more parsimonious explanation is that each has retained the adaptation (a conservative pattern) from the time when their ancestors belonged to a common species. Thus, the first mothlike insect was probably a nectar feeder. Species that evolved from this ancestral population retained the trait because there were many sources of nectar that could be exploited by different species.

If some members of a cluster of related species possess distinctive behavior patterns, it is probable that these traits arose more recently in evolutionary time. The more uncommon a character (as measured against the presumptive ancestral pattern), the more likely it is to have been a recent modification of a less radically altered version of the original behavior (Figure 6). The basic sucking proboscis and digestive system of Lepidoptera need be little modified to enable feeding on fruit juices in place of flower nectar. A modest mutation of these adaptations could result in fruit-scraping and then fruit-piercing. The kind of proboscis capable of stabbing through an orange could also readily be employed to penetrate the skin of mammals to drink blood.

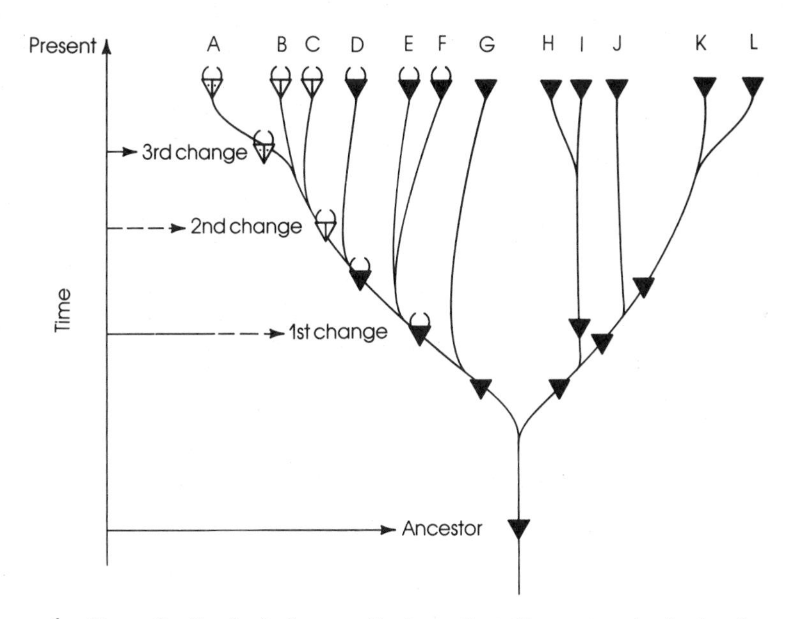

6 **Hypothetical phylogenetic tree** that illustrates the logic of the comparative approach to reconstruction of a historical pathway. Traits that arose in the distant past have the chance to be retained through many speciation events and so may appear in many currently existing species. More recently evolved traits should appear in fewer modern species.

If enough comparative material exists, it may be possible to develop a plausible series of adaptational changes that could easily follow one after the other, with no step requiring a massive alteration of the structural, physiological, or behavioral abilities of a species (Figure 7). Scenarios of this sort are consistent with the generally accepted notion that evolution proceeds gradually through a series of small changes rather than by drastic leaps (but see [273]).

The History of a Fly's Courtship Signals

The comparative method has helped solve the problem of how an exceptionally unusual courtship signal evolved in a small fly, *Hilara sartor* [376]. The male of this species constructs a delicate silk balloon that is approximately as large as he is. He then flies to a swarm composed of other balloon-carrying males, there to circle about until a female arrives (Figure 8). She selects a mate from the swarm, accepts his balloon, and the two drift away from the group to copulate.

When these flies were first observed in the nineteenth century, all that was known was that a group of males gathered together carrying silk balloons. Naturally this prompted a certain amount of bewildered speculation about the function of the silken ball. Even the discovery that the female accepts the balloon as a condition for mating hardly dispels the mystery. Why should copulation depend upon receipt of an empty ball of silk? Why does the male make it in the first place? Without comparative data these puzzles would remain unsolved. There are, however, literally thousands of species of empidid flies (the family to which *H. sartor* belongs); and although few have been studied in any detail, enough is known to suggest a solution to the riddle of the balloon fly. E. L. Kessel has summarized this information and shown that empids fall into at least eight categories [376].

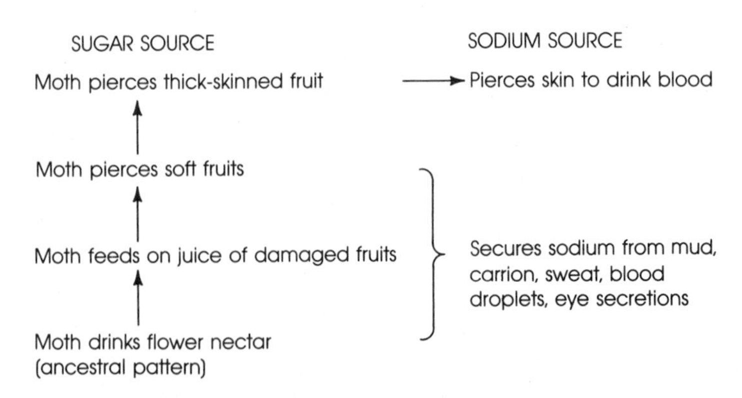

7 **Historical pathway** that might have led to the evolution of blood-sucking and tear-drinking in certain moths. The pathway has been derived through a comparison of the behavior of currently existing species.

1. Flies in this group and all others, except for those in groups 7 and 8, are at least partly carnivorous, hunting for other small flies, including mosquitoes, midges, crane flies, and even other members of their own species on occasion. Type 1 courtships are those in which the male searches for a female and courts her in isolation from other individuals.
2. The male of species in this group captures prey before locating a female. She takes the food prior to copulation and consumes it during mating.
3. In still other species, males with captured prey form groups. A female attracted to the swarm selects a male, receives the prey, and mates (Figures 9 and 10).
4. The behavior of group 4 flies is very similar to that of group 3, but here the male applies some strands of silk to the prey prior to joining the swarm.
5. The same as in group 4, except that the male wraps the prey entirely in a heavy silk bandage before offering it to a female.
6. The same as in group 5, except that the male removes the juices from his offering prior to wrapping it. As a result the female receives a nonnutritious husk.
7. These species feed only on nectar. However, prior to courtship the male finds a dried insect fragment and uses it as a foundation for the construction of a large balloon, which he then presents to a female before copulating.
8. *Hilara sartor* and a few other species omit an insect fragment as the starting point for balloon construction.

As the entomologist Harold Oldroyd points out, it is as if one first presented a fiancée with elegant diamonds, then plied her with diamonds in an exquisite case, and finally offered her an elaborate, but totally empty,

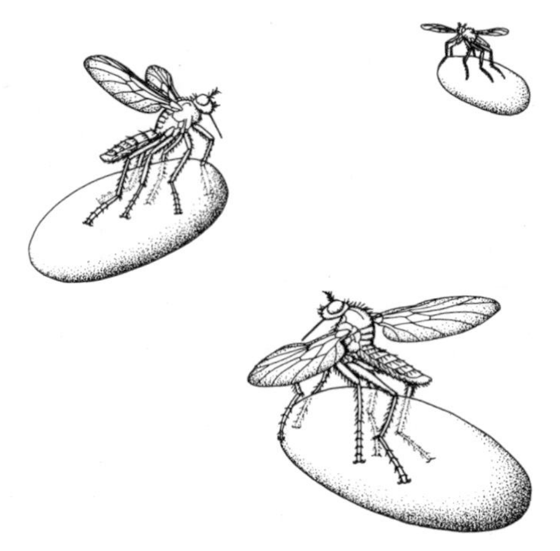

8 **Swarming balloon flies** display their empty silken balloons. In this species, females require a balloon before permitting copulation to occur even though the balloon contains no nutritional benefit.

9 **Male empidid fly** with an edible nuptial gift. Females of this species require a food present before they will copulate. Photograph by the author.

gift-wrapped box [546]. One can speculate about the impetus for each evolutionary change, but the point is that the sequence outlined by Kessel is an eminently logical one. It is reasonable that the most ancient pattern is the one shared by the majority of insects, with the male searching out the female, courtship occurring without the unusual addition of a prey offering, and so on. For logical reasons, type 5 courtships could hardly have preceded type 4 in evolutionary time; and type 8 is surely an elaboration founded on type 7.

Kessel and others have suggested that the initial "wedding gift" invention is advantageous because of the predatory nature of the fly and the risk of cannibalism. The present can be viewed as a distraction for the female so that the male can get on with the serious business of reproducing—safely. An alternative and not mutually exclusive hypothesis is that the gift of prey gives the females a high-protein meal that helps her eggs to mature rapidly [696]. Thus, it is to her advantage to choose a present-giving male if one is available (Chapter 12), and it may be to the male's advantage as well, if the prey assists in the development and survival of eggs he has fertilized. The continued existence of type 1 courtships in empidids could stem simply from the failure of any variant male in these populations to offer a female a fly while courting. Or there may be something about the ecology of these species that reduces the risk of cannibalism or lowers the advantage to males of giving the female an energy boost.

The same kind of speculation can be applied to every other change.

10 **Mated pair** of *Empis barbatoides*. The male is above the female, who is consuming prey given to her by her mate. Photograph by the author.

Collecting in swarms may evolve if females tend to refuse to mate with solitary males. It may be to their advantage to have a chance to compare males (in a swarm) and to choose to copulate with the male carrying the best present (type 3). The silk may serve to restrain prey initially (type 4) and then to make it look larger and more attractive than it really is (type 5). By this stage the balloon has become the releaser of mate choice for the female, and the prey itself is no longer the initial basis for the union of the sexes. A male can "cheat" the female (if he gains some reproductive advantage from doing so) and remove the juices of the prey (type 6). Upon adoption of a purely nectivorous feeding niche (instead of feeding only occasionally on nectar), capture of prey serves no nutritional function and is really a waste of time and energy. However, because the releaser of balloon building is a prey item, the male may continue to collect a piece of an insect around which the balloon is built (type 7). Finally, this behavioral component (which is time consuming) is lost and the balloon constructed from scratch (type 8). Thus, thanks to the abundance of comparative material, one need not envision large and unusual changes in behavior in order to account for the gradual evolution of the remarkable activities of *H. sartor*.

The History of a Bowerbird's Display

Fascinating though balloon flies are, not even they can equal the golden bowerbird of Australia for strangeness of courtship display [260]. Males of this species build a maypole bower consisting of two towers of sticks that may be as much as 2.5 meters high. The two towers are connected at the base by a bridge of twigs, which are decorated with pale green items such as lichens, mosses, certain flowers, and fruits (Figure 11). Females visit these display grounds and listen to the male's complex and varied vocalizations, after which mating may occur. This display pattern, although unique in its details, is reminiscent of that of the manakins and three-wattled bellbirds (Chapter 12). Golden bowerbirds are fruit-eating animals whose females can care for the young without male assistance. Mate choice therefore seems to have shifted away from male parental qualities to display ability, a characteristic presumably related to the value of the male's genes. But how did the particular display behavior of the golden bowerbird evolve?

Once more we resort to comparisons with relatives of this species [178]. There are other bowerbirds, some of which build equally elaborate structures of different designs and some of which build quite simple display bowers. The simple bowers have more than a passing resemblance to a nest, which gives us a potential clue to the origins of the trait. For many birds, display at a nest site is a fundamental part of courtship. Frequently, males of these birds employ signals that are related to nest building as indicators of the suitability of the territory for a nest (see Chapter 7 on kittiwake behavior). This behavior gets carried one step further in those species (among them the familiar North American house wren) whose males actually construct several nests in their territories, directly advertising the nest-building capacity of the male. In species of this sort, the nest itself takes on the properties of a courtship signal, stimulating the female to ovulate and become receptive. Bowerbirds conceivably evolved from birds in which the visual stimuli provided by a nest served this courtship function. Currently, inasmuch as males of many bowerbirds are completely exempt from parental duties, the bower is never used as a functional nest. Female choice, however, could favor ever more elaborate demonstrations of male construction abilities. Perhaps these activities are correlated with male physiological condition or other heritable attributes of utility to a female's offspring.

We can apply the same kind of analysis to the origins of the decorations that golden bowerbirds apply to their towers. A possible clue is provided by the occurrence of fruits as ornaments of its bower, and this is true for many other bowerbirds [260]. In fact, a male of one maypole builder, *Amblyornis flavifrons,* places mounds of fruits on a mossy display platform at the base of its tower and holds a fruit in its beak while displaying to a visiting female (Figure 12). As noted earlier, courtship feeding occurs in many birds, and thus it is plausible that bower decorating is derived from this more widespread avian trait [178]. Because female golden bowerbirds feed upon highly abundant foods, they would gain little nutritional benefit if fed by a mate. Freed from its original function, courtship feeding has taken on a purely symbolic and ever more elaborate nature. Perhaps females prefer males that demonstrate their foraging skill in gathering large

11 **Bower of the golden bowerbird,** perhaps the most elaborate display platform created by an animal. The drawing shows the bird close-up at the center of the bower. The photograph shows the two towers of twigs on either side of the central display area.

quantities of difficult-to-locate items of the same color as a once-preferred fruit used originally in nuptial feedings.

Thus, the bowers of male bowerbirds are probably not so very different from the silken balloons built by some male empidids. In both cases, these structures may be derived from once-utilitarian actions that have secondarily taken on a pure courtship function capable of stimulating females to copulate.

12 **Bower of the yellow-fronted gardener bowerbird.** This bower, although still complex, may more closely resemble the presumably simple bowers built by ancestors of the related golden bowerbird. Courtesy of Jared Diamond.

The History of Honeybee Communication

We return again to the insects to trace the pathway that might have lead to the evolution of one of the most famous of all behavioral traits, the dance of the honeybees. Karl von Frisch won the Nobel prize for his research on the dance "language" of the bees. He concluded (after 20 years of work) that the dance movements contained symbolic information about the distance and direction to food sources, information that other bees could use to locate supplies of pollen and nectar [743–745].

The dances take place on the vertical surface of a comb in the hive of a colony of bees. When a forager bee has found a rich new food source (a patch of flowers, a watch glass filled with honey), it will return to the hive and perform elaborate maneuvers on the comb. If it has collected high-quality nectar within roughly 80 meters of the hive, it will execute a ROUND DANCE (Figure 13). Because it is normally dark inside the hive, other bees do not watch a dancer from a distance but instead may follow it about on the comb as well as sensing vibrations produced by it. The followers may fly out of the hive to search more or less randomly for the nectar or pollen, keeping relatively close to the hive. The search will not be entirely random because they will have tasted the nectar regurgitated to them by the recruiter and will also have detected odor cues adhering to the body of the dancer. These clues enable identification of the flower that produced the food. Moreover, experienced bees learn the locations of specific patches of flowers whose nectar supply fluctuates. A worker following a recruiter that bears the odor and nectar associated with a familiar food source will make a beeline to this location. However, for our purposes, what is most interesting is that an inexperienced worker just joining the forager force will "know" not to wander too far from the hive if recruited by a round dancer.

If the bee has found a rich food source more than 80 meters from the hive, it will perform a WAGGLE DANCE (Figure 14). This action conveys information about the relative distance of the food source from the hive in the range of roughly 80 to 600 meters. The information is coded (1) in the number of times the bee runs through a complete dance circuit in a unit of time, (2) in the number of abdomen waggles given during the straight-run portion of the dance, and (3) in the frequency with which sound bursts are produced while dancing. The adaptiveness of providing recruits with several redundant cues becomes apparent when one observes the crowded chaos in the hive that inevitably distorts the movement of a dancer.

The energy expended in going to a food source on the *outward* flight determines the nature of the dance. If a bee flies into a headwind or up a hill, it will perform fewer waggles and a slower dance compared with one executed by a bee flying the same distance with a tailwind or along level ground. The less energy expended to get to a particular location, the more animated the dance—more waggles, more revolutions per unit of time, and a higher frequency of sound bursts.

Figure 15 shows the results of an experimental test of the ability of a forager to recruit others to a specific food source. It is apparent that one bee can transfer information about the distance to a resource to others. This is not to say that a recruit learns the *exact* distance it must fly. However, the very imprecision of the information conveyed in a dance is possibly adaptive. The bee's natural food source consists of patches of flowers. These are rarely concentrated in a tiny area but tend to be scattered

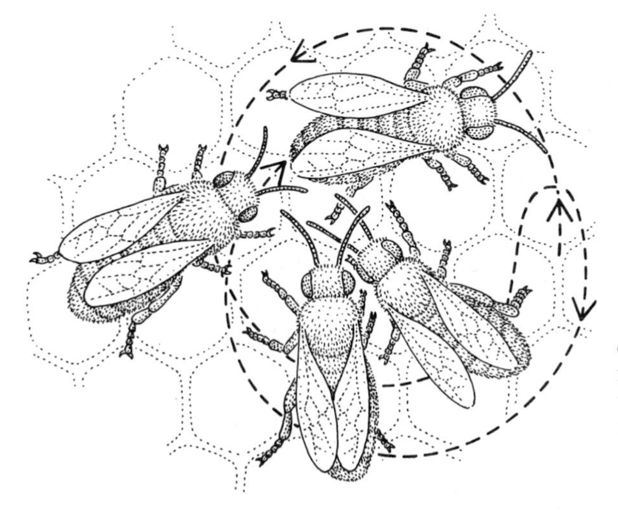

13 **Round dance of the honeybee.** The dancer (the uppermost bee) is followed by three workers, who may acquire the information that a profitable food source is located within 80 meters of the hive.

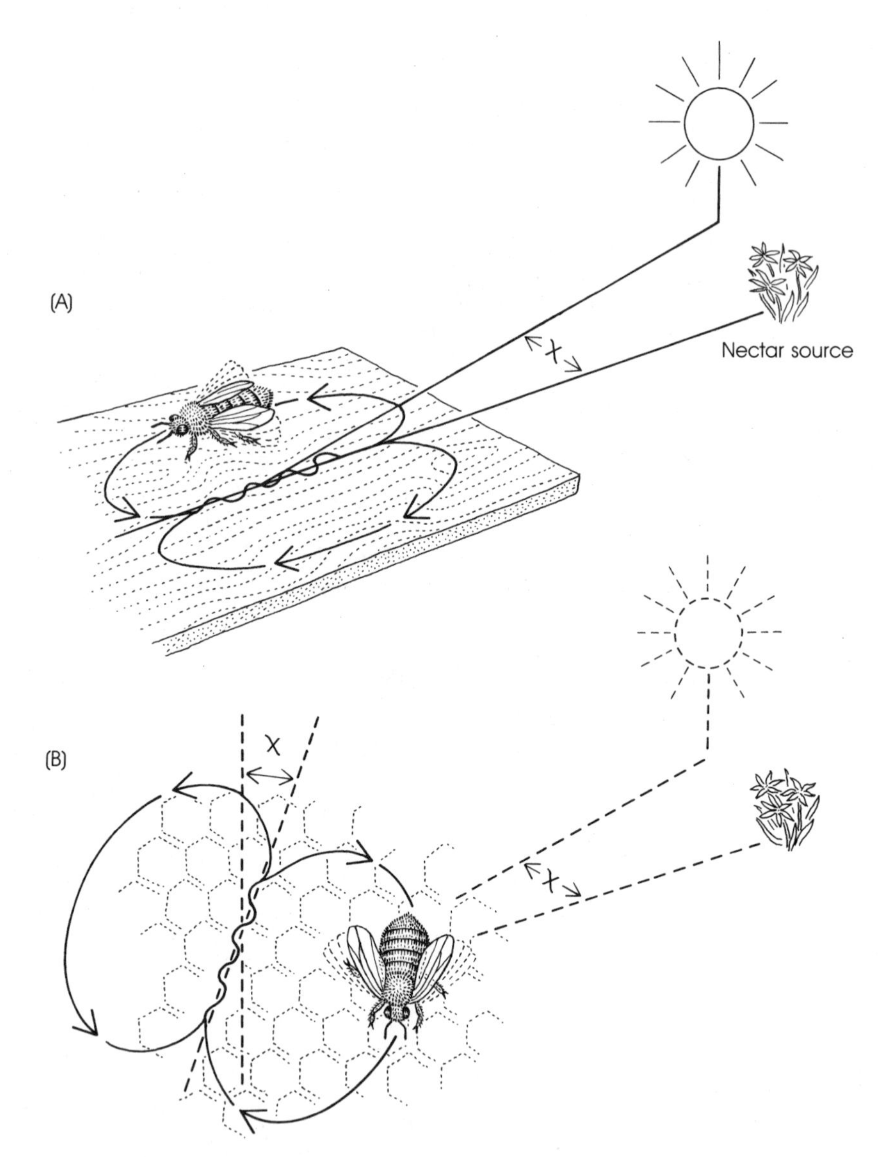

14 **Waggle dance of the honeybee.** As the insect performs the straight run, her abdomen waggles; the number of waggles and the orientation of the straight run contain information about the distance and direction to a food source (see text). (A) The directional component of the dance is most obvious when the display is performed outside the hive on a horizontal surface. Here the bee runs directly at the food source. (B) On the comb, inside the dark hive, the same dance is oriented with respect to gravity so that the displacement of the straight run from vertical equals the displacement of the location of the food site from a line running between the hive and the sun. In the illustration, workers attending to the dancer learn that food may be found by flying 20° to the right of the sun when they leave the hive.

about. A bee that is directed to the general area will probably enjoy foraging success and may find subsidiary sources unknown to the recruiter itself [743].

Direction Communication

Honeybee dances indicate not only how far food is from the hive but in what direction it lies. A forager on the way to a discovered food source notes the angle between the food, hive, and sun. During a waggle dance (not during round dances), it transposes this information onto the vertical surface of the comb when it performs the straight-run portion of the dance (Figure 14). If the bee walks waggling straight up the comb, the nectar or pollen will be found by flying directly toward the sun. If the bee

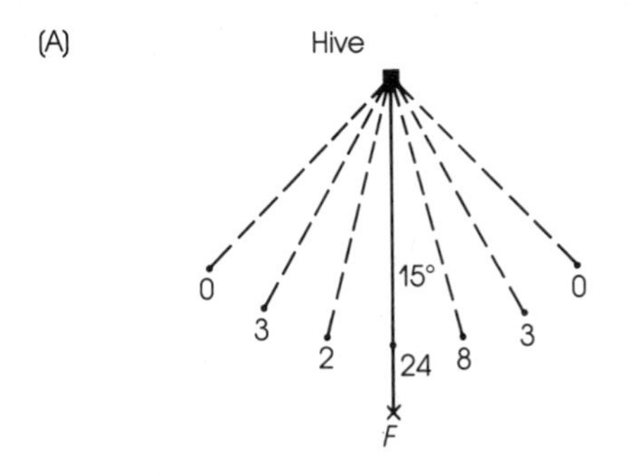

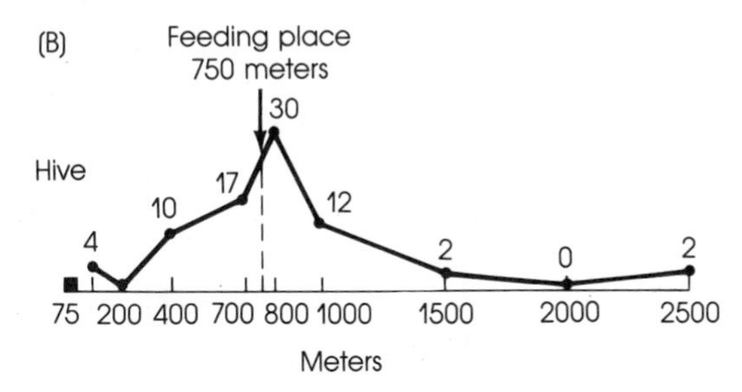

15 **Ability of worker bees to communicate** with other workers about (A) the direction and (B) the distance to a food source. (A) A "fan" test for directional communication. After training a number of recruiters to come to a food dish at F, all newcomers are then collected at seven tables with equally attractive sugar water. Most new bees arrive at the site on line with F. (B) A test for distance communication. Some workers are trained to visit a feeding place 750 meters from the hive. Thereafter, all newcomers are collected at feeding tables placed at various distances from the hive. Many more bees are taken at the tables closest to the training table, again indicating that transfer of information from recruiters to other workers had taken place in the hive.

waggles straight down the comb, the food is located directly away from the sun. A patch of nectar-producing flowers positioned 90° to the right of a line between the hive and the sun triggers waggle runs oriented at 90° to the right of vertical on the comb, and so on. In other words, sun-compass information is converted into a code based on gravity. Recruits following a dancer determine the angle of its movement with respect to vertical with special sensory devices [428] and then use this information when leaving the hive in search of food. Again the dance does not provide absolutely precise directional information but narrows the field of search considerably.

Von Frisch and others [428] have tested the directional component of the dance information by training recruiters to come to one feeding station. They then place a series of feeders equidistant from the hive but in various directions from it. By collecting each honeybee as it arrives at one of these feeders, one can determine the effectiveness of the recruiters in transmitting directional information to other workers. Feeding stations with recruiter bees have many more visitors than those without them (Figure 15).

But von Frisch's work did not categorically rule out an alternative hypothesis for the ability of recruited bees to find food sources advertised by other foragers. Adrian Wenner and his colleagues argued that odor cues from the source adhering to the body of the recruiter were sufficient to identify a feeding site [770]. Other bees could find the site simply by tracking familiar scents in the neighborhood of the hive. It remained for J. L. Gould to confirm that the dance language really did have meaning for the bees that attended a dancer [266, 269, 270]. To do this he tricked his bees, getting a forager to give a dance that was misinterpreted by the recruited bees. He took advantage of the ability of honeybees to orient their dances directly to the sun if they can see it (Figure 14). Only when foragers cannot see the sun (or a substitute light) will they transpose their directional information into a gravity-based code. If one paints over the three simple light-receiving organs (the ocelli) on the head of a bee, the insect becomes much less sensitive to light even though its two much larger compound eyes are not blocked. By placing a weak light as an artificial sun near a comb, Gould provided a cue that recruits could see but that the ocelli-blocked dancers could not. Thus, when the foragers returned from a rich food source set up by the experimenters, they oriented their waggle runs with respect to gravity. But the bees that followed them could see the "sun," so they interpreted the dance as if the waggles were oriented with respect to light, not gravity (Figure 16). Off they went in a direction that did not lead them to the feeder found by the dancing foragers. Because Gould could predict where the (marked) recruited bees would go, he was at this spot to record their arrival. He proved that workers can be directed purely by the symbolic information contained in the dances of their fellow bees.

The Evolution of the Dance Displays

Honeybee dancing is such a highly specialized and unusual behavior that its evolution has understandably been the topic of considerable interest and speculation [428]. Martin Lindauer has used the comparative method creatively to analyze the evolution of this behavior. Because all populations of the honeybee perform round and waggle dances [222], the logical group to examine first is the genus *Apis,* to which the honeybee (*Apis mellifera*) belongs.

The three other *Apis* species are tropical bees, and all except *A. florea* exhibit the capacity to perform dance displays on vertical surfaces. This small bee dances on the horizontal surface of a comb built in the open over a limb (see Figure 10, Chapter 10). To indicate the direction to a food source, *A. florea* simply orients her waggle run directly at it. Because this is a less sophisticated maneuver than the transposed pointing done on vertical surfaces, it is probably a relatively ancient form of dance communication. (Moreover, *A. florea* builds a simpler nest—a single comb—instead of the multiple combs of the other *Apis* bees.) The transition to vertical surface dancing with coding of sun-compass information in terms of gravity may be favored initially because there is more vertical surface area on a

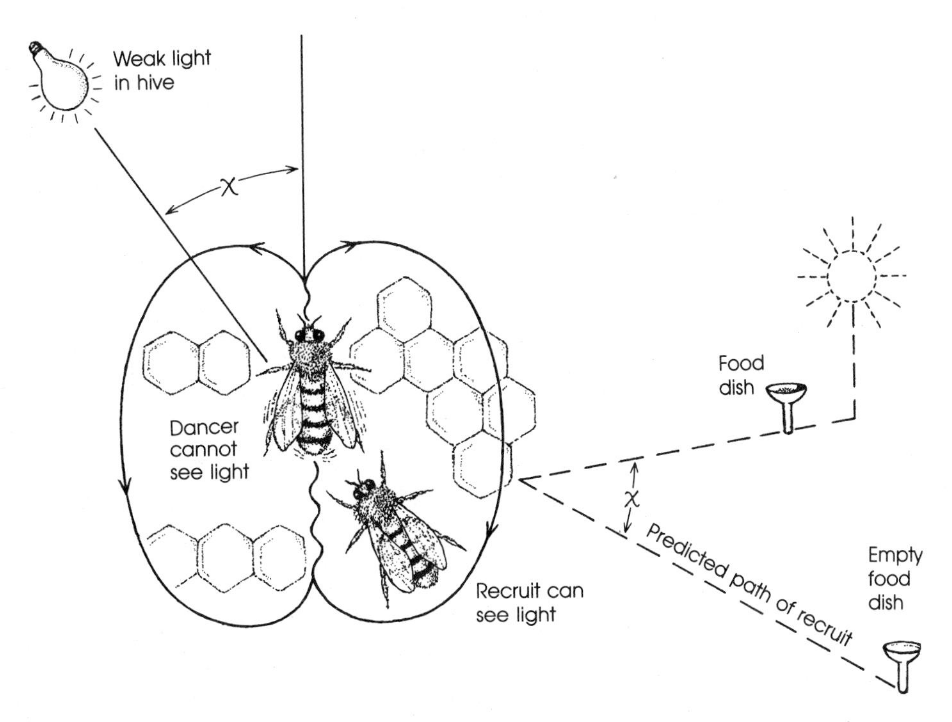

16 **Experiment that demonstrates communication** occurs through the dance movements of honeybees. The ocelli-blocked dancer cannot see the weak light above the comb; it therefore dances using gravity to indicate that food is located directly toward the sun. The recruits, however, can see the light in the hive and they interpret the dance to mean that they should fly x degrees to the right of the sun.

comb to accommodate more recruiters and their followers. In the case of *A. mellifera,* the close packing of vertical combs in a tree trunk or hole in the ground leaves little horizontal surface.

Just how the first dancers happened to transpose a sun-oriented dance into an accurate gravity-oriented dance is somewhat puzzling. However, it is known that a number of other insects make crude conversions of this sort. Moreover, moving upward is moving roughly toward the sun; moving downward leads the bee roughly away from the sun (especially in the tropics). A pioneer bee dancing on a vertical surface might, therefore, be reasonably expected to adopt a code similar to, but obviously not nearly as detailed as, the modern displays of honeybees.

In any event, because all *Apis* species have both round and waggle dance displays, one has to look elsewhere for other patterns possibly similar to ancestral behaviors. Although not all social bees communicate about foraging sites, the stingless bees of the subfamily Meliponini do exhibit an exciting range of communication systems.

A worker of the species *Trigona droryana* that has found high-quality nectar returns to the nest and there performs unstructured excited movements coupled with a high-pitched "humming." This arouses its hivemates, who request food samples from the "dancer" and presumably detect the odor of the source on the dancer's body. With this information, they leave the nest and search for similar odors. The actions of the recruiter do *not* offer any specific cues about the direction and distance to the desirable food, as experiments have proved.

Trigona postica workers, on the other hand, do convey this information, although in a way quite different from the dances of the honeybee. A worker that makes a substantial find marks the area with a pheromone produced by its mandibular glands. As the bee returns to the hive, it continues to chew on grasses and rocks every few yards. At the hive entrance there may be a group of bees waiting to be recruited. The forager cooperates by leading these individuals back along the trail it has marked. Martin Lindauer and his associates tested the importance of the scent guides by placing a hive on one side of a pond and then gradually training some foragers to come to a food supply directly on the other side (Figure 17).

> Now the bees always flew across the water, where they were of course unable to place any odor marks. They did try it, by flying again and again very low over the water surface. They then repeatedly returned to the shore, where they generously bestowed scent marks all over the feeding table, and the notebook and hat of the observer.
>
> During the whole feeding time of two days, not a single newcomer appeared at the feeding place. Then we strung a rope across the water, and hung a few twigs at 2–3 meter intervals. This now served as a bridge for the newcomers. Within 35 minutes, 37 newcomers reached the goal and we could conclusively observe how all the marked collector bees were eagerly placing odor marks on the twigs. [428]

Trigona postica lives in tropical areas with dense vegetation, thick tree trunks, and other obstacles between the hive and a source of nectar. Personal leading is a very effective way to overcome these obstacles as well as a useful strategy for attracting newcomers to nectar high in flower-bearing tropical trees. It is instructive that honeybees have difficulty in communicating that a food source is upward rather than outward (but see [764]). Von Frisch trained some would-be recruiters to come to a dish of sugar water up in a tower. Although the trained bees returned to the hive to dance, no recruits ever came to the food source. Instead, they moved *parallel* to the ground the "correct" distance from the hive rather than flying up to the top of the tower. Honeybees have evolved in a more open habitat, where food is generally found close to the ground and rarely in high trees.

A number of species belonging to the genus *Melipona* separate distance and direction information, unlike *T. postica*. A dancing forager communicates that food is near by producing pulses of sound of short duration. The longer the pulse of sound, the farther away the food is. In order to transmit directional information, the recruiter leaves the nest with a number of followers and performs a short zigzag flight that is oriented toward the

17 **Information transfer in the bee** ***Trigona postica*****:** Martin Lindauer's experiment. As soon as the line with attached branches was strung across the lake, recruits began to appear at the food table on the other side, having followed a trail pheromone applied to the shrubbery by the trained workers. Photograph by Martin Lindauer.

source of nectar. The lead bee returns and repeats the flight a number of times before flying straight off to the nectar site with the novice bees in close pursuit. Evidently the zigzag flights alert the recruits to the general direction of the upcoming straight flight.

Trends in Bee Communication

With the comparative evidence, it is possible to suggest an evolutionary sequence of events leading to the honeybee dances [428, 793]. Communication about the *distance* to a food source by an ancestor of the honeybee probably first involved relatively unorganized, agitated movements on a comb. A worker that had found food nearby might tend to remain highly active back at the nest as it prepared to return to the site it had found. Selection would favor colonies with workers that happened to be easily socially stimulated to leave the hive. This would lead workers out to forage at propitious times rather than remaining engaged in hive-related activities. Nectar and pollen close to a nest are desirable foods. Hives that collect it quickly enjoy a large net energy gain because foraging costs are relatively low and because, by going in groups, the bees gather it before other animals do.

In some populations selection presumably acted to stereotype the action, as it has on so many initially variable behaviors with communication value. Selection may have at first had the effect of standardizing the sound components of the display (as in *Melipona*), and secondarily, led to an increasing stereotypy of the movements of the excited bee. The selection process culminated in the waggle dance of modern *Apis*. A reduction in behavioral variation permits the transfer of more information with less ambiguity. In the honeybee hive, the presence of many dancers actually enables a worker to choose among various available foraging sites or new nest locations (Chapter 8), selecting the best possible one.

Communication about the *direction* to a food source appears to have originated with personal leading, a worker guiding a group of recruits directly to a nectar-rich area. Subsequently, in some populations selection may have favored less and less complete performance of the guiding movement—in effect, selecting queens that produced workers with a tendency to perform incomplete leading. At first this may have taken the form of partial leading (as in *Melipona*) and then just pointing in the proper direction with a very restricted movement toward the feeding place (as in *A. florea*). The culmination of this trend has been the transposed pointing of *A. mellifera,* in which directional cues based on a sun compass are converted into cues based on gravity (Table 1).

Thus, a comparative approach makes the dances of honeybees somewhat more comprehensible. Like all other behavior, the dances have an evolutionary past, a gradual history that can be tentatively reconstructed given the necessary comparative material. This does not, however, make the dances of the bees any less wonderful examples of the adaptive effects of natural selection.

The examples of behavioral evolution presented in the preceding pages are sufficient to illustrate how one can make reasonable guesses about the history of a behavior with no fossil evidence at hand. In a sense, inquiry into the origins and subsequent evolution of a behavior is detective work. It involves the determined collection of pieces of evidence from many living species and arrangement of the data in a logical sequence until finally a plausible historical pattern emerges. This is satisfying and creative work, even though the hypotheses generated may not be easy to test.

The development of a historical perspective makes one aware that there are constraints on evolutionary pathways imposed by the previous history of a species. In their analyses of the possible adaptive value of a trait, behavioral ecologists often pay little attention to this point, assuming that whatever is most adaptive will eventually evolve with the occurrence of the appropriate mutation. But past history limits what mutations can become established within a species [275]. In principle, balloon fly males might enhance the effectiveness of their displays with elaborate acoustical displays. However, the small size and delicate construction of empidids probably places physical limitations on their ability to generate long-dis-

Table 1
Summary of the Possible Evolution of Honeybee Displays

Stage	Information about Direction to Food	Information about Distance to Food
1	Generalized agitated movements upon returning to hive with high-quality food; no direction or distance information conveyed by dance	
2	In addition to excited movements, recruiter personally leads recruits to site	
3	Dance plus partial leading; recruits started off in right direction	Activity and sounds produced by dancer proportional to value of resource to colony; food that is close to hive is economically valuable and stimulates much activity, rapid buzzing
4	Pointing in dance; incomplete leading in the straight run of the waggle dance (a ritualized version of the outward journey)	Increasing stereotypy of displays with distance information accurately coded in speed of dance, number of waggles in straight run, and sound production
5	Transposed pointing on vertical surface of comb	

tance acoustical signals. Thus, mutants that employed silken nuptial advertisements were possible, but stridulating males were unlikely to be effective signalers even if one did arise by chance. The study of the effect of historical limitations or PHYLOGENETIC INERTIA on the course of evolution has been relatively neglected to date and should receive more attention in the future.

The Evolution of Communication Systems

Having explored some specific examples of communication by empidid flies, bowerbirds, and honeybees, it is appropriate here to examine some general evolutionary aspects of communication systems. Let us begin at the beginning with a definition of a communication signal. How do we know when animals are truly communicating with one another? Most of us would suspect that communication has occurred if a signaler provides a stimulus that causes another animal (the receiver) to react voluntarily in a well-defined way. Thus, when a foraging bee dances, the bees that follow the dancer will subsequently behave in a predictable fashion, flying out to the food source the forager had discovered (unless tricked by a human experimenter).

But what about this case? A female douglas-fir beetle produces feces (frass) that contain volatile components when she feeds upon the tissues of a tree. Other females looking for a site in which to build their egg-laying galleries are attracted to these odors and may establish their burrows near the original colonist. Under certain circumstances this reduces the reproductive success of the original colonist because her progeny will face heightened competition for the food contained within the tree [595]. But the attracted females may gain because they can reproduce rather than having to keep searching for a rare, safe oviposition site. Is the original colonist communicating with her competitors? And what about the firefly *femme fatale* or the bola spider that produces a stimulus that attracts males of other species to their doom? Is this communication?

Most evolutionary biologists argue that true communication occurs only when both signaler and receiver enjoy fitness benefits from their interaction [438]. "Deceit" and "eavesdropping" can occur, with negative reproductive consequences for either signaler or receiver (Table 2). If the fitness effects of giving a signal or responding to it are negative significantly more often than they are positive, then selection will inevitably eliminate the tendency to send the message or to react to it. This is a critical point because it suggests that communication systems cannot originate unless they raise the fitness of both signaler and receiver.

Now this does not mean that, once established, communication signals cannot be exploited (Figure 18). A female bark beetle may release a sex pheromone to attract a mate who will help her defend her burrow and remove frass from the egg-laying galleries as they are constructed. This is to the mutual benefit of both male and female. But other females may exploit the system to locate a host tree whose chemical defenses have been weakened by the beetles already established in the tree. These ILLEGITIMATE

Table 2
Effects of Information Transfer on the Fitness of a Signaler and Receiver

Fitness Effects on		
Signaler	*Receiver*	**Label**
+	+	*Communication* between legitimate signaler and legitimate receiver
−	+	*Eavesdropping* on signaler by an illegitimate receiver
+	−	*Deceitful signaling* by an illegitimate signaler

RECEIVERS reduce the overall benefits associated with release of a sex pheromone whose adaptive function is to attract a mate.

Likewise firefly *femme fatales* exploit an evolved male–female communication system by being ILLEGITIMATE SIGNALERS. The probability that a male *Photinus* will respond to a predator female's mimetic signal reduces the average gain derived from reacting to a particular kind of light flash. The system can persist, however, if alternative communication channels provide lower fitness benefits on average to male *Photinus* than the current flash signaling system.

The Origins of Communication Signals

Even if we accept the argument that communication cannot evolve unless it raises the reproductive success of some signalers and their receivers, we are still left with the question of how mutually beneficial signals might originate. The douglas-fir bark beetle provides a clue. Just by passing the tissue of a tree through her digestive tract, a female beetle generates some incidental by-products that include volatile scents. Originally, males that happened to detect these chemicals and tracked them to their source would have derived a benefit and so would the receptive females that secured a helpful partner in this manner.

Once a communication system has been established between male and female, subsequent selection can favor mutant signalers that happen to produce more pheromone or a message that is more distinctive, more easily dispersed, or harder to exploit. It may be significant that the sex pheromone currently used by bark beetles is often a highly complex combination of chemicals [70, 662]. Likewise, mutant receivers that are better able to detect and track the sex pheromone may enjoy an advantage. To this end, silk moth males have evolved the ultimate in sensitivity to the female's scent—the capacity to detect a single molecule of the substance [637]!

There are many other animals in addition to bark beetles that use digestive waste products to communicate with conspecifics [596]. Thus, wolves, hyenas, lions, and tigers specifically mark the boundaries of their territories with mounds of feces or sprayed urine (or both). Moreover, hyenas and tigers possess special secretory glands whose products supple-

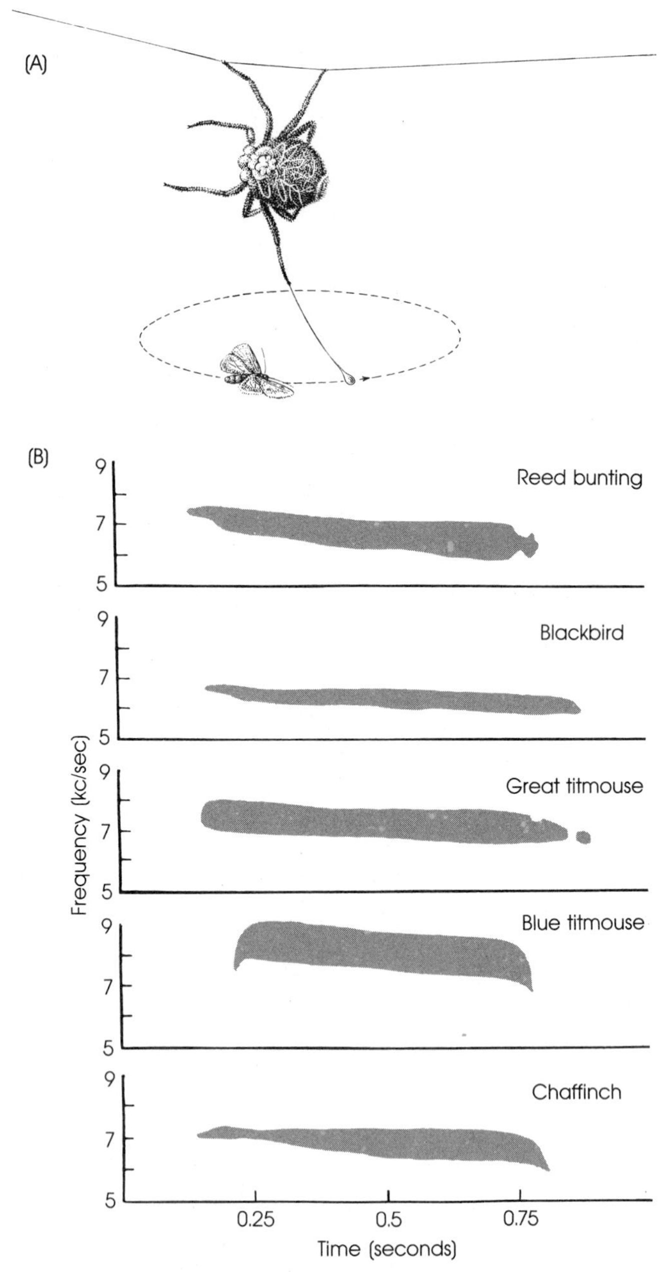

18 **Illegitimate signalers and illegitimate receivers** pose problems for communicating animals. (A) The predatory bola spider lures males of certain moths to their death by mimicking the sex pheromone signal of the female moths. (B) The alarm calls of some European song birds are a thin whistled "seet"; the narrow band of frequencies in the call may make it difficult for an illegitimate receiver, a predator, to locate the sender. Courtesy of Peter Marler.

ment the chemicals in waste materials, elaborating on the communication function served by feces or urine [82, 397].

Another rich source for incipient communication signals are the actions of animals that are placed in situations that elicit two opposing tendencies [203]. The European ethologists labeled these activities "conflict behavior" (Figure 19). Thus, two rival herring gulls that meet at the border of territories may attempt to attack *and* to escape from their opponent. The resulting conflict may cause a male to raise his head back as if he were about to peck at his opponent without actually carrying out the action. Movements of this sort that are made in preparation for a response are called "intention movements." Alternatively, the male may attack a clump of grass rather than his neighbor (redirecting his aggression onto a safe object rather than the stimulus that aroused aggression in the first place). Still another frequently observed response is an apparently irrelevant behavior unrelated to either attack or escape motivation, such as preening of the feathers (a so-called "displacement activity"). Each of the ambivalent movements may convey a subtly different message to a rival. If certain actions deter one male from entering another's territory and both individuals gain from this (the deterred male avoids a beating, the resident saves his energy), then communication has taken place.

Ethologists have long argued that conflict behavior provides the raw material for many communication signals that, once variable, gradually evolved into ritualized and stereotyped DISPLAYS. If the value of a communication signal lies in its ability to transfer information from a signaler to a particular receiver, then whatever constitutes a less ambiguous message should be selectively advantageous. This may favor individuals that reliably produce a distinctive and unchanging signal in a particular context. The more exaggerated the behavior and the more it involves conspicuous

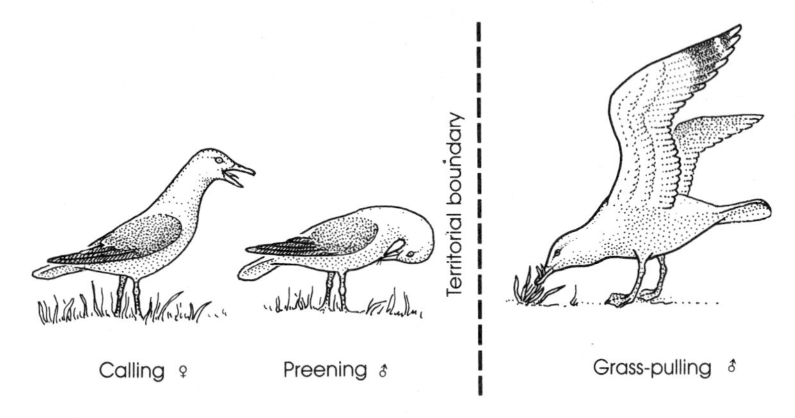

19 **Conflict behavior in herring gulls.** Two males, at the boundary of their territories, are torn between tendencies to attack an opponent and to remain in the safety of their homesites. One bird responds by pulling grass; the other, by preening its feathers. The female mate of one male calls behind him.

structural traits, the less likely it is to be misunderstood. This is the kind of argument that has been used, for example, to explain the complex dance displays of honeybees (Table 1).

Deceitful "Communication"

The reduction-of-ambiguity hypothesis is based on a benign view of animal interactions that emphasizes the cooperative consequences of information transfer. But although communication systems cannot evolve unless there is the potential for mutual benefit between signaler and receiver, there is no requirement that every potentially communicative interaction must be based on cooperation. It is one thing to apply the assumption of cooperation to honeybee workers (which really do gain from the transfer of accurate information between foragers and recruits) and another when the interactants are two aggressive herring gull males fighting for a territory or a male and a female bowerbird each attempting to achieve different sexual goals. In many circumstances animals may be expected to employ their signals in ways that exploit the individual with whom they are "communicating." Thus, female langurs that are pregnant by one male may enter into a false estrus when their band is taken over by a new male. They signal a readiness to copulate even though they have no egg to fertilize. By permitting the new harem master to mate with her, a pregnant female may reduce the chances that this new male will kill her offspring when it is born [350] (Chapter 1).

Another possible case of deception in signaling involves the use of song repertoires by male birds [390, 810]. In many territorial birds (and some lizards [369]), there has been selection for individual recognition based on distinctive songs or displays. This potentially benefits both display giver and receiver; if established territory owners can recognize each other, neighbors need not waste time and energy reacting to each other's songs or displays but will retain the ability to respond strongly to new intruders. John Krebs has suggested that the ability of birds to recognize an individual's song has been exploited through the evolution of complex song repertoires. The male of the great tit has from two to eight different song types; he tends to change his territorial song when he changes his perch [398]. This may help convince a territory-hunting intruder listening in the woods that there are more individuals in the area than there really are. This could cause him to leave the "saturated" woodlot without competing for a territory with the singing male.

Possible support for the Beau Geste* hypothesis comes from an experimental study of red-winged blackbirds, whose males also employ a complex song repertoire and change song types when they change perches. An empty

* The Beau Geste hypothesis is named for P. C. Wren's adventure novel about the French Foreign Legion. The book was named after its hero. Beau Geste employed deception to thwart an attack on an undermanned fortress by propping up dead men in view of the enemy to create the false impression that there was a larger number of defenders than there actually was.

territory site occupied by a speaker broadcasting a variety of song types was significantly less often invaded by trespassing males (other than neighboring territory owners) than an empty territory with a speaker that played only a single song type [810]. Ken Yasukawa points out, however, that this result is also compatible with another hypothesis, which is that males signal their territory-holding ability through their songs. If males that have a complex song repertoire are older and more powerful than males with a less varied song, intruders might avoid the versatile songster, not because they are deceived about the number of occupants in the area, but because they can accurately assess that they would not fare well if they were to trespass against such a singer.

Another test of the communication-as-manipulation hypothesis comes from a study of cricket song. Richard Dawkins and John Krebs proposed that male singing crickets have evolved the ability to induce females to do the risky and energetically more costly task of traveling to them, rather than the other way around [170]. Leaving aside the question of relative risk of predation on singing versus traveling individuals, a study of the energetics of these two activities by Bradford Rence indicates that singing males actually expend more metabolic energy than walking females. This result does not confirm the exploitation argument. Again this is not to suggest that manipulation and deceit do not sometimes occur in communication, nor that signaler and receiver are trying at all costs to be helpful, only that communication has as its evolutionary basis the cooperative transfer of useful information between some individuals.

Channels of Communication

Leaving aside the as yet incompletely resolved problem of how much deception can exist in an evolved communication system, let us turn to another evolutionary puzzle about communication. Why is it that bark beetles employ odors to attract mates, whereas empidid flies use visual displays, and bowerbirds and gulls rely heavily on acoustical signals? As indicated earlier, accident, chance, and especially the constraints imposed by past evolution play a role in determining the nature of an animal's communication abilities. This does not mean, however, that natural selection has not shaped animal signals. If communication systems have evolved in response to particular environmental pressures, we should be able to perceive general correlations between the ecology and the signal mode of a species.

In order to determine if some signals have been selected for their ability to overcome obstacles to information transfer imposed by a particular environment, it is useful to consider the factors that influence the fitness value of a signal to a sender. These include

1. How effectively the signal reaches desired receivers.
2. The amount of information encoded in the signal.
3. The cost to the sender of producing and broadcasting the signal.

4. The ease with which the sender can be located by a receiver, if this is a goal of the sender.
5. The risk that the signal will be detected by an illegitimate receiver and used against the sender.

These parameters of a message vary, depending on the environment in which they are used. Obviously, visual signals, except for self-manufactured bioluminescent flashes, cannot be used at night and will not pass around physical barriers, unlike chemical and auditory messages. Moreover, visual signals may be readily detected by the sender's predators. On the other hand, visual communication enables instantaneous localization of the sender if the receiver gets the message, unlike most other communication modes. Pheromones, for example, must be tediously tracked by a receiver. Moreover, these long-distance chemical signals are dependent upon wind conditions. Too little wind, and the signal goes nowhere; too much and the pheromone may be blown erratically hither and yon, a result that makes it difficult for receivers to find the sender. In addition, because a chemical message lingers in the air a long time, it is probably difficult (but not impossible [139]) for an animal to send out a rapid-fire series of pheromonal messages, each conveying a new bit of information. This lowers the rate of information transfer by chemical means relative to auditory and visual signals, which usually have a rapid fadeout time and can be quickly replaced with new messages [794].

Thus, the utility of a communication channel depends on the ecology of the species and what kind of message the animal "wishes" to convey to its receivers. Because moths can fly to a perch exposed to night breezes, they can use pheromonal messages to attract mates from a great distance. The chemicals used for this purpose tend to be moderately long-chain carbon molecules (Figure 20), small enough to be lightweight and volatile, but large enough to be distinctive in structure [792]. The great diversity of pheromones makes it possible for females to attract males of their species whose olfactory apparatus is specifically designed for the detection of the female signal.

Crickets rely primarily on the auditory channel to draw conspecifics to them from considerable distances [9]. Constant chirping may be more energetically expensive than the release of a minute amount of chemical and may also provide some male competitors and predators with information damaging to the chirper. (Because pheromones are released in such small quantities and are so distinctive, predators must be highly specialized to use pheromonal messages to track down their prey.) The risk of exploitation associated with a loud calling song may be one reason why male crickets that have succeeded in attracting a female switch to a soft song in the later phases of courtship. Frogs do this as well [767].

On the plus side, auditory messages are less subject to environmental vagaries than olfactory signals. Moreover, because crickets live in and under grasses and debris, an auditory signal can usually be broadcast more

Bombykol (silkworm moth)

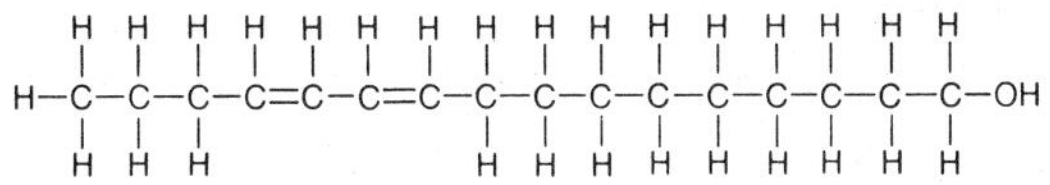

Disparlure (gypsy moth)

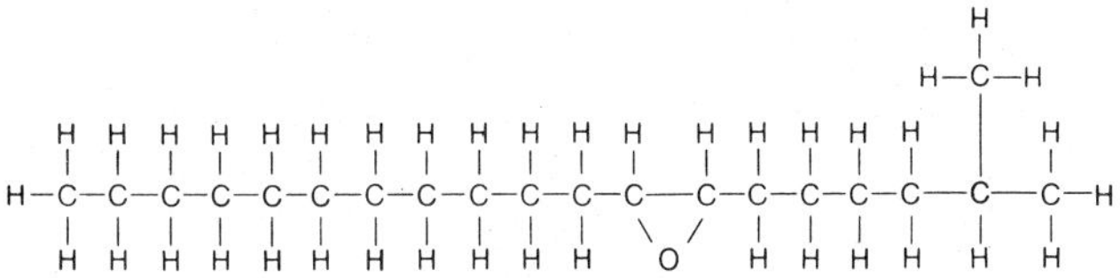

2,3-Dihydro-7-methyl-1H-pyrrolizin-L-one
(Queen butterfly)

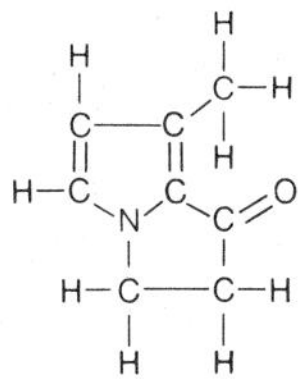

Honeybee queen substance

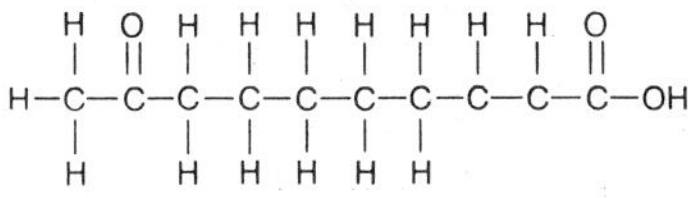

20 **Molecular structures of four insect pheromones.** These fairly simple, but distinctive, hydrocarbons help individuals recognize members of their own species.

effectively than an olfactory one. A moth perched high on a tree is exposed to wind currents; a cricket under a log is not. The species-specificity of the signal is produced by varying the temporal spacing of the components of the chirping or trilling song (Figure 7, Chapter 4).

Fireflies are nocturnal insects that manufacture their own visual signals with the aid of photochemical equipment at the tip of their abdomens [438]. This would seem to make them highly vulnerable to predators, perhaps more so than signaling moths and crickets. However, fireflies are probably poisonous [207] and may rarely be attacked (lizards find them nauseating). Thus, they are able to use a highly conspicuous, easy-to-locate, and economical visual communication channel. The specificity of the message is achieved through the evolution of unique patterns of light pulses, which vary from species to species in duration, intensity, frequency of occurrence, and color [436] (Figure 21).

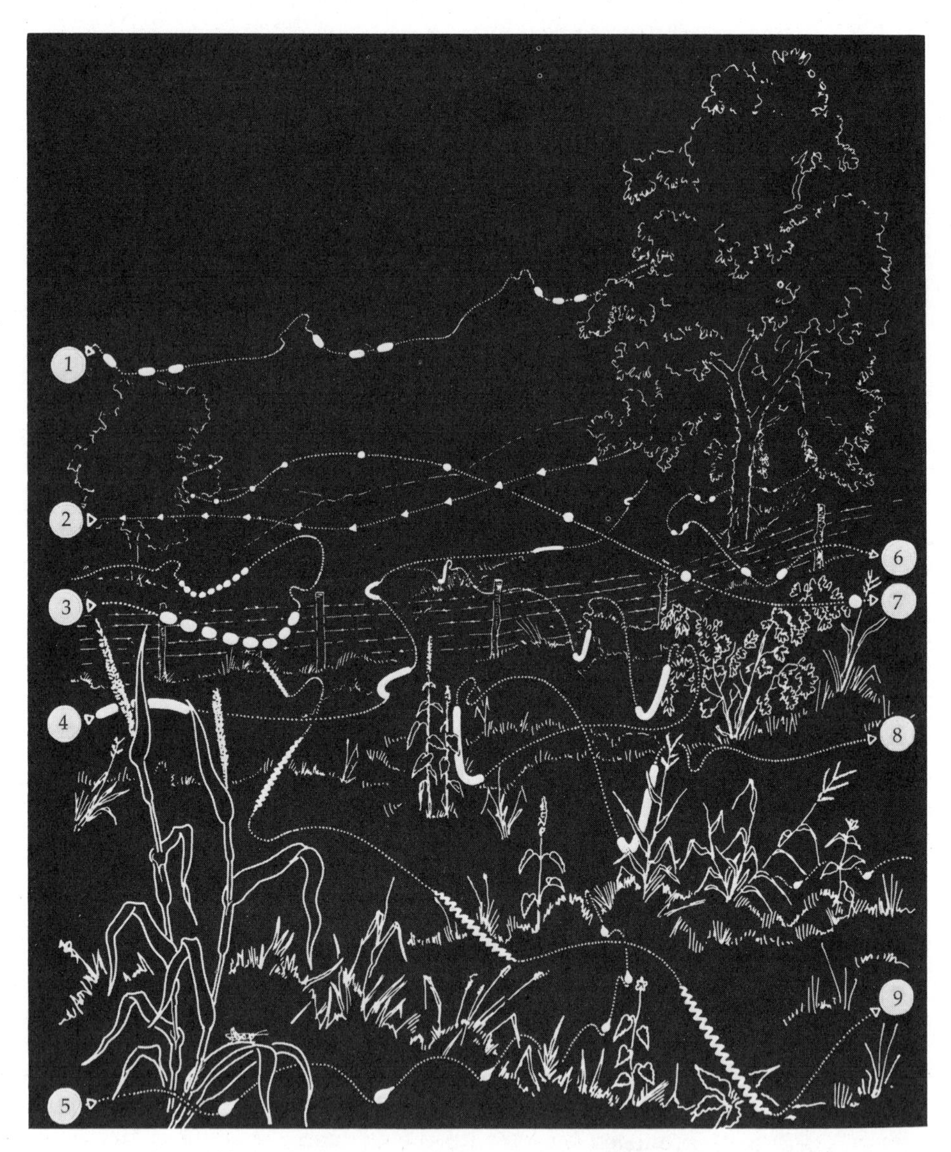

21 **Patterns of light flashes** and flight paths of nine species of fireflies, illustrating the diversity of male courtship signals within this group of insects. Courtesy of James E. Lloyd.

Therefore, although the mate-locating signals of moths, crickets, and fireflies all are the product of selection for species-specific advertisements, each group relies heavily on a different channel of communication appropriate for its ecology. This reflects the different costs and benefits of communication modes in different environments (Table 3).

An individual need not limit itself to a single channel for each message.

Some male crickets use olfactory signals in addition to auditory ones to induce a female to mate [557]. Figure 22 shows a vocalizing female chimpanzee engaged in a visual display (the facial grimace) with a tactile component (touching the male). The context in which each signal element is given often greatly affects its meaning. Therefore, by combining different actions, animals are potentially able to generate many subtly different messages. More often, multichannel communication appears to be redundant, involving the use of several different signals that mean roughly the same thing and that reinforce the main message of the display. This is true for honeybees, for which wing vibrations, odor cues, and the tactile information in the dance all combine to direct a recruit to the right location. Likewise, when humans speak, considerable information is carried on the auditory channel, but facial displays, hand movements, shifts of body position, and tactile signals may all complement the spoken message and make the intent of the speaker clearer. Written transcripts of spoken interactions often are nearly complete gibberish to a reader, whereas there was probably little or no confusion between the signaler and receiver thanks to the nonverbal communication that accompanied the words.

Diversity within Communication Channels

Additional support for the hypothesis that communication signals help individuals solve specific ecological problems is provided by the diversity in signal structure that exists within any one communication

Table 3
Comparisons among the Major Channels of Communication

	Ability of Signal to Reach Receiver			
Channel	*Range*	*Rate Transmission*	*Flow around Barrier?*	*Night Use?*
Chemical	Long	Slow	Yes	Yes
Auditory	Long	Fast	Yes	Yes
Visual	Medium	Fast	No	No[a]
Tactile	Very short	Fast	No	Yes

	Information Available		Cost to Sender	
Channel	*Fadeout Time*	*Locatability of Sender*	*Broadcast Expense*	*Risk of Exploitation*
Chemical	Slow	Difficult	Low	Low
Auditory	Fast	Fairly easy	High	Medium
Visual	Fast	Easy	Low-moderate	High
Tactile	Fast	Easy	Low	Low

[a] Except bioluminescent signals.

22 **Multichannel communication.** The female chimpanzee simultaneously provides tactile, visual, and auditory signals to the male on the right. The function of her behavior is to determine whether she will be permitted to remain near the male. Photograph by Leanne Nash.

channel. For example, different pheromones function as alarm signals, sex attractant messages, or territorial markers. Each function favors chemicals of rather different design. We have already indicated that there is a tradeoff between molecular distinctiveness (which is best served through molecules of high weight) and volatility (which is best served by molecules of low weight). Thus, sex attractants generally sacrifice a certain amount of volatility in order to achieve a unique structure and thereby convey unambiguously the species-membership of the signaler [793]. An alarm pheromone is under no such constraint. If a worker ant or termite is to arouse other colony members to repel an attacker, the primary requirement is that the message be broadcast as quickly and widely as possible. Alarm pheromones of insects therefore tend to be extremely lightweight and convergently similar in structure from species to species [793]. In contrast, territorial markers should be durable if they are to be effective, yet volatile enough to be easily detected by intruders. This favors relatively large molecules that can nonetheless be carried on air currents. A further refinement is in the delivery system employed by the tiger (and perhaps some other species). The tiger applies the marker scent in a fatty base, which acts to retard the evaporation of the odorous components; a territorial boundary point sprayed with the marker can still smell of tiger weeks later [82].

Similar correlations between signal type and environmental problems have been reported for electrical [342, 343], visual [404], and acoustical signals [523]. Perhaps the most intriguing of these examples involves the interpretation of song variation among birds. The ease with which sounds of different frequencies are transmitted over long distances depends upon the acoustical properties of the environment. Acoustical signals may be absorbed by vegetation quickly, reducing the number of receivers they may reach. Eugene Morton broadcast sounds of different frequencies in the forest and grassland habitats of Panama [523]. Tape recorders placed at varying distances from the sound producer made a record of what frequencies reached them. Morton found that sounds of low and high frequency were relatively quickly absorbed in the forest, but sounds between 1500 and 2500 hertz were much less severely attenuated. The bird species typically found in the Panamanian forest rarely produce songs with frequencies higher than 3500 hertz, and the majority signal within the 1500- to 2500-hertz band. These songs can be projected the greatest distance with the least energetic expense (Figure 23).

In grassland habitats the scatter of frequencies used by songbirds is much greater. Here Morton's test recordings revealed that no range of sounds was relatively safe from absorption (although high-frequency sounds were somewhat more attenuated than low-frequency ones).

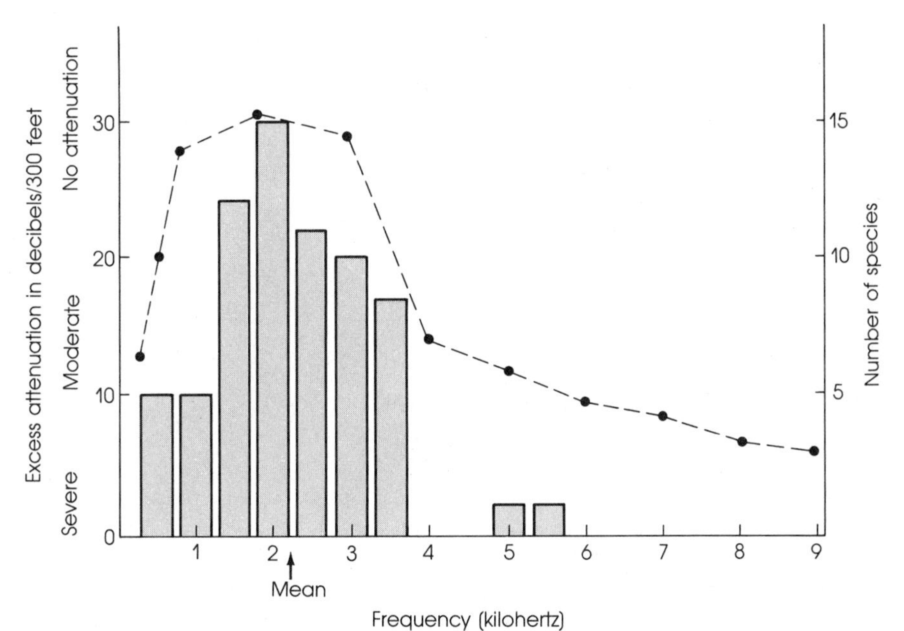

23 **Song frequencies of bird species** found in lowland forest habitats of Central America. The histograms show the number of species employing various frequencies; the mean frequency for the group as a whole is a little over 2000 hertz (2 kilohertz). The dotted line shows the attenuation effect of the habitat on sounds of different frequencies. Sounds of about 2000 hertz are relatively free from attenuation.

The hypothesis that attenuation problems are the basis for the difference between grassland and forest bird songs has been challenged on a variety of grounds [470, 471]. An alternative hypothesis is that the key problem for birds is one of song distortion or degradation rather than song attenuation. In forests, the difficulty is that sounds echo back from leaves and limbs; the echoes create competing noises that may damage the integrity of a song. In grasslands, song distortion may be produced by heating and wind effects, which create a turbulent air zone close to the earth. Forest birds may therefore sing low-frequency songs, not so much to project a message a great distance, but to avoid deflection by small objects such as leaves and branches, which greatly interfere with high-frequency sounds. In grasslands, the solution may be to escape the distortion zone by singing from elevated perches or by flying up to sing 20 feet or more above the ground. Also, in this environment there appears to be greater emphasis on repeating song elements in trills; a listener that hears even a part of a repetitive song during a lull in winds can probably identify the singer. Trills are rarely employed by forest birds because of the disruptive effect of echoes on sounds of this structure.

The advocates of the antidegradation hypothesis point to cases of variation in the songs of birds of the same species that happen to live in different habitats. In the portions of their range where they inhabit open country, the rufous-collared sparrow uses a song with a prominent rapid trill component. This is largely missing from the songs of birds that occupy forest habitat [540]. On the other hand, Morton has evidence from Carolina wrens that is consistent with the antiattenuation hypothesis (Figure 24). If one plays the songs of wrens from various geographic regions in a given habitat, the song that carries the farthest is the one sung by the local birds [261]. Actually it would not be too surprising if both attenuation and distortion problems played a role in shaping the evolution of bird song and other acoustical signals. The key point is that even populations of the same bird may differ in their songs in predictable ways. This suggests that communication signals rapidly evolve to match the properties of the particular environments in which they are broadcast.

Predation, Competition, and Communication

The environment of most signaling species not only provides physical barriers to signal transmission but also contains other obstacles in the form of illegitimate receivers and interfering competitors for signal space. These features have been shown to influence not only the design of a signal but also how it is delivered, as can be illustrated by examining the communication systems of two frogs.

Males of the tropical frog *Physalaemus pustulosus* have a two-part call that sounds like a "whine" followed by "chuck." Sometimes males give just the whine, despite the fact that females are preferentially attracted to complete calls (because only the chuck contains clear information about male body size and females prefer large males [627]). The chuck component

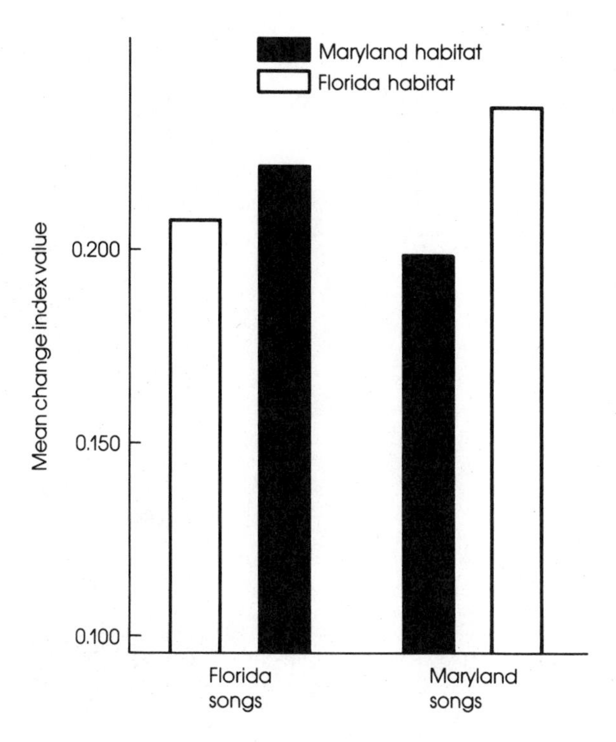

24 **Carolina wren songs** retain their signal characteristics better (change less over distance) in the habitat in which the singer actually lived. The "mean change index value" is a complex measure of the degradation of tape-recorded songs over a standard distance.

also probably helps announce the location of the calling male better than the whine. This would seem to be advantageous to a male attempting to attract a female; but unfortunately (from the frog's perspective), the calls can also attract predators, notably a Panamanian bat that specializes in tracking down singing frogs [726] (Figure 25). This may explain why males that are calling alone omit the chuck and give only the more ventriloquistic whine, compromising mate attraction because of the high risk of predation. But when chorusing groups form, calling males sing their complex songs. Not only are males safer when in a group (Chapter 10), but competition with other callers forces a frog to give his complete advertisement call. If he does not, females will ignore him in favor of one that does [599].

Competition for broadcast space as well as predation appears to have played a strong role in the evolution of calls and their delivery in some frogs. The Puerto Rican treefrog *Eleutherodactylus coqui* gives its call, "co-qui," in a rainforest filled with loudly calling rivals and members of other species. Peter Narins has suggested that the following features of the coqui's call help males transmit their messages effectively to appropriate receivers [531]:

1. The call is very loud, 90–100 decibels at 50 centimeters from the male.

25 **Frog-eating bat** about to capture a male frog that the predator has located by listening to the frog's calls.

2. Males call from elevated perches and orient themselves to face skyward, the better to propagate the song in the horizontal plane [533].
3. The song's energy is focused in a band of frequencies not used by other species of frogs in the forest.
4. Males are remarkably adept at producing a call almost instantly after a nearby rival's "co-qui" has ended. This enables an individual to inject his signal in the relatively quiet interval between neighbors' calls [532].

All four attributes of the male's calling behavior appear to facilitate the transmission of a message in a particularly competitive and noisy environment, an observation that again supports the hypothesis that communication signals reflect the particular selective pressures acting on the signaler.

Summary

1 Understanding the history of a behavior pattern contributes to answering a major ultimate question about the causation of behavior. Historical aspects of behavior can sometimes be inferred by matching the fossils left by extinct animals with comparable traits in living species.
2 The more generally useful method of tracing the evolutionary pathway to an unusual behavior pattern is to compare the behavior of closely related species for possible clues about intermediate stages. The central assumption is that behavior patterns that are widespread in a complex of related species are probably similar to the original characteristic in an ancestral population that gave rise to the present-day complex. Unusual behavior for a group is likely to represent a relatively recent modification of the ancestral behavior.
3 Comparing clusters of related species has produced plausible historical scenarios for a number of complex communication signals. Communication originates from actions that incidentally happen to convey mutually beneficial information from one individual to another. Once originated, further selection may result in the extreme elaboration of the system and its exploitation by illegitimate signalers and receivers.
4 Different channels of communication have different advantages and disadvantages in various environments. The correspondence between ecological problems and the kind of signal used suggests that selection has promoted effectiveness of signal transmission, detection, and response over evolutionary time.

Suggested Reading

Hans Bänziger's article on blood-sucking moths [38] is a pleasure to read. James Gould's *Ethology* [269] has a succinct review chapter on honeybee social behavior and communication. Martin Lindauer's *Communication Among Social Bees* [428] is a classic book that outlines the evolutionary pathways to honeybee dances.

More general discussions of the evolution of animal communication can be found in Daniel Otte's review article [556], several major chapters in E. O. Wilson's *Sociobiology* [794], and the book *How Animals Communicate*, edited by Thomas Sebeok [645].

Stephen Jay Gould and Richard Lewontin have been leading advocates of the argument that historical constraints greatly influence and limit the evolution of the characteristics of living species [275].

Suggested Films

Evolution of Nests of the Weaverbird (8780). Color, 24 minutes. An application of the comparative method to analyze the evolution of the complex nest-building behavior of weaverbirds.

Life Cycle of the Honeybee (06005). Color, 12 minutes. A well-made film that illustrates, among other things, the dances of the honeybee.

Signals for Survival. Color, 50 minutes. Much of this film on the lesser black-backed gull is devoted to description of the courtship signals of the birds.

The preceding chapter stressed that behavior patterns have a historical basis that can be reconstructed through comparative procedures. The fact that some behavioral traits appear to be relatively recently derived from older ones does not, however, mean that they are better in some absolute sense than the characteristics from which they have evolved. One assumes only that the more recent innovation is better suited for the changed environment that the species now happens to occupy. This is a particularly important point to keep in mind when analyzing the evolution of social behavior. There is an unfortunate tendency to think that the highly complex societies of some vertebrates, most conspicously our own human societies, represent a crowning achievement of evolution. To counteract this mistaken impression, Chapter 14 begins with a discussion of the costs, as well as the benefits, of social living. Only under certain ecological conditions will the benefits of sociality exceed the costs and selection favor the evolution of societies. Otherwise, solitary living is the more adaptive mode of existence. After establishing this point, we shall examine one of the most intriguing aspects of some social species, the evolution of altruism. Here we shall present a variety of examples of the phenomenon, including its most extreme expression in the reproductive self-sacrificing actions of some animals. There are various hypotheses to account for the evolution of helping, and the chapter will examine the different proposals and will show that there is more than one way in which cooperation among social creatures can arise.

CHAPTER 14

The Evolution of Societies

The Costs and Benefits of Social Living

In recent years there has been a remarkable proliferation of studies on the behavior of social animals ranging from reef fishes [602] and frogs [765] to bats [85] and marmots [41], and from ants [793] to birds [212]. These studies have been inspired in part by the recognition that there are major disadvantages as well as advantages to living in groups (Table 1) [11]. This point has been beautifully illustrated in a comparative study by Mart Gross and Anne MacMillan on the bluegill sunfish (*Lepomis macrochirus*) and its relative, the pumpkinseed sunfish (*Lepomis gibbosus*) [289]. Bluegills are generally social; pumpkinseeds are solitary. The social nature of bluegills is expressed during the breeding season when groups of as many as 50–100 males may build their nests (depressions in a sandy lake bottom) side-by-side to form a dense

Table 1
Some Major Advantages and Disadvantages of Sociality[a]

Advantages
Reduction in predator pressure by improved detection or repulsion of enemies
Improved foraging efficiency for large game or clumped ephemeral food resources
Improved defense of limited resources (space, food) against other groups of conspecific intruders
Improved care of offspring through communal feeding and protection
Disadvantages
Increased competition within the group for food, mates, nest sites, nest materials, or other limited resources
Increased risk of infection by contagious diseases and parasites
Increased risk of exploitation of parental care by conspecifics
Increased risk that conspecifics will kill one's progeny

[a] For solitary species, certain of the disadvantages outweigh any benefits of social living; for social species, the costs are more than matched by certain of the advantages of sociality.

colony (Figure 1). The primary benefit to these males is almost certainly a reduction in predator pressure on the eggs deposited in the nests by spawning females. Largely through a selfish herd effect (Chapter 10), males whose nests are surrounded by those of their neighbors are less likely to experience predation by egg-eating catfish or snails. In addition, clumped nests may swamp the feeding capacity of the local predators, and the males, by defending overlapping territories, "cooperate" in the repulsion of certain predators, such as the catfish (Table 2).

But colonial nesting bluegills do not get their antipredator benefits for free (nor do the many other social animals that band together to defeat their enemies [11]). In particular, an individual that nests in a group must contend with the tendency of his neighbors (and other nonnesting bluegills attracted to the group) to consume the eggs in his nest, which he has fertilized. Sexual interference in courtship and spawning is another problem that occurs because of the sneaker males (Chapter 11) that gather at large colonies. Moreover, fungi that destroy eggs may be transmitted from nest to nest in a dense colony. These costs reduce the net benefit enjoyed by social bluegills but do not eliminate it entirely.

Pumpkinseed sunfish do not breed in colonies apparently because they derive few antipredator benefits from group living. Pumpkinseeds are much

1 **Colonial nesting fish.** Each male bluegill places his nest close to those of other males, creating a cluster of roughly hexagonal territories. Two colonies of nesting males are shown. Predation pressure provided by bass (above), bullhead catfish (left), snails, pumpkinseed sunfish (right foreground), and other bluegills may have led to the evolution of bluegill societies. Courtesy of Mart Gross.

Table 2
Predation Pressure on Bluegill and Pumpkinseed Sunfish

Position of Nest	Mean Number of Snails per Nest	Mean Number of Chases[a] per Nest	Group Response[b] (%)
BLUEGILL			
Central	6.9	1.5	50
Peripheral	13.7	8.7	8.2
Solitary	29.7	10.4	0
PUMPKINSEED			
Solitary	0.4	—	—

Source: Gross and MacMillan [289].
[a] Chases involve reaction by nest defender to a potential predator like another bluegill or bullhead catfish
[b] Group response is shown as the percentage of all chases at a nest in which two or more males simultaneously pursued the predator

less affected by predators than bluegills because they have mouthparts adapted for picking up, crushing, and consuming heavy-bodied, molluscan prey. Bluegills, on the other hand, have delicate, small mouths designed for "inhaling" small, soft-bodied, insect larvae. Thus, although a bluegill cannot pick up a snail and cart it away from the nest, pumpkinseeds are easily able to do this (and may consume their enemy to boot). In addition, a bluegill can be ignored by a nest-raiding bullhead catfish, but a pumpkinseed's attack has considerably more bite to it. Because pumpkinseeds are relatively free from nest predation, there are insufficient advantages to social breeding to compensate for the costs of this lifestyle [289]. Pumpkinseed sunfish are in no way inferior or less well-adapted than bluegills because they are solitary; they simply face different ecological circumstances, for which colonial nesting would yield reduced individual fitness.

Degrees of Sociality in Prairie Dogs

A cost–benefit analysis of sociality has also been applied by John Hoogland in his studies of prairie dogs [337, 338]. Prairie dogs are similar to bluegills in that antipredator benefits are likely to have been the major impetus for the evolution of their colonial way of life. Hoogland tested the response of prairie dogs to potential predators by pulling a stuffed badger or weasel toward and through a colony to his blind at a programmed rate of speed. He recorded the length of the interval between the moment at which the "predator" was activated and the first alarm call. For both the white-tailed prairie dog (*Cynomys leucurus*) and the black-tailed prairie dog (*C. ludovicianus*), the smaller the colony the longer before the enemy was detected and an alarm sounded (Figure 2). Animals that hear an alarm call scamper back to their burrows, where they are relatively safe from

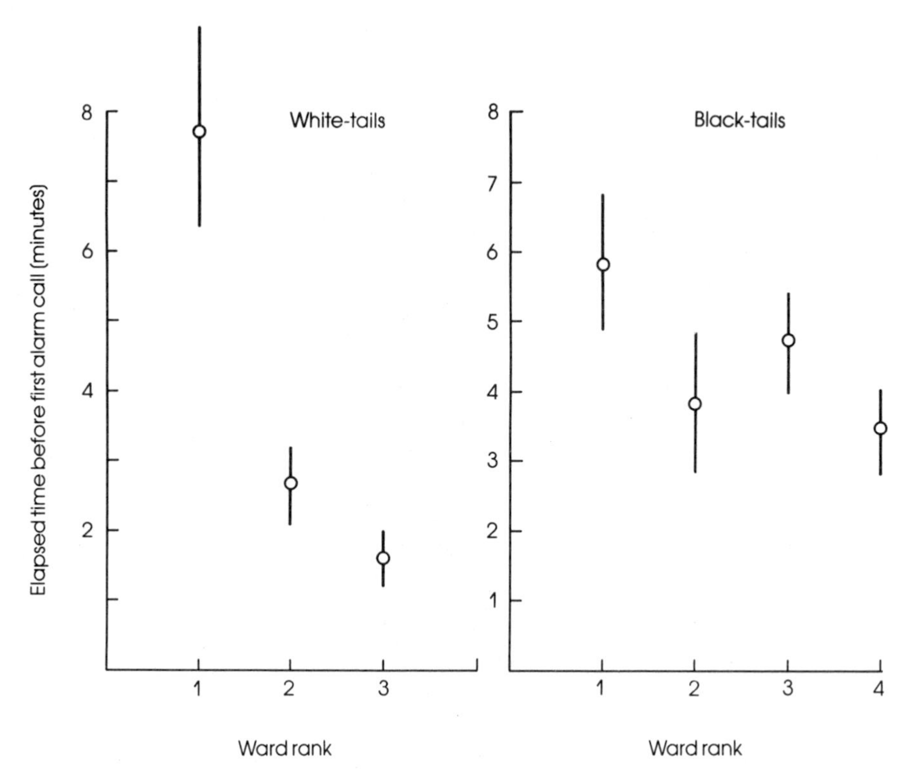

2 **A benefit of sociality.** For two species of prairie dogs, an experimentally introduced predator is more quickly detected by large groups than by smaller ones. Wards (groups) are ranked by numbers living together with rank 1 the smallest group.

many of their predators. Because black-tails tend to live in larger groups than white-tails, they generally react with alarm sooner than their relatives. Moreover, black-tailed prairie dogs can afford to devote significantly less time (35 percent of their average daily time budget) to scanning for predators than can white-tails (43 percent).

The question becomes, Why don't white-tailed prairie dogs live in large groups, too, so that individuals can enjoy the multiple benefits that come from coping with predators efficiently? As with bluegills, however, there are costs to individuals that live in large groups. Although a prairie dog may spend less time looking about for danger in a large group, it will also spend more time dealing with territorial intrusions from its neighbors (Figure 3). In addition, fleas that bear sylvatic plague, a devastating disease, are more abundant in large colonies. In dense black-tail groups Hoogland counted an average of 3.3 fleas per burrow entrance, whereas in white-tail colonies Hoogland counted an average of only 0.8 flea per burrow entrance. Because black-tails pay a greater price in terms of disease risks and social interference, they presumably are under greater predation pressure than white-tails.

White-tailed prairie dogs do appear to have an environmental advantage in dealing with predators because they live in areas with significantly more woody shrubs than the plains-dwelling black-tails. These shrubs provide cover for foraging white-tails, who can better hide from their enemies than their more exposed relatives. (Hoogland documented that a significantly smaller proportion of feeding white-tails are visible at any moment to human observers than are black-tails.) This suggests that, whereas black-tails rely on one another to detect their predators, white-tails can use naturally occurring shrubs as cover and therefore gain less from group living. As a result, benefit-to-cost ratios for living in large, dense colonies are lower for white-tails and may account for the different degrees of sociality in two otherwise very similar prairie dog species [337, 338].

Intraspecific Variation in Group Size

We have compared a social fish with a solitary fish and a highly social prairie dog with a moderately social relative. There are some animal species whose members can be either social or solitary. These animals provide an ideal opportunity to test hypotheses about the correlation between key environmental pressures and social living. For example, a lion has the capacity to live alone or join a large band of other lions (the pride), a band that may consist of as many as 15 adults and numbers of younger individuals [635]. What factors promote the formation of lion prides?

Unlike bluegills and prairie dogs, group living in lions appears to have

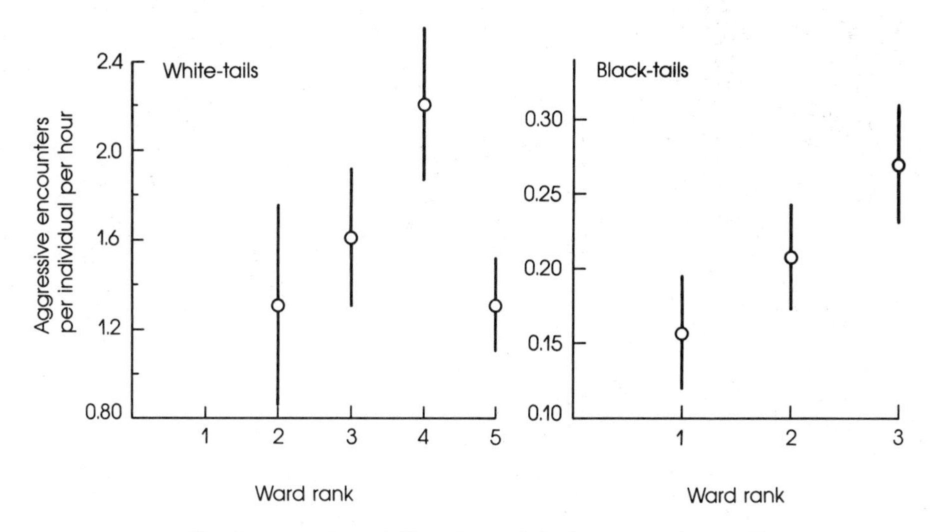

3 **A cost of sociality.** In prairie dogs, members of large groups spend more time and energy in aggressive encounters with intruders than do members of small groups.

little to do with foiling predators. As you might imagine, lions have relatively few enemies that dare to attack them. In this case, sociality seems to be related to improved territorial defense and increased success in hunting [635]. As discussed in Chapter 9, lions are able to kill unusually large herbivores through cooperation. Moreover, the members of a pride also work together to defend a foraging territory against other lions and clans of hyenas. Still other benefits of social living include mutual defense of captured prey against competitors (Figure 4) and the joint protection and feeding of offspring of the pride by its members.

Given this array of benefits, why do lions ever live alone? The answer appears to be related to seasonal fluctuations in food supply [125]. The primary prey of lions in the Serengeti consists of the relatively small, but abundant, Thomson's gazelles and the much larger wildebeests and zebras. The large herbivores are migratory, moving from the plains to the woodland areas of the Serengeti in the dry season as grasses disappear. They return to the plains during the wet season to harvest the new crop of grasses. Lions that do not migrate with the wildebeests and zebras are forced to prey upon the resident gazelles, each of which provides about 11 kilograms of edible meat. A lion needs 6 kilograms of meat per day to maintain itself. George Schaller observed that hunting lions manage about three stalks per day [635]. For a solitary lion hunting gazelles, the chance of success per hunt is 15 percent as opposed to 30 percent if two or more lions cooperate in a stalk. Therefore, if two lions hunt and feed together, they can average

4 **Two lionesses defending their kills** against a clan of hyenas. One of the benefits of sociality in lions may be improved defense of captured prey. Photograph by Norman Myers.

about one gazelle captured per day, which is just enough to support them. If more than two lions forage together, their chances of a kill do not improve, and the 11-kilogram gazelle caught each day must be shared among three or more consumers. Each lion's intake of food will fall below the 6-kilogram requirement. Schaller recorded an average of about 1.5 lions at gazelle kills during the season when lions were restricted to these prey. When gazelles are the staple food item, it is advantageous for lions to disperse; and therefore the prides break up.

During the time of year in which wildebeests or zebras are available, lions largely ignore the smaller gazelles. The capture rate of lions hunting zebras increases from 15 percent for solitary hunters to over 40 percent for groups of five. A zebra carcass can feed a group for three days. The calculations of Thomas Caraco and Larry Wolf reveal that a group of two or three lions could achieve an average daily food intake of 8 to 10 kilograms of meat, whereas a group of four or five will secure the minimum daily requirement [125]. Schaller found an average of four or five lions feeding at each wildebeest or zebra victim. (In areas with exceptionally large concentrations of prey or exceptionally vulnerable prey, prides may be larger still.) Lions form groups larger than two or three because the quantity of food captured per individual is not the only determinant of gene-copying success [607]. Nevertheless, food supply is the basic limiting factor that regulates the social life of lions. When ecological conditions (the abundance of large game) raise the benefits of sociality by increasing the energy that can be gained by social foraging, prides form. When conditions change and the foraging gains of large groups fall, prides disband.

The Evolution of Cooperation

Animals that live in groups often appear to help one another, as when two or three bluegills jointly attack a bullhead, or when a prairie dog gives an alarm call after it spots a badger, or when a pride of lionesses overwhelms a dangerous Cape buffalo. Until recently, biologists were not especially surprised by this sort of thing because they assumed that the members of a group should help each other for the benefit of the species as a whole. But with the recognition that group selection is less potent than individual selection (Chapter 1), helpful actions become considerably more interesting [786], particularly if they appear to reduce a helpful individual's reproductive success while raising that of the animal it helped (Table 3).

There is, however, no requirement that an individual lower its personal fitness by helping another. In many cases, helping may lead to immediate gains for both helper and helpee, as for example, when one lioness drives a wildebeest toward her fellow pride members. If the antelope is killed by the ambushers, the driver will feed upon the prey as well. Likewise, if several male bluegills succeed in fending off a bullhead catfish that has entered the space they mutually claim and defend, the eggs guarded by all these males are at once safer as a result. If it can be demonstrated that both helper and recipient enjoy reproductive gains through cooperation,

Table 3
Reproductive Consequences of Helpful Actions for the Helper and the Recipient of the Assistance

Effect on Individual Reproductive Success (Personal Fitness)		
Actor	*Recipient*	**Category**
+	+	Mutualism
		Short-term cooperation
		Long-term reciprocity
–	+	Altruism
+	–	Pseudocooperation

then there is no special evolutionary puzzle about the evolution of these helpful tendencies.

In addition, it is possible that animals can raise their reproductive success by participating in reciprocal cooperation in which individuals take it in turns to help one another. Robert Trivers invented the label RECIPROCAL ALTRUISM for helpful actions that result in *deferred* gain for the helper [715]. He pointed out that if an individual could help another now at relatively little cost and later receive valuable "repayment" from the helped animal, the original helper would experience a net reproductive benefit from its original reproductive self-sacrifice or "altruism." Critics of this notion, however, raised questions about the evolutionary stability of reciprocal altruism. What would prevent a recipient from gratefully accepting help but later conveniently refusing to repay its helper? The problem of "cheating" would seem to make reciprocal arrangements difficult to enforce and therefore difficult to evolve.

However, Robert Axelrod and W. D. Hamilton have shown that if there is the possibility of a long series of interactions between potential cooperators, then individuals that use the simple behavioral rule "Do unto individual X as he did unto you the last time you met" can reap greater fitness gains than individuals that always try to accept assistance while never helping in return [30]. The strategy of taking advantage of a cooperator, although possibly providing a large return for the cheater in that one interaction, in the long haul may yield less than a strategy of cooperating with those individuals that have in the past cooperated with you.

Imagine a group of lionesses that excluded the female that drove the game toward them from a kill that they made as a result of the driver's help. The cheaters would enjoy the fruit of the driver's labor at her expense. But if this caused the driver to leave the group, the exploiters would have taken a short-term gain at the cost of losing the long-term help of the departed individual. Thus, cheating could be a short-sighted strategy that would yield less than permitting the driver to share in the kill, an action that would ensure her cooperation in future hunts.

Altruism and Kin Selection

In cases of profitable mutualism, whether short term or long term, individuals experience a net increase in their reproductive success. To demonstrate that a helpful action is truly altruistic, one must show that the individual really does lose reproductively over the long haul as a result of its help. It is not enough to determine that an individual fails to gain from one interaction because at a later time it may be "repaid" by the animal that it helped. Moreover, one must also show that the receiver truly benefits from the "assistance" that it receives, because there is always the chance that the recipient is really being exploited by an apparent helper. Cases of pseudocooperation may occur in which false helpers manipulate others in ways that raise the manipulator's reproductive success. Like short-term cooperation and long-term reciprocity, these actions can evolve through individual (natural) selection.

But helpful actions that permanently reduce a helper's reproductive success cannot evolve via individual selection. The puzzle is that examples of altruism exist. The solution to the puzzle comes from the realization that if a helper directs its aid to its genetic relatives it may be more than compensated for a reduction in its own fitness by an increase in the reproductive success of the related individuals [300]. Remember that from an evolutionary viewpoint the point of reproduction is to propagate one's alleles. Personal reproduction is an admirably direct way to achieve this ultimate goal. But helping relatives survive to reproduce provides an indirect route to this same end.

To understand why this is so, we must discuss the concept of degrees of relatedness. There are ways to determine the probability that any two individuals will share a particular allele as a result of having inherited it from a common ancestor. For a parent with a genotype (*amy 1,amy 2*), the probability that its offspring will inherit, say, the *amy 1* allele, is 0.5 (Chapter 2). There is one chance in two that the egg or sperm that the parent produced that helped form the offspring will contain the *amy 1* allele. The probability of possessing an allele by common descent is the COEFFICIENT OF RELATEDNESS or *r*. Between nonrelatives, *r* is essentially 0; between parents and offspring, *r* is 0.5.

The coefficient of relatedness can be calculated for any pair of individuals that are descended from a common ancestor. The value of *r* reflects the number of meiotic events that separate the two individuals. Table 4 presents a sample of *r* values. With this information we can determine the impact of a helpful, self-sacrificing action on the transmission of an allele that predisposes its bearer to behave in an altruistic manner. The key question is whether an individual can leave more copies of an altruism-promoting allele by helping relatives reproduce or by reproducing personally. Let us say that the animal could potentially have an average of two offspring of its own *or* give up reproduction, invest entirely in relatives, and help create an average of three siblings that would not have survived without its help. Offspring share half their genes with a parent; siblings share half their genes with each other. Therefore, in this example, the cost of abandoning personal reproduction is ($r \times 2 = 0.5 \times 2 = 1$); the benefit in terms of genes passed on indirectly in the bodies of relatives is ($r \times 3 = 0.5 \times 3 = 1.5$). The altruistic route therefore offers greater genetic success than the "selfish" personal reproduction route.

If the cost of an altruistic act were the loss of one offspring ($C = 0.5 \times 1$) but if the altruistic act led to the survival of three nephews that would have otherwise perished ($B = 0.25 \times 3$), the action would also create a net genetic gain for the altruist. B must be greater than C if the action is to increase the frequency of the allele underlying the helpful act. When one thinks in these terms, it becomes clear that there can be a form of selection that occurs when genetically different individuals differ in their effects on the reproductive success of relatives. This form of selection is usually called KIN SELECTION, and it can lead to the evolution of genetically profitable altruism [773, 794].

W. D. Hamilton [300] recognized that an individual's total impact on evolution via transmission of genes to a subsequent gene pool required a *combined measure* of the animal's direct and indirect contributions of genes. Therefore, he invented the term INCLUSIVE FITNESS as a quantitative measure of total genetic success via personal reproduction and effects on rela-

Table 4
The Coefficient of Relatedness of Certain Relatives and Inclusive Fitness

Relationship	Coefficient of Relatedness (r)	Relationship	Coefficient of Relatedness (r)
Parent–offspring	0.5	Full siblings	0.5
Uncle–nephew	0.25	Half siblings[a]	0.25
Aunt–niece	0.25	Grandparent–grandchild	0.25
Cousins	0.125		

Inclusive fitness = Personal fitness + kinship component
Personal fitness = r × number of surviving offspring personally produced by the individual
Kinship component = r × number of surviving relatives that the individual helped produce or keep alive

Example: Individual has one surviving offspring, saves one full brother's life, and helps his sister rear two nephews to the age of reproduction

$$\text{Inclusive fitness} = \underbrace{(0.5 \times 1)}_{\text{Personal fitness}} + \underbrace{[(0.5 \times 1) + (0.25 \times 2)]}_{\text{Kinship component}} = 1.5$$

[a] Half-siblings are brothers and sisters that have only one parent in common.

tives' reproduction (Table 4). As we have seen, it is theoretically possible for an individual under the right circumstances to have higher inclusive fitness by allocating his resources to help relatives rather than by investing all in attempts to reproduce personally.

But what evidence is there that altruism actually has evolved through kin selection? One frequently mentioned candidate for beneficial self-sacrificing actions directed at relatives is the alarm call given by various animals. But as we have already noted in our discussion of tail-flagging by white-tailed deer, the issue is a complex one because there are many potential competing hypotheses to account for specific cases of apparent alarm calls.

How Might Alarm Calls Evolve?

Many social animals, like prairie dogs and ground squirrels (Figure 5) give special vocalizations in the face of danger [e.g., 465, 654, 693]. Although these superficially appear to be truly altruistic, to demonstrate unequivocally that this is true one must show that (1) the call really does help other individuals and (2) that the caller really does reduce its chances for survival and reproductive success by giving the signal. The following list provides a spectrum of possible advantages to a caller, many of which could evolve via individual selection rather than kin selection [655].

1. *Individual selection hypotheses:* The caller enhances its personal chances for reproductive success by giving an "alarm" signal.
   a. *The predator deterrence hypothesis:* Although one might think that the caller risks drawing attention to itself and should slip off quietly when it spots danger, actually predators that know they have been detected are likely to give up the hunt. The caller signals to communicate with the predator to save his own skin. Any ben-

5 **Alarm calling by Belding's ground squirrel.** In this animal, females are more likely to give alarm calls than males. Photograph by George Lepp.

efits enjoyed by others are purely incidental and not the evolved function of the alarm call.

b. *The short-term mutualism hypothesis:* The call's function is to alert the caller's neighbors so that they together can adopt a cohesive escape formation that better foils the predator and improves the chances of survival of the caller. Because the signaler knows where the predator is but the other animals do not, the caller can position itself within the group in the safest possible position. The others respond because they are better off fleeing together than fleeing separately [133].

c. *The reciprocal altruism hypothesis:* A caller may give the signal at some immediate risk to itself, but later it will be more than repaid by others in the group when they return the favor. This presupposes a stable relationship between at least some group members who have the capacity to withhold help if they do not receive it from their fellows.

2. *Kin selection hypothesis:* The caller reduces its lifetime chances for reproductive success by sounding the alarm, but the altruism nevertheless raises its inclusive fitness. Parents, aunts, uncles, brothers, sisters, or cousins are alerted by the signal, and the increase in reproductive output by these relatives is great enough that the sacrifice of the altruist is genetically advantageous [486].

All of these hypotheses presume that the actions of the alarm giver do *not* reduce the propagation of the genetic basis of the behavior. If alarm giving resulted in a decreased representation of the allele(s) responsible for the action, the trait would disappear from the population in subsequent generations.

There have been few thorough tests of any of the hypotheses that account for the evolution of alarm calls. For some birds, the short-term mutualism hypothesis appears most plausible [526, 558]. Many sandpipers and plovers give a loud and easily located alarm call when they spot an aerial predator. These birds forage in loose, mixed-species flocks on migration, and the calls draw together unrelated individuals. Once the flock has formed, its members may all enjoy the benefit of not being an isolated and easily captured prey.

Alarm Calls and Ground Squirrels

But in the best-studied case of alarm calls, Paul Sherman has made a convincing argument that the kin selection hypothesis is most likely [655]. Belding's ground squirrels, like prairie dogs, form breeding colonies with many burrows aggregated in grassy mountain meadows. The squirrels sometimes produce an excited chattering call when they spot a predatory mammal and a high whistle when an aerial predator sails overhead. Other squirrels stop what they are doing when they hear the call and either scan the environment for the enemy or duck into their burrows. There is no group mobbing or formation of a defensive herd, nor are predators deterred from a hunt when they hear a squirrel call. In fact, Sherman regularly saw weasels, badgers, and coyotes stalk alarm callers and sometimes kill them. Because the squirrels were marked and because the colony had been followed for several years, Sherman knew who was related to whom. This permitted him to discover which animals gave alarm calls in response to predators. But over 3000 hours of observation were required in order to secure just 100 records of predators in the presence of marked squirrels!

As noted in Chapter 7, the females in an area tend to be sedentary and therefore are related to one another, whereas the males, which move each year, are not. The kin selection hypothesis predicts, therefore, that females should give more alarm calls than males because the risk to a female caller might be compensated by the improved chances of survival of her nearby relatives. A male giving an alarm call would usually risk his life for nonrelatives, not a strategic move in evolutionary terms. Sherman found that females are in fact much more likely to give alarm calls when they have an opportunity to do so than are males. An even more gratifying finding was the discovery that females known to have relatives living nearby called more frequently than females without living relatives as neighbors [655]. This is powerful support for the operation of kin selection in the evolution of alarm calls in this species (Table 5), especially because the probability that an individual will perform certain altruistic acts is not correlated with familiarity or length of association between actor and beneficiary. This result decreases the probability that alarm calling is maintained through reciprocal altruism among relatives [657].

Kin selection may also be at the root of the alarm signals given by a number of other ground squirrels, although there are interesting differences among the many species in this group. Alarm calling is much reduced in the more widely dispersed thirteen-lined ground squirrel and is restricted primarily to mothers with recently fledged young. In this species and the round-tailed ground squirrel, juvenile males do occasionally give alarm calls, but they do so *only* before leaving their mother's home area, that is, when there are littermates living with them [189, 643]. Adult males and juvenile males that have left their relatives behind and emigrated to a new area remain quiet when a predator approaches. This provides further support for the hypothesis that animals give costly alarm calls only when they have relatives near them to benefit from their altruism.

Additional evidence for the impact of kin selection in promoting helpful behavior in Belding's ground squirrels comes from studies by Warren Holmes and Paul Sherman, studies that show that mothers, daughters, sisters, and half-sisters can recognize one another [336]. As noted before, females in the wild appear to help their close female relatives in territorial conflicts with intruders (see Figure 15, Chapter 7). The proximate mechanism for this cooperation lies in the ability of females to identify their kin. In order to explore how this mechanism works, Warren Holmes arranged to have some pregnant females captured and shipped to his laboratory.

Table 5
Who Gives Alarm Calls among Belding's Ground Squirrels?

Category of Squirrel	Exposure to Predator[a]	Squirrels Observed to Give Alarm Call		Number of Squirrels Expected to Call If Alarms Were Given Randomly
		Number	*Percentage*	
Males one year old or older	67	12	18	19
Females one year old or older with living daughters or granddaughters, mothers, or sisters	190	75	39	53
Females one year old or older without living descendants, mothers, or sisters	168	31	18	46

[a] Number of times ground squirrels in each category were present when a terrestrial predator appeared at Sherman's study site, 1974–1977. Data courtesy of Paul Sherman.
"G" statistic = 16.57 (observed vs. expected); $P < .01$.

When two females gave birth close in time, Holmes switched some of the pups to create four classes of juveniles: (1) siblings reared apart, (2) siblings reared together, and (3) nonsiblings reared apart and (4) together. The foster pups were readily accepted by the adult females (presumably because in nature the chance of litter substitutions involving the helpless newborn pups is nil).

When the juveniles had reached the postweaning stage, pairs were placed in an arena and given a chance to interact. Animals that were reared together, whether siblings or not, generally treated each other nicely, whereas animals that had been reared apart were likely to react aggressively to one another. But the most exciting finding was that biological sisters *reared apart* engaged in a significantly lower rate of aggressive interactions than nonsiblings reared apart. In other words, sisters have some way of recognizing one another that is not dependent on the shared experience of growing up together. But brothers do not direct significantly fewer aggressive acts to their siblings (male or female) than to nonsiblings (Figure 6). Because mothers, daughters, and sisters often live in close association, they have the opportunity to provide help to one another and do so because they can make the appropriate discrminations. Under natural conditions, Belding's ground squirrel males are unlikely ever to interact with their close genetic relatives, and they lack the capacity to identify these individuals [657]. The evolution of sex-linked kin identification mechanisms provides an additional line of support for Sherman's interpretation of alarm calling by females as an example of altruism toward genetic relatives (see also [503]).

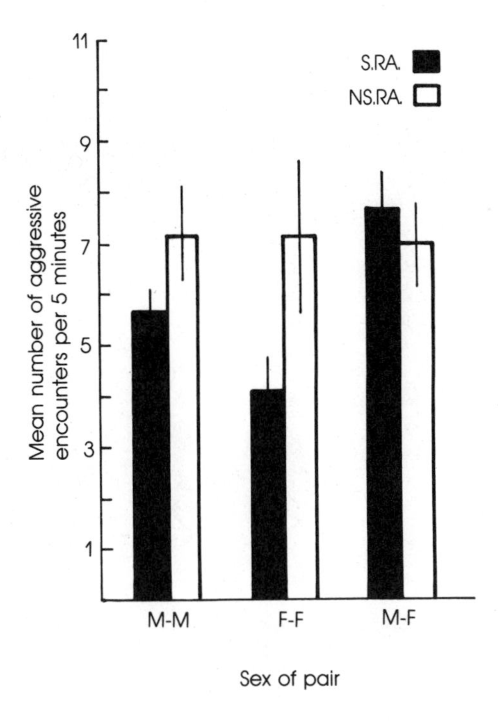

6 **Experimental test of sibling recognition** in Belding's ground squirrels. Sisters reared apart exhibit significantly less aggression to each other than do genetic strangers reared apart. All other combinations of siblings reared apart (S.RA.) are as aggressive to one another when they meet in an experimental chamber as nonsiblings reared apart (NS.RA.).

Cooperation in Mate Acquisition

Kin selection may also play a role in the evolution of behavior other than alarm calling and recognition of relatives. For example, despite the typically ruthless nature of male competition for access to mates, in some populations of wild turkeys, males actually help one another in mating displays [759]. The adult males form groups of two or three and congregate in open meadows where females come to choose a partner on the basis of male displays, which involve much wing drooping, tail spreading, and gobbling (Figure 7). Well before the first females are receptive, male groups compete with one another and in the process establish a dominance hierarchy. When the breeding season begins, females invariably choose to mate with one particular member of the dominant group. If the helper male(s) in this dominant group are consistently deprived of the opportunity to mate, why do they persist in helping?

The answer to this question requires an analysis of the composition of display groups. Robert Watts and Allen Stokes believed that the male members of a single brood (and there may be six to eight male progeny produced by a female in one clutch) remain together as a single unit throughout life [759]. Each male competes with his siblings for the position of dominant male. When this has been settled, the top male and his brothers compete with other display groups to become the dominant band at the lek. If they are successful as a group, the dominant male may mate many times,

7 **Male cooperation in mating attempts** by turkeys. Two brother–brother display groups and one single male are competing for the attention of the female feeding across the road. Photograph by C. Robert Watts.

but the subordinates do not. Nevertheless, a subordinate almost surely improves his fitness by helping his brother because by himself he would have almost no chance to reproduce. If a male cannot secure dominance over his brother, he is very unlikely to be attractive to females who prefer the very largest, most dominant males. But a subordinate can still enjoy a degree of gene-copying success by helping his brother dominate other male groups, thereby raising the reproductive potential of his dominant brother and his own inclusive fitness in the bargain.

A similar case involves the cooperation shown by male lions in defense of a pride [68, 117]. Just as in turkeys, there is little or no fighting over females among the males within a group, and yet some individuals enjoy greatly disproportionate reproductive success. In one group of six lions that owned a pride, the top male copulated 3.5 times as frequently as the lowest male in the hierarchy. In this case, however, the males were a cluster of brothers, half-brothers, and cousins that were expelled from their natal pride at the same time and then remained together to compete with other male groups for access to a harem of females. In Brian Bertram's terms, each male was "reproducing by proxy through his companions" because they shared a high proportion of the same alleles. As a result, in the six-male pride, the *genetic* success of the least sexually active male was about 67 percent that of the most successful male, despite the great differences in personal reproductive output.

Cooperation in Mate Acquisition: Individual Selection

The kin selection hypothesis cannot plausibly explain all cases of cooperation in mate acquisition. In other prides of lions, for example, it has been shown that harem masters are *not* related to one another and yet they still cooperate in defense of their females [562]. If a male does not have brothers or half-brothers to assist him, his options are to try to control a cluster of females by himself *or* to form a coalition with other males, using their combined force to repel opponents. Inasmuch as the probability that a solitary male will be able to subdue two or more cooperating rivals is vanishingly small, the second alternative is likely to be the more productive. Even if a male in a coalition of nonrelatives is low lion on the dominance totem pole and therefore gets to mate relatively little, a few copulations are better than none at all.

The same argument may be valid for some groups of cooperating male turkeys, for here, too, there is evidence that not all coalitions are composed of brothers. Moreover, there is some indication that subordinate males may be able to mate occasionally, thereby deriving a personal fitness gain from association with a more powerful nonrelative [37].

The kin selection hypothesis also seems unlikely to apply to long-tailed manakins, whose males have an even more elaborate communal display than male turkeys [239]. Two males defend an isolated display site together by giving a ringing, synchronous call over and over again. When a female arrives, they perform a variety of mutual displays, including the spectacular cartwheel display. With a female sitting at the end of the display branch, the male closest to her leaps into the air with a cry and then drops down to where the other male had been perched (Figure 8). His companion, having slipped forward, is where the first male had been prior to his jump. The birds may carry out 100 cartwheels in a row while the female watches them, presumably in amazement. By marking both members of several pairs, Mercedes Foster found that one of the males was dominant to the other and this individual secured *all* the copulations. It is unlikely that the helper male is the brother of the helpee, given the small clutch size of manakins (one or two eggs). This would require that the female produce two sons simultaneously and that they both survive for the three or four years needed to achieve adult plumage. Thus, kin selection probably is not responsible for the evolution of mutual display in manakins. Therefore, we must explain the helper's actions in terms of their contribution to his personal reproductive success, something that would seem to be hard to do given the failure of the helper to copulate.

One can speculate, however, that the lifetime mating success of the subordinate manakin is improved through his cooperation. Because the display sites of this species are separated visually, a female must visit

8 **Cooperative display by long-tailed manakins.** The two males are in the midst of the cartwheel display in which they change perches on the limb, leaping over one another to induce the female to mate.

several places to compare the behavior of males prior to selecting one as a mate. A female might prefer males that display together, as this permits her to compare more than a single male at a time. If a counterselection pressure of some sort prevents the addition of still more males at a display court, the stage is set for the evolution of mutual displays. Presumably the two birds at an arena compete for the dominant position. But the subordinate does not benefit by leaving after dominance relations have been established because (1) his probability of becoming dominant at another site would be very low and (2) his chances of mating as a solo bird would be nil. Instead, it might be to his advantage to remain. It could even be in his interest to try to make the dominant male "look good" to females in order to attract a regular clientele to the site. (Alan Lill found that, in white-bearded manakins, females do form preferences for certain display arenas and will return to them repeatedly [424].) When the dominant male dies, the original subordinate is in a position to become the dominant male of a new pair (a new male might join him for the opportunity to become the sexually active male eventually). Females that have formed an attachment to the display area might well recognize him and therefore could judge that he had survived the long apprenticeship period. This might make him a more desirable mate than a male of unknown age. Thus, cooperative behavior in long-tailed manakins could possibly be the product of individual selection [239].

Cooperation in Mate Acquisition: Reciprocal Cooperation

In lions and manakins, males do not take turns copulating; but in the olive baboon, cooperating males do reciprocate in this manner. Male baboons regularly form political alliances to dominate a high-ranking competitor who has formed a consort relationship with an estrous female [560] (Figure 9). Another closely matched male may threaten the consort male and may attempt to enlist the aid of another baboon in this task. If he is successful, the helper male will respond to his head-turning movements and the two monkeys will together drive the consort male from the female. If this happens, the male that enlisted the support of the helper always takes the female. The helper baboon runs a small risk of being attacked by the consort male, who is naturally not enthusiastic about relinquishing a fertilizable female. What does the helper gain for his risk?

9 **Two male baboons** on the right have formed a "political alliance" in a dominance contest with the male on the left. They are threatening this male by staring at him and by raising the skin above their eyes, exposing a whitish patch that is normally concealed. Photograph by Leanne Nash.

The adult males in a troop of *Papio anubis* generally transfer into the group from many other bands and are therefore unlikely to be close relatives. This reduces the likelihood that kin selection is responsible for the cooperating male's "altruism." Instead, this appears to be a genuine case of reciprocal altruism [560]. A helper is likely to receive assistance from the baboon he helped when he wishes to form an alliance to steal a female from another male. Males that participate in mutual threat displays often have favorite partners that take it in turn to help and be helped. Males that fail to give assistance when solicited are far less likely to receive aid than individuals that will help in threatening an opponent. Reciprocity can readily evolve in a baboon troop because males are able to recognize each other as individuals and because male baboons have good memories and well-developed learning abilities. The cost of the altruistic act is small because it is rare that a single male dares withstand a threatening cooperating pair. The benefit of the act for the helpful male may be great if he eventually is repaid and inseminates a female that would otherwise have mated with another male.

Helpers at the Nest

The message that comes from the studies described thus far is that there is no single evolutionary basis for alarm calling or male cooperation in mating in animal species. This conclusion is further reinforced by investigations of another major category of "helpful" behavior, assistance given by nonbreeding birds to the offspring of other individuals [212].

In the large majority of birds, parents defend by themselves an all-purpose territory in which they rear their progeny without help from other individuals. But in a substantial minority, there are helpers-at-the-nest about whom the following questions can be asked [214]:

1. Does a helper raise the reproductive success of the birds it appears to be helping?
2. What is the helper's degree of relatedness, if any, to the pair that it helps?
3. Does a helper benefit by helping? Does helping increase an individual's inclusive fitness via the kinship component or does the behavior contribute eventually to increased personal reproductive success for the helper?

Let us ask these questions of the Florida scrub jay, a population of which has been studied in remarkable detail by Glen Woolfenden [680, 803–806]. A breeding pair of jays may have as many as six other individuals living with them in their territory. These nonbreeding adults may be two or three years old, or even older; they are physiologically capable of producing offspring of their own but instead defend the breeding pair's territory, feed the nestlings, and repel predators (Figure 10). If one measures the number of offspring fledged by pairs with and without helpers, one finds an increase, albeit a modest one (Table 6), for the pair with helpers.

Interestingly a reproductive increase of almost identical magnitude has been reported for the totally unrelated gray-crowned babbler, a communal breeding bird of Australia. In carefully designed experiments, some helpers were removed from some groups of this species. This change produced a decline in the reproductive output of the deprived breeding pairs, a finding that confirms that helpers-at-the-nest really do benefit the helped parent birds [107].

By color-banding large numbers of young scrub jays and observing their behavior in succeeding years, Woolfenden showed that helper jays are the adult offspring of the pair they help. Thus, potentially their helpful behavior could be the product of kin selection because they help produce genetically similar individuals (siblings) that have a high probability of carrying the genetic basis for their altruistic behavior.

But is helping the best way for a young adult scrub jay to propagate its genes; and if so, why? If one compares the genetic success of jays that are breeding for the first time with those that are helping rear additional brothers and sisters, it would appear that breeding is by far the superior option (Table 6). A novice breeder, even without any helpers, can produce an average of one fledgling. A helper jay never contributes more than 0.60 siblings through its assistance at a nest. First-time breeders can reproduce, however, only if they secure a territory in which to nest. In order to determine the productivity of the personal reproduction option, one must therefore consider how easy it is for young jays to find an unoccupied site. Woolfenden found that his population was stable and that established pairs occupied almost all the habitat suitable for breeding. If this applies generally to Florida scrub jays, a young adult has little chance to secure a place in which to breed. If it sets out on its own, it risks being forced into a marginal environment where it may well die without ever reproducing successfully. Under these conditions, a young jay may often do better by deferring reproductive attempts until an opening is available in an area near its natal territory.

Helper jays do not sign a life-long pledge of celibacy, and they are quick to take advantage of vacancies in neighboring territories, provided they have secured a high dominance position in the hierarchy of helpers at their parents' nest [805]. A bird that succeeds in becoming a dominant assistant (subordinate only to the breeding male and female) may even inherit a portion of its parents' territory. Low-ranking jays may never have a chance to reproduce personally, but if they were to leave their parents' site they would forfeit the genetic benefits of helping produce additional brothers and sisters, some of which may breed and pass on shared genes.

Helping and Habitat Saturation

Among acorn woodpeckers there are populations in which helping is standard practice and others in which it is not [382, 679, 714]. When one compares the different populations, one finds a strong correlation between the frequency of helpers-at-the-nest and the stability of territory ownership [215]. In a Californian site, not 1 of 20 territories became vacant

during a three-year study. Here average group size was large and most pairs had helpers living with them. But in the Chiricahua Mountains of Arizona, acorn supplies are highly variable from year to year, not constant as in the Californian research area. Almost all adult pairs in the Chiricahuas dispersed from their territories after breeding and only a few pairs

Table 6
Effect of Helpers at the Nest on the Reproductive Success of Their Parents and Their Own Inclusive Fitness

	Inexperienced Pairs[a]	Experienced Pairs
Average number of fledglings produced with no helpers	1.03	1.62
Average number of fledglings produced with helpers	2.06	2.20
Average number of helpers	1.7	1.9
Increase due to help	1.03	0.58
Indirect contribution per helper	0.60	0.30

Source: Emlen [212].
[a] Inexperienced pairs are ones in which one or both members of the pair are breeding for the first time.

10 **Cooperation among scrub jay relatives.** Several helpers at the nest are assisting their parents in rearing more siblings, which they do by feeding and defending nestlings.

acquired a helper (Table 7). Thus, juvenile acorn woodpeckers, like the Florida scrub jay, tend to remain with their parents to help them rear more siblings when habitat saturation is high and the probability of finding a vacant territory low. But when openings in territory sites are numerous, young adults leave home and attempt to breed as soon as possible.

Table 7
Habitat Saturation and Altruism in Acorn Woodpeckers

Location	Percentage of Territories That Are Vacant from Year to Year	Average Group Size	Percentage of Groups with a Helper(s)
California	0	5.1	70
New Mexico	19	3.0	59
Arizona			
Huachucas	—	2.8	56
Chiricahuas	93	2.2	16

Source: Emlen [215].

Australian blue wrens provide a variation on this theme. Individuals of the superb blue wren seem to be able to assess their chances for successful personal reproduction and to use this information to determine whether to help or attempt to breed on their own [212, 620]. Adult males of these minute, but beautifully plumaged, birds outnumber females in most populations but to an extent that varies from year to year. When females are relatively abundant in a population, males disperse from their parents' territory to seek out mates and to breed. When potential mates are scarce, young males more often remain at home as nonbreeding helpers (Figure 11).

Reciprocity and Communal Nesting in Green Woodhoopoes

In all the birds discussed thus far, helpers tend to be the offspring of the individuals they help and they live in stable populations in which territorial turnover is much reduced. But these are not the only ecological conditions correlated with communal breeding [215]. In the green woodhoopoe of Kenya, flocks consist of as many as 16 birds and include a single breeding pair; some members of a flock are close relatives of the pair,

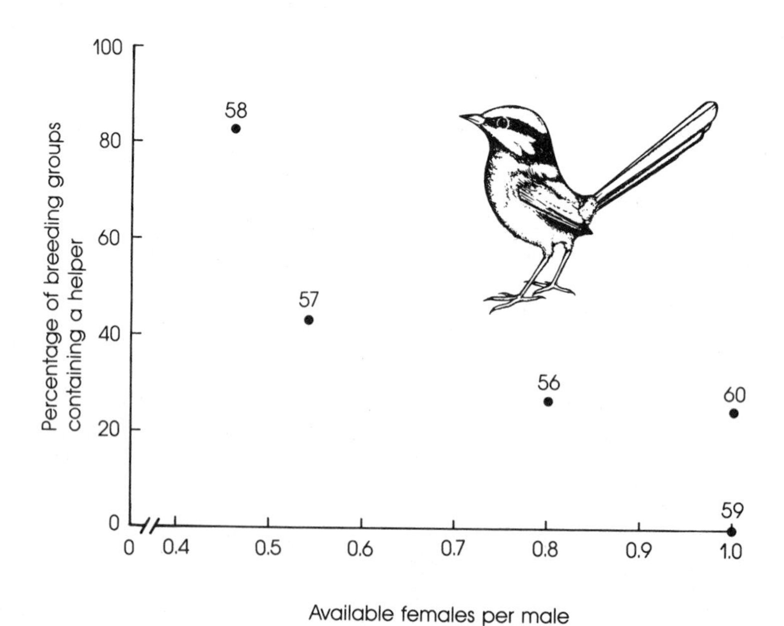

11 **Helping at the nest versus personal reproduction** in blue wrens. In the superb blue wren, males are more likely to help their parents in years in which opportunities for personal reproduction are slight because of a shortage of available females. The number above each point refers to the year (1956–60) in which the results were collected.

whereas others may be unrelated outsiders. By studying a marked population of these exotic birds for many years, David and Sarah Ligon have shown that nonrelatives are likely to be accepted into a breeding band that has been recently reduced in number [423]. Disasters strike green woodhoopoe flocks regularly. Breeding females, in particular, are often killed because they nest in holes in trees and have predators that attack at night when an incubating bird has little chance of escaping. Moreover, the insect productivity of the woodlands in which they live fluctuates wildly from year to year; at times food is so scarce that birds may die of starvation. The overall result is that 40 percent of the population dies per year, creating frequent vacancies in communally defended territories as entire flocks disappear or become too small to defend their home base.

You will recall the argument that environmental stability created conditions that favored helping at the nest in birds like the acorn woodpecker and scrub jay. The case of the green woodhoopoe obviously does not conform to the environmental stability hypothesis because this species occupies a social and physical environment of great instability. Stephen Emlen has pointed out, however, that there is a common problem facing would-be breeders in both very stable *and* extremely erratic environments [215]. At both ends of this spectrum, the probability of successful reproduction for a young bird is unusually low. In a saturated environment, the rarity of vacant territories makes it unlikely that a first-time breeder will find a suitable place in which to reproduce. In a strongly fluctuating environment, the vagaries of shifting food supplies and the high rate of mortality for incubating adults means that the probability of rearing offspring to maturity for a single pair is low. In a saturated environment, young adults often help their parents while waiting for a territorial opening. In highly unpredictable environments, young adults join their parents to form a breeding flock that is buffered to some extent against food shortages and accidents that cause the death of some members. If a parent dies, the other birds in the flock can perhaps compensate for its loss by providing the incubation, feeding, and territorial defense services that are necessary if the eggs, nestlings, or fledglings are to survive. Moreover, with the death of a breeding adult, younger helpers may ascend to breeding status with a ready-made clan of helpers to assist them.

It is the possibility of inheriting a flock of helpers that makes it attractive for some nonrelatives to attempt to join an established group. These individuals come from very large flocks, where their chances of living long enough to graduate to breeding status are low; or they may be the last survivors of a group diminished and dispersed by starvation or predation. They may be accepted into certain bands that are just below the optimal size for the reproductive success of the breeding pair. The result is a coalition of nonrelatives in which the newcomers provide help for genetic strangers in return for a small, but real, chance to reproduce in the future. Thus, long-term reciprocal arrangements favored by individual selection can contribute to the evolution of communal breeding in some circumstances as well as kin selection [215, 216]. The critical point is that if a young bird's chances of successful independent breeding are close to nil, it does not have to have a high probability of becoming a breeder in a communal flock in order to make the helper-at-the-nest option the genetically advantageous decision.

The Evolution of Eusocial Insects

Although a Martian ethologist would surely admire the complex social lives of green woodhoopoes and olive baboons, he or she would be even more impressed upon being stung by a honeybee. What is most impressive about this event is not the pain caused by the sting, but the realization that the worker bee dies when her stinger catches in the skin of an enemy [743] (Chapter 10). This action exemplifies the extreme altruism that is commonplace in honeybees and some other social insects. Colonies of termites, ants, some bees, and some wasps may contain hundreds or tens of thousands of workers that *cannot* reproduce but instead sacrifice themselves in many ways for the welfare of others in their group. One does not have to be a Martian to appreciate that the evolution of sterility and a readiness for suicidal self-sacrifice by a worker caste pose an intriguing evolutionary problem. In order to understand how these eusocial insects might have evolved, one can try to reconstruct a historical pathway that runs from a solitary life history pattern to one characterized by colonial living with a nonreproducing worker group and a few reproductively active individuals. The wasps are a good choice for a comparative historical study because there are many solitary, some moderately social, and a few highly social species. Howard Evans and Mary Jane Eberhard have used this diversity of social organization to construct an evolutionary sequence that leads to eusociality [226].

The variation in the social life of wasps is clear when one compares a wasp like *Ammophila novita* (Chapter 1) with the social paper wasps of the genus *Polistes*. The solitary *Ammophila* captures a single large prey, digs a nest, deposits the prey in the nest, lays an egg on the victim, and closes the nest as she leaves. The solitary female has no contact with her progeny and receives no assistance in her reproductive efforts from any other individual.

A totally different pattern occurs in the familiar *Polistes* paper wasps. Almost everybody living in the Americas has probably been stung by a *Polistes* at least once (and a most unpleasant experience it is) because paper wasps regularly build their nests (Figure 12) in the shelter provided by an eave on a house. Despite their intimidating stings, *Polistes* wasps are well worth getting to know bcause they are remarkable social beings [226]. Females capable of reproduction emerge from cells in the nest late in the breeding season (in a number of temperate-zone species). They mate with unrelated males, which are also being produced at this time, and then spend the winter hibernating in a sheltered spot. In the spring, they rouse themselves and start a nest, which is constructed of chewed plant fibers. The nest contains a series of cells, each of which receives a single egg from

12 **Paper wasp nest.** One of the four adult *Polistes* females on the nest is the queen; the others are helping her rear the translucent larvae in the brood cells. An egg laid by the queen can be seen in the central cell; the capped cells contain pupae that will join the work force as sterile females when they emerge. Photograph by the author.

the new queen. A foundress female may be joined by other overwintering females; this generates dominance contests and the formation of a hierarchy, with the dominant female reproducing and the subordinate(s) helping her to rear the larvae and protect the nest against predators and parasites.

The eggs that are laid early in the season are destined to become daughter wasps; the female controls the sex of her offspring by deciding whether to fertilize an egg with sperm she has stored from last fall's copulation or to lay an unfertilized egg. Young wasps reared from fertilized eggs become females; those that develop from unfertilized eggs will be males.

As the eggs hatch, the newborn larvae are fed water, nectar, and fragments of insect prey. When the first females emerge, they assist their mother in raising still more daughters (their sisters) rather than flying off to start new colonies in which to rear their own offspring. Several broods are produced in this fashion during the colony's life, with more and more females joining the work force. They increase the size of the nest, add cells, feed the larvae, detect and drive off parasites, and descend en masse on potential predators, stinging them into retreat. However, as the summer progresses in temperate regions, the queen produces increasing numbers of females that do not join the workers but instead lounge about on the nest, appropriating food from their working sisters. Later in the summer males emerge for the first time and these, too, do little to aid the welfare of the colony. Activity at the nest dramatically decreases at this time, the "lazy" females and males fly off, mating occurs with members of other colonies, the males die, and the mated females—future colony foundresses—hibernate through the winter months to resume the cycle the succeeding spring [226].

By employing the comparative method, Evans and Eberhard were able to suggest how a purely solitary wasp might over evolutionary time gradually become transformed into a species rather like *Polistes*. The sequence they suggest has seven steps.

1. Some wasps merely sting a prey and lay an egg on the victim; by this act they terminate their investment in the care of their offspring.
   Sequence: Prey, egg
2. The next stage appears to involve dragging captured prey to a crevice, laying an egg on it, and leaving both prey and egg in the protected site.
   Sequence: Prey, natural burrow, egg
3. Some species (e.g., *A. novita*) capture an insect, dig a burrow, deposit the prey in the nest, and lay an egg on it, closing the nest as they leave.
   Sequence: Prey, nest, egg, closure
4. The same as 3, except that the nest is dug prior to prey capture. This prevents parasites from exploiting the paralyzed prey while the nest is being built.
   Sequence: Nest, prey, egg, closure
5. The fifth stage involves bringing many prey to a larva in a burrow over a period of days. This permits the exploitation of relatively small prey, any one of which would be insufficiently large to rear the larva.
   Sequence: Nest, prey, egg, progressive provisioning, closure
6. The same as 5, except that the wasp builds and provisions several nests or cells at once.
   Sequence: Multiple nests, prey, egg, progressive provisioning, closure
7. The same as 6, except that the eggs are laid before the first prey are brought to the cells. This may be advantageous if parasite pressure is severe; food will not be brought to the nest until the wasp's grubs are ready to consume it immediately.
   Sequence: Multiple cells, eggs, prey, progressive provisioning, closure

System 7 is obviously close to that of *Polistes*. If some daughters are born before others and if the mother is long-lived, the stage is set for the evolution of helpful behavior by firstborn daughters, provided they can render useful services by collecting food or protecting the colony.

The Role of Kin Selection in Eusociality

The by-now familiar question is, Why should a daughter forego reproduction in order to help rear additional sisters? In *Polistes* and many other highly social insects, daughter workers may not have the option to reproduce personally should a suitable opportunity arise. They are chained to their helper roles for life, unlike the helper scrub jay. This is altogether more extreme than the reproductive cooperation exhibited by social vertebrates, although there are indications that there may be an effectively sterile worker caste in colonies of the bizarre naked mole rat, an African burrowing rodent [362].

The fact that obligatorily sterile workers are known only from certain Hymenoptera (ants, bees, and wasps) and the termites reinforces the unusual nature of the eusocial phenomenon [793]. The link between the Hymenoptera and eusociality led W. D. Hamilton to consider the genetic consequences of the peculiar mode of sex determination practiced by this group [300]. As indicated already, males are haploid, having been produced from unfertilized eggs, whereas females are diploid, with two sets of chromosomes, one from the egg and the other from the sperm. If a female hymenopteran mates with just one male, all the sperm she has received will be identical (the male having only one set of chromosomes, which he copies when making gametes). Therefore, all the daughters a mother produces will carry the same set of male genes. Any one daughter will share these genes (50 percent of her total genotype) with all her sisters. The other set of chromosomes comes from her mother. The foundress' eggs are not identical genetically because the mother is diploid; gamete formation in animals with two sets of chromosomes involves producing a cell with just one set. The statistically average egg made by a female wasp, bee, or ant will have 50 percent of the same alleles carried in her other eggs. Thus, when eggs unite with identical sperm, genotypes are created that are identical for an average of 75 percent of their alleles—50 percent of those alleles carried in common are from the father and 0 to 50 percent are from the mother (Figure 13). The probability that an allele present in one female will also be carried by her sister is 0.75; $r = 0.75$.

Because of the haplodiploid nature of sex determination in the Hymenoptera, sisters may be genetically very similar to one another, more so than a mother to her daughters and sons, with which she has only half her genotype in common ($r = 0.50$). Because sisters are so closely related in the Hymenoptera, Hamilton suggested, and many others have agreed, that kin selection is the driving force behind the evolution of the social insects [300, 793]. The high proportion of identical alleles increases the probability that a helpful act between sisters will elevate the inclusive fitness of the altruist. In support of the kin selection hypothesis, observers have noted that the workers of social Hymenoptera are sisters. Males are not as highly related to each other and have on average only 25 percent of their alleles in common with their sisters. They show no special cooperative interactions with siblings of either sex. Moreover, in one case in which *larvae* of both sexes contribute silk for nest construction, the males have smaller silk glands and provide less material than their sisters [796].

The essence of the Hamiltonian kin selection hypothesis is that female wasps are really helping their reproductively competent sisters (future queens) and only incidentally assisting their mother. That is, the genetic goal of their behavior is to increase the chances of survival of their own genes by aiding very closely related siblings. The advantages of altruism are so great in this case that it can favor females that, so to speak, put all their eggs (alleles) in a sister's basket and give up personal reproduction entirely.

That sterility can arise through kin selection is supported by the occurrence of nonreproducing soldier aphids that defend the reproducing members of their clone, which are genetically identical to them [23]. A sterile

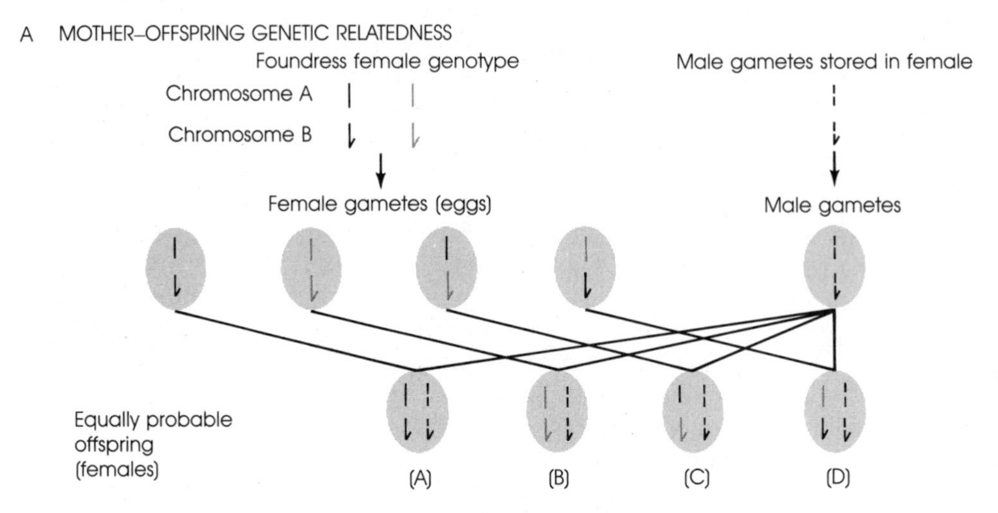

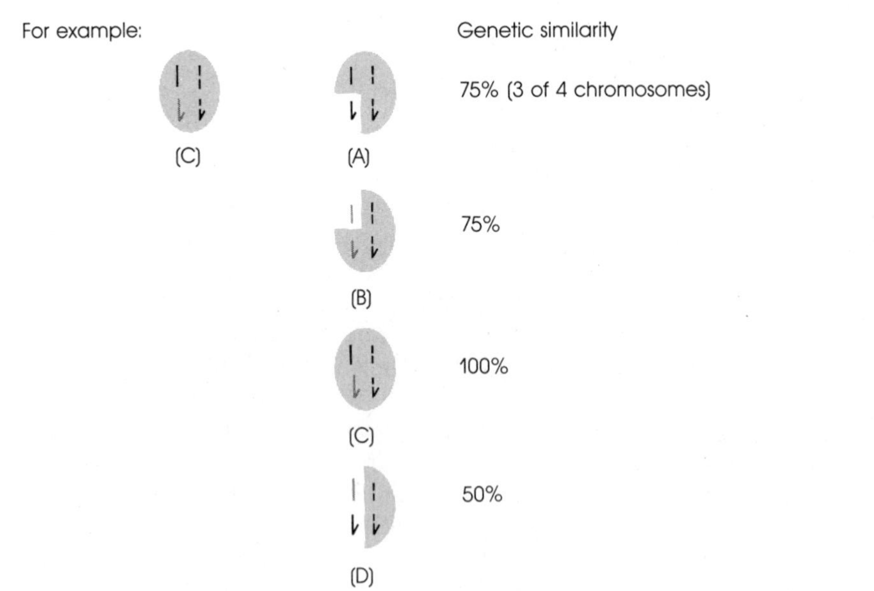

13 **Degree of genetic relatedness** of a female wasp to her offspring (A) and among sisters (B). For the sake of simplicity only two chromosomes are considered. Sisters may be more closely related to one another than they would be to their own progeny (see text).

soldier caste has also evolved in a parasitic wasp whose females lay numerous eggs within a single host. Some members of a female's brood develop into large-jawed larvae (Figure 14) that within the host attack and dispatch other parasitic competitors that endanger the siblings of the defender form. The large-jawed types never metamorphose into reproducing adults, sacrificing their reproductive chances to help their close relatives survive [151].

These cases do not prove, however, that workers in the social Hymenoptera have evolved via kin selection. To test this hypothesis, Robert Trivers and Hope Hare looked at the sex ratio of colonies of eusocial ants [719]. The queen of a colony has 50 percent of her alleles in common with both her male and female progeny; she presumably gains no genetic advantage by investing more resources in the production of females than males or vice versa. Her daughters, however, should benefit if more sisters are produced than brothers because $r = 0.75$ for sisters whereas $r = 0.25$ between sister and brother.

Thus, there is a potential conflict between the queen and the workers. (This is merely one of many potential disputes between parents and offspring of sexually reproducing species [216, 717]. These conflicts arise because although an animal and its progeny are genetically similar, they are not identical and therefore may have different ways in which to maximize the survival of the alleles in their unique genotypes.) Many offspring must be cared for and fed by their sisters. Workers could materially influence the weight and even survival of the male larvae in a colony by withholding food from their brothers and channeling it instead to their sisters. If female Hymenoptera behave in ways that benefit their alleles rather than their mother's alleles, we should expect that the weight of all the female reproductives produced by a colony should exceed the combined weight of male reproductives by 3 to 1 (sisters having 75 percent of their alleles in common with other sisters and only 25 percent with their brothers). Acquiring the

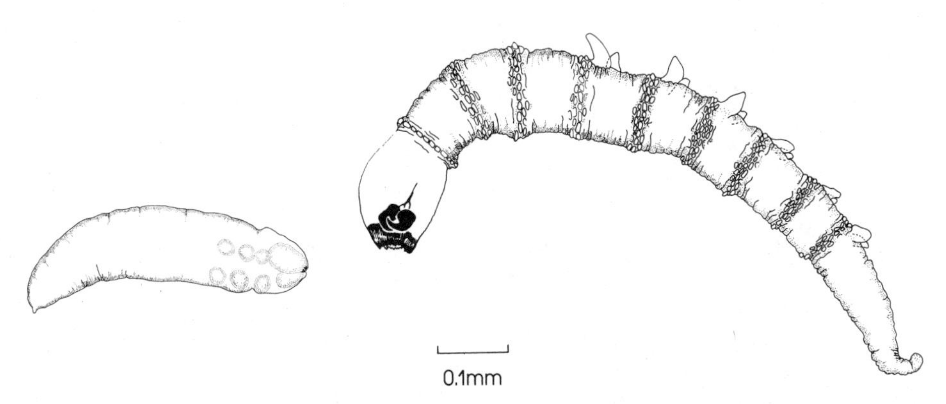

14 **Altruistic soldier larvae** of the parasitic wasp *Pentalitomastix* (*right*) have jaws, but they never mature into adults. Instead they protect their brothers and sisters (*left*), which live in the same host, by dispatching other parasites in the host.

necessary data to test this prediction is difficult, but Trivers and Hare's review indicated a strong skewing of investment toward females (but see [537]). Although some biologists feel that this test conclusively supports the kin selection hypothesis, others have argued with equal conviction that there may be alternative explanations for the skewed sex ratio data in particular [15] and for the origins of insect sociality in general [225].

Individual Selection and the Evolution of Eusociality

Few skeptics of the kin selection hypothesis as applied to social insects fail to admire Hamilton's genius for having uncovered the correspondence among the haplodiploid system of sex determination, the close genetic relatedness of sisters, and the evolution of sterile female castes in the Hymenoptera. An ingenious and plausible idea, however, may still be incorrect or at least only part of the story. Some critics feel that the reliance on kin selection as the force behind the evolution of all social behavior has become too great [225, 427]. They point to the termites, which are highly social despite the fact that males and females are diploid. They note that if a queen bee, wasp, or ant mates more than once, there will be more than one kind of male gamete available for the fertilization of her eggs. The greater the number of male mating partners, the greater the genetic diversity among sperm and the lower the average degree of genetic relatedness among sisters. Multiple mating by females is common in the social Hymenoptera; for example, a honeybee queen usually copulates with more than one dozen males on her nuptial flights [563]. The skeptics point, moreover, to the widespread occurrence of joiner females in various wasps, including *Polistes,* that become workers at an established colony, although they are probably not helping produce relatives. (In *Polistes,* however, there is some evidence that joiners may be genetically related to the foundresses they help, perhaps because sisters return, after hibernating, to nest at the site where their mother's nest was placed the previous year [509, 510].)

But joiner workers need not be related to the foundress in order to benefit from helping. As in the case of the long-tailed manakins or green woodhoopoes, what is required is that an individual have a very low probability of reproducing successfully on its own [215, 773]. For example, in areas with a high density of *Polistes* females, a single foundress nest may have little chance of producing offspring for the foundress because other females will raid the nest and take it over when she is off collecting food or nest material. Under these conditions, a female may be better off joining a nest and protecting it against conspecific usurpers in order to have a chance to inherit an established productive nest later in the season, should the foundress die [245]. Or to take another example, a female that happens to emerge from hibernation later than most may have very low reproductive potential because of the reduced opportunity to find an unoccupied prime nesting site [773]. If this is true, a late-emerging female need have only a small chance of reproducing if she joins a successful nest to make the joining option selectively advantageous. Marcia Litte has shown that in the social wasp *Mischocyttarus mexicanus* a dominance hierarchy is formed at the nest among the foundress and joiners, as in *Polistes*. When the dominant female (the queen) is experimentally removed, the next highest ranking individual invariably takes over and becomes reproductively active [434].

But how do proponents of individual selection account for the daughter workers that slave away for the colony after they emerge rather than establishing their own nest? They, like Trivers, envision a conflict of interest between parents and offspring; but instead of the offspring winning, they see the worker caste as evidence that the queen has won [11, 15]. If a queen mother carries genes that predispose her to generate some sterile progeny that do not reproduce but instead help her make reproductively competent individuals, the queen's fitness may be elevated. There are two general strategies a queen could follow to achieve this end. One would be to produce offspring that, regardless of their genes, were coerced to behave in ways that benefited the queen's genes [11, 88]. For example, the queens of some social species lay eggs for the express purpose of feeding them to other progeny (Figure 15). The trophic eggs contain genotypes whose chances for personal reproduction are sacrificed entirely when they are eaten by a fellow colony member, but this might be a way for the queen to raise her own genetic success. For adherents of the parental manipulation hypothesis, workers are merely grown-up trophic eggs that may be forced to do things that will lower their inclusive fitness but increase that of their mother [773].

Parental manipulation of offspring need not take so dramatic a form. Instead, a mother hymenopteran might produce some small progeny who, in order to maximize their inclusive fitness, would have to do just those things that also raised their mother's fitness [773]. The parent would be taking advantage of her offspring's genotypic capacity to make the best of a bad situation. When the mother directly feeds her offspring (as she would have done in the early stages of the evolution of insect colonies), she can control the diet and size of her progeny. The animal's genotype might have the capacity to direct the individual to reproduce personally only if it were large and, therefore, likely to compete successfully for nest sites and other resources essential for reproduction. If the animal were small, its genotype might instruct it to remain at the natal colony and attempt to rear large, reproductively successful sisters. In this way its inclusive fitness might be positive, whereas if it were to try to reproduce itself it might never succeed. If the helper strategy benefits the queen, she could make it genetically profitable for some of her (small) offspring to give up reproduction and help her by regulating the allocation of food to her brood. In this regard, it is suggestive that workers in social insects are usually considerably smaller than the queen.

The distinction between the kin selection and the individual selection (or parental manipulation) hypotheses is subtle, and the definitive test of the alternatives has yet to be performed. This has not prevented, and may even have encouraged, the strong feelings and statements by adherents of opposing viewpoints. It will be surprising if the argument can be resolved to the complete satisfaction of one camp or the other because it is improbable that there was a single pathway to insect sociality. Complex societies are believed to have arisen independently at least eleven times in the Hymenoptera alone. Instead, there may be many different blends of individual and kin selection working in concert to produce and maintain the remarkable behavioral adaptations of the members of these and other animal societies.

15 **Trophic eggs of the bulldog ant.** A worker ant with a trophic egg that she has just laid. Larvae await a feeding on her left (*top left*). Workers feeding trophic eggs to the offspring of their mother. Queens sometimes produce trophic eggs that are distributed to workers for use in feeding their progeny (*top right*). A queen accepting a trophic egg from a worker. She may consume it or feed it to a larva (*bottom*). Photographs by Jenny Barnett.

Summary

1 In animal societies, individuals tolerate the close presence of other conspecifics despite the increased competition for limited resources and the heightened risk of disease and intraspecific exploitation that this entails. Under some ecological circumstances, the advantages of sociality (usually improved defense against predators and sometimes improved foraging efficiency) are great enough to outweigh the costs of social living. If ecological conditions affecting a species are variable, its members may be able to adjust their degree of sociality to changing conditions.

2 A central problem for evolutionary biology is to explain the evolution of helpful behavior in which an individual sacrifices a portion of its personal fitness to elevate the reproductive success of another. This behavior is far from rare in animal societies and includes the alarm calls of some birds and mammals, assistance given by one male that helps another acquire a mate, help given to rear offspring that do not belong to the helper, and, most amazing of all, the complete rejection of reproduction in favor of helping other individuals reproduce.

3 Several alternative hypotheses apply to the evolution of helpful behaviors. Some cooperative acts may immediately elevate the personal reproductive success of the cooperator. Other actions may require that costly help be provided now in order to receive reciprocal assistance later that will advance the reproductive chances of the helper. Both these categories of cooperation can, therefore, be the product of individual (natural) selection. If a helper really does *permanently* reduce its personal reproduction by helping, it may nevertheless raise its inclusive fitness (total genetic propagation) if its altruism is directed to its relatives. Costly helpful acts that increase the fitness of a relative can be the product of kin selection.

4 To discriminate between the alternative hypotheses for a given case, some kinds of information are especially useful—in particular, the degree of genetic relatedness between the helper and the individual(s) it helps and the impact of the helpful act on the reproductive success of the interacting animals. Even if this information is available, however, the relative importance of kin selection and individual selection in the evolution of a trait may not be entirely clear, as illustrated by the continuing debate on the evolutionary basis for the sterile castes of some social insects.

Suggested Reading

An especially large number of significant, readable articles and books on social behavior have been written recently. Edward O. Wilson's *Sociobiology* [794] is a masterful compendium of material on all aspects of the evolution of social behavior and societies. It is useful as a general reference and yet can be read with enjoyment as well. I like David Barash's *Sociobiology and Behavior* [43] and Richard Dawkins's *The Selfish Gene* [168] as shorter distillations of the key concepts in *Sociobiology*. J. F. Wittenberger's *Animal Social Behavior* is also recommended [801].

Key theoretical articles on the evolution of sociality include W. D. Hamilton's work [299, 300], which helped initiate the revolution in animal behavior, and more recent reviews by Richard Alexander [11], Mary Jane West Eberhard [773], Stephen Emlen [215, 216], and Christopher Starr [681].

Especially good descriptions and analyses of the social behavior of different groups of animals are provided by E. O. Wilson [793] (on Hymenoptera and termites), by George Schaller [635] and Brian Bertram [68] (on lions), by Paul Sherman [655–658] (on ground squirrels), by John Hoogland [337–338] (on prairie dogs), and by Mart Gross and Anne MacMillan [289] (on sunfish).

Suggested Films

Baboon Behavior. Color, 31 minutes. A film that illustrates baboon social life.

The Biology of Polistes. Color, 17 minutes. The social behavior of a species of paper wasp is featured, with sections on the various kinds of social interactions that occur at the nest.

Castles of Clay. Color, 54 minutes. A beautiful film on certain termites of Africa with footage on the dwarf mongoose, a social mammal.

The Wild Dogs of Africa. Color, 51 minutes. On the social life of a pack of wild dogs, with dramatic evidence of the conflict of reproductive interests that can exist within social units.

Life Cycle of the Honeybee (06005). Color, 12 minutes. A sketch of honeybee social organization.

The Social Cat. Color, 25 minutes. Lion social behavior is described with reference to apparently altruistic behavior of pride members.

The Termite Colony. Color, 30 minutes. This film shows some aspects of the complex social structure of various species of termites.

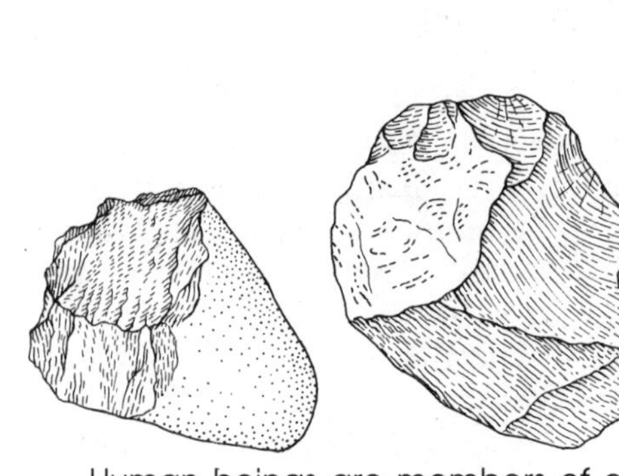
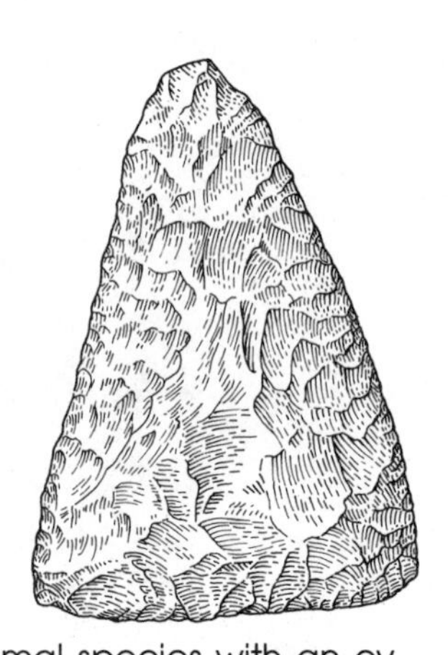

Human beings are members of an animal species with an evolutionary history. It is true that we are an unusual species, but so too is an amoeba or a flea or a kittiwake, each of which has unique and wonderful adaptations. An evolutionary approach predicts that our behavior, like that of an amoeba and a kittiwake, will tend to maximize the inclusive fitness of individual human beings. This is a bold and controversial hypothesis that has generated vigorous debate in recent years. This chapter briefly reviews the controversy and considers why it is that social scientists, among others, are reluctant to accept the possibility that our behavior fosters gene-copying success. At the heart of the matter is the problem of the relation between genes and culturally transmitted behavioral practices. Evolutionary theory requires that our genes constrain the development of our brains in ways that "encourage" us to adopt only those cultural practices that will tend to raise our inclusive fitness. The challenge for both proponents and opponents of this approach is to test whether there is a genetically sensible correlation between ecological and behavioral variation in human societies. The first step in this process is to develop hypotheses about human behavioral attributes that are consistent with individual (or kin) selection. Here I shall present a sample of these hypotheses, some more speculative than others, some better tested than others. None of these ideas has been exhaustively tested against a full range of alternative hypotheses. My goal is simply to demonstrate as a conceptual exercise that it is possible to formulate plausible evolutionary hypotheses and testable predictions about a variety of elements of human behavior, including a number that appear superficially to be genetically maladaptive. I do not present these examples as the Final Truth on the matter; but I do want to suggest how an evolutionary approach might eventually contribute to a better, fuller understanding of ourselves.

CHAPTER 15

An Evolutionary Approach to Human Behavior

The Sociobiology Controversy

Although any number of biologists had already made the case for an evolutionary analysis of social behavior, E. O. Wilson's *Sociobiology* [794] brought this approach to the attention of a broad audience. Because the book was carefully written and exhaustively documented, its section on human behavior could not be dismissed as mere biological popularizing, as some earlier books had been. And unlike many of these earlier books, which were founded on group selectionist thinking, *Sociobiology* showed that the social behavior of animals, from colonial coelenterates to honeybees and elephants, appears to promote the survival of the genes within individuals. Wilson cautiously argued that human behavior has an evolutionary foundation and could be interpreted in a similar manner.

In response to the publication of *Sociobiology,* a number of scientists and laymen in the Boston area belonging to a group called Science for the People launched a widely publicized attack on Wilson and the sociobiological approach to human behavior. In a letter to the *New York Review of Books,* Wilson's ideas were labeled irresponsible, racist, and genocidal [17]. Prominent among the SFTP group were two Harvard colleagues of Wilson, Richard Lewontin and Stephen Jay Gould [18, 272, 274], a fact that heightened public interest in the attack and resulted in a burst of news articles on sociobiology [418, 484, 750].

Lewontin and his companions were apparently motivated by a concern that *Sociobiology* would provide what they considered unjustified scientific support for immoral social policies. They felt that *Sociobiology* would be read as saying that human behavior is both genetically determined and biologically adaptive and therefore *cannot* and *should not* be changed. Noting that racist and fascist demagogues have misused biological theories in the past to promote evil political programs, they were eager to prevent the sociobiological perspective from gaining public acceptance, fearing its social implications [18].

One can sympathize with this concern (although the biological hypotheses promoted by the SFTP group are also potentially liable to political misuse). However, the opponents based their criticisms of *Sociobiology* on two fundamental errors of interpretation [795]. They argued that if one claims that human behavior is adaptive for an individual's genes, then one is also claiming that the behavior must be morally or socially desirable. But evolutionary theory provides a tool for understanding the evolution of the traits of an organism. Sociobiology is a discipline that attempts to explain why traits exist, *not to justify them*. This distinction is easily understood in other fields of biology. For example, biologists who have studied the bacterium that causes pneumonia have found that this organism possesses a number of adaptive traits that result in the infection and illness of people. As David Barash has pointed out, these researchers have not been accused of approving of the bacterium or the disease [43]. Everyone recognizes that they are simply seeking to understand the organism. Therefore to say that something is biologically or genetically adaptive means only that the something tends to elevate the inclusive fitness of individuals—nothing more.

Second, an evolutionary hypothesis does not require that behavior be rigidly determined by one's genes, only that there be a correlation between genotype and behavioral abilities [10, 12, 13] (Chapter 2). It is not biological determinism to hypothesize that there is a connection between genetic information and the development of specific traits. Indeed, by increasing our understanding of how specific alleles predispose some humans to certain illnesses, it has been possible to devise environmental prescriptions (e.g., a controlled diet) that help the individual avoid development of the trait altogether. There is behavior because genes have relinquished direct control of their destinies and have taken up residence in animalian survival machines whose reproductive success and ability to help relatives determine genetic duplication and propagation. In multicelled animals such as ourselves, behavioral control is the responsibility of the nervous system [168]. This system's properties are influenced and constrained by the developmental process, which is genetically regulated. But once constructed, our brains are partly independent of continuing genetic manipulation. As a result, our central nervous systems even have abilities that have enabled us to discover that in evolutionary terms we exist solely to propagate the genes within us. Because our genes have to operate by proxy through our brains, we may decide to thumb our noses at what our genes ultimately "want" us to do and do the exact opposite. Although we are an extreme case, the partial independence of the brain from genes, of phenotype from genotype, is true for all organisms and recognized as such by modern evolutionary biologists.

Genes, Culture, and Behavior

The fact that sociobiologists understand that animals do not have within them a gene or genes ordering the individual to be automatically aggressive or to copulate once a day no matter what does not satisfy many social scientists. These students of human behavior often argue that there is not even an indirect connection between human genes and human behavior. In my few attempts to present to social anthropologists the idea that our genes may influence our behavior through regulation of the development of our nervous systems, I have been greeted with reactions ranging from incredulity to indignation (vehemently expressed). This reaction is perfectly captured in *The Use and Abuse of Biology* [629], an outraged critique of evolutionary theory, especially as applied to human behavior, by the distinguished anthropologist Marshall Sahlins.

For Sahlins and many others, two aspects of human behavior are taken as convincing evidence that human behavior has not evolved to promote the survival of the genes of individual human beings. First, the critics note that if you were to ask any human why he did something, the person would never respond that he performed the action to elevate his inclusive fitness. But if a baby cuckoo could talk, it too surely would not tell an interrogator that it rolled its host's eggs out of the nest "because my genes made me do it to raise my inclusive fitness." No animal need possess a conscious awareness of the ultimate reasons for its activities. It is enough that its proximate mechanisms predispose it to behave in ways that help its genes survive. On the proximate level we enjoy eating sweet foods, we fall in love, we want to be liked by our friends and relatives, and we learn a language, all because we possess physiological mechanisms that make it easy for us to do these things. We do not have to be aware of the caloric content of honey and its contribution to cell growth and maintenance in order to eat it. It is sufficient that our brains tell us that honey tastes good [43]. Thus, evolutionary theory does not require that an animal know the genetic consequences of its acts, only that animals with certain abilities or preferences will tend to replace those with other aptitudes.

The second objection is much more challenging. Social anthropologists have documented that human cultures are amazingly diverse. To them this suggests that we are the completely flexible product of whatever social environment we choose to invent. How can it be said that humans tend to behave in ways that elevate their fitness when humans around the world behave in such different ways? There are polyandrous, polygynous, and monogamous societies; cultures in which females make important political decisions and others in which males dominate the females. There are human groups for which warfare is a constant fact of life and other groups that never fight openly with one another. The list of cultural peculiarities is seemingly endless; depending on where you were born, you might be allowed to marry your first cousin or not, you might be allowed to look at your mother-in-law or not, you might be forced to have your penis cut from stem to stern at adolescence or not. Moreover, cultures are not only enormously variable, but they have the capacity to change rapidly, as seen in the transformation of human societies over the past 10,000 years from predominantly hunter–gatherer groups to the industrial, high-density, high-technology civilizations of today.

Some authors have interpreted this diversity to mean that many aspects of the differences among cultures are simply the incidental by-products of the evolution of a large complex brain [274]. According to this hypothesis, once our brain reached a certain size and we evolved the capacity for language (by far our most remarkable evolutionary "achievement"), humans became capable of creating an almost infinite variety of social environments and cultural practices. Advocates of this view argue that to search for adaptation in the fondness of some humans for a Bach cantata or for the Boston Red Sox or for Buddhist religious beliefs is futile because such characteristics are simply random artifacts of our intellect.

But where does one draw the line with this argument? An interest in good food, economic achievement, and sex are all encouraged in a host of ways by most (or all?) cultures. Shall we agree that these are potentially adaptive aspects of culture but leave the hard cases without testing to some other nonadaptationist explanation? Is it not fairer, more challenging, and more consistent to accept as a *working hypothesis* the position that human behavioral traits, no matter how diverse or flexible, rest on evolved abilities that have been naturally selected because of their contribution to the genetic success of individuals in the past? This does not require us to argue that to be attached to the Boston Red Sox is to advance one's inclusive fitness. But it does demand that we accept the possibility that the ability to identify emotionally with a sporting team (something that billions of humans do) has an evolutionary, adaptive foundation. Perhaps an emotional capacity for group identification has helped cement cooperative alliances among humans in the past (and present) for economic advancement, defense of land, or other mutually advantageous endeavors—particularly in opposition to other groups. This is a hypothesis only, but at least it is the basis for the creation of testable predictions.

The point is that we are an animal species. We do have within us genes. Our brains require genetic information for their development. Our genes (and brains) should therefore promote our inclusive fitness. Opponents of sociobiology could demolish not only sociobiology as applied to human beings but evolutionary theory in general if they could show that any basic human ability consistently reduces the inclusive fitness of individuals [13]! Darwin recognized this general point, writing: "If it could be proved that any part of the structure of any one species had been formed for the exclusive good of another species, it would annihilate my theory, for such could not have been produced through natural selection" [160]. In modern terms, Darwin's challenge could be rephrased to read that current evolutionary theory would be annihilated if it could be shown that a trait originated exclusively for the genetic benefit of individuals that did not share genes with the practitioners of the trait. There are certainly some possible candidates for genetically altruistic human traits that by evolutionary terms should not have evolved if they really do result in lowered inclusive fitness: risking death in warfare, the use of birth control measures, and the practice of infanticide, for example. I hope, however, to show how these and other similar phenomena are all explicable by an evolutionary approach and that we can, at least for the moment, retain our confidence in evolutionary theory as applied to animals in general and humans in particular. But first let us consider why it is that cultural diversity may have evolved.

The Evolution of Cultures

Earlier chapters in this book presented evidence in support of two principles that are essential for an understanding of the possible evolutionary bases of human cultures. We begin with the realization that cultures are the product of neural machinery within the human skull. Cultural practices are learned behaviors, transmitted from one generation to the next through tradition (teaching). One basic biological principle relevant to an understanding of cultures is that all learning abilities have a genetic foundation of some sort and, therefore, can and do evolve. I will not repeat the key arguments here (see Chapter 4), but simply note that in order for cultural practices to be acquired by each new generation, young humans must have brains that are capable of storing the information contained in tradition. We can predict that to the extent culturally acquired information elevates individual fitness there will be selection in favor of humans who are able to absorb this information more completely, rapidly, or with less expenditure of effort than others. Conversely, if a cultural practice reduces inclusive fitness, individuals who are *unable* to learn it should enjoy a selective advantage.

Human brains appear to have properties that facilitate certain kinds of learning. The classic example is language learning. There is considerable evidence that the brain contains neural mechanisms that facilitate language acquisition and use (see Chapter 3). Simply because there is cultural variation in the languages spoken around the world is not proof of the independence of this behavior from genetic influence. Genes are deeply involved in the production of the neural foundation of language, and it seems inescapable that alleles have been selected on the basis of how well "their" brains have been able to contribute to language use. If selection can produce humans with great skill at learning a language, it should also be able to shape the design of the human brain so that humans can readily learn which individuals are relatives, how to use tools, how to behave in certain social settings, how to extract economic gain from the environment, and so on [13]. As in the case of language, the successful genes need not promote the development of neural mechanisms that are restrictively designed to permit only the learning of a specific cultural trait (such as the Turkestan or Jivaro or English language). Instead, these genes may be involved in more open developmental systems (Chapter 4), permitting a flexible (but still not completely plastic) accommodation to whatever cultural variables dominate an individual's social environment. By this view, cultural evolution involves selection for various learning abilities that enable individuals to adopt the adaptive cultural practices of their societies. Thus, genetic evolution and cultural evolution complement rather than compete with one another [13, 191].

One of the obvious advantages of cultural adaptation is the rapidity with which a cultural adjustment to ecological variation can be made. If an evolutionary perspective is correct, our brains should have evolved properties that make it easier to accept novel cultural practices that raise individual inclusive fitness while rejecting those that would reduce the genetic success of individuals. That cultural change is *not* random or arbitrary is indicated by the spread of tool use and the perfection of tool manufacture during the course of human evolution. These traditional practices are clearly related to improved resource acquisition and the ability to produce and support more offspring. Moreover, two major cultural innovations of the past 10,000 years—the invention of agriculture and the development of the use of fossil fuels as an energy source—have (1) spread extremely rapidly around the world and (2) been associated with a phenomenal increase in the human population. Both inventions, in effect, have given participating individuals access to more calories and nutrients for conversion into offspring than nonparticipating individuals.

Do Cultures Change Adaptively?

But not all culturally accepted changes appear so plainly adaptive. Many "primitive" societies, once exposed to Western culture, have rapidly dropped their customs in favor of Western clothing and foods. To some anthropologists, these changes appear to be the arbitrary result of the weaker society slavishly (and often maladaptively) adopting the superficial trappings of the more dominant group. But a recent comparative study of the degree to which four Brazilian tribes have taken to Western goods and crop-growing practices suggests that people adjust their cultures in ways that make economic sense to individuals [287]. Some tribes now live in areas where the soil is poor and the forests depleted of wild foods. These groups have made the greater investment of time and energy in attempting to grow foods, like rice, for exchange with goods from Western markets. The harder it is to make a living with traditional agricultural techniques, the more likely a tribal group is to switch to an alternative mode of producing food in order to maintain themselves and their families.

Let's examine another example of a cultural practice that at first glance seems to be maladaptive: the practice of cooking corn in a 5 percent solution of lime before making tortillas from the cooked mash. Once invented, this tradition spread to every South and Central American culture that relies on maize as the basic foodstuff. Alkali cooking of corn is a time-consuming procedure and leads to the loss of some of the thiamine, niacin, nitrogen, and fat content of the corn. Although it would be easy to interpret this method of corn preparation as maladaptive, S. H. Katz and his colleagues have shown that alkali cooking actually improves the nutritional quality of cornmeal (Table 1). It does so by breaking down the indigestible portions of the corn kernel and thereby freeing for human use substantial quantities of certain essential amino acids, including lysine, which is deficient in those parts of the corn seed that can be digested without this system of preparation [371]. A diet that is dependent on cornmeal as a staple would be dangerous without alkali cooking; we lack the metabolic pathways to manufacture lysine, an amino acid that appears in many vital proteins.

Another cultural trait that has baffled outside observers is the sacred-cow tradition, which forbids or greatly restricts the sale or slaughter of cows for meat. In India the Hindu reverence for cows is such that they are

Table 1
Effect of Alkali Cooking on the Availability of Certain Essential Amino Acids

	Milligrams of Amino Acid per Gram of	
Amino Acid	*Uncooked Corn*	*Alkali-Treated Corn*[a]
Histidine	0.012	0.028
Isoleucine	0.074	0.158
Leucine	0.309	0.358
Lysine	0.045	0.126
Threonine	0.145	0.381
Ratio: leucine/isoleucine	4.2	2.3

Source: Katz, Hediger, and Valleroy [731].
[a] Corn subjected to enzymatic digestion for 12 hours.

allowed to roam freely, interfering with traffic, and that they are almost never killed, even in times of famine. But Marvin Harris has argued persuasively that far from being a liability, cow love is almost certainly to the economic benefit of the Indian peasants who maintain the tradition [310]. Emaciated or not, cows provide their owners with a certain amount of milk and with quantities of dung, which is used for fertilizer, cooking, and construction; most importantly, they occasionally produce offspring, the males of which can be converted into oxen. Oxen are the Indian farmer's tractor, and without them he faces economic doom. This is powerful incentive not to dispose of your cow, even if she is a decrepit beast. The sacred-cow tradition is a proximate mechanism that encourages individuals to do those things that are economically and biologically adaptive.

If this argument is correct, we should expect to find that other groups that rely on cattle for their subsistence have a similar attitude toward cows. The cattle-owning Karamojong of Uganda and Masai of Kenya have exactly the same view [194]. The tribesmen try to accumulate large numbers of cattle, which provide their owners with milk and blood (Figure 1), important elements of the owner's diet. Despite pressure in recent times to sell portions of the herds to prevent overgrazing, the Karamojong and Masai are extremely reluctant to do so, in part because cattle exchanges are currency for forming protective alliances with other men and for the purchase of wives. The herdsmen feel a strong affinity for the cattle they own and absolutely refuse to trade or kill female animals, which may provide them with additional herd members. Thus, although Western observers of Indian and African cultures may feel that these peoples' attitudes are irrational, from the perspective of individual cattle owners their practices are entirely sensible. Examples of this sort persuade me that proponents of the hypothesis that human cultures are filled with irrational, random, or maladaptive practices must do more than merely assert that this is true. People seem to have an uncanny ability to employ tradition that will raise rather than lower their genetic success.

1 **Group of Masai men** preparing to open a vein in the neck of an ox to collect some blood for food, an important souce of protein for several cattle-herding tribes. Photograph by Anthro-Photo.

The Evolution of Human Warfare

The argument that cultural innovations spread and will be adopted to the extent that they contribute to individual inclusive fitness can be further tested by an analysis of one of the most dramatic of all human cultural activities, violent group aggression. There are great differences among cultures in the use of war, ranging from the essentially pacifist societies of some Eskimos and Pygmies to some extremely combative New Guinean tribes and the hyperaggressive societies of the United States and Russia (Figure 2). This variation has been interpreted by some as evidence that war is largely a capricious feature of certain cultural environments and could easily be eradicated [519]. Certainly the evidence indicates that the genotypes of humans do not blindly program them to engage in warfare. But this does not mean that our genes have not endowed our nervous systems with special properties that provide humans with the *capacity* to engage in group aggression under *certain conditions*. From an evolutionary perspective, it is possible that some ecological circumstances may so elevate the benefits of warfare to individuals that they exceed the costs to individual inclusive fitness. An argument that the abilities that underlie the capacity for warfare are genetically advantageous does not imply either that warfare is inevitable or that it is morally

2 **Men at war.** Two groups of warriors are fighting in a valley in New Guinea (*top*). Photograph by Robert Gardiner, Peabody Museum. A patrol sweep in Vietnam by members of the First Cavalry Division (*bottom*). This U.S. Army photograph is entitled "Mission—Search and Destroy the Vietcong." Photograph by SFC Jack H. Yamaguchi.

desirable. My personal belief, for what it is worth, is that wars and "defense" establishments are morally indefensible and socially disastrous. This does not prevent me from accepting as a hypothesis that the traits that may be employed in group aggression can help maximize the inclusive fitness of the humans that possess these traits.

An evolutionary approach suggests that humans will engage in warfare to the extent that they can defend or gain valuable resources from cooperative aggression [192]. An essentially similar hypothesis has proved useful for analyzing the behavior of animals other than humans. For example, one of the reasons why scrub jays and green woodhoopoes form communal breeding flocks is to defend a breeding site against attacking strangers [423, 805]. In these cases, groups can better maintain or claim a valuable territory than pairs of birds. Other examples, and particularly relevant ones, are provided by lions and hyenas. Within these species there is variation in the degree to which individuals engage in "warfare" (here defined as communal defense of a foraging territory against communal attack by another group). Prides of lions and clans of hyenas only form in areas that contain unusually dense populations of big game. Prides and clans defend these rich foraging locations, but otherwise lions and hyenas tend to be solitary and to avoid aggression with others of their species. This illustrates the basic evolutionary principle that the benefits of group aggression must outweigh its obvious costs for individuals if the capacity for group territoriality is to evolve. Furthermore, it supports the argument that humans are not the only animals that have the flexibility to adjust their aggressive behavior in a potentially adaptive fashion.

But is it legitimate to compare our behavior with that of lions or green woodhoopoes? A common complaint is that sociobiologists make capricious use of comparative material, selecting whatever species help support the hypotheses they favor [758]. There are, however, two kinds of disciplined, rational comparisons we can make among animal species to gain insight into the historical and ecological bases of a behavioral characteristic of interest. First, we can look at closely related species to determine which of their behaviors are similar to our own. This information can help us determine the history of a behavior pattern. Second, we can examine unrelated species that experience similar ecological pressures to determine if they have evolved convergent behavioral solutions to these shared problems (Chapter 7).

If we were to select animals most likely to have evolved adaptations convergent with our own, we would select the social carnivores. A great deal of evidence from anthropology and paleontology indicates that humans have for hundreds of thousands of years (1) lived in bands and (2) hunted large animals [123]. The increase in tool quality and diversity over the past one million years of human evolution (Figure 3) is correlated with an increasing reliance on big game as a major food item in the diet of our ancestors. Campsites that are 300,000 years old contain great quantities of the bones of the largest mammals then alive: mastodons, rhinoceroses, and cave bears. Big-game hunting may have been the central adaptation of most human populations from that time until the recent past. Paul Martin has advocated the hypothesis that humans were such successful hunters that they may have been responsible for the mass extinctions of big-game species that occurred first in Africa and then later in North America [472,

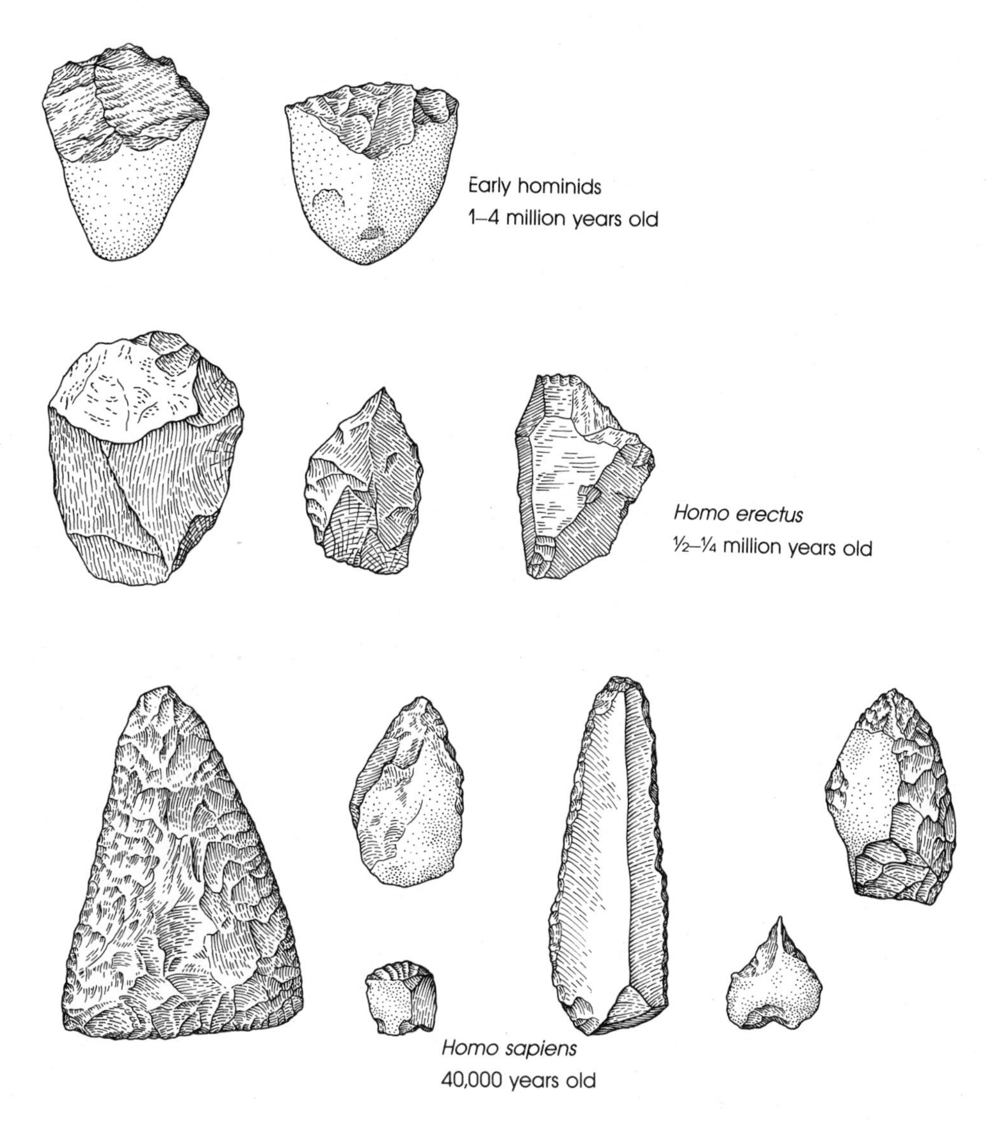

3 **The evolution of tools.** The earliest man-made tools were simple, crude, pebble choppers. The tools constructed by more recent hominids have increased in diversity and quality over evolutionary time.

473]. Although his position is controversial and is contradicted by some evidence [277], there is little doubt that humans have been remarkably effective social predators throughout much of their evolution.

Therefore, it is accurate to characterize our ancestors as having adopted the life of a social carnivore to a large degree. As predicted, we are in many respects more similar behaviorally to lions, hunting dogs, and hyenas than we are to our closest primate relatives, the great apes (Table 2). Male chimpanzees (the species to which we are most closely related [813]) do cooperate in attacking outsiders in their territory, primarily to defend sexual access to females whose home ranges lie within the male's area [264, 725]. But the degree of coordination in the aggression of chimps is much less than that exhibited in a clash of hyena clans or human bands. Both the social carnivores and hominids have an unusually great capacity for group cooperation, an obviously adaptive skill when hunting large and dangerous game on an African plain. In addition, both the social carnivores and our ancestors developed powerful weapons that help individuals dispatch large prey safely and quickly. Cooperation and weapons can also be used aggressively against members of the same species (Figure 2). But because attacking conspecifics is a risky business, we can predict that humans, as well as lions and green woodhoopoes, will engage in group conflict only when something of great fitness value is to be gained by the winners, thereby compensating for the costs of the attack [192].

Testing the Ecological Hypothesis

If an evolutionary cost–benefit approach is applicable to human warfare, we should be able to detect a pattern in the use of group aggression through a comparison of cultures that vary in their readiness to fight with other groups. One such comparison contrasts the highly aggressive coastal tribes of northwestern North America with their much more peaceful neighbors to the interior. The tribal territories of the coast produced an abundance of food (fish, shellfish, and marine mammals including whales), and this food occurred reliably in relatively small strips

Table 2
Comparison of Major Human Traits with Behavior of the Chimpanzee and the Social Carnivores (Lion, Hunting Dog, Wolf, and Hyena)

Human Behavioral Adaptations	Chimpanzee	Social Carnivores
Bipedal locomotion	Rare	No
Tool making and using	Rudimentary	No
Degree of intelligence	High	Moderate
Prolonged maternal care and family formation	Yes	Moderate
Sexual division of labor (beyond child rearing)	No	Rudimentary
Prolonged pair bond between mates	No	Yes
Highly structured social organization	No	Yes
Cooperation in hunting and food sharing	Rare	Yes
Group territoriality under some circumstances	Yes	Yes
Intense aggression toward conspecifics and other competitors	Rare	Yes

of coastline and adjacent ocean. Thus, these locations were both worth defending and sufficiently restricted in size to be energetically defendable. Inland areas lacked the concentrated productivity of the coastal zone, and warfare was correspondingly much reduced among groups living in this environment [192].

The western Shoshone and Paiute Indians offer another instructive comparison. Both tribes lived in the Great Basin area of western North America, but they differed strongly in social organization and in a readiness to employ aggression in defense of food resources. Both groups relied heavily on grass seeds as a food staple during summer. But the Paiute lived along watercourses where the seed-producing grasses grew predictably and in abundance. These Indians formed small villages and cooperated in the defense of a territory along a stretch of river. The Shoshone, in contrast, lived in dry short-grass prairies in regions without permanent streams and in an environment where rainfall was scarce and patchy. As a result, the seed-producing plants that they searched for were scattered, ephemeral, and unpredictable in location. These Indians lived in small family groups that wandered widely and did not engage in warfare with other bands living in the prairie [213].

Just as divergent cultural evolution in neighboring tribes provides a test of the ecological correlates of warfare, so too convergent cultural practices in geographically distinct people is helpful evidence on the economic rationale for group territoriality. W. H. Durham notes that in tropical South America, Borneo, and New Guinea, warfare of exceptional ferocity was commonplace within historical times [192]. These aggressive cultures engaged in intensive agricultural production of a carbohydrate-rich, but protein-poor, food staple such as manioc or sweet potato. Protein was either secured as wild game from forests adjacent to villages or grown in the form of pigs. Warfare seems to have been correlated with population growth and the associated pressure placed on the protein-producing sector of the environment.

In the Mundurucú of Brazil, for example, villages apparently began to carry out head-hunting raids when their preferred animal prey, the wild peccary, became scarce in their traditional hunting areas. A decrease in peccaries would occur when neighboring groups grew in size and sent hunting expeditions farther and farther from home villages, depressing the population of peccaries on which other groups depended. The warfare that broke out between neighboring groups had as its proximate goal the collection of the severed head of an enemy tribesman. But the Mundurucú seem to have realized that there was an ultimate connection between successful warfare and the eventual improvement of peccary hunting because they called a warrior that came home, head in hand, a "mother of the peccary." (Similarly, the head-hunting Ibans of Borneo believed that the heads of members of opposing tribes contained magical powers that would make the rice grow well, the forests more productive for their hunters, and their wives more fertile [736]!) A Mundurucú village that staged a murderously effective raid on its neighbors could reduce the competition for peccaries by killing some competitors outright and by encouraging the defeated enemies to emigrate to a more distant village site. Warfare among the Ibans [736], the Yanomamö of tropical Venezuela [132], and various New Guinean tribes [192] had the functional effect for the victors of driving their opponents away, thereby freeing more land for the production of wild game or for agricultural purposes.

Relatives, Reciprocity, and the Costs of Warfare

The advantages of going to war apply only to certain patterns of competition and resource distribution. Other conditions may make the peaceful coexistence of neighboring bands an adaptive strategy for all concerned. In harsh ecosystems, periodic famines may have held human populations below the carrying capacity of the land, reducing the likelihood that desirable materials would become limited. This may have promoted peaceful relations among the widely dispersed groups of Eskimos living in central Arctic regions and among the scattered Bushmen bands of the Kalahari Desert of southern Africa [192]. In addition, if neighboring units consist of individuals who are related to one another, any reproductive gains to the successful aggressors of one band would be reduced to some extent by the loss of fitness experienced by relatives, the loss being proportional to the degree of relatedness among the aggressors and their victims.

It is a general point that people are much more likely to murder nonrelatives than genetically related individuals. In their study of a sample of more than 500 homicides committed in Detroit, Martin Daly and Margo Wilson showed that when people kill other people that they know and live with, a nonrelative is about ten times more likely to be the victim than a relative of the offender (Table 3). Moreover, using samples drawn from eight different cultures, they demonstrated that in every case the proportion of genes shared by descent between victim and offender was lower, usually a great deal lower, than the average genetic relatedness of people who collaborated in a murder [157].

If it is usually true that kin cooperate in attacking nonrelatives, we can expect that warfare will be less frequent or absent when contiguous tribal units contain close relatives. The pacific Bushmen and Eskimos were in fact tribes consisting of an extended network of relatives divided into numerous small groups. Even highly warlike peoples were sensitive to the question of relatedness, and this factor influenced their selection of victims. The Tiv of Nigeria always attacked the neighboring group that was most distantly related to the attacking village [192]. Likewise, the extraordinarily warlike Yanomamö moderate their aggressivity somewhat when dealing with villages that contain their relatives [132] (Figure 4).

In addition to the possibility that one may be immediately reducing one's inclusive fitness by assaulting a group with relatives, there is the risk that an attack will destroy the possibility of mutually beneficial arrangements between groups, whether related or not. Reciprocal sharing of resources

Table 3
Risk of Homicide by Relationship (Cohabitants)

Average Person > 14 Years Old Lives with 3.0 People	**Number of Victims**		**Relative Risk (observed/expected)**
	Observed	*Expected*	
0.6 Spouses	65	20	3.32
0.1 Nonrelatives	11	3	3.33
0.9 "Children"	8	29	0.27
0.4 "Parents"	9	13	0.69
1.0 Other "relatives"[a]	5	33	0.15

Source: Daly and Wilson [157].
[a] Other relatives are almost all siblings.

and assistance can produce significant benefits to individuals. For example, the large herbivorous prey of the Bushmen and the high Arctic Eskimo provide a treacherously unpredictable resource base for a family group. Local droughts in the Kalahari may result in the absence of game from much of a band's foraging range; a shift of even a few miles in the migratory route of a caribou herd one year can spell hunting failure for a group of

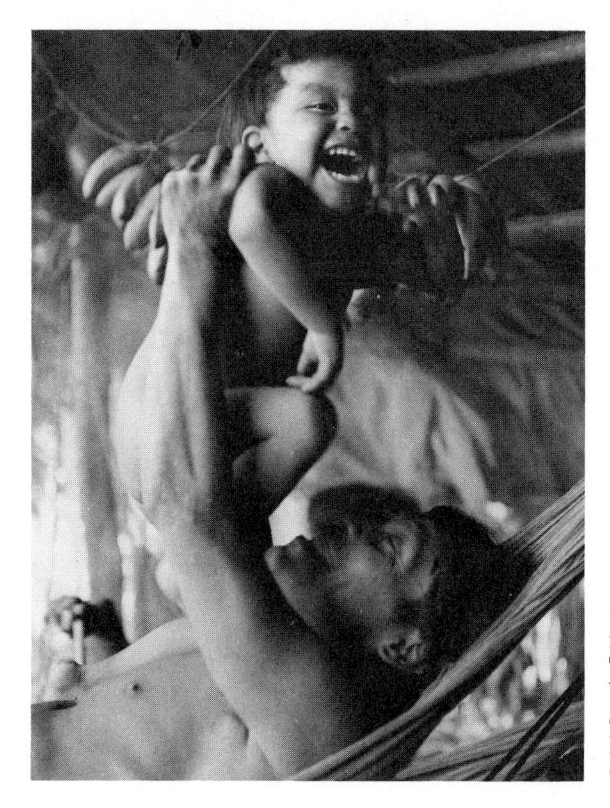

4 **Yanomamö man playing with his nephew** in a village he is visiting. The man's life is in danger because the two villages are on the verge of hostilities. Photograph by Napoleon Chagnon.

Eskimos. A group of Bushmen or Eskimos weakened by starvation would have little chance to oust a better nourished neighboring band from its still productive hunting grounds. Instead, an agreement between bands to provide assistance to one another during capriciously unpredictable hard times may be the superior strategy to protect individuals from reproductive disaster [192].

Reciprocity among related human beings has great potential for genetic benefit because a participant may gain by elevating his own reproductive success, and he may also increase the fitness of a relative [13, 715]. This does not mean that blind altruism among relatives will always be advantageous. In reciprocal altruism among relatives, the reproductive gain for the initial helper (as opposed to the genetic gain from elevating a relative's fitness) comes some time after the helpful act. If not repaid for his kindness, the altruist's inclusive fitness may be lowered because the genetic cost of the act may not be adequately compensated solely by the improvement in the reproductive success of the relative. Related persons share some, but not all, their genes in common. This might favor individuals who were careful about helping another person (and of accepting help from others as well), even if they were related and especially if they were not.

There is considerable evidence that human beings are remarkably aware of social debts owed them and social obligations of their own [715]. Giving someone anything usually obligates the other person to match or exceed the present, a fact exploited in extreme fashion by the Indians of coastal northwestern North America in the potlatch, an extravagant party given by one family for a rival group. The intent was to give so many things that the competitor would be socially destroyed by his inability to exceed the performance in a counter-potlatch [138].

R. L. Trivers has argued persuasively that the origins of the emotions that underlie gift giving and receiving stem from the risk that the initial giver will be cheated [715]. Selection favors individuals who are alert to cheating, so that they can cut their losses when it becomes apparent they are unlikely to be repaid. Similarly, humans with a sense of guilt, who avoid incurring large social obligations that they may not be able to repay, also should enjoy higher fitness. Social cheaters face the penalties of ostracism and the refusal of others to help when it is really needed. Suspicion, hostility toward social cheaters, and concern about a person's reputation for generosity might have had their origins in selection for reciprocal altruism among relatives. Once established, however, these psychological traits might be used to regulate reciprocal altruism among nonrelatives as well. In some societies, a visitor need simply ask another for something that person has (e.g., food or a tool) and the owner will relinquish it immediately [36, 695]. Behavior of this sort has been consistently misinterpreted (and abused) by persons from Western cultures, who have believed that these "primitive" peoples have no sense of property. This is untrue because each gift is noted and remembered by both the giver and the receiver. He who takes is obligated to reciprocate in equal or greater measure at some future date. The helper is entitled to deferred payment at a moment of his choosing (when he goes visiting and is hungry or in danger).

Reciprocity among Nonrelatives

Thus, members of groups of unrelated humans may potentially gain through reciprocal agreements. In the past, the highly aggressive Yanomamö and New Guinean tribes expended as much time and effort in attempts to form cooperative alliances with other groups as they did fighting with each other [132] (Figure 5). (Admittedly, the usual goal of these alliances was to join forces to slaughter or drive off the members of another village. If successful, two previously cooperating groups might then turn on one another in a manner reminiscent of modern nations.)

Because of the possible kin ties between allied villages of Yanomamös and others, it is hard to know whether completely unrelated groups of these peoples form reciprocal alliances. However, the pacific Pygmies in their dealings with neighboring Bantu villages offer an instructive example of reciprocal altruism among totally unrelated peoples [724]. The Pygmies and Bantus do not fight with one another, although members of each tribe heartily despise members of the other, as Colin Turnbull reports in his marvelous book *The Forest People*. The Bantus have convinced themselves and outsiders that they "own" the Pygmies, who provide them with meat and forest products for practically nothing. The Pygmies, on the other hand, encourage the Bantu in their self-deception and feel, with considerable justification and pleasure, that it is they who are exploiting the Bantu and not the other way around. In reality, individuals of both groups benefit by exchanging their own surplus goods for materials they could not easily get otherwise. Because individuals of the two tribes exploit largely different habitats and resources, they have more to gain by maintaining the reciprocal arrangement than they would by fighting, which would mean the end of trading. This most assuredly does not mean that they trust or like one

5 **Formation of an alliance.** Two Yanomamö men from previously hostile groups have engaged in ritual chants and have exchanged gifts to help cement a cooperative alliance between them and their villages. Photograph by Napoleon Chagnon.

another or that they do not try to get the best of every bargain. The point is that the potential advantages of reciprocity must be considered in decisions about whether to go to war; and, if the losses are sufficiently great, they may outweigh any benefits to be gained by group aggression.

The cost–benefit equation of warfare is a complex one, dependent upon the potential gains, the substantial risks of injury taken by warring groups, the potential loss of inclusive fitness if relatives are killed or injured in attacks, and the potential loss of the benefits of reciprocity. All these factors can and do vary, affecting the net genetic gain of individuals behaving in different ways. Here we have made the suggestion that humans are sensitive to these costs and benefits. Although many more analyses of group aggression in relation to ecological conditions are required, the little evidence reviewed here is consistent with the prediction that a broad spectrum of human abilities is employed in ways that tend to elevate the inclusive fitness of some participants in aggressive or cooperative interactions between groups.

The Evolution of Human Reproductive Behavior

We are all aware that cultural tradition has an enormous influence, not only on the practice of warfare by a society, but also on the nature of its sexual mores. Nevertheless, men and women generally behave differently with respect to reproductive tactics [155]. In this section, I shall examine some of these differences and shall argue that cultural traditions consistently reinforce biologically adaptive decision-making by men and women, an outcome that refutes the hypothesis that cultural evolution has become completely independent of our biology.

Table 4 lists a set of differences that occur in the sexual behavior of men and women of our own Western societies and that appear commonly in other cultures as well. As Donald Symons has argued, differences of this sort are entirely in line with predictions from evolutionary theory [690]. As we saw in Chapters 11 and 12, males and females are not expected to have identical strategies for reproductive success. Human males, like males of most other animals, have the capacity to produce such vast quantities of sperm that an individual could, at little physiological expense, father great numbers of offspring. (The record for a man is 888 [155].) To achieve extensive fatherhood, however, requires that a male inseminate more than one female because each woman's reproductive potential is severely limited. Not only does a human female produce a relatively small total number of fertilizable eggs over her lifetime, but a woman almost never has more than a single mature egg available for fertilization and then only for a few days each month. When she is pregnant or nursing a baby, eggs are no longer prepared for fertilization and this further limits the total number of children a female can have. (The record is 69 for a woman who specialized in having triplets [155].)

The Desire for Sexual Variety

On the one hand, a male's genetic success could easily be positively correlated with the number of different females he inseminates (the more copulations with the greater number of females, the more eggs he is likely to fertilize). On the other hand, there is no obvious reason why a female's fitness should increase as she mates with more men (but see [352]). Her genetic success will depend far more on the quality of a husband rather than the quantity of her mates. This simple evolutionary argument helps us interpret a large number of sexual differences between men and women [690]. In particular, males appear to find copulation more rewarding and desirable as an end in itself than women. Males have a lower threshold for sexual arousal and achieve orgasm more predictably than females. These characteristics motivate males to engage in copulation, reinforce them for success, and ensure that even a brief copulation results in the transfer of sperm to a female. Evidence that males, as predicted, find a diversity of partners more desirable than females comes from the widespread reports that men are more likely to engage in adultery than women. Moreover, the Coolidge effect (Chapter 11) applies to humans as well as to many other mammals [690].

Donald Symons has also noted that the behavior of male homosexuals, in Western society, is very different from that of female lesbians. Not only is male homosexuality much more common than the female variety, but males typically have a progression of partners, whereas their female counterparts have much more long-lasting, stable relationships. Symons's argument is that male homosexuals, although not necessarily advancing their genetic interests, are expressing the proximate mechanisms that motivate males to try to achieve sexual variety [690]. These same mechanisms often drive heterosexual males to seek out multiple sexual partners and thereby increase their egg fertilization opportunities.

Finally, the fact that males sometimes force women to engage in sexual acts against their wishes can be interpreted as still more evidence that males are more strongly motivated to copulate than females. Rape is a rare,

Table 4
Differences in Male and Female Sexual Physiology and Behavior

Trait	Male	Female
Threshold for sexual arousal	Low	High
Orgasm during copulation	Almost always	Not always
Desire for sexual variety	High	Lower
Adultery	More frequent	Less frequent
Rape	Occasional	Almost never
Concern for mate's sexual fidelity	Very high	Moderate

but regular, phenomenon in most societies. At the proximate level it may reflect the greater ease with which men become sexually aroused than women. An ultimate explanation for the behavior is that it enables some men to fertilize eggs that they would otherwise never have a chance to fertilize. This hypothesis was vigorously rejected by Susan Brownmiller in her influential and favorably reviewed book *Against Our Will* [110]. There she seriously advanced the thesis that rapists were agents of psychological terror, acting on behalf of all men to subjugate all women. The fact that this more than mildly fevered notion was plausible to many readers indicates the degree to which conflict and suspicion characterize the interactions between the sexes.

Randy and Nancy Thornhill have recently used an evolutionary approach to test the hypothesis that rape is a "conditional strategy" that some men may use to increase their fitness [701]. For example, if rape has a reproductive function, then rapists should tend to force copulation on women of fertilizable age. In fact, raped women do not constitute a random sample of the population; women in the years of peak fertility have an unusually high probability of being raped (Figure 6). Moreover, if there were no sexual motive to rape but only the intent to inflict violence on women, then rapists should select the same population of women as murderers. But the age distribution of raped females is significantly skewed toward the younger age classes when compared with female murder victims. Finally, young men are overrepresented in the rapist category, as predicted from an evolutionary perspective. These males are likely to be relatively low in economic status and thus not as acceptable to females as older, better established males. Just as *Panorpa* scorpionflies that do not have resources to offer females engage in forced copulation attempts, so too human males excluded from "legitimate" avenues of reproductive competition may be more likely to engage in a low-return, high-risk option. (By this time my readers presumably do not have to be reminded that describing human rape as a "conditional strategy" does not imply moral approval of the behavior. But just in case, let me state again that evolutionary hypotheses represent an attempt to explain biological phenomena and not to justify them in social terms.)

Female Choice and the Incest Taboo

The fact that rapes occur shows that opportunities for males to mate with hosts of females, although potentially rewarding at both the proximate and ultimate levels, are few and far between for many men in most human societies. Women tend to be resistant to casual sex. They, like all female mammals, make a great energetic investment in each offspring while it is in the embryonic stage and during the early years of life of a child when it is nourished at the breast. Therefore, if there is variation in the "quality" of potential partners from a female's perspective, she should refuse to mate with males that will do less to advance her fitness than other individuals. Human males do vary in both the genetic quality of their gametes and the material benefits they can offer their mates.

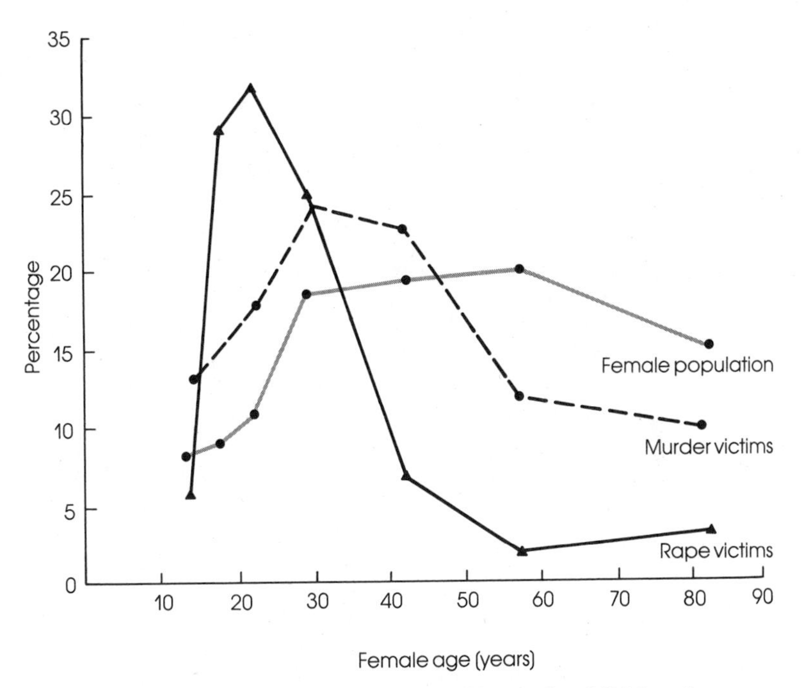

6 **Victims of rape** are more likely to be in the child-bearing years than are females that are murdered; this conclusion is based on United States crime data collected in 1971. This result suggests a reproductive basis for rape.

For example, human females would do well to avoid copulating with close relatives. Inbreeding in our species produces offspring that tend to have deleterious recessive alleles in double dose. Humans from incestuous matings exhibit a substantially higher rate of genetic defects and congenital abnormalities than members of outbred populations [648]. In addition to the genetic benefits of outbreeding, marrying a nonrelative promotes the formation of alliances between different genetic lineages. The progeny produced in a marriage between two separate family groups creates a tie that may increase cooperation between the two bands. Without a mutual interest in the children of the marriage, the two groups might have no other shared concerns and no motivation to help one another.

Given these arguments, it is striking that incest taboos appear to be a general feature of human cultures [730]. Parents and their children are expressly forbidden to copulate, and people are usually not permitted to marry their siblings (or even their first cousins). Incest taboos are sometimes culturally extended to forbid marriage between individuals belonging to particular categories, such as the same clan. These rules simply reinforce a "natural" tendency of individuals that have been reared together to avoid having sexual relations. The strength of this tendency has been demonstrated by the refusal of children reared together in Israeli kibbutzim to marry one another [13]. When the kibbutz system was devised, the social scientists in charge envisioned that communally raised children would eventually marry within the kibbutz. Such marriages were actually encouraged, and yet the children of one kibbutz almost without exception fall in love with members of another kibbutz. It is as if the brain had a program that reads: "Do not become sexually attracted to humans with whom you have been reared." (Note that this "rule" contrasts sharply with the ability of humans to fall in love with complete strangers [13].) Over evolutionary time, a program of this sort would have usually prevented siblings or other very close relatives from mating. It is interesting that the cultural rules against incest are most strongly developed in societies in which siblings are regularly reared apart. Under these conditions, individuals might make reproductive errors were it not for societal proscription against sexual relations between brothers and sisters [730].

The Loss of Estrus and the Evolution of Monogamy

Men vary in the quality of the resources they can transfer to a mate as well as in the compatibility of their genes. In most societies, some men help in rearing their wives' children. Women should, therefore, prefer males willing and able to be good parents and good providers. But from a male's perspective, investments in one female and her progeny reduce his ability to search for and achieve copulations with many other females. Given all that has been said about the reproductive advantages of a male strategy that emphasizes sexual variety, how can we explain the high rate of monogamous marriage and male parental care in human cultures?

There have been a number of different hypotheses advanced to account for the peculiar institution of helpful marriage, which is so different from the typical primate arrangement. I present the scenario that I find most plausible [13]. Readers interested in alternative viewpoints are referred to [115, 690].

The scenario begins with the assumption that early in hominid evolution there was a division of labor in which males hunted animal prey more than females. Because the pelvis of women is modified to enable childbirth, in general females are slightly less efficient runners than men. In addition, women tend to be somewhat smaller and less muscular than men. They therefore would have been better suited for the collection of vegetable foods than for the pursuit of big game.

The second assumption is that female protohominids, like all modern primates except humans, at one time had a well-defined estrus phase in which they were sexually receptive. During estrus, a female might attract a consort male who would guard her during the time she was most likely to ovulate. (Mate guarding is common in extant social primates.)

During the consort phase, a male might share prey he had recently captured with the female because it would be likely that the progeny produced by the female would bear his genes; protein gifts to the female might, therefore, yield a genetic gain for the male. (As usual, there is no requirement that individuals had to be aware of the genetic consequences of their acts.) If gift giving became an established component of courtship and consort guarding, then a female might gain if she could extend the period of guarding and thus receive more presents. Instead of blatantly advertising the time of ovulation by becoming receptive just prior to this event, female protohominids might have partly hidden the time of ovulation by lengthening the period of receptivity. The advertisement-of-ovulation strategy may be practiced by some female mammals as an incitation display to encourage competition among males for access to them, thus increasing the probability of mating with a dominant male. But if material benefits are offered by a male, the quality of this investment may become more important to a female than male genetic quality.

By making it difficult for a male to determine when his chances to fertilize an egg were best, females that retained their receptivity could make it advantageous for at least some males to practice prolonged guarding. The more hidden the ovulation time and the more prolonged the period of receptivity, the longer a male must remain with a female (copulating with her at regular intervals) if he is to increase his chances of fertilizing her eggs when they do become available for fertilization. A widespread use of a strategy of permanent guarding (marriage) reduces the availability of fertilizable females and reduces the time a married male has available to find these females (because of the guarding time required to protect a mate). This being so, it may become advantageous to be a devoted husband and father and to expend resources on one's presumptive offspring, which are highly likely to bear one's genes. This expenditure may increase the offsprings' chances of surviving and reproducing, thereby providing a genetic gain that may often (but not always) exceed a Casanova strategy of attempts at multiple sexual conquests.

Female Choice and Economics

If the above scenario is correct, the advantages of gaining a male's paternal investment and valuable resources were a major factor in the evolution of the loss of estrus by female humans. We have predicted that as soon as some males offered useful material benefits in return for copulation, selection would favor women that evaluated a male's economic status in making mating decisions. In this context it is surely significant that courting males in many societies, our own included, are expected to provide gifts of food (e.g., meals at French restaurants) and valuables (e.g., diamond rings) to either the female or her family. Among the Bushmen, a male was not eligible for marriage until he had killed a large antelope. By imposing economic tests on a partner, a female (and her family) have a direct way of assessing the material benefits she is likely to derive from a potential husband. This, in turn, is related to a female's reproductive chances, as William Irons has shown in his study of the Yomut society (Figure 7). The wives of males of the wealthier one-half of the population have substantially more descendants than the wives of males in the poorer half [358].

Given that men can and do sometimes provide resources of value to their

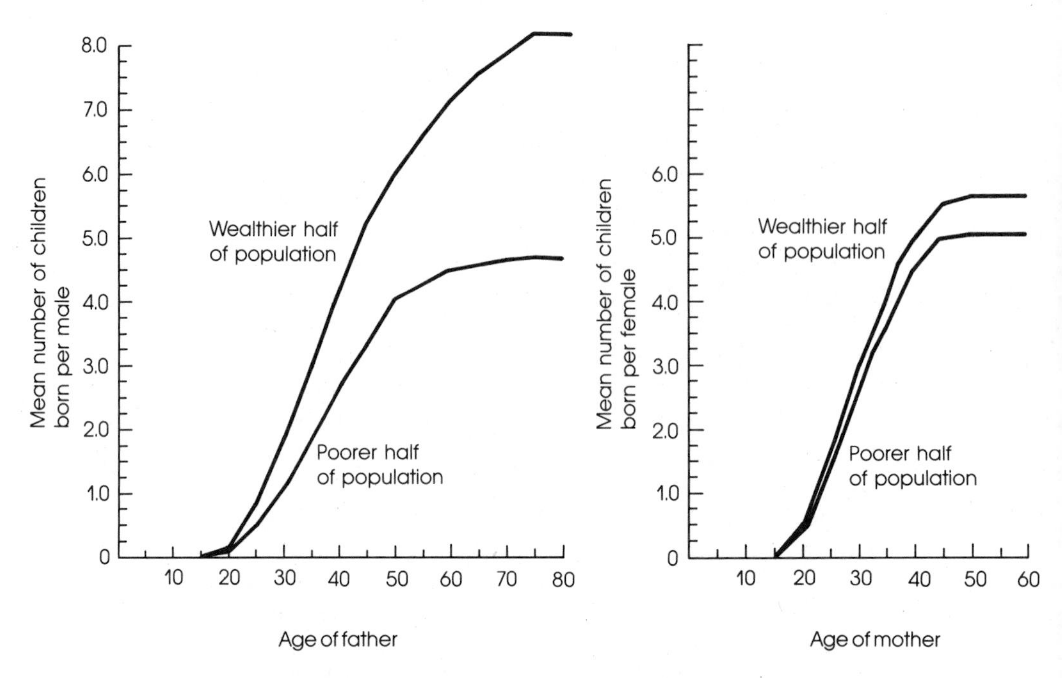

7 **Reproductive success and wealth** in a polygynous society, the Yomuts of Turkey. As predicted, both males and females in the wealthier half of the population have higher reproductive success than individuals in the poorer half of the Yomut culture.

mates, one might predict that women should always prefer to have an unmarried male as a husband in order not to have to share the material gains derived from marriage. But although the vast majority of human marriages have surely been monogamous, in most cultures a few males have more than one wife (Figure 8). Why?

Perhaps human males are rather like red-winged blackbird males (Chapter 12). Most human societies offer opportunities for truly exceptional performance in various fields, with commensurate material rewards. Outstanding leaders, hunters, weapon makers, and healers have probably always secured large amounts of food directly or indirectly. Because the difference between successful and unsuccessful men is fairly obvious, marriageable women have probably had no difficulty discriminating between the two [534]. Given a choice between an outstanding provider and an inferior one, it may be to the women's advantage to choose the former, even though this means entering into a polygamous relationship.

Sexual Jealousy and Male Parental Care

Females should choose mates that will help them rear children, as this enhances female fitness. By the same token, one can predict that males that intend to make a commitment of parental care and wife support will choose females that will raise *their* reproductive success. The

8 **A polygynous household.** A Yanomamö headman, Kaobawa, with two of his three wives. Photograph by Napoleon Chagnon.

reproductive value of a woman varies considerably in relation to her age, and this helps explain why men consistently prefer young women as wives in many cultures. Although it is common for an older man to marry a much younger woman, the reverse is very rare. Because young women have greater expected reproductive output than older women, a male can expect to father more children by a young wife than by an older one [690].

But the fecundity of a woman is not the only determinant of her contribution to her husband's fitness. If she engages in extramarital copulations, the children she bears may be fathered by men other than her husband. The genetic consequences of cuckoldry are severe. A male that unknowingly provides costly parental care for the offspring of a competitor not only fails to promote his own reproductive success but elevates the fitness of a rival. Female humans on the other hand cannot be cuckolded. If a woman gives birth to a baby, it will have her genes with complete certainty. But her husband can never be asolutely sure that the baby is truly his child.

This fundamental difference between men and women in the reliability of parenthood seems clearly related to the higher level of sexual jealousy on the part of males and their fear of cuckoldry, characteristics that are evident in all cultures studied to date [159]. Even in societies said to be sexually permissive, anthropological interviews indicate that men regard the sexual favors of their wives as their "property" and consider adultery by a wife to be a serious offense against them. Sexual jealousy focused on the right to copulate is an emotion that motivates males to guard their mates and prevent other males from cuckolding them.

Adultery (or suspicion of it) often precipitates a violent response by the offended husband against his wife and her lover. Even among the relatively peaceful Bushmen of the Kalahari Desert, most murders are committed by men in disputes over women—particularly if adultery is involved [408]. In many other societies as well, sexual jealousy is a major cause of violent crime by men [157]. Violence by a cuckolded individual is, moreover, often considered partly or wholly justified in many legal codes. It is striking in how many cultures laws have independently been formulated in which a married woman's adultery is identified as a legal offense against her husband and severe punishment specified for the offenders [159]. (In contrast, adultery by the male is usually treated, not as a crime against his wife, but as a crime against the male he cuckolds.) This is a classic example of how cultural tradition is constrained by (and reinforces) evolved attributes of the human brain that advance individual fitness.

The typical concern of a male about the paternity of his children extends even to cases in which his partner has been forced to copulate against her will by another man. Frequently the husband of a rape victim is so profoundly disturbed by the event that he may ultimately divorce his wife, a reaction accepted or even encouraged by a diversity of religious groups and legal codes [110].

The fear of loss of a husband's support may be why married women that have been raped are more reluctant to report the crime than unmarried females [701]. In addition, contrary to all expectation based on sociological theory, raped married women that have not been beaten or otherwise marked with obvious injury are every bit as psychologically damaged as those women that have been cut or brutalized during a rape. But if a woman has not violently resisted her rapist (and been injured in the attempt), her husband is perhaps more likely to suspect that he has been the victim of adultery than if his wife has been obviously hurt. The desire of a woman to allay these suspicions may be at the root of the psychological distress of the raped, but otherwise unhurt, woman [701].

There is another response of women that appears to be an effort to reduce a husband's sexual jealousy and fear of cuckoldry. Martin Daly and Margo Wilson have shown that after a child has been born in our society, the mother and her relatives are significantly more likely to claim that the child resembles the father more than the mother [158]. Indeed, pregnant women apparently often fantasize that their child will look like their husband. The desire for a baby, male or female, to resemble the father need not involve conscious deceit or manipulation on the part of a woman and her kin. But whether conscious or not, the response reflects the psychological importance for a husband of establishing confidence of paternity if he is to remain with his wife and care for her children.

Are There Maladaptive Aspects of Human Reproductive Behavior?

The protectiveness of husbands about their wives and their reluctance to continue to provide for adulterous (or raped) wives and their progeny seem to confirm evolutionary predictions about how males should behave to maximize their inclusive fitness. But what about wife sharing, a practice not unknown in our society and fully sanctioned in certain other cultures? A speculative interpretation of wife sharing is that it may be an extreme example of reciprocal altruism. If the sharing is truly equal, each husband stands the same chance as the other of eventually providing support for offspring fathered by the other individual. Therefore, neither male necessarily reduces his fitness if he sees to it that his generosity is fully reciprocated. Moreover, if the exchange of wives helps cement between the males an alliance that enables them to cooperate more effectively in the acquisition of resources, the fitness of both males might be elevated in the long run. Males that exchange the right to copulate with their mates do not lack the emotion of jealousy or the sense of possessiveness about their wives, as is shown by the violent treatment or death that adulterous (nonreciprocating) males receive in some Eskimo groups and other tribes in which wife sharing occurs.

Any number of cultures are far more sexually permissive than our own, and in them "adulterous" sexual relationships may be commonplace. The sexually open South Sea Island societies of the Pacific are a classic case. In human groups of this sort, the reliability of paternity may be very low. The standard practice within such cultures is for a male to withhold parental care from the children of his wife and instead to adopt the father role for the children of his sister. From an evolutionary perspective, parents may offer their offspring love and care for many years because these individuals usually have a high proportion of their genes. But a male whose wife's children were fathered by other males does not share genes in common with these offspring. If most or all of his wife's children are genetic strangers, it may become adaptive to share resources with his relatives instead of his wife and her progeny. Richard Alexander has shown that if a male can expect to father only one of four children produced by his wife, he would gain genetically by diverting his parental care to the children of his sister (Figure 9). Under these conditions, a male shares $5/32$ of his genes with his nephews and nieces, but only $4/32$ with his wife's children. Importantly, in this society, three times out of four a male helping his wife's offspring will be assisting an individual with no shared genes at all. In contrast, by becoming an uncle–father, he will be helping a sister's progeny, all of whom have at least some of his genes by common descent [13].

Because humans everywhere display an enormous interest in and aptitude for learning "who is related to whom and by how much," they are in a position to regulate the distribution of their resources and services in ways that could help them maximize their inclusive fitness. The ability to assume the uncle–provider role in some situations is merely one example of the genetic utility of knowing who is related to you.

Adoption

Two aspects of human reproductive behavior provide what may be the most difficult challenges to an evolutionary approach to our behavior. They are the practices of adoption and birth control. We all know people in our society who have adopted into their families the child of a

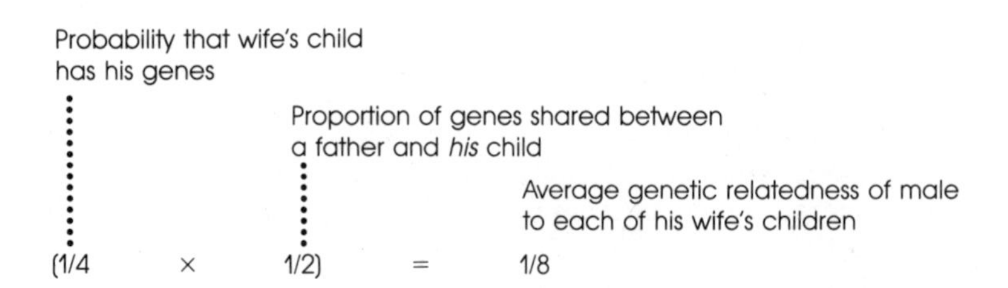

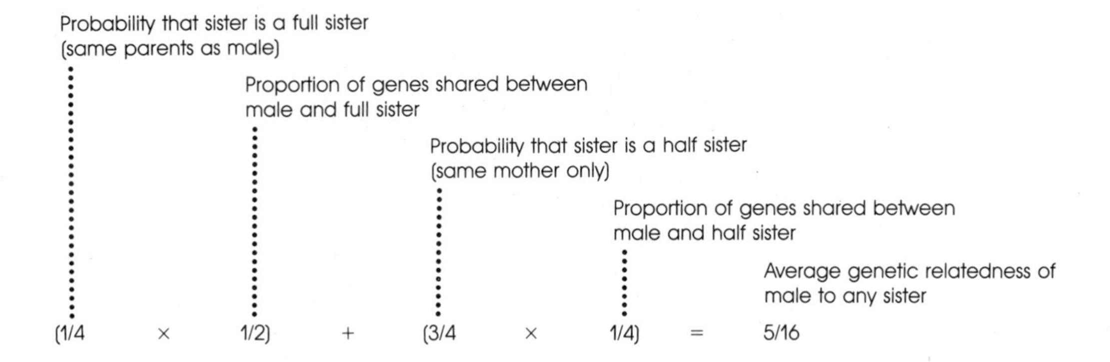

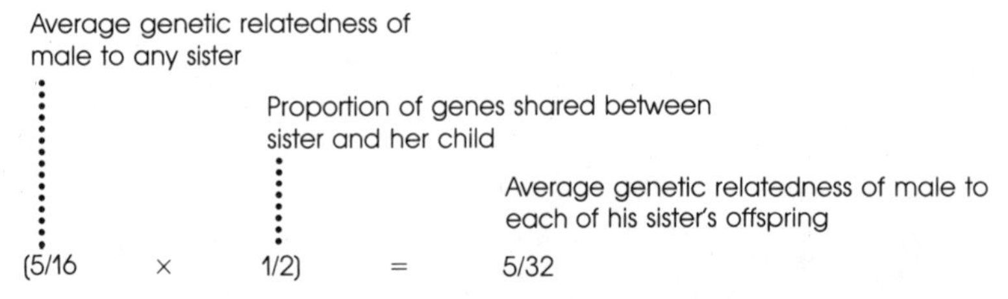

9 **The average degree of genetic relatedness** between a man and an offspring of his wife may be lower than the genetic relatedness of the man to an offspring of his sister. In a society in which the reliability of paternity is low, a male can therefore propagate his genes more effectively by helping his sister's progeny than by providing care for the children of his wife.

complete stranger, which they treat with the same affection that is customarily offered to one's biological children. These adoptive parents are, in effect, subsidizing competitor genes. How is it possible to account for such a maladaptive response in evolutionary terms?

Advocates of an evolutionary approach tackle problems of this sort in a special way. If the action is highly unlikely to increase inclusive fitness, as is the case for the adoption of a complete stranger, there are two alternative evolutionary hypotheses that may help explain the behavior:

1. The action is one of the costly effects of an ability whose benefits generally exceed the costs.

2. The action is currently disadvantageous solely because of the radically altered setting in which it is now performed; in the past, the disadvantageous response would not have occurred.

These explanations for traits that seem to harm an individual's chances for genetic success have been used throughout this text. For example, the flight of a moth to a bola spider was held to be a disadvantageous aspect of the generally adaptive ability of these male moths to find receptive females of their species. A hypothesis of the second type was employed to explain why robins sometimes hurl their banded offspring from the nest. The silvery band stimulates the nest-cleaning response, which is used in a way that would never have occurred in the past because human bird banders are a recent, novel addition to the bird's environment.

Similar hypotheses may help explain the modern practice of the adoption of strangers [43]. First, adopting a nonrelative may be a genetic cost of the usually adaptive desire of adult humans to have children and raise a family. Although it is true that adults who adopt strangers are reducing their fitness, the urge to have a family and the love of children that cause them to behave this way are on average beneficial. Because these traits usually elevate fitness, they are therefore maintained within human populations. The point is that all adaptations have costs and benefits, and one cost is that they may be misused, from a genetic perspective, especially in rare or unusual circumstances (Figure 10). In Joel Welty's *The Life of Birds,* there is a photograph of a cardinal feeding a beakful of insects to a goldfish that has lunged out of the water, mouth gaping, to receive the food [769]. This act certainly could not benefit the cardinal's genes. The bird, however, had recently lost its brood and was employing its food gathering and brood feeding in a rare and inappropriate manner. The ability to collect food for young and the strong desire to feed them surely raises the fitness of most parent cardinals.

The misplaced-parental-care hypothesis generates a testable prediction, which is that husbands and wives who fail to produce children themselves should be especially prone to adopt strangers. Their frustrated desire to be parents could find inappropriate expression (from an evolutionary perspective) in the adoption of a biological-child substitute, such as an animal pet or an unrelated infant. However, the risk that parental drive will be "misused" is relatively slight, and the overall genetic benefit of the eagerness to have children is obvious [773].

Another prediction is that if the adopting parents proceed to have children of their own (not an uncommon occurrence), then they will be more likely to withhold care and resources from an adopted child of strangers than from their own biological children. One way to test this prediction would be to examine the victims of child abuse in a family with both natural and adopted children.

The second basic alternative hypothesis is that the adoption of strangers is a novel, recent phenomenon that occurs primarily because of changes in the typical human environment over the past several hundred years. In

10 **Two emperor penguins** struggling for "possession" of an infant penguin. "Adoption" of this sort may arise when one individual has lost its chick or because one adult is mistaken about the identity of the chick. Photograph by Yvon Le Maho.

large industrial societies, children are made available to people who do not know the parents of the adoptee. Throughout the vast majority of human evolution, during which people lived in small bands or villages, this event would be exceptionally rare. Instead, adoption in this environment may have been practiced regularly to the genetic benefit of the adopter.

1. Almost certainly acceptance of the children of relatives was the most common kind of adoption in the past [773]. A direct way to elevate the fitness of a relative, and thereby perhaps to raise one's own inclusive fitness, would be to assist the children of that relative. In effect, males that practice the uncle–father strategy are adopting their sister's children as their own. Adoptive assistance given to related persons might also occur frequently when the biological parents died or experienced some disastrous turn of fortune that greatly reduced their ability to care for their children. By stepping in to help, an individual could salvage nieces and nephews whose reproductive potential would otherwise be nil or very low.

2. In cases in which an individual adopted a stranger, the adoptee might be a captured child from an enemy group. The adopting family would probably not treat the child in an altruistic fashion, but instead would exploit him or her as a servant or slave.
3. There are some well-documented cases of self-sacrificing adoption in preindustrial societies of individuals known not to be relatives of the adopters. Marshall Sahlins, who believes this phenomenon cannot be explained by recourse to evolutionary theory, also points out that "adoption is like marriage, a mode of alliance between groups," thereby providing a possible evolutionary explanation for adoption of this sort [629]. A reciprocal altruism hypothesis predicts that individuals adopting nonkin, although temporarily lowering their fitness, gain by establishing a "mode of alliance" with the parents (and kin) of the adopted child. In the Pacific Island cultures where this practice was widespread, warfare was also highly developed. Adoption among nonrelatives may have increased the possibility of alliances between villages for mutual benefit in defensive or offensive war maneuvers. Such alliances are often critical in determining the survival of a group in cultures where warfare is a major activity. Adoptive practices may also have decreased the probability that allies would turn on each other during or after the war because they would have a direct genetic stake in the survival of the other group. Finally, if one could adopt the child of an enemy tribe, one might reduce the likelihood of continuing aggression between the opponents and so end or moderate hostilities. In this context it is significant that the warlike Tahitians were willing to adopt the child of a person they had killed in a battle or a child of a close relative of the deceased person [629].

Altruistic adoption of a nonrelative essentially converts that individual to kin and is an extreme example of a common practice of human beings: attributing kinship status to nonrelatives. My children call an unrelated family friend "Aunt Beth," and I have been called "Brother" by persons who were not tenth cousins of mine. This procedure is sometimes taken to mean that humans really do not appreciate the distinction between relative and nonrelative, but as Richard Alexander has argued, just the opposite is true. Giving the title of a relative to someone you both know perfectly well is not related to you is a signal that you desire friendly relations with that person because it is assumed that relatives will interact in mutually helpful ways [13].

Birth Control

How is it possible to reconcile the contention that people love children so much that they will sometimes voluntarily adopt the infant of a stranger with the widespread current and ancient practices of birth control, including abortion and infanticide, all of which prevent people from rearing children? An evolutionary approach to this problem begins with the understanding that one can sometimes maximize fitness by not reproducing under certain conditions. The reproductive physiology of humans reflects this principle [485]. Rose Frisch has shown that food intake helps control both the onset of menstruation in women and their fertility during the childbearing years [242, but see 78]. In fact, a minimum of 16 kilograms of stored body fat is required for maintenance of menstrual cycling. This reserve contains 144,000 calories, an amount that can sustain the development of a typical fetus and three months of lactation for the infant after its birth. A sharp weight loss (or emotional stress) can cause the physiological blockage of ovulation and, therefore, the prevention of pregnancy. The mechanisms, hormonal and otherwise, that make this happen are performing unconscious birth control for the person, preventing the formation of a zygote that has a poor chance of surviving and a high probability of simply draining resources from a mother with no genetic return.

Conscious behavioral decisions about when to produce children can also have the effect of raising rather than lowering fitness by increasing the total number of reproducing offspring created by a parent during his or her entire lifetime. It is possible to make some predictions about when people in many cultures will practice birth control, if the ultimate function of this behavior is to maximize individual gene-copying success.

Prediction 1. Unmarried women in societies in which the biological father normally assists in child rearing should be more likely to practice birth control, including abortion and infanticide, than married women. The chances of success in child-rearing would have historically been much less for a single parent than for a married woman.

Prediction 2. Married couples will practice birth control primarily with the goal of spacing the births of their children at intervals of several years. In many cultures, a male is forbidden to have sexual intercourse with his wife for months after the birth of a child. As you might suspect, this reduces the probability that the new mother will become pregnant, an event that would divert metabolic resources from milk production to embryonic development. Such a diversion would often endanger the survival of the living infant. If a newborn can survive the first few years of life, it stands a better chance of surviving to reproduce and, therefore, becomes increasingly valuable and deserving (in genetic terms) of continued investment. In hunter–gatherer cultures, nursing continues for up to five years after the birth of a child [383]. Nursing activates hormonal systems that help prevent ovulation; cultural proscriptions against copulation between new parents reinforce the ultimate effect of the antiovulation mechanisms, which is to space births widely.

There is some peripheral evidence that suggests that humans are designed to have children at intervals of three to four years. In modern cultures, some people are able to have many children with very short periods between each birth. Psychologists have found that the intelligence of children (which may influence their potential reproductive success and thus the fitness of the parents) is significantly influenced by both birth interval and family size [59, 187]. Later children in large families have considerably lower intelligence scores on the average than earlier-born progeny. Likewise, the longer the interval between births (up to four or five years), the higher the scores of children on intelligence tests. These results may indicate that human developmental systems are designed to operate in a social environment with siblings four years or so older or younger or that parents may be better able to assist in the intellectual development of their children if they are widely spaced.

Despite hormonal and behavioral adaptations designed to keep a mother with a very young child from becoming pregnant again, it may still happen. Under these circumstances, aborting the fetus or killing the newborn may in evolutionary terms represent ending a genetically unwise investment in reproduction earlier rather than later, thus saving energy to devote to living offspring and to promote the long-term reproductive success of the parent.

Prediction 3. In cases in which infanticide does not have the effect of spacing attempts to rear offspring, it will result in the removal of progeny whose potential for genetic success is relatively low [11]. Thus, we can predict infanticide to occur with higher-than-average frequency when the newborn has a congenital defect apparent at birth or when twins are born. (Because of their low weight and because their mother's milk must be shared between them, twins run a higher risk of mortality than single children.)

It is possible to interpret the selective infanticide of healthy newborn females in some cultures in the light of this hypothesis. Female infanticide is practiced primarily in highly warlike cultures in which male mortality produces a skewed adult sex ratio. In the absence of female infanticide, the sex ratio would become even more strongly unbalanced in favor of females. Because such societies usually exhibit a polygynous mating system, a surviving male child in a population with many females may have far higher potential for genetic success than a female whose reproductive success is limited by the relatively small number of children she can bear. This favors couples who produce male infants. If a mother rears a female newborn, there will be a long interval (three to six years) before she can expect to conceive again. By killing her female infant, a woman can quickly become pregnant again and perhaps this time produce a genetically more valuable male child [180].

Prediction 4. Life-long celibacy will almost never be practiced by children that are the sole offspring of their parents. Individuals who never engage in sexual intercourse are rare but do occur regularly in human societies. Females in this category "should" devote themselves to their relatives' children, preferably their nieces and nephews, who share a substantial proportion of genes in common with them. The tradition of the old maid aunt is well established in our society and others. Celibate males are usually found in the priesthood. We might predict that they will occur more frequently in polygynous cultures than monogamous ones, with the celibate role being adopted by males who could judge (probably unconsciously) that their chances of competing with other males for mates were poor. By proclaiming their rejection of personal reproduction, they might be better tolerated by reproductively active males (perhaps sometimes giving them a chance to deceive these males). In the role of religious leaders, they might also be in the position to donate goods received for their services from nonrelatives to their relatives, but only if they had relatives to help. Thus, a male, if he were an only child, would generally have less to gain genetically by joining a celibate priesthood than if he were to reproduce personally. A male from a large family might have more opportunities to elevate his inclusive fitness as a celibate priest than by trying to form a family of his own.

We have argued that birth control can be and often has been used in ways that contribute to individual genetic success. It is indisputable, however, that in some modern societies birth control is used by many couples to reduce their reproductive potential. To explain this misapplication of birth control (from the gene's standpoint), it may be sufficient to note that reliable means of preventing pregnancy (other than abstinence, which has never been popular) have become available only in the last few years. This provides couples with the unique freedom to copulate without the subsequent burdens and demands of parenthood.

It is striking, however, that almost all husbands and wives in Western society, even with access to nearly perfect birth control measures, still choose to have several children. However, they tend to draw the line at two or three, probably because of the great time and energy requirements of child rearing as well as the enormous economic costs. Humans value freedom from economic constraints and freedom to enjoy the pleasurable aspects of life. In the past, efforts to secure these freedoms have almost inevitably been associated with producing a family with as many surviving children as possible. In the present, modern birth control technology enables us to satisfy our proximate drives without experiencing their ultimate consequences.

Human Behavior and Evolutionary Theory

We are at the threshold of an exciting new phase of the analysis of human behavior, and a new generation of biologists and social scientists has an unparalleled opportunity to help develop this phase. Evolutionary theory has just begun to be used to explain why we might behave the way we do. As we have seen, evolutionary thinking generates a host of predictions about how humans should behave if the ultimate, unconscious goal of individuals is to maximize their inclusive fitness. The extraordinary differences in the behavior of humans of different cultures could potentially be used to test directly whether there is in fact a correlation among cultural strategy, the ecology of the society, and the inclusive fitness of members of the culture. Weighed against human cultural diversity, the variation in the social behavior of other species is vanishingly small. It is unfortunate that much of this diversity is disappearing before it can be thoroughly described and before new information can be collected that would permit scientists to test evolutionary predictions about human behavior. Nevertheless, there is still time for a productive union of anthropology, psychology, political science, economics, sociology, and evolutionary biology, although forming such a union will be a challenging task. We have a great chance to increase our knowledge of ourselves; I hope we have the willingness to do so.

Summary

1 Human beings are an evolved animal species. An evolutionary approach predicts that our behavior, like the behavior of other animals, will tend to maximize the survival of the genes within us.
2 This is a controversial proposition for a variety of reasons, not the least of which is the great flexibility of human behavior, as shown by the bewildering array of cultural practices and the rapid change in culturally transmitted patterns of behavior within historical times. The plasticity of human behavior has persuaded many social scientists that humans have "escaped" biological constraints on their behavior and can adopt almost any imaginable behavioral option.
3 Variation in social organization and social behavior is, however, not unique to humans, but occurs in a number of animals that encounter variable ecological conditions. The ability to adopt the practices of a culture may therefore be a biological adaptation to variation in the environmental pressures affecting humans in different regions of the world. If this is true, we should expect cultural traditions to help raise the inclusive fitness of the members of a culture, and there is evidence that this is generally true.
4 By identifying the costs and benefits of a trait, whether culturally acquired or not, and by determining how the benefit:cost ratio should be affected by variable ecological conditions, one can predict under what circumstances a trait should occur. A preliminary analysis of this sort for human warfare supports the view that human beings engage in war only when it is likely to elevate their inclusive fitness.
5 Likewise, most aspects of human reproductive behavior are easily interpreted from an evolutionary perspective. Men and women differ in several aspects of sexual physiology and behavior that appear to be linked to the greater reproductive advantages for men of indiscriminate copulations. Because pregnancies are always very costly for a woman, females are expected to choose mates that are likely to assist them in child-rearing, and they appear to be able to do so. Human males often sacrifice opportunities to fertilize many females in order to be a good parent and provider. These men are under selection pressure to increase the probability that the offspring they help actually carry their genes and not those of a competitor male. Male sexual jealousy and mate guarding are proximate mechanisms that help achieve this ultimate goal. If reliability of paternity is very low, however, males may withhold assistance from

their wives' offspring and invest in the children of female relatives, a culturally sanctioned option in some societies.

6 Two human practices related to reproduction—adoption and birth control—appear to violate the evolutionary hypothesis. On closer inspection, however, there are several complementary explanations for these behaviors that are completely consistent with evolutionary theory and that can be tested.

7 Opportunities to devise evolutionary predictions about human behavior and to test them using anthropological data have hardly begun to be tapped, despite their potential importance for an understanding of evolutionary processes and humankind.

Suggested Reading

Readers interested in the conflict over *Sociobiology* should begin by reading the last chapter in E. O. Wilson's book [794] and then the exchange between Wilson and Science for the People that appears in *BioScience* [18, 795].

W. G. Durham has written two valuable articles on the evolution of culture and warfare from an evolutionary perspective [191, 192].

Richard Alexander has made a number of important contributions to the development of an evolutionary approach to human behavior that deal head-on with the typical objections of social scientists [10–13].

The psychologists Martin Daly and Margo Wilson and the anthropologist Donald Symons have shown just how profitable it can be for social scientists to adopt evolutionary principles. Their work on human sexual behavior is both entertaining and instructive [155–159, 690].

There are several excellent accounts of "primitive" cultures that make superb reading for their own sake but can also be read for examples to test evolutionary predictions about human behavior. They include Napoleon Chagnon's *Yanomamö, The Fierce People* [132], R. A. Gould's *Yiwara, Foragers of the Australian Desert* [271], Peter Mathiessen's *Under the Mountain Wall* [481], Kenneth Read's *The High Valley* [601], Elizabeth Marshall Thomas's *The Harmless People* [695], Colin Turnbull's *The Forest People* [724], and Laurence van der Post's *The Lost World of the Kalahari* [731].

Suggested Films

The Ax Fight. Color, 30 minutes. One of a series of films made by Napoleon Chagnon and Timothy Asch on the Yanomamö Indians of South America. This film illustrates the escalation of violence in a conflict between two villages and the role of kinship relations in structuring the interactions among the participants. An extremely dramatic film.

The Desert People. Black and white, 51 minutes. A film on Australian hunter–gatherers.

Feeding and Food Sharing (05987). Color, 23 minutes. The feeding behavior of wild chimpanzees.

The Hunters, Parts I and II. Color, 36 minutes each. A superb film on Bushmen hunter–gatherers.

Miss Goodall and the Wild Chimpanzees. Color, 28 minutes. A film showing elements of chimpanzee behavior.

Bibliography

1 Abele, L. G., and S. Gilchrist. 1977. Homosexual rape and sexual selection in acanthocephalan worms. *Science 197*:81–83.

2 Alatalo, R. V., A. Carlson, A. Lundberg, and S. Ulfstrand. 1981. The conflict between male polygamy and female monogamy: The case of the pied flycatcher *Ficedula hypoleuca. American Naturalist 117*:738–753.

3 Alcock, J. 1972. The evolution of the use of tools by feeding animals. *Evolution 26*:464–474.

4 Alcock, J. 1976. The behavior of the seed-collecting larvae of a carabid beetle (Coleoptera). *Journal of Natural History 10*:367–375.

5 Alcock, J. 1979. Multiple mating in *Calopteryx maculata* (Odonata: Calopterygidae) and the advantage of non-contact guarding by males. *Journal of Natural History 13*:439–446.

6 Alcock, J., and D. W. Pyle. 1979. The complex courtship behavior of *Physiphora demandata* (F.) (Diptera: Otitidae). *Zeitschrift für Tierpsychologie 49*:352–362.

7 Alcock, J., G. C. Eickwort, and K. R. Eickwort. 1977. The reproductive behavior of *Anthidium maculosum* and the evolutionary significance of multiple copulations by females. *Behavioral Ecology and Sociobiology 2*:385–396.

8 Alcock, J., C. E. Jones, and S. L. Buchmann. 1977. Male mating strategies in the bee *Centris pallida* Fox (Hymenoptera: Anthophoridae). *American Naturalist 111*:145–155.

9 Alexander, R. D. 1961. Aggressiveness, territoriality, and sexual behavior in field crickets (Orthoptera: Gryllidae). *Behaviour 17*:130–223.

10 Alexander, R. D. 1971. The search for an evolutionary philosophy of man. *Proceedings of the Royal Society of Victoria 84*:99–120.

11 Alexander, R. D. 1974. The evolution of social behavior. *Annual Review of Ecology and Systematics 5*:325–383.

12 Alexander, R. D. 1977. Evolution, human behavior and determinism. *Philosophy of Science Association 2*:3–21.

13 Alexander, R. D. 1979. *Darwinism and Human Affairs.* University of Washington Press, Seattle.

14 Alexander, R. D., and G. Borgia. 1978. On the origin and basis of the male–female phenomenon. In *Sexual Selection and Reproductive Competition in Insects.* M. F. Blum and A. Blum (eds.). Academic Press, New York.

15 Alexander, R. D., and P. W. Sherman. 1977. Local mate competition and parental investment in social insects. *Science 196*:494–500.

16 Alexander, R. D., J. L. Hoogland, R. D. Howard, K. M. Noonan, and P. W. Sherman. 1979. Sexual dimorphism and breeding systems in pinnipeds, ungulates, primates and humans. In *Evolutionary Biology and Human Social Behavior: An Anthropological Perspective.* N. A. Chagnon and W. Irons (eds.). Duxbury Press, North Scituate, Massachusetts.

17 Allen, G. E., et al. 1975. Against "sociobiology." Letter to the Nov. 13 issue, *New York Review of Books.*

18 Allen, G. E., et al. 1976. Sociobiology—Another biological determinism. *BioScience 26*:183–186.

19 Andersson, M. 1982. Female choice selects for extreme tail length in a widowbird. *Nature 299*:818–820.

20 Andersson, M., F. Götmark, and C. G. Wiklund. 1981. Food information in the black-headed gull, *Larus ridibundus. Behavioral Ecology and Sociobiology 9*:199–202.

21 Angst, W., and D. Thommen. 1977. New data and a discussion of infant killing in old world monkeys and apes. *Folia Primatologica 27*:198–229.

22 Anonymous. 1973. A conversation with Jerome Kagan. *Saturday Review of Education 1*(Mar. 10):41–43.

23 Aoki, S. 1977. *Colophina clematis* (Homoptera: Pemphigidae), an aphid species with "soldiers." *Kontyu 45*:276–282.

24 Appleby, M. C. 1982. The consequences and causes of high social rank in red deer stags. *Behaviour 80*:259–273.

25 Arms, K., P. Feeny, and R. C. Lederhouse. 1974. Sodium: Stimulus for puddling behavior by tiger swallowtail butterflies, *Papilio glaucus. Science 185*:372–374.

26 Arnold, S. J. 1976. Sexual behavior, sexual interference and sexual defense in the salamanders *Ambystoma maculatum, Ambystoma tigrinum* and *Plethodon jordani. Zeitschrift für Tierpsychologie 42*:247–300.

27 Arnold, S. J. 1980. The microevolution of feeding behavior. In *Foraging Behavior: Ecological, Ethological and Psychological Approaches.* A. Kamil and T. Sargent (eds.). Garland Press, New York.

28 Arnold, S. J. 1982. A quantitative approach to antipredator performance: Salamander defense against snake attack. *Copeia 1982*:247–253.

29 Arnold, S. J., and R. J. Wasserburg. 1978. Differential predation on metamorphic anurans by garter snakes (*Thamnophis*): Social behavior as a possible defense. *Ecology 59*:1014–1022.

30 Axelrod, R., and W. D. Hamilton. 1981. The evolution of cooperation. *Science 211*:1390–1396.

31 Baerends, G. P., et al. 1971. Reviews of *Animal Behavior* by R. A. Hinde. *Animal Behaviour 19*:791–818.

32 Baker, M. C. 1975. Song dialects and genetic differences in white-crowned sparrows, *Zonotrichia leucophrys. Evolution 29*:226–241.

33 Baker, M. C., and L. R. Mewaldt. 1978. Song dialects as barriers to dispersal in white-crowned sparrows, *Zonotrichia leucophrys nuttalli. Evolution 32*:712–722.

34 Baker, M. C., K. J. Spitler-Nabors, and D. C. Bradley. 1982. The response of female mountain white-crowned sparrows to songs from their natal dialect and an alien dialect. *Behavioral Ecology and Sociobiology 10*:175–179.

35 Baker, R. R. 1978. *The Evolutionary Ecology of Animal Migration.* Hodder & Stoughton, London.

36 Balickci, A. 1970. *The Netsilik Eskimo.* Doubleday (Natural History Press), Garden City, New York.

37 Balph, D. F., G. S. Innis, and M. H. Balph. 1980. Kin selection in Rio Grande turkeys: A critical reassessment. *Auk 97*:854–860.

38 Bänziger, H. 1971. Bloodsucking moths of Malaya. *Fauna 1*:4–16.

39 Bänziger, H. 1975. Skin-piercing blood-sucking moths. I. Ecological and ethological notes on *Calpe eustrigata* (Lepid, Noctuidae). *Acta Tropica 32*:125–144.

40 Bänziger, H. 1980. Skin-piercing blood-sucking moths. III: Feeding act and piercing mechanism of *Calyptra eustrigata* (Hmps.) (Lep., Noctuidae). *Bulletin de la Societé Entomologique Suisse 53*:127–142.

41 Barash, D. P. 1974. The evolution of marmot societies: A general theory. *Science 185*:415–420.

42 Barash, D. P. 1977. Sociobiology of rape in mallards (*Anas platyrhynchos*): Response of the mated male. *Science 197*:788–789.

43 Barash, D. P. 1982. *Sociobiology and Behavior,* 2nd edition. Elsevier, New York.

44 Barfield, R. J., and L. A. Geyer. 1972. Sexual behavior: Ultrasonic post-ejaculatory song of the male rat. *Science 176*:1349–1350.

45 Barlow, G. W. 1977. Modal action patterns. In *How Animals Communicate.* T. A. Sebeok (ed.). Indiana University Press, Bloomington.

46 Barrows, E. M. 1975. Individually distinctive odors in an invertebrate. *Behavioral Biology 15*:57–64.

47 Bastock, M. A. 1956. A gene mutation which changes a behavior pattern. *Evolution 10*:421–439.

48 Bateson, P. P. G. 1976. Rules and reciprocity in behavioural development. In *Growing Points in Ethology.* P. P. G. Bateson and R. A. Hinde (eds.). Cambridge University Press, Cambridge.

49 Bateson, P. P. G. 1982. Preferences for cousins in Japanese quail. *Nature 295*:236–237.

50 Bateson, P. P. G., W. Lotwick, and D. K. Scott. 1980. Similarities between the faces of parents and offspring in Bewick's swans and the differences between mates. *Journal of Zoology 191*:61–74.

51 Baylis, J. R. 1981. The evolution of parental care in fishes, with reference to Darwin's rule of male sexual selection. *Environmental Fish Biology 6*:223–251.

52 Beach, F. 1976. Sexual attractivity, proceptivity, and receptivity in female mammals. *Hormones and Behavior 7*:105–138.

53 Beecher, M. D. 1982. Signature systems and kin recognition. *American Zoologist 22*:477–490.

54 Beecher, M. D., and I. M. Beecher. 1979. Sociobiology of bank swallows: Reproductive strategy of the male. *Science 205*:1282–1285.

55 Beecher, M. D., I. M. Beecher, and S. Lumpkin. 1981. Parent–offspring recognition in bank swallows (*Riparia riparia*): I. Natural History. *Animal Behaviour 29*:86–94.

56 Bell, G. 1978. The handicap principle in sexual selection. *Evolution 32*:872–885.

57 Bell, G. 1982. *The Masterpiece of Nature: The Evolution and Genetics of Sexuality.* University of California Press, Berkeley, California.

58 Bellugi, U. 1970. Learning the language. *Psychology Today* (Dec.):32ff.

59 Belmont, L., and F. A. Marolla. 1973. Birth order, family size and intelligence. *Science 182*:1096–1101.

60 Belovsky, G. E. 1981. Food plant selection by a generalist herbivore: The moose. *Ecology 62*:1020–1030.

61 Bentley, D., and R. R. Hoy. 1972. Genetic control of the neuronal network generating cricket (*Teleogryllus gryllus*) song patterns. *Animal Behaviour 20*:478–492.

62 Bentley, D., and R. R. Hoy. 1974. The neurobiology of cricket song. *Scientific American 231*(Aug.):34–44.

63 Benzer, S. 1973. Genetic dissection of behavior. *Scientific American 229* (Dec.):24–37.

64 Berens von Rautenfeld, D. 1978. Bemerkungen zur Austauschbarkeil von Küken der Silbermöve (*Larus argentatus*) nach der ersten Lebenswoche. *Zeitschrift für Tierpsychologie 47*:180–181.

65 Berger, P. J., E. H. Sanders, P. D. Gardner, and N. C. Negus. 1977. Phenolic plant compounds functioning as reproductive inhibitors in *Microtus montanus*. *Science 195*:575–577.

66 Bernstein, H., G. S. Byers, and R. E. Michod. 1981. Evolution of sexual reproduction: Importance of DNA repairs, complementation and variation. *American Naturalist 117*:537–549.

67 Bernstein, I. L. 1978. Learned taste aversion in children receiving chemotherapy. *Science 200*:1302–1303.

68 Bertram, B. C. R. 1976. Kin selection in lions and evolution. In *Growing Points in Ethology*. P. P. G. Bateson and R. A. Hinde (eds.). Cambridge University Press, New York.

69 Bill, R. G., and W. F. Herrnkind. 1976. Drag reduction by formation movement in spiny lobsters. *Science 193*:1146–1148.

70 Birch, M. C. 1978. Chemical communication in pine bark beetles. *American Scientist 66*:409–419.

71 Birkhead, T. R. 1978. Behavioural adaptations to high density nesting in the common guillemot *Uria aalge*. *Animal Behaviour 26*:321–331.

72 Birkhead, T. R. 1979. Mateguarding in the magpie *Pica pica*. *Animal Behaviour 27*:866–874.

73 Black, A. H. 1971. The direct control of neural processes by reward and punishment. *American Scientist 59*:236–245.

74 Blest, A. D. 1957. The evolution of protective displays in the Saturnoidea and Sphingidae (Lepidoptera). *Behaviour 11*:257–309.

75 Blest, A. D. 1957. The function of eye-spot patterns in the Lepidoptera. *Behaviour 11*:209–256.

76 Blumer, L. S. 1979. Male parental care in the bony fishes. *Quarterly Review of Biology 54*:149–161.

77 Bolles, R. C. 1973. The comparative psychology of learning: The selective association principle and some problems with "general" laws of learning. In *Perspectives on Animal Behavior*. G. Bermant (ed.). Scott, Foresman & Company, Glenview, Illinois.

78 Bongaarts, J. 1980. Does malnutrition affect fecundity? A summary of the evidence. *Science 208*:564–569.

79 Borgia, G. 1980. Sexual competition in *Scatophaga stercoraria*: Size- and density-related changes in male ability to capture females. *Behaviour 75*:185–206.

80 Bouchard, T. J., Jr., and M. McGue. 1981. Familial studies of intelligence: A review. *Science 212*:1055–1059.

81 Bowman, R. I. 1961. Morphological differentiation and adaptation in the Galapagos finches. *University of California Publications in Zoology 58*:1–302.

82 Brachmachary, R. L., and J. Dutta. 1981. On the pheromones of tigers: Experiments and theory. *American Naturalist 118*:561–567.

83 Bradbury, J. W. 1977. Lek mating behavior in the hammer-headed bat. *Zeitschrift für Tierpsychologie 45*:225–255.

84 Bradbury, J. W. 1982. The evolution of leks. In *Natural Selection and Social Behavior*. R. D. Alexander and D. W. Tinkle (eds.). Chiron Press, New York.

85 Bradbury, J. W., and S. L. Vehrencamp. 1977. Social organization and foraging in emballonurid bats. III. Mating systems. *Behavioral Ecology and Sociobiology 2*:1–17.

86 Brady, R. O. 1976. Inherited metabolic diseases of the nervous system. *Science 193*:733–739.

87 Branch, G. M. 1975. Mechanisms reducing intraspecific competition in *Patella* spp.: Migration, differentiation and territorial behaviour. *Journal of Animal Ecology 44*:575–600.

88 Breed, M. D., and G. J. Gamboa. 1977. Behavioral control of workers by queens in primitively eusocial insects. *Science 195*:694–696.

89 Breven, K. A. 1981. Mate choice in the wood frog, *Rana sylvatica*. *Evolution 35*:707–722.

90 Bristowe, W. S. 1958. *The World of Spiders*. Collins, London.

91 Brockmann, H. J. 1979. Nest-site selection in the great golden digger wasp, *Sphex ichneumoneus* L. (Sphecidae). *Ecological Entomology 4*:211–224.

92 Brockway, B. F. 1964. Social influences on reproductive physiology and ethology of budgerigars (*Melopsittacus undulatus*). *Animal Behaviour 12*:493–501.

93 Brockway, B. F. 1965. Stimulation of ovarian development and egglaying by male courtship vocalizations in budgerigars (*Melopsittacus undulatus*). *Animal Behaviour 13*:575–578.

94 Brodie, E. D. 1977. Hedgehogs use toad venom in their own defense. *Nature 268*:627–628.

95 Brodie, E. D., Jr., and E. D. Brodie, III. 1980. Differential avoidance of mimetic salamanders by free-ranging birds. *Science 208*:181–182.

96 Brodie, E. D., Jr., J. L. Hensel, Jr., and J. A. Johnson. 1974. Toxicity of urodele amphibians *Taricha, Notophthalmus, Cynop,* and *Paramesotriton* (Salamandridae). *Copeia 1974*:506–511.

97 Brower, J. V. 1958. Experimental studies of mimicry in some North American butterflies. I. The monarch, *Danaus plexippus,* and viceroy, *Limenitis archippus archippus*. *Evolution 12*:3–47.

98 Brower, J. V., and L. P. Brower. 1962. Experimental studies of mimicry. 6. The reaction of toads (*Bufo terrestris*) to honeybees (*Apis mellifera*) and their dronefly mimics (*Eristalis vinetorum*). *American Naturalist 96*:297–307.

99 Brower, L. P. 1969. Ecological chemistry. *Scientific American 220*(Feb.):22–29.

100 Brower, L. P. 1971. Prey coloration and predator behavior. In *Topics in the Study of Life: The BIO Source Book, Section 6, Animal Behavior*. V. G. Dethier (ed.). Harper & Row, New York.

101 Brower, L. P. 1977. Monarch migration. *Natural History 86*(June–July):40–53.

102 Brower, L. P., J. V. Brower, and F. P. Cranston. 1965. Courtship behavior of the queen butterfly, *Danaus gilippus berenice* (Cramer). *Zoologica 59*:1–39.

103 Brower, L. P., and S. C. Glazier. 1975. Localization of heart poisons in the monarch butterfly. *Science 188*:19–25.

104 Brower, L. P., W. N. Ryerson, L. L. Coppinger, and S. C. Glazier. 1968. Ecological chemistry and the palatability spectrum. *Science 161*:1349–1351.

105 Brown, J. H., and G. A. Lieberman. 1973. Resource utilization and coexistence of seed eating desert rodents in sand dune habitats. *Ecology 54*:788–797.

106 Brown, J. L. 1975. *The Evolution of Behavior*. Norton, New York.

107 Brown, J. L., and E. R. Brown. 1981. Kin selection and individual selection in babblers. In *Natural Selection and Social Behavior*. R. D. Alexander and D. W. Tinkle (eds.). Chiron Press, New York.

108 Brown, J. L., and G. H. Orians. 1970. Spacing patterns in mobile animals. *Annual Review of Ecology and Systematics 1*:239–262.

109 Brownell, P. H. 1977. Compressional and surface waves in sand: Used by desert scorpions to locate prey. *Science 197*:479–482.

110 Brownmiller, S. 1975. *Against Our Will: Men, Women and Rape*. Simon & Schuster, New York.

111 Buchler, E. R., T. B. Wright, and E. D. Brown. 1981. On the functions of stridulation by the passalid beetle *Odontotaenius disjunctus* (Coleoptera: Passalidae). *Animal Behaviour 29*:483–486.

112 Buchsbaum, M. S., R. D. Coursey, and D. L. Murphy. 1976. The biochemical high-risk paradigm: Behavioral and familial correlates of low platelet monamine oxidase activity. *Science 194*:339–341.

113 Bunge, R., M. Johnson, and C. D. Ross. 1978. Nature and nurture in the development of the autonomic neuron. *Science 199*:1409–1416.

114 Burk, T. 1982. Evolutionary significance of predation on sexually signaling males. *Florida Entomologist 65*:90–104.

115 Burley, N. 1979. The evolution of concealed ovulation. *American Naturalist 114*:835–858.

116 Bush, G. L., R. W. Neck, and G. B. Kitto. 1976. Screwworm eradication: Inadvertent selection for noncompetitive ecotypes during mass rearing. *Science 193*:491–493.

117 Bygott, J. D., B. C. R. Bertram, and J. P. Hanby. 1979. Male lions in large coalitions gain reproductive advantage. *Nature 282*:839–841.

118 Cade, W. 1975. Acoustically orienting parasitoids: Fly phonotaxis to cricket song. *Science 190*:1312–1313.

119 Cade, W. 1980. Alternative male reproductive strategies. *Florida Entomologist 63*:30–45.

120 Cade, W. 1981. Alternative male strategies: Genetic differences in crickets. *Science 212*:563–564.

121 Callahan, J. R. 1981. Vocal solicitation and parental investment in female *Eutamias*. *American Naturalist 118*:872–875.

122 Calvert, W. H., L. E. Hedrick, and L. P. Brower. 1979. Mortality of the monarch butterfly (*Danaus plexippus* L.): Avian predation at five overwintering sites in Mexico. *Science 204*:847–851.

123 Campbell, B. 1966. *Human Evolution*. Aldine, Chicago.

124 Caraco, T. 1981. Energy budgets, risk and foraging preferences in dark-eyed juncos (*Junco hyemalis*). *Behavioral Ecology and Sociobiology 8*:213–217.

125 Caraco, T., and L. L. Wolf. 1975. Ecological determinants of group sizes of foraging lions. *American Naturalist 109*:343–352.

126 Carey, M., and V. Nolan, Jr. 1975. Polygyny in indigo buntings: A hypothesis tested. *Science 190*:1296–1297.

127 Carey, M., and V. Nolan, Jr. 1979. Population dynamics of indigo buntings and the evolution of avian polygyny. *Evolution 33*:1180–1192.

128 Carr, A. 1967. Adaptive aspects of the scheduled travel of *Chelonia*. In *Animal Orientation and Navigation*. R. M. Storm (ed.). Oregon State University Press, Corvallis.

129 Carr, A. 1967. *So Excellent a Fish*. Anchor Books, Garden City, New York.

130 Carter-Saltzman, L. 1980. Biological and sociocultural effects on handedness: Comparison between biological and adoptive families. *Science 209*:1263–1265.

131 Catchpole, C. K. 1980. Sexual selection and the evolution of complex songs among European warblers of the genus *Acrocephalus*. *Behaviour 74*:149–166.

132 Chagnon, N. 1968. *Yanomamö, The Fierce People*. Holt, Rinehart & Winston, New York.

133 Charnov, E. L., and J. R. Krebs. 1975. The evolution of alarm calls: Altruism or manipulation? *American Naturalist 109*:107–112.

134 Chapman, M., and G. Hausfater. 1979. The reproductive consequences of infanticide in langurs: A mathematical model. *Behavioral Ecology and Sociobiology 5*:227–240.

135 Chen, J.-S., and A. Amsel. 1980. Recall (versus recognition) of taste and immunization against aversive taste anticipations based on illness. *Science 209*:831–833.

136 Clutton-Brock, T. H., and S. D. Albon. 1979. The roaring of red deer and the evolution of honest advertisement. *Behaviour 69*:145–170.

137 Clutton-Brock, T. H., S. D. Albon, R. M. Gibson, and F. E. Guinness. 1979. The logical stag: Adaptive aspects of fighting in red deer. *Animal Behaviour 27*:211–225.

138 Codere, H. 1950. *Fighting with property*. American Ethnological Society Monograph, No. 18.

139 Conner, W. E., T. Eisner, R. K. Vander Meer, A. Guerrero, D. Ghiringelli, and J. Meinwald. 1980. Sex attractant of an arctiid moth (*Utetheisa ornatrix*): A pulsed chemical signal. *Behavioral Ecology and Sociobiology 7*:55–63.

140 Constanz, G. D. 1975. Behavioral ecology of mating in the male Gila topminnow, *Poeciliopsis occidentalis* (Cyprinodontiformes: Poeciliidae). *Ecology 56*:966–973.

141 Cook, S. B. 1969. Experiments on homing in the limpet *Siphonaria normalis*. *Animal Behaviour 17*:679–682.

142 Cooper, R. A., and J. R. Uzmann. 1971. Migrations and growth of deep-sea lobsters, *Homarus americanus*. *Science 171*:288–290.

143 Cott, H. 1940. *Adaptive Coloration in Animals*. Methuen, London.

144 Cox, C. R., and B. J. LeBoeuf. 1977. Female incitation of male competition: A mechanism in sexual selection. *American Naturalist 111*:317–335.

145 Crews, D. 1975. Psychobiology of reptilian reproduction. *Science 189*:1059–1065.

146 Crews, D. 1977. The annotated anole: Studies on the control of lizard reproduction. *American Scientist 65*:428–434.

147 Crews, D. 1979. The hormonal control of behavior in a lizard. *Scientific American 241*(Aug.):180–187.

148 Crews, D. 1980. Interrelationships among ecological, behavioral, and neuroendocrine processes in the reproductive cycle of *Anolis carolinensis* and other reptiles. *Advances in the Study of Behavior 11*:1–74.

149 Crews, D., and N. Greenberg. 1981. Function and causation of social signals in lizards. *American Zoologist 21*:273–294.
150 Cronin, E. W., Jr., and P. W. Sherman. 1977. A resource-based mating system: The orange-rumped honey guide. *Living Bird 15*:5–32.
151 Cruz, Y. P. 1981. A sterile defender morph in a polymorphic hymenopterous parasite. *Nature 294*:446–447.
152 Cuellar, O. 1977. Animal parthenogenesis. *Science 197*:837–843.
153 Cullen, E. 1957. Adaptations in the kittiwake to cliff nesting. *Ibis 99*:275–302.
154 Curio, E. 1976. *The Ethology of Predation.* Springer-Verlag, New York.
155 Daly, M., and M. Wilson. 1983. *Sex, Evolution and Behavior,* 2nd edition. Willard Grant Press, Boston.
156 Daly, M., and M. I. Wilson. 1981. Abuse and neglect of children in evolutionary perspective. In *Natural Selection and Social Behavior.* R. D. Alexander and D. W. Tinkle (eds.). Chiron Press, New York.
157 Daly, M., and M. Wilson. 1982. Homicide and kinship. *American Anthropologist 84*:372–378.
158 Daly, M., and M. I. Wilson. 1982. Whom are newborn babies said to resemble? *Ethology and Sociobiology 3*:69–78.
159 Daly, M., M. Wilson, and S. J. Weghorst. 1982. Male sexual jealousy. *Ethology and Sociobiology 3*:11–27.
160 Darwin, C. 1859. *On the Origin of Species.* Murray, London.
161 Davidson, D. W. 1977. Foraging ecology and community organization in desert seed-eating ants. *Ecology 58*:725–737.
162 Davies, N. B. 1977. Prey selection and social behaviour in wagtails (Aves: Motacillidae). *Journal of Animal Ecology 46*:37–57.
163 Davies, N. B., and T. R. Halliday. 1978. Deep croaks and fighting assessment in toads *Bufo bufo. Nature 274*:683–685.
164 Davies, N. B., and A. I. Houston. 1981. Owners and satellites: The economics of territory defense in the pied wagtail, *Motacilla alba. Journal of Animal Ecology 50*:157–180.
165 Davis, W. J., G. J. Mpitsos, and J. M. Pinneo. 1974. The behavioral hierarchy of the mollusk *Pleurobranchaea.* II. Hormonal suppression of feeding associated with egg-laying. *Journal of Comparative Physiology 90*:225–243.
166 Davis, W. J., G. J. Mpitsos, J. M. Pinneo, and J. L. Ram. 1977. Modification of the behavioral hierarchy of *Pleurobranchaea.* I. Satiation and feeding motivation. *Journal of Comparative Physiology 117*:99–125.
167 Davison, J. 1976. *Hydra hymanae*: Regulation of the life cycle by time and temperature. *Science 194*:618–620.
168 Dawkins, R. 1977. *The Selfish Gene.* Oxford University Press, New York.
169 Dawkins, R. 1980. Good strategy or evolutionarily stable strategy? In *Sociobiology: Beyond Nature/Nurture?* G. Barlow and J. Silverberg (eds.). Westview Press, Boulder, Colorado.
170 Dawkins, R., and J. R. Krebs. 1978. Animal signals: Information or manipulation? In *Behavioural Ecology, An Evolutionary Approach.* J. R. Krebs and N. B. Davies (eds.). Blackwell, Oxford.
171 Delcomyn, F. 1980. Neural basis of rhythmic behavior in animals. *Science 210*:492–498.
172 Dethier, V. G. 1962. *To Know A Fly.* Holden-Day, San Francisco.
173 Dethier, V. G. 1976. *The Hungry Fly, A Physiological Study of the Behavior*

Associated with Feeding. Harvard University Press, Cambridge, Massachusetts.

174 Dethier, V. G., and D. Bodenstein. 1958. Hunger in the blowfly. *Zeitschrift für Tierpsychologie 15*:129–140.

175 Devine, M. C. 1975. Copulatory plugs in snakes: Enforced chastity. *Science 187*:844–845.

176 Dewsbury, D. A. 1982. Ejaculate cost and male choice. *American Naturalist 119*:601–610.

177 Dewsbury, D. A., and D. Q. Estep. 1975. Pregnancy in cactus mice: Effects of prolonged copulation. *Science 187*:552–553.

178 Diamond, J. M. 1982. Rediscovery of the yellow-fronted gardener bowerbird. *Science 216*:431–434.

179 Diamond, J. M., E. Cooper, C. Turner, and L. Macintyre. 1976. Trophic regulation of nerve sprouting. *Science 193*:371–377.

180 Dickemann, M. 1979. Female infanticide and reproductive strategies of stratified human societies: A preliminary model. In *Evolutionary Biology and Human Social Behavior*. N. A. Chagnon and W. Irons (eds.). Duxbury Press, North Scituate, Massachusetts.

181 Dilger, W. C. 1962. The behavior of lovebirds. *Scientific American 206*(Jan.):88–98.

182 Dobzhansky, T. 1962. *Mankind Evolving*. Yale University Press, New Haven.

183 Dolhinow, P. C. 1977. Letter to the editor. *American Scientist 65*:266.

184 Dominey, W. J. 1980. Female mimicry in male bluegill sunfish—A genetic polymorphism? *Nature 284*:546–548.

185 Downes, J. A. 1973. Lepidoptera feeding at puddle-margins, dung, and carrion. *Journal of the Lepidopterists' Society 27*:89–99.

186 Downes, J. A. 1978. Feeding and mating in the insectivorous Ceratopogoninae (Diptera). *Memoirs of the Entomological Society of Canada 104*:1–62.

187 Downhower, J. F., and K. B. Armitage. 1971. The yellow-bellied marmot and the evolution of polygamy. *American Naturalist 105*:355–370.

188 Duffey, S. S. 1970. Cardiac glycosides and distastefulness: Some observations on the palatability spectrum of butterflies. *Science 169*:78–79.

189 Dunford, C. 1977. Kin selection for ground squirrel alarm calls. *American Naturalist 111*:782–785.

190 Dunn, J. 1976. How far do early differences in mother–child relations affect later development? In *Growing Points in Ethology*. P. P. G. Bateson and R. A. Hinde (eds.). Cambridge University Press, Cambridge.

191 Durham, W. H. 1976. The adaptive significance of cultural behavior. *Human Ecology 4*:89–121.

192 Durham, W. H. 1976. Resource competition and human aggression. *Quarterly Review of Biology 51*:385–415.

193 Dyer, F. C., and J. L. Gould. 1981. Honey bee orientation: A backup system for cloudy days. *Science 214*:1041–1042.

194 Dyson-Hudson, R., and N. Dyson-Hudson. 1969. Subsistence herding in Uganda. *Scientific American 220*(Feb.):76–89.

195 Eberhard, W. G. 1974. The natural history and behaviour of the wasp *Trigonopsis cameronii* Kohl (Sphecidae). *Transactions of the Royal Entomological Society of London 125*:295–328.

196 Eberhard, W. G. 1977. Aggressive chemical mimicry by a bolas spider. *Science 198*:1173–1175.

197 Eckert, R., and Y. Naitoh. 1972. Bioelectric control of locomotion in the ciliates. *Journal of Protozoology 19*:237–243.

198 Edmunds, G. F., Jr., and D. N. Alstad. 1978. Coevolution in insect herbivores and conifers. *Science 199*:941–945.

199 Edmunds, M. 1974. *Defence in Animals*. Longman Group, Harlow, Essex.

200 Edwards, J. S. 1966. Observations on the life history and predatory behaviour of *Zelus exsanguis* (Stål) (Heteroptera: Reduviidae). *Proceedings of the Royal Entomological Society of London* (A) *41*:21–24.

201 Ehrlich, P., and P. H. Raven. 1967. Butterflies and plants. *Scientific American 216*(June):104–113.

202 Ehrman, L., and P. A. Parsons. 1976. *The Genetics of Behavior*. Sinauer Associates, Sunderland, Massachusetts.

203 Eibl-Eibesfeldt, I. 1975. *Ethology, The Biology of Behavior*, 2nd edition. Holt, Rinehart, & Winston, New York.

204 Eimas, P. D. 1975. Speech perception in early infancy. In *Infant Perceptions from Sensation to Cognition*. L. B. Cohen and P. Salapatek (eds.). Academic Press, New York.

205 Eisner, T. E. 1966. Beetle spray discourages predators. *Natural History 75*(Feb.):42–47.

206 Eisner, T. E. 1970. Chemical defense against predation in arthropods. In *Chemical Ecology*. E. Sondheimer and J. B. Simeone (eds.). Academic Press, New York.

207 Eisner, T., D. F. Weimer, L. W. Haynes, and J. Meinwald. 1978. Lucibufagins: Defense steroids from the fireflies *Photinus ignitus* and *Photinus marginellus* (Coleoptera, Lampyridae). *Proceedings of the National Academy of Sciences 75*:905–908.

208 Elliott, P. F. 1975. Longevity and the evolution of polygamy. *American Naturalist 109*:281–287.

209 Emlen, J. T., and R. L. Penney. 1966. The navigation of penguins. *Scientific American 215*(Oct.):104–113.

210 Emlen, S. T. 1975. The stellar-orientation system of a migratory bird. *Scientific American 223*(Aug.):102–111.

211 Emlen, S. T. 1975. Migration: Orientation and navigation. In *Avian Biology*, Vol. 5. D. S. Farner and J. R. King (eds.). Academic Press, New York.

212 Emlen, S. T. 1978. Cooperative breeding. In *Behavioural Ecology, An Evolutionary Approach*. J. R. Krebs and N. B. Davies (eds.). Blackwell, Oxford.

213 Emlen, S. T. 1980. Ecological determinism and sociobiology. In *Sociobiology: Beyond Nature/Nurture?* G. W. Barlow and J. Silverberg (eds.). Westview Press, Boulder, Colorado.

214 Emlen, S. T. 1981. Altruism, kinship and reciprocity in the white-fronted bee-eater. In *Natural Selection and Social Behavior*. R. D. Alexander and D. W. Tinkle (eds.). Chiron Press, New York.

215 Emlen, S. T. 1982. The evolution of helping. I. An ecological constraints model. *American Naturalist 119*:29–39.

216 Emlen, S. T. 1982. The evolution of helping. II. The role of behavioral conflict. *American Naturalist 119*:40–53.

217 Emlen, S. T., and L. W. Oring. 1977. Ecology, sexual selection and the evolution of mating systems. *Science 197*:215–223.

218 Emlen, S. T., W. Wiltschko, N. J. Demong, R. Wiltschko, and S. Bergman. 1976. Magnetic direction finding: Evidence for its use in migratory indigo buntings. *Science 193*:505–508.

219 Englemann, F. 1970. *The Physiology of Insect Reproduction.* Pergamon Press, Oxford.
220 Epstein, R., R. P. Lanza, and B. F. Skinner. 1980. Symbolic communication between two pigeons (*Columba livia domestica*). *Science 207*:543–545.
221 Erickson, C. J., and P. G. Zenone. 1977. Courtship differences in male ring doves: Avoidance of cuckoldry? *Science 192*:1353–1354.
222 Esch, H. 1967. The evolution of bee language. *Scientific American 216*(Apr.):96–104.
223 Evans, H. E. 1966. *The Comparative Ethology and Evolution of the Sand Wasps.* Harvard University Press, Cambridge, Massachusetts.
224 Evans, H. E. 1966. *Life on a Little Known Planet.* Dell, New York.
225 Evans, H. E. 1977. Extrinsic and intrinsic factors in the evolution of insect sociality. *BioScience 27*:613–617.
226 Evans, H. E., and M. J. W. Eberhard. 1970. *The Wasps.* University of Michigan Press, Ann Arbor.
227 Evans, H. E., and K. M. O'Neill. 1978. Alternate mating strategies in the digger wasp *Philanthus zebratus. Proceedings of the National Academy of Science 75*:1901–1903.
228 Ewert, J.-P. 1974. The neural basis of visually guided behavior. *Scientific American 230*(Mar.):34–42.
229 Ewert, J.-P. 1980. *Neuro-Ethology.* Springer-Verlag, New York.
230 Ewing, A. W. 1963. Attempts to select for spontaneous activity in *Drosophila melanogaster. Animal Behaviour 11*:369–378.
231 Farentinos, R. C., P. J. Capretta, R. E. Kepner, and V. M. Littlefield. 1981. Selective herbivory in tassel-eared squirrels: Role of monoterpenes in ponderosa pines chosen as feeding trees. *Science 213*:1273–1275.
232 Farlow, J. O. 1981. Estimates of dinosaur speeds from a new trackway site in Texas. *Nature 294*:747–748.
233 Farner, D. S. 1964. Time measurement in vertebrate photoperiodism. *American Naturalist 98*:375–386.
234 Farner, D. S., and R. A. Lewis. 1971. Photoperiodism and reproductive cycles in birds. *Photophysiology 6*:325–370.
235 Feduccia, A. 1980. *The Age of Birds.* Harvard University Press, Cambridge, Massachusetts.
236 Feduccia, A., and H. B. Tordoff. 1970. Feathers of *Archeopteryx*: Asymmetric vanes indicate aerodynamic function. *Science 203*:1021–1022.
237 Fink, L. S., and L. P. Brower. 1981. Birds can overcome the cardenolide defence of monarch butterflies in Mexico. *Nature 291*:67–70.
238 Fisher, R. A. 1930. *The Genetical Theory of Natural Selection.* Clarendon Press, Oxford.
239 Foster, M. S. 1977. Odd couples in manakins: A study of social organization and cooperative breeding in *Chiroxiphia linearis. American Naturalist 11*:845–853.
240 Franklin, W. L. 1974. The social behaviour of the vicuña. In *The Behaviour of Ungulates and Its Relation to Management.* Vol. 1. V. Giest and F. Walther (eds.). IUCN Publications, Morges, Switzerland.
241 Freeland, W. J. 1981. Parasitism and behavioral dominance among male mice. *Science 213*:461–462.
242 Frisch, R. E. 1978. Population, food intake and fertility. *Science 199*:22–29.
243 Gadgil, M. 1972. Male dimorphism as a consequence of sexual selection. *American Naturalist 106*:574–580.

244 Galaburda, A. M., M. LeMay, T. L. Kemper, and N. Geschwind. 1978. Right–left asymmetries in the brain. *Science 199*:852–856.

245 Gamboa, G. J. 1978. Intraspecific defense: Advantage of social cooperation among paper wasp foundresses. *Science 199*:1463–1465.

246 Garcia, J., and F. R. Ervin. 1968. Gustatory–visceral and telereceptor–cutaneous conditioning: Adaptation in internal and external milieus. *Communications in Behavioral Biology* (A) *1*:389–415.

247 Garcia, J., W. G. Hankins, and K. W. Rusiniak. 1974. Behavioral regulation of the milieu interne in man and rat. *Science 185*:824–831.

248 Gardner, R. A., and B. T. Garnder. 1969. Teaching sign language to a chimpanzee. *Science 165*:664–672.

249 Garrick, L. D., and J. W. Lang. 1977. The alligator revealed. *Natural History 87*(June–July):54–61.

250 Garstka, W. R., and D. Crews. 1981. Female sex pheromone in the skin and circulation of a garter snake. *Science 214*:681–682.

251 Geschwind, N. 1970. The organization of language and the brain. *Science 170*:940–944.

252 Getting, P. A. 1975. *Tritonia* swimming: Triggering of a fixed action pattern. *Brain Research 96*:128–133.

253 Ghiselin, M. T. 1974. *The Economy of Nature and the Evolution of Sex*. University of California Press, Berkeley.

254 Ghysen, A. 1978. Sensory neurones recognize defined pathways in *Drosophila* central nervous system. *Nature 274*:869–872.

255 Gibbons, J. A., and H. B. Lillywhite. 1981. Ecological segregation, color matching and speciation in lizards of the *Amphibolurus decresii* species complex (Lacertilia: Agamidae). *Ecology 62*:1573–1584.

256 Gilbert, L. E. 1976. Postmating female odor in *Heliconius* butterflies: A male-contributed antiaphrodisiac? *Science 193*:419–420.

257 Gill, F. B., and L. L. Wolf. 1975. Economics of feeding territoriality in the golden-winged sunbird. *Ecology 56*:333–345.

258 Gill, F. B., and L. L. Wolf. 1975. Foraging strategies and energetics of East African sunbirds at mistletoe flowers. *American Naturalist 109*:491–510.

259 Gill, F. B., and L. L. Wolf. 1978. Comparative foraging efficiencies of some montane sunbirds in Kenya. *Condor 80*:391–400.

260 Gilliard, E. T. 1969. *Birds of Paradise and Bower Birds*. Natural History Press, Garden City, New Jersey.

261 Gish, S. L., and E. S. Morton. 1981. Structural adaptations to local habitat acoustics in Carolina wren songs. *Zeitschrift für Tierpsychologie 56*:74–84.

262 Gittleman, J. L., and P. H. Harvey. 1980. Why are distasteful prey not cryptic? *Nature 286*:149–150.

263 Goldin-Meadow, S., and H. Feldman. 1977. The development of language-like communication without a language model. *Science 197*:401–403.

264 Goodall, J. 1968. The behaviour of free-living chimpanzees in the Gombe Stream Reserve. *Animal Behaviour Monographs 1*:165–301.

265 Goodman, C. S., and N. C. Spitzer. 1979. Embryonic development of identified neurones: Differentiation from neuroblast to neurons. *Nature 280*:208–214.

266 Gould, J. L. 1975. Honey bee recruitment. *Science 189*:685–693.

267 Gould, J. L. 1980. The case for magnetic field sensitivity in birds and bees. *American Scientist 68*:256–267.

268 Gould, J. L. 1982. Why do honey bees have dialects? *Behavioral Ecology and Sociobiology 10*:53–56.

269 Gould, J. L. 1982. *Ethology, The Mechanisms and Evolution of Behavior.* Norton, New York.

270 Gould, J. L., M. Henerey, and M. C. MacLeod. 1970. Communication of direction by the honeybee. *Science 169*:544–554.

271 Gould, R. A. 1969. *Yiwara, Foragers of the Australian Desert.* Scribner, New York.

272 Gould, S. J. 1974. This view of life: The nonscience of human nature. *Natural History 83*(Apr.):21–25.

273 Gould, S. J. 1977. Evolution's erratic pace. *Natural History 86*(May):12–16.

274 Gould, S. J. 1980. Sociobiology and the theory of natural selection. In *Sociobiology: Beyond Nature/Nurture?* G. Barlow and J. Silverberg (eds.). Westview Press, Boulder, Colorado.

275 Gould, S. J., and R. Lewontin. 1979. The spandrels of San Marco and the Panglossian paradigm: A critique of the adaptationist programme. *Proceedings of the Royal Society of London* (B) *205*:581–598.

276 Gouzoules, H. 1974. Harassment of sexual behavior in the stumptail macaque *Macaca arctoides. Folia Primatologica 22*:208–217.

277 Grayson, D. K. 1977. Pleistocene avifaunas and the overkill hypothesis. *Science 195*:691–693.

278 Greene, H. W. 1973. Defensive tail display by snakes and amphisbaenians. *Journal of Herpetology 7*:143–161.

279 Greene, H. W., and J. A. Campbell. 1972. Note on the use of caudal lures by arboreal green pit vipers. *Herpetologica 28*:32–34.

280 Greenfield, M. D. 1981. Moth sex pheromones: An evolutionary perspective. *Florida Entomologist 64*:4–17.

281 Greenstone, M. H. 1979. Spider feeding behavior optimises dietary essential amino acid composition. *Nature 282*:501–503.

282 Griffin, D. R. 1958. *Listening in the Dark.* Yale University Press, New Haven.

283 Griffin, D. R. 1958. More about bat "radar." *Scientific American 199*(July):40–44.

284 Griffin, D. R., F. A. Webster, and C. R. Michael. 1960. The echolocation of flying insects by bats. *Animal Behaviour 8*:141–154.

285 Grohmann, J. 1939. Modifikation oder Funktionsreifung? *Zietschrift für Tierpsychologie 2*:132–144.

286 Grosberg, R. K. 1981. Competitive ability influences habitat choice in marine invertebrates. *Nature 290*:700–702.

287 Gross, D. R., G. Eiten, N. M. Flowers, F. M. Leoi, M. L. Ritter, and D. W. Werner. 1979. Ecology and acculturation among native peoples of central Brazil. *Science 206*:1043–1050.

288 Gross, M. R., and E. L. Charnov. 1980. Alternative life histories in bluegill sunfish. *Proceedings of the National Academy of Sciences 77*:6937–6940.

289 Gross, M. R., and A. M. MacMillan. 1981. Predation and the evolution of colonial nesting in bluegill sunfish (*Lepomis macrochirus*). *Behavioral Ecology and Sociobiology 8*:163–174.

290 Gross, M. R., and R. Shine. 1981. Parental care and mode of fertilization and ectothermic vertebrates. *Evolution 35*:775–793.

291 Gurney, M. E., and M. Konishi. 1980. Hormone-induced sexual differentiation of brain and behavior in zebra finches. *Science 208*:1380–1383.

292 Gwinner, E., and W. Wiltschko. 1980. Circannual changes in migratory orientation of the garden warbler, *Sylvia borin. Behavioral Ecology and Sociobiology 7*:73–78.

293 Gwynne, D. T. 1981. Sexual difference theory: Mormon crickets show role reversal in mate choice. *Science 213*:779–780.

294 Hailman, J. P. 1967. The ontogeny of an instinct. *Behaviour Supplements 15*:1–159.

295 Hailman, J. P. 1977. *Optical Signals: Animal Communication and Light.* Indiana University Press, Bloomington.

296 Hall, J. C. 1977. Portions of the central nervous system controlling reproductive behavior in *Drosophila melanogaster. Behavior Genetics* 7:291–312.

297 Hall, J. C., R. J. Greenspan, and W. A. Harris. 1982. *Genetic Neurobiology.* MIT Press, Cambridge, Massachusetts.

298 Hall, K. R. L., and G. B. Schaller. 1964. Tool-using behavior of the California sea otter. *Journal of Mammalogy 45*:287–298.

299 Hamilton, W. D. 1963. The evolution of altruistic behavior. *American Naturalist 97*:354–356.

300 Hamilton, W. D. 1964. The evolution of social behavior. *Journal of Theoretical Biology* 7:1–52.

301 Hamilton, W. D. 1970. Selfish and spiteful behavior in an evolutionary model. *Nature 228*:1218–1220.

302 Hamilton, W. D. 1971. Geometry for the selfish herd. *Journal of Theoretical Biology 31*:295–311.

303 Hamilton, W. D. 1975. Gamblers since life began: Barnacles, aphids, elms. *Quarterly Review of Biology 50*:175–180.

304 Hamilton, W. D., P. A. Henderson, and N. A. Moran. 1981. Fluctuation of environment and coevolved antagonist polymorphism as factors in the maintenance of sex. In *Natural Selection and Social Behavior.* R. D. Alexander and D. W. Tinkle (eds.). Chiron Press, New York.

305 Hamilton, W. J., and G. H. Orians. 1965. Evolution of brood parasitism in altricial birds. *Condor 67*:361–382.

306 Handford, P., and F. Nottebohm. 1976. Allozymic and morphological variation in population samples of rufous-collared sparrow, *Zonotrichia capensis,* in relation to vocal dialects. *Evolution 30*:802–817.

307 Harcourt, A. H., P. H. Harvey, S. G. Larson, and R. V. Short. 1981. Testis weight, body weight and breeding system in primates. *Nature 293*:55–57.

308 Harlow, H. F., and M. K. Harlow. 1962. Social deprivation in monkeys. *Scientific American 207*(Nov.):136–146.

309 Harlow, H. F., M. K. Harlow, and S. J. Suomi. 1971. From thought to therapy: Lessons from a primate laboratory. *American Scientist 59*:538–549.

310 Harris, M. 1977. *Cannibals and Kings: The Origins of Cultures.* Random House, New York.

311 Hartung, J. 1981. Genome parliaments and sex with the Red Queen. In *Natural Selection and Social Behavior.* R. D. Alexander and D. W. Tinkle (eds.). Chiron Press, New York.

312 Haseltine, F. P., and S. Ohno. 1981. Mechanisms of gonadal differentiation. *Science 211*:1272–1277.

313 Hausfater, G. 1975. Dominance and reproduction in baboons (*Papio cynocephalus*): A quantitative analysis. *Contributions in Primatology* 7:1–150.

314 Hay, R. L., and M. D. Leakey. 1982. The fossil footprints of Laetoli. *Scientific American 246*(Feb.):50–57.

315 Heinrich, B. 1976. The foraging specializations of individual bumblebees.

Ecological Monographs 46:105–128.

316 Heinrich, B. 1979. *Bumblebee Economics.* Harvard University Press, Cambridge, Massachusetts.

317 Heinrich, B. 1979. Foraging strategies of caterpillars: Leaf damage and possible predator avoidance strategies. *Oecologia 42*:325–337.

318 Heinrich, B., and S. L. Collins. In press. Caterpillar leaf damage, and the game of hide-and-seek with birds. *Ecology.*

319 Hennessy, D. F., and D. H. Owings. 1978. Snake species discrimination and the role of olfactory cues in the snake-directed behavior of the California ground squirrel. *Behaviour 65*:115–124.

320 Hennessy, D. F., D. H. Owings, M. P. Rowe, R. G. Coss, and D. W. Leger. 1981. The information afforded by a variable signal: Constraints on snake-elicited tail flagging by California ground squirrels. *Behaviour 78*:188–226.

321 Henry, C. S. 1972. Eggs and rapagula of *Ulolodea* and *Ascaloptynx* (Neuroptera: Ascalaphidae): A comparative study. *Psyche 79*:1–22.

322 Henson, O. W., Jr. 1970. The ear and audition. In *Biology of Bats,* Vol. II. W. A. Wimsatt (ed.). Academic Press, New York.

323 Heston, L. 1970. The genetics of schizophrenic and schizoid disease. *Science 167*:248–255.

324 Hill, J. L. 1974. *Peromyscus*: Effect of early pairing on reproduction. *Science 186*:1042–1044.

325 Hinde, R. A. 1970. *Animal Behavior,* 2nd edition. McGraw-Hill, New York.

326 Hirth, D. H., and D. R. McCullough. 1977. Evolution of alarm signals in ungulates with special reference to white-tailed deer. *American Naturalist 111*:31–42.

327 Holden, C. 1980. Identical twins reared apart. *Science 207*:1323–1328.

328 Hölldobler, B. 1971. Communication between ants and their guests. *Scientific American 224*(Mar.):86–95.

329 Hölldobler, B. 1974. Communication by tandem running in the ant *Camponotus sericeus. Journal of Comparative Physiology 90*:105–127.

330 Hölldobler, B. 1975. Home range orientation and territoriality in harvester ants. *Proceedings of the National Academy of Science 71*:3274–3277.

331 Hölldobler, B. 1976. Recruitment behavior, home range orientations and territoriality in harvester ants, *Pogonomyrmex. Behavioral Ecology and Sociobiology 1*:3–44.

332 Hölldobler, B. 1979. Territories of the African weaver ant (*Oecophylla longinoda* (Latreille)), a field study. *Zeitschrift für Tierpsychologie 51*:201–213.

333 Hölldobler, B. 1979. Territoriality in ants. *Proceedings of the American Philosophical Society 123*:211–218.

334 Hölldobler, B. 1980. Canopy orientation: A new kind of orientation in ants. *Science 210*:86–88.

335 Hölldobler, B., and C. J. Lumsden. 1980. Territorial strategies in ants. *Science 210*:732–739.

336 Holmes, W. G., and P. W. Sherman. 1982. The ontogeny of kin recognition in two species of ground squirrels. *American Zoologist 22*:491–517.

337 Hoogland, J. L. 1979. Aggression, ectoparasitism and other possible costs of prairie dog (Sciuridae, *Cynomys* spp.) coloniality. *Behaviour 69*:1–35.

338 Hoogland, J. L. 1981. The evolution of coloniality in white-tailed and black-tailed prairie dogs (Sciuridae: *Cynomys leucurus* and *C. ludovicianus*). *Ecology 62*:252–272.

339 Hoogland, J. L. 1982. Prairie dogs avoid extreme inbreeding. *Science 215*:1639–1641.
340 Hoogland, J. L., and P. W. Sherman. 1976. Advantages and disadvantages of bank swallow (*Riparia riparia*) coloniality. *Ecological Monographs 46*:33–58.
341 Hopkins, C. D. 1972. Sex difference in electric signalling in an electric fish. *Science 176*:1035–1037.
342 Hopkins, C. D. 1974. Electric communication in fish. *American Scientist 62*:426–437.
343 Hopkins, C. D. 1980. Evolution of electric communication channels of mormyrids. *Behavioral Ecology and Sociobiology 7*:1–13.
344 Horner, J. R., and R. Makela. 1979. Nest of juveniles provides evidence of family-structure among dinosaurs. *Nature 282*:296–298.
345 Hotta, Y., and S. Benzer. 1979. Courtship in *Drosophila* mosaics: Sex-specific foci for sequential action patterns. *Proceedings of the National Academy of Science 73*:4154–4158.
346 Howard, R. D. 1978. The evolution of mating strategies in bullfrogs, *Rana catesbiana*. *Evolution 32*:850–871.
347 Howard, R. D. 1980. Mating behaviour and mating success in woodfrogs, *Rana sylvatica*. *Animal Behaviour 28*:705–716.
348 Howard, R. R., and E. D. Brodie, Jr. 1973. A Batesian mimetic complex in salamanders: Response of avian predators. *Herpetologica 29*:33–41.
349 Howard, R. W., C. A. McDaniel, and G. J. Blomquist. 1980. Chemical mimicry as an integrating mechanism: Cuticular hydrocarbons of a termitophile and its host. *Science 210*:431–433.
350 Hrdy, S. B. 1977. *The Langurs of Abu*. Harvard University Press, Cambridge, Massachusetts.
351 Hrdy, S. B. 1977. Infanticide as a primate reproductive strategy. *American Scientist 65*:40–49.
352 Hrdy, S. B. 1979. The evolution of human sexuality: The latest word and the last. *Quarterly Review of Biology 54*:309–314.
353 Hubel, D. H., and T. N. Wiesel. 1965. Receptive fields and functional architecture in two non-striate visual areas (18 and 19) of the cat. *Journal of Neurophysiology 28*:229–289.
354 Hubel, D. H., and T. N. Wiesel. 1979. Brain mechanisms of vision. *Scientific American 241*(Mar.):150–162.
355 Hubert, H. B., R. R. Fabsitz, M. Feinleib, and K. S. Brown. 1980. Olfactory sensitivity in humans: Genetic versus environmental control. *Science 208*:607–608.
356 Immelmann, K. 1969. Song development in the zebra finch and other estrildid finches. In *Bird Vocalizations*. R. A. Hinde (ed.) Cambridge University Press, Cambridge.
357 Inglis, I. R., and J. Lasarus. 1981. Vigilance and flock size in brent geese: The edge effect. *Zeitschrift für Tierpsychologie 57*:193–200.
358 Irons, W. 1979. Cultural and biological success. In *Evolutionary Biology and Human Social Behavior*. N. Chagnon and W. Irons (eds.). Duxbury Press, North Scituate, Massachusetts.
359 Jaenike, J. 1978. An hypothesis to account for the maintenance of sex within populations. *Evolutionary Theory 3*:191–194.
360 Janzen, D. H. 1972. Protection of *Barteria* (Passifloraceae) by *Pachysima* ants (Pseudomyrmecinae) in a Nigerian rain forest. *Ecology 53*:885–892.

361 Jarman, M. V. 1979. Impala social behaviour. Territory, hierarchy, mating and use of space. *Fortschritte Verhaltensforschung 21*:1–92.

362 Jarvis, J. U. M. 1981. Eusociality in a mammal: Cooperative breeding in naked mole-rat colonies. *Science 212*:571–573.

363 Jaycox, E. R., and S. G. Parise. 1980. Homesite selection by Italian honey bee swarms, *Apis mellifera ligustica* (Hymenoptera: Apidae). *Journal of the Kansas Entomological Society 53*:171–178.

364 Jaycox, E. R., and S. G. Parise. 1981. Homesite selection by swarms of black-bodied honey bees, *Apis mellifera caucasia* and *A. m. carnica*. *Journal of the Kansas Entomological Society 54*:697–703.

365 Jeanne, R. L. 1970. Chemical defense of brood by a social wasp. *Science 168*:1465–1466.

366 Jen, P. H.-S., and N. Suga. 1976. Coordinated activities of middle-ear and laryngeal muscles in echolocating bats. *Science 191*:950–952.

367 Jenni, D. A. 1974. Evolution of polyandry in birds. *American Zoologist 14*:129–144.

368 Jenni, D. A., and G. Collier. 1972. Polyandry in the American jacana. *Auk 89*:743–765.

369 Jenssen, T. A. 1977. Evolution of anoline lizard display behavior. *American Zoologist 17*:302–315.

370 Kagan, J., and R. E. Klein. 1973. Cross-cultural perspectives on early development. *American Psychologist 28*:947–961.

371 Katz, S. H., M. L. Hediger, and L. A. Valleroy. 1974. Traditional maize processing techniques in the New World. *Science 184*:765–773.

372 Keeton, W. T. 1974. The mystery of pigeon homing. *Scientific American 231*(Dec.):96–107.

373 Keeton, W. T. 1974. The orientational and navigational basis of homing in birds. *Advances in the Study of Behavior 5*:47–132.

374 Kenagy, G. J. 1972. Saltbush leaves: Excision of hypersaline tissue by a kangaroo rat. *Science 178*:1094–1096.

375 Kenward, R. E. 1978. Hawks and doves: Factors affecting success and selection in goshawk attacks on wild pigeons. *Journal of Animal Ecology 47*:449–460.

376 Kessel, E. L. 1955. Mating activities of balloon flies. *Systematic Zoology 4*:97–104.

377 Kettlewell, H. B. D. 1955. Selection experiments on industrial melanism in the Lepidoptera. *Heredity 9*:323–343.

378 Kimura, D. 1973. The asymmetry of the human brain. *Scientific American 228*(Mar.):70–78.

379 King, A. P., and M. J. West. 1977. Species identification in the North American cowbird: Appropriate responses to abnormal song. *Science 195*:1002–1004.

380 King, M. C., and A. C. Wilson. 1975. Evolution at two levels in humans and chimpanzees. *Science 188*:107–116.

381 Kirk, V. M., and B. J. Dupraz. 1972. Discharge by a female ground beetle, *Pterostichus lucublandus* (Coleoptera: Carabidae), used as a defense against males. *Annals of the Entomological Society of America 65*:513.

382 Koenig, W. D. 1981. Reproductive success, group size and the evolution of cooperative breeding in the acorn woodpecker. *American Naturalist 117*:421–443.

383 Kolata, G. B. 1974. !Kung hunter-gatherers: Feminism, diet and birth control. *Science 185*:932–934.

384 Konishi, M. 1965. The role of auditory feedback in the control of vocalization in the white-crowned sparrow. *Zeitschrift für Tierpsychologie 22*:770–783.

385 Konishi, M., and F. Nottebohm. 1969. Experimental studies on the ontogeny of avian vocalization. In *Bird Vocalization*. R. A. Hinde (ed.). Cambridge University Press, New York.

386 Koshland, D. E. 1977. A response regulator model in a simple sensory system. *Science 196*:1055–1063.

387 Kovac, M. P., and W. J. Davis. 1977. Behavioral choice: Neural mechanisms in *Pleurobranchaea*. *Science 198*:632–634.

388 Krebs, J. R. 1971. Territory and breeding density in the great tit, *Parus major* L. *Ecology 52*:2–22.

389 Krebs, J. R. 1976. Sexual selection and the handicap principle. *Nature 261*:192.

390 Krebs, J. R. 1977. The significance of song repertoires: The Beau Geste hypothesis. *Animal Behaviour 25*:475–478.

391 Krebs, J. R. 1978. Optimal foraging: Decision rules for predators. In *Behavioural Ecology, An Evolutionary Approach*. J. R. Krebs and N. B. Davies (eds.). Blackwell, London.

392 Krebs, J. R., and N. B. Davies. 1981. *An Introduction to Behavioural Ecology*. Sinauer Associates, Sunderland, Massachusetts.

393 Krebs, J. R., R. Ashcroft, and M. Webber. 1978. Song repertoires and territory defense in the great tit. *Nature 271*:539–542.

394 Kroodsma, D. 1976. Reproductive development in a songbird: Differential stimulation by quality of male song. *Science 192*:574–575.

395 Kroodsma, D. 1978. Aspects in the ontogeny of bird song: Where, from whom, when, how many, which and how accurately? In *The Development of Behavior*. G. M. Burghardt and M. Bekoff (eds.). Garland STPM Publishing, New York.

396 Kruuk, H. 1964. Predators and anti-predator behaviour of the black-headed gull *Larus ridibundus*. *Behaviour Supplements 11*:1–129.

397 Kruuk, H. 1972. *The Spotted Hyena*. University of Chicago Press, Chicago.

398 Kuffler, S. W., and J. G. Nicholls. 1976. *From Neuron to Brain*. Sinauer Associates, Sunderland, Massachusetts.

399 Kung, C., S.-Y. Chang, Y. Satow, J. van Houten, and H. Hansma. 1975. Genetic dissection of behavior in *Paramecium*. *Science 188*:898–904.

400 Labov, J. B. 1980. Factors influencing infanticidal behavior in wild male housemice (*Mus musculus*). *Behavioral Ecology and Sociobiology 6*:297–303.

401 Labov, J. B. 1981. Pregnancy blocking in rodents: Adaptive advantages for females. *American Naturalist 118*:361–371.

402 Lack, D. 1947. *Darwin's Finches*. Cambridge University Press, New York.

403 Lack, D. 1968. *Ecological Adaptations for Breeding in Birds*. Methuen, London.

404 Lall, A. B., H. H. Seliger, W. H. Biggley, and J. E. Lloyd. 1980. Ecology of colors of firefly bioluminescence. *Science 210*:560–562.

405 LeBoeuf, B. J. 1974. Male–male competition and reproductive success in elephant seals. *American Zoologist 14*:163–176.

406 LeCroy, M. 1981. The genus *Paradisaea*—Display and evolution. *American Museum Novitates 2714*:1–52.

407 Lederhouse, R. C. 1982. Territorial defense and lek behavior of the black swallowtail butterfly, *Papilio polyxenes*. *Behavioral Ecology and Sociobiology 10*:109–118.

408 Lee, R. B. 1979. *The !Kung San*. Cambridge University Press, Cambridge.

409 Lee, T., P. Seeman, W. A. Toutellotte, I. J. Farley, and O. Hornykeiwicz. 1978. Binding of ^{3}H-neuropeptides and ^{3}H-apomorphine in schizophrenic brains. *Nature 274*:897–900.

410 Lehrman, D. S. 1953. A critique of Konrad Lorenz's theory of instinctive behavior. *Quarterly Review of Biology 28*:337–363.

411 Lehrman, D. S. 1970. Semantic and conceptual issues in the nature–nurture problem. In *Development and Evolution of Behavior*. L. R. Aronson, E. Tobach, D. S. Lehrman, and J. S. Rosenblatt (eds.). W. H. Freeman, San Francisco.

412 Lenington, S. 1980. Female choice and polygyny in redwinged blackbirds. *Animal Behaviour 28*:347–361.

413 Lenington, S. 1981. Child abuse: The limits of sociobiology. *Ethology and Sociobiology 2*:17–29.

414 Lenneberg, E. H. 1968. *The Biological Foundations of Language*. Wiley, New York.

415 Leopold, A. S. 1977. *The California Quail*. University of California Press, Berkeley.

416 Levick, M. G. 1914. *Antarctic Penguins, A Study of Their Social Habits*. William Heineman, London.

417 Levine, S. 1966. Sex differences in the brain. *Scientific American 214*(Apr.):84–90.

418 Lewin, R. 1976. The course of a controversy. *New Scientist 70*(13 May):344–345.

419 Li, S. K., and D. H. Owings. 1978. Sexual selection in the three-spined stickleback. I. Normative observations. *Zeitschrift für Tierpsychologie 46*:359–371.

420 Licht, P. 1973. Influence of temperature and photoperiod on the annual ovarian cycle of the lizard *Anolis carolinensis*. *Copeia 1973*:465–472.

421 Lightcap, J. L., J. A. Kurland, and R. L. Burgess. 1982. Child abuse: A test of some predictions from evolutionary theory. *Ethology and Sociobiology 3*:61–67.

422 Ligon, J. D. 1978. Reproductive interdependence of piñon jays and piñon pines. *Ecological Monographs 48*:111–126.

423 Ligon, J. D., and S. H. Ligon. 1982. The cooperative breeding behavior of the green woodhoopoe. *Scientific American 247*(July):126–135.

424 Lill, A. 1974. Social organization and space utilization in the lek-forming white-bearded manakin, *M. manacus trinitatis* Hartert. *Zeitschrift für Tierpsychologie 36*:513–530.

425 Lill, A. 1974. Sexual behavior of the lek-forming white-bearded manakin (*Manacus manacus trinitatis* Hartert). *Zeitschrift für Tierpsychologie 36*:1–36.

426 Lill, A. 1974. The evolution of clutch size and male "chauvinism" in the white-bearded manakin. *The Living Bird 13*:211–231.

427 Lin, N., and C. D. Michener. 1972. Evolution of sociality in insects. *Quarterly Review of Biology 47*:131–159.

428 Lindauer, M. 1961. *Communication Among Social Bees*. Harvard University Press, Cambridge, Massachusetts.

429 Linsenmair, K. E. 1972. Die Bedeutung familienspezifischer "Abseichen" für den Familienzusammenhalt bei der sozialen Wüstenassel *Hemilepistus reaumuri* Audouin u. Savigny (Crustacea, Isopoda, Oniscoidea). *Zeitschrift für Tierpsychologie 31*:131–162.

430 Linsenmair, K. E., and C. Linsenmair. 1971. Paarbildung und Paarzusammenhalt bei der monogamen Wüstenassel *Hemilepistus reaumuri* (Crustacea, Isopoda, Oniscoidea). *Zeitschrift für Tierpsychologie 29*:134–155.

431 Lipetz, V. E., and M. Bekoff. 1982. Group size and vigilance in pronghorns. *Zeitschrift für Tierpsychologie 58*:203–216.

432 Lissmann, H. W. 1958. On the function and evolution of electric organs in fish. *Journal of Experimental Biology 35*:156–191.

433 Lissmann, H. W. 1963. Electric location by fishes. *Scientific American 208*(Mar.):50–59.

434 Litte, M. 1977. Behavioral ecology of the social wasp, *Mischocyttarus mexicanus*. *Behavioral Ecology and Sociobiology 2*:229–246.

435 Lloyd, J. E. 1965. Aggressive mimicry in *Photuris*: Firefly femmes fatales. *Science 149*:653–654.

436 Lloyd, J. E. 1966. Studies on the flash communication system in *Photinus* fireflies. *Miscellaneous Publications of the Museum of Zoology, University of Michigan 130*:1–95.

437 Lloyd, J. E. 1975. Aggressive mimicry in *Photuris* fireflies: Signal repertoires by femmes fatales. *Science 197*:452–453.

438 Lloyd, J. E. 1977. Bioluminescence and communication. In *How Animals Communicate*. T. A. Sebeok (ed.). Indiana University Press, Bloomington.

439 Lloyd, J. E. 1979. Mating behavior and natural selection. *Florida Entomologist 62*:17–34.

440 Lloyd, J. E. 1980. Insect behavioral ecology: Coming of age in bionomics or compleat biologists have revolutions too. *Florida Entomologist 63*:1–4.

441 Lockard, R. B. 1978. Seasonal change in the activity pattern of *Dipodomys spectabilis*. *Journal of Mammalogy 59*:563–568.

442 Lockard, R. B., and D. H. Owings. 1974. Seasonal variation in moonlight avoidance by bannertail kangaroo rats. *Journal of Mammalogy 55*:189–193.

443 Loher, W. 1972. Circadian control of stridulation in the cricket *Teleogryllus commodus* Walker. *Journal of Comparative Physiology 79*:173–190.

444 Loher, W. 1979. Circadian rhythmicity of locomotor behavior and oviposition in female *Teleogryllus commodus*. *Behavioral Ecology and Sociobiology 5*:383–390.

445 Loher, W., and B. Rence. 1978. The mating behavior of *Teleogryllus commodus* (Walker) and its central and peripheral control. *Zeitschrift für Tierpsychologie 46*:225–259.

446 Lombardi, J. R., and J. G. Vandenbergh. 1977. Pheromonally induced sexual maturation in females: Regulation by the social environment of the male. *Science 196*:545–546.

447 Lore, R., and K. Flannelly. 1977. Rat societies. *Scientific American 236*(May):106–116.

448 Lorenz, K. Z. 1952. *King Solomon's Ring*. Crowell, New York.

449 Lorenz, K. Z., 1965. *Evolution and Modification of Behavior*. University of Chicago Press, Chicago.

450 Lorenz, K. Z. 1969. Innate bases of learning. In *On the Biology of Learning*. K. H. Pribram (ed.). Harcourt Brace Jovanovich, New York.

451 Lorenz, K. Z. 1970. *Studies on Animal and Human Behavior*, Vols. 1 and 2. Harvard University Press, Cambridge, Massachusetts.

452 Lorenz, K. Z. 1970. Companions as factors in the bird's environment. In

Studies on Animal and Human Behavior. K. Z. Lorenz (ed.). Harvard University Press, Cambridge, Massachusetts.

453 Lott, D. F. 1979. Dominance relations and breeding rate in mature male American bison. *Zeitschrift für Tierpsychologie 49*:418–432.

454 Low, B. S. 1978. Environmental uncertainty and the parental strategies of marsupials and placentals. *American Naturalist 112*:197–213.

455 Luling, K. H. 1963. The archer fish. *Scientific American 209*(July):100–109.

456 MacLean, S. F., Jr., and T. R. Seastedt. 1979. Avian territoriality: Sufficient resources or interference competition. *American Naturalist 114*:308–312.

457 MacLusky, N. J., and F. Naftolin. 1981. Sexual differentiation of the central nervous system. *Science 211*:1294–1302.

458 Macrides, F., A. Bartke, and S. Kalterio. 1975. Strange females increase plasma testosterone levels in male mice. *Science 189*:1104–1105.

459 Malcolm, W. M., and J. P. Hanks. 1973. Landing-site selection and searching behaviour in the micro-lepidopteran *Agonopteryx pulvipennella*. *Animal Behaviour 21*:45–48.

460 Mallory, F. F., and R. J. Brooks. 1978. Infanticide and other reproductive strategies in the collared lemming, *Dicrostonyx groenlandicus*. *Nature 273*:144–146.

461 Manning, A. 1961. The effect of artificial selection for mating speed in *Drosophila melanogaster*. *Animal Behaviour 9*:82–92.

462 Markl, H., and J. Tautz. 1975. The sensitivity of hair receptors in caterpillars of *Barathra brassicae* L. (Lepidoptera, Noctuidae) to particle movement in a sound field. *Journal of Comparative Physiology 99*:79–87.

463 Markow, T. A. 1982. Mating systems of cactophilic *Drosophila*. In *Ecological Energetics and Evolution: The Cactus–Yeast–Drosophila Model System*. J. S. F. Barker and W. T. Starmer (eds.). Academic Press, New York.

464 Markow, T. A., M. Quaid, and S. Kerr. 1978. Male mating experience and competitive courtship success in *Drosophila melanogaster*. *Nature 276*:821–822.

465 Marler, P. 1959. Developments in the study of animal communication. In *Darwin's Biological Work*. P. R. Bell (ed.). Cambridge University Press, New York.

466 Marler, P. 1970. Birdsong and speech development: Could there be parallels? *American Scientist 58*:669–673.

467 Marler, P., and W. J. Hamilton. 1966. *Mechanisms of Animal Behavior*. Wiley, New York.

468 Marler, P., and S. Peters. 1981. Sparrows learn adult song and more from memory. *Science 213*:780–782.

469 Marler, P., and M. Tamura. 1964. Culturally transmitted patterns of vocal behavior in sparrows. *Science 146*:1483–1486.

470 Marten, K., and P. Marler. 1977. Sound transmission and its significance for animal vocalization. I. Temperate habitats. *Behavioral Ecology and Sociobiology 2*:271–290.

471 Marten, K., D. Quine, and P. Marler. 1977. Sound transmission and its significance for animal vocalization. II. Tropical forest habitats. *Behavioral Ecology and Sociobiology 2*:291–302.

472 Martin, P. S. 1967. Pleistocene overkill. *Natural History 76*(Dec.):32–38.

473 Martin, P. S. 1973. The discovery of America. *Science 179*:969–974.

474 Marx, J. L. 1980. Ape-language controversy flares up. *Science 207*:1330–1333.

475 Maschwitz, U. 1975. Old and new trends in the investigation of chemical recruitment in ants. In *Proceedings of the Symposium of the International Union for the Study of Social Insects*. Ch. Noirot, P. E. Howse, and G. le Masne (eds.). Université de Dijon, Dijon, France.

476 Maschwitz, U., and E. Maschwitz. 1974. Platzende Arbeiterinnen: Eine neue Art der Feindabwehr bei sozialen Hautflüglern. *Oecologia 14*:289–294.

477 Mason, W. A. 1968. Early social deprivation in non-human primates: Implications for human behavior. In *Biology and Behavior: Environmental Influences*. D. C. Glass (ed.). Rockefeller University Press, New York.

478 Mason, W. A. 1978. Social experience and primate cognitive development. In *The Development of Behavior: Comparative and Evolutionary Aspects*. G. M. Burghardt and M. Bekoff (eds.). Garland STPM Press, New York.

479 Massey, A., and J. G. Vandenbergh. 1980. Puberty delay by a urinary cue from female house mice in feral populations. *Science 209*:821–822.

480 Masters, W. M. 1979. Insect disturbance stridulation: Its defensive role. *Behavioral Ecology and Sociobiology 5*:187–200.

481 Mathiessen, P. 1962. *Under the Mountain Wall*. Random House (Ballantine Books), New York.

482 Matsubara, J. A. 1981. Neural correlates of a nonjammable electrolocation system. *Science 211*:722–724.

483 Maxson, S. J., and L. W. Oring. 1980. Breeding season time and energy budgets of the polyandrous spotted sandpiper. *Behaviour 74*:200–263.

484 May, R. M. 1976. Sociobiology: A new synthesis and an old quarrel. *Nature 260*:390–392.

485 May, R. M. 1978. Human reproduction reconsidered. *Nature 272*:491–495.

486 Maynard Smith, J. 1965. The evolution of alarm calls. *American Naturalist 94*:59–63.

487 Maynard Smith, J. 1974. The theory of games and the evolution of animal conflict. *Journal of Theoretical Biology 47*:209–221.

488 Maynard Smith, J. 1977. Parental investment: A prospective analysis. *Animal Behaviour 25*:1–9.

489 Maynard Smith, J. 1977. Why the genome does not congeal. *Nature 268*:693–696.

490 Maynard Smith, J. 1978. *The Evolution of Sex*. Cambridge University Press, Cambridge, Massachusetts.

491 Maynard Smith, J., and G. R. Price. 1973. The logic of animal conflict. *Nature 246*:15–18.

492 Mayr, E. 1963. *Animal Species and Evolution*. Harvard University Press, Cambridge, Massachusetts.

493 Mayr, E. 1974. Behavior programs and evolutionary strategies. *American Scientist 62*:650–659.

494 Mayr, E. 1977. Darwin and natural selection. *American Scientist 65*:321–327.

495 McCann, T. S. 1981. Aggression and sexual activity of male southern elephant seals, *Mirounga leonina*. *Journal of Zoology 195*:295–310.

496 McCloskey, L. R. 1971. A marine farmer: The rod-building amphipod. *Fauna 1*:20–25.

497 McCosker, J. E. 1977. Flashlight fishes. *Scientific American 236*(Mar.):106–115.

498 McCracken, G. F., and J. W. Bradbury. 1981. Social organization and kinship in the polygynous bat *Phyllostomus hastatus*. *Behavioral Ecology and Sociobiology 8*:11–34.

499 McEwen, B. S. 1981. Neural gonadal steroid actions. *Science 211*:1303–1311.

500 McGregor, P. K., and J. R. Krebs. 1982. Mating and song types in the great tit. *Nature 297*:60–61.

501 McGregor, P. K., J. R. Krebs, and C. M. Perrins. 1981. Song repertoires and lifetime reproductive success in the great tit (*Parus major*). *American Naturalist 118*:149–159.

502 McKinney, F., and P. Stolen. 1982. Extra-pair-bond courtship and forced copulation among captive green-winged teal (*Anas crecca carolinensis*). *Animal Behaviour 30*:461–474.

503 McLean, I. G. 1982. The association of female kin in the Arctic ground squirrel *Spermophilus parryi*. *Behavioral Ecology and Sociobiology 10*:91–99.

504 McNicol, D., Jr., and D. Crews. 1979. Estrogen/progesterone synergy in the control of female sexual receptivity in the lizard, *Anolis carolinensis*. *General and Comparative Endocrinology 38*:68–74.

505 Mech, L. D. 1970. *The Wolf: The Ecology and Behavior of an Endangered Species*. Doubleday (Natural History Press), Garden City, New York.

506 Meeuse, B. J. D. 1961. *The Story of Pollination*. Ronald Press, New York.

507 Mendlewicz, J., and J. D. Ranier. 1977. Adoption study supporting genetic transmission of manic-depressive illness. *Nature 268*:327–329.

508 Menge, J. L. 1974. Prey selection and foraging period of the predaceous rocky intertidal snail, *Acanthina punctulata*. *Oecologia 17*:293–316.

509 Metcalf, R. A., and G. S. Whitt. 1977. Intra-nest relatedness in the social wasp *Polistes metricus*. *Behavioral Ecology and Sociobiology 2*:339–352.

510 Metcalf, R. A., and G. S. Whitt. 1977. Relative inclusive fitness in the social wasp *Polistes metricus*. *Behavioral Ecology and Sociobiology 2*:353–360.

511 Meyerriecks, A. J. 1960. Comparative breeding behavior of four species of North American herons. *Nuttall Ornithological Club Pubs. No. 2*:1–158.

512 Michener, C. D. 1974. *The Social Behavior of the Bees*. Harvard University Press, Cambridge, Massachusetts.

513 Miller, L. A. 1975. The behaviour of flying green lacewings, *Chrysopa carnea*, in the presence of ultrasound. *Journal of Insect Physiology 21*:205–219.

514 Miller, L. A., and J. Olesen. 1979. Avoidance behavior in green lacewings. I. Behavior of free flying green lacewings to hunting bats and ultrasound. *Journal of Comparative Physiology 131*:113–120.

515 Miller, P. L. 1977. Neurogenic pacemakers in the legs of *Opiliones*. *Physiological Entomology 2*:213–224.

516 Mock, D. W. 1984. Infanticide, siblicide, and avian nestling mortality. In *Infanticide: Comparative and Evolutionary Perspectives*. G. Hausfater and S. B. Hrdy (eds.). Aldine, Chicago.

517 Möglich, M., U. Maschwitz, and B. Hölldobler. 1974. Tandem calling: A new kind of signal in ant communication. *Science 186*:1046–1047.

518 Monahan, M. W. 1977. Determinants of male pairing success in the red-winged blackbird (*Agelaius phoeniceus*): A multivariate, experimental analysis. Ph.D. thesis, Indiana University.

519 Montagu, A. 1976. *The Nature of Human Aggression*. Oxford University Press, New York.

520 Moore, M. C. 1983. Effect of female sexual displays on the endocrine physiology and behaviour of male white-crowned sparrows, *Zonotrichia leucophrys*. *Journal of Zoology 199*:137–148.

521 Morse, D. H. 1977. Resource partitioning in bumble bees: The role of behavioral factors. *Science 197*:678–680.

522 Morse, D. H. 1980. *Behavioral Mechanisms in Ecology.* Harvard University Press, Cambridge, Massachusetts.

523 Morton, E. S. 1975. Ecological sources of selection on avian sounds. *American Naturalist 109*:17–34.

524 Morton, E. S. 1977. On the occurrence and significance of motivation-structural rules in some bird and mammal sounds. *American Naturalist 111*:855–869.

525 Moskowitz, B. A. 1978. The acquisition of language. *Scientific American 239*(Nov.):92–108.

526 Moynihan, M. 1962. The organization and probable evolution of some mixed flocks of neotropical birds. *Smithsonian Miscellaneous Collection 143*(7):1–140.

527 Munger, J. C., and J. H. Brown. 1981. Competition in desert rodents: An experiment with semipermeable enclosures. *Science 211*:510–512.

528 Murray, M. G. 1982. The rut of the impala: Aspects of seasonal mating under tropical conditions. *Zeitschrift für Tierpsychologie 59*:319–337.

529 Naftolin, F., and E. Butz (eds.). 1981. Sexual dimorphism. *Science 211*:1263–1324.

530 Nakatsuru, K., and D. L. Kramer. 1982. Is sperm cheap? Limited male fertility and female choice in the lemon tetra (Pisces, Characidae). *Science 216*:753–755.

531 Narins, P. 1982. Effects of masking noise on evoked calling in the Puerto Rican coqui (Anura: Leptodactylidae). *Journal of Comparative Physiology 147*:439–446.

532 Narins, P. 1982. Behavioral refractory period in Neotropical treefrogs. *Journal of Comparative Physiology 148*:337–344.

533 Narins, P., and D. D. Hurley. 1982. The relationship between call intensity and function in the Puerto Rican coqui (Anura: Leptodactylidae). *Herpetologica 38*:287–295.

534 Neel, J. V. 1970. Lessons from a "primitive" people. *Science 170*:815–822.

535 Negus, N., and P. J. Berger. 1977. Experimental triggering of reproduction in a natural population of *Microtus montanus. Science 196*:1230–1231.

536 Nisbet, I. C. T. 1973. Courtship-feeding, egg size and breeding success in common terns. *Nature 241*:141–142.

537 Noonan, K. M. 1978. Sex ratio of parental investment in colonies of the social wasp *Polistes fuscatus. Science 199*:1354–1356.

538 Norris, K. S. 1967. Some observations on the migration and orientation of marine mammals. In *Animal Orientation and Navigation.* R. M. Storm (ed.). Oregon State University Press, Corvallis.

539 Nottebohm, F. 1970. Ontogeny of bird song. *Science 167*:950–956.

540 Nottebohm, F. 1975. Continental patterns of song variability in *Zonotrichia capensis*: Some possible ecological correlates. *American Naturalist 109*:605–624.

541 Nottebohm, F. 1981. A brain for all seasons: Cyclical anatomical changes in song control nuclei of the canary brain. *Science 214*:1368–1370.

542 Nottebohm, F., and A. P. Arnold. 1976. Sexual dimorphism in vocal control areas of the songbird brain. *Science 194*:211–213.

543 Nottebohm, F., and M. E. Nottebohm. 1978. Relationship between song repertoire and age in the canary, *Serinus canarius. Zeitschrift für Tierpsychologie 46*:298–305.

544 Oglesby, J. N., D. L. Lanier, and D. A. Dewsbury. 1981. The role of prolonged copulatory behavior in facilitating reproductive success in male Syrian golden hamsters (*Mesocricetus auratus*) in a competitive mating situation. *Behavioral Ecology and Sociobiology 8*:47–54.

545 Ohmart, R. D. 1969. Physiological and ethological adaptations of the rufous-winged sparrow (*Aimophila carpalis*) to a desert environment. Unpublished Ph.D. thesis, University of Arizona.

546 Oldroyd, H. 1964. *The Natural History of Flies*. Norton, New York.

547 Olson, S. L., and Y. Hasegawa. 1979. Fossil counterparts of giant penguins from the North Pacific. *Science 206*:688–689.

548 O'Neill, W. E., and N. Suga. 1979. Target range-sensitive neurons in the auditory cortex of the mustache bat. *Science 203*:69–72.

549 Orians, G. H. 1962. Natural selection and ecological theory. *American Naturalist 96*:257–264.

550 Orians, G. H. 1969. On the evolution of mating systems in birds and mammals. *American Naturalist 103*:589–603.

551 Orians, G. H. 1980. *Adaptations of Marsh-Nesting Blackbirds*. Princeton University Press, Princeton, New Jersey.

552 Oring, L. W., and M. L. Knudson. 1973. Monogamy and polyandry in the spotted sandpiper. *The Living Bird 11*:59–73.

553 Oring, L. W., and D. B. Lank. 1982. Sexual selection, arrival times, philopatry, and site fidelity in the polyandrous spotted sandpiper. *Behavioral Ecology and Sociobiology 10*:185–192.

554 Orr, R. T. 1970. *Animals in Migration*. Macmillan, New York.

555 Ostrom, J. H. 1974. *Archeopteryx* and the origin of flight. *Quarterly Review of Biology 49*:27–47.

556 Otte, D. 1974. Effects and functions in the evolution of signaling systems. *Annual Review of Ecology and Systematics 5*:385–417.

557 Otte, D., and W. Cade. 1976. On the role of olfaction in sexual and interspecies recognition in crickets (*Acheta* and *Gryllus*). *Animal Behaviour 24*:1–6.

558 Owens, N. W., and J. D. Goss-Custard. 1976. The adaptive significance of alarm calls given by shorebirds on their winter feeding grounds. *Evolution 30*:397–398.

559 Owings, D. H., and R. G. Coss. 1977. Snake mobbing by California ground squirrels: Adaptive variation and ontogeny. *Behaviour 62*:50–69.

560 Packer, C. 1977. Reciprocal altruism in *Papio anubis*. *Nature 265*:441–443.

561 Packer, C. 1979. Inter-troop transfer and inbreeding avoidance in *Papio anubis*. *Animal Behaviour 27*:1–36.

562 Packer, C., and A. E. Pusey. 1982. Cooperation and competition within coalitions of lions: Kin selection or game theory? *Nature 296*:740–742.

563 Page, R. E., Jr. 1980. The evolution of multiple mating behavior by honey bee queens (*Apis mellifera* L.). *Genetics 96*:263–273.

564 Palmer, J. D. 1974. *Biological Clocks in Marine Organisms: The Control of Physiological and Behavioral Tidal Rhythms*. Wiley, New York.

565 Palmer, J. D. 1976. *An Introduction to Biological Rhythms*. Academic Press, New York.

566 Parker, G. A. 1970. Sperm competition and its evolutionary consequences in the insects. *Biological Reviews 45*:525–567.

567 Parker, G. A. 1974. Assessment strategy and the evolution of fighting behaviour. *Journal of Theoretical Biology 47*:223–243.

568 Parker, G. A., R. R. Baker, and V. G. F. Smith. 1972. The origin and evolution of gamete dimorphism and the male–female phenomenon. *Journal of Theoretical Biology 36*:529–553.

569 Partridge, B. L. 1982. The structure and function of fish schools. *Scientific American 246*(June):114–123.

570 Partridge, L. 1974. Habitat selection in titmice. *Nature 247*:573–574.

571 Partridge, L. 1976. Field and laboratory observations on the foraging and feeding techniques of blue tits (*Parus caeruleus*) and coal tits (*Parus ater*) in relation to their habitats. *Animal Behaviour 24*:534–544.

572 Payne, R. B., and K. Payne. 1977. Social organization and mating success in local song populations of village indigo birds, *Vidua chalybeata. Zeitschrift für Tierpsychologie 45*:113–173.

573 Pengelley, E. T., and S. J. Asmundson. 1974. Circannual rhythmicity in hibernating animals. In *Circannual Clocks*. E. T. Pengelley (ed.). Academic Press, New York.

574 Perrill, S. A., H. C. Gerhardt, and R. Daniel. 1978. Sexual parasitism in the green tree frog (*Hyla cinerea*). *Science 200*:1179–1180.

575 Pierce, N. E., and P. S. Mead. 1981. Parasitoids as selective agents in the symbiosis between lycaenid butterfly larvae and ants. *Science 211*:1185–1187.

576 Pietrewicz, A. T., and A. C. Kamil. 1977. Visual detection of cryptic prey by blue jays (*Cyanocitta cristata*). *Science 195*:580–582.

577 Pietsch, T. W., and D. B. Grobecker. 1978. The compleat angler: Aggressive mimicry in the antennariid anglerfish. *Science 201*:369–370.

578 Pitcher, T. 1979. He who hesitates lives. Is stotting antiambush behavior? *American Naturalist 113*:453–456.

579 Pleszycynska, W. K. 1978. Microgeographic prediction of polygyny in the lark bunting. *Science 201*:935–937.

580 Pleszczynska, W., and R. I. C. Hansell. 1980. Polygyny and decision theory: Testing a model in lark buntings (*Calamospiza melanocorys*). *American Naturalist 116*:821–830.

581 Plomin, R., and D. C. Rowe. 1978. Genes, environment and development of temperament in young human twins. In *The Development of Behavior*. G. M. Burghardt and M. Bekoff (eds.) Garland STPM Press, New York.

582 Pollak, G., D. Marsh, R. Bodenhamer, and A. Souther. 1977. Echo-detecting characteristics of neurons in inferior colliculus of unanesthetized bats. *Science 196*:675–677.

583 Pough, F. H. 1972. Newts, leeches and agriculture. *New York's Food and Life Sciences Quarterly 5*:4–7.

584 Powell, G. V. N. 1974. Experimental analysis of the social value of flocking by starlings (*Sturnus vulgaris*) in relation to predation and foraging. *Animal Behaviour 22*:501–505.

585 Premack, D. 1971. Language in the chimpanzee? *Science 172*:808–822.

586 Pritchard, P. C. H. 1976. Post-nesting movements of marine turtle (Cheloniidae and Dermochelyidae) tagged in the Guianas. *Copeia 1976*:749–754.

587 Provine, R. R. 1981. Wing-flapping develops in chickens made flightless by feather mutations. *Developmental Physiology 14*:481–486.

588 Purcell, J. E. 1980. Influence of siphonophore behavior upon their natural diets: Evidence for aggressive mimicry. *Science 209*:1045–1047.

589 Pyburn, W. F. 1980. The function of eggless capsules in leaf nests of the frog *Phyllomedusa hypochondrialis* (Anura: Hylidae). *Proceedings of the Biological*

Society of Washington 93:153–167.
590 Pyke, G. 1979. The economics of territory size and time budget in the golden-winged sunbird. *American Naturalist 114*:131–145.
591 Pyle, D. W., and M. H. Gromko. 1978. Repeated mating by female *Drosophila melanogaster*: The adaptive importance. *Experientia 34*:449–450.
592 Pyle, D. W., and M. H. Gromko. 1981. Genetic basis for repeated mating in *Drosophila melanogaster. American Naturalist 117*:133–146.
593 Quinlan, R. J., and J. M. Cherrett. 1977. The role of substrate preparation in the symbiosis between the leaf-cutting ant *Acromyrmex octospinosus* (Reich) and its food fungus. *Ecological Entomology 2*:161–170.
594 Quinn, W. G., P. P. Sziber, and R. Booker, 1979. The *Drosophila* memory mutant *amnesiac. Nature 277*:212–214.
595 Raffa, K. F., and A. A. Berryman. 1983. The role of host resistance in the colonization behavior and ecology of bark beetles (Coleoptera: Scolytidae). *Ecological Monographs 63*:27–49.
596 Ralls, K. 1971. Mammalian scent marking. *Science 171*:443–449.
597 Ralls, K., K. Brugger, and J. Ballou. 1979. Inbreeding and juvenile mortality in small populations of ungulates. *Science 206*:1101–1103.
598 Ram, J. L., S. R. Salpeter, and W. J. Davis. 1977. *Pleurobranchaea* egg-laying hormone: Localization and partial purification. *Journal of Comparative Physiology 199*:171–194.
599 Rand, A. S., and M. J. Ryan. The adaptive significance of a complex vocal repertoire in a Neotropical frog. *Zeitschrift für Tierpsychologie 57*:209–214.
600 Ray, T. S., and C. C. Andrews. 1980. Antbutterflies: Butterflies that follow army ants to feed on antbird droppings. *Science 210*:1147–1148.
601 Read, K. E. 1965. *The High Valley.* Scribner, New York.
602 Reese, E. S. 1975. A comparative field study of the social behavior and related ecology of reef fishes of the family Chaetodontidae. *Zeitschrift für Tierpsychologie 37*:37–61.
603 Reichman, J. 1979. Subtly suited to a seedy existence. *New Scientist 81*(Mar.):658–660.
604 Ridley, M. 1978. Paternal care. *Animal Behaviour 26*:904–932.
605 Rijksen, H. D. 1981. Infant killing: A possible consequence of a disputed leader role. *Behaviour 78*:138–167.
606 Robbins, R. K. 1981. The "false head" hypothesis: Predation and wing pattern variation of lycaenid butterflies. *American Naturalist 118*:770–775.
607 Rodman, P. S. 1981. Inclusive fitness and group size with a reconsideration of group sizes in lions and wolves. *American Naturalist 118*:275–283.
608 Roeder, K. D. 1963. *Nerve Cells and Insect Behavior.* Harvard University Press, Cambridge, Massachusetts.
609 Roeder, K. D. 1965. Moths and ultrasound. *Scientific American 212*(Apr.):94–102.
610 Roeder, K. D. 1970. Episodes in insect brains. *American Scientist 58*:378–389.
611 Roeder, K. D., and A. E. Treat. 1961. The detection and evasion of bats by moths. *American Scientist 49*:135–148.
612 Rosenblatt, J. S., H. I. Siegel, and A. D. Mayer. 1979. Progress in the study of maternal behavior in the rat: Hormonal, nonhormonal, sensory, and developmental aspects. *Advances in the Study of Behavior 10*:225–331.
613 Rosenthal, G. A., C. G. Hughes, and D. H. Janzen. 1982. L-Canavanine, a dietary nitrogen source for the seed predator *Caryedes brasiliensis* (Bruchidae). *Science 217*:353–355.

614 Rosenzweig, M. L. 1973. Habitat selection experiments with a pair of coexisting heteromyid rodent species. *Ecology 54*:111–117.

615 Ross, D. M. 1971. Protection of hermit crabs (*Dardanus* spp.) from octopus by commensal sea anemones (*Calliactis* spp.). *Nature 230*:401–402.

616 Roth, L. M. 1981. The mother–offspring relationship of some blaberid cockroaches (*Dictyoptera*: Blattaria: Blaberidae). *Proceedings of the Entomological Society of Washington 83*:390–398.

617 Rothenbuhler, W. C. 1964. Behavior genetics of nest cleaning in honey bees. IV. Responses of F_1 and backcross generations to disease-killed brood. *American Zoologist 4*:111–123.

618 Rothstein, S. I. 1975. An experimental and teleonomic investigation of avian brood parasitism. *Condor 77*:250–271.

619 Routtenberg, A. 1978. The reward system of the brain. *Scientific American 239*(Nov.):154–165.

620 Rowley, I. 1965. The life history of the superb blue wren, *Malurus cyaneus*. *Emu 64*:251–297.

621 Rumbaugh, D. (ed.) 1977. *Language Learning by a Chimpanzee: The Lana Project*. Academic Press, New York.

622 Rusak, B., and G. Groos. 1982. Suprachiasmatic stimulation phaseshifts rodent circadian rhythms. *Science 215*:1407–1409.

623 Rusiniak, K. W., W. G. Hankins, J. Garcia, and L. P. Brett. 1979. Flavor-illness aversions: Potentiation of odor by taste in rats. *Behavioral and Neural Biology 25*:1–17.

624 Rutowski, R. L. 1980. Courtship solicitation by females of the checkered white butterfly, *Pieris protodice*. *Behavioral Ecology and Sociobiology 7*:113–117.

625 Rutowski, R. L. 1982. Epigamic selection by males as evidenced by courtship partner preferences in the checkered white butterfly (*Pieris protodice*). *Animal Behaviour 30*:108–112.

626 Rutowski, R. L., C. E. Long, L. D. Marshall, and R. S. Vetter. 1981. Courtship solicitation by *Colias* females (Lepidoptera: Pieridae). *American Midland Naturalist 105*:334–340.

627 Ryan, M. J. 1980. Female mate choice in a neotropical frog. *Science 209*:523–525.

628 Ryan, M. J., M. D. Tuttle, and L. K. Taft. 1981. The costs and benefits of frog chorusing behavior. *Behavioral Ecology and Sociobiology 8*:273–278.

629 Sahlins, M. 1976. *The Use and Abuse of Biology*. University of Michigan Press, Ann Arbor.

630 Sargent, T. D. 1966. Background selection of geometrid and noctuid moths. *Science 154*:1674–1675.

631 Sargent, T. D. 1969. Behavioural adaptations of cryptic moths. III. Resting attitudes of two bark-like species, *Melanolophis canadaria* and *Catocala ultronia*. *Animal Behaviour 17*:670–672.

632 Sargent, T. D. 1976. *Legion of Night: The Underwing Moths*. University of Massachusetts Press, Amherst.

633 Savage-Rumbaugh, E. S., D. M. Rumbaugh, and S. Boysen. 1980. Do apes use language? *American Scientist 68*:49–61.

634 Schaller, G. B. 1964. *The Year of the Gorilla*. University of Chicago Press, Chicago.

635 Schaller, G. B. 1972. *The Serengeti Lion*. University of Chicago Press, Chicago.

636 Schmidt-Koenig, K., and W. T. Keeton (eds.). 1978. *Animal Migration, Navi-*

gation and Homing. Springer-Verlag, Berlin.

637 Schneider, D. 1969. Insect olfaction: Deciphering system for chemical messages. *Science 163*:1031–1036.

638 Schneider, D. 1974. The sex-attractant receptor of moths. *Scientific American 231*(July):28–35.

639 Schoener, T. W. 1982. The controversy over interspecific competition. *American Scientist 70*:586–595.

640 Scholz, A. T., R. M. Horrall, J. C. Cooper, and A. D. Hasler. 1976. Imprinting to chemical cues: The basis for home stream selection in salmon. *Science 192*:1247–1249.

641 Schuckit, M. A., and V. Rayses. 1979. Ethanol ingestion: Differences in blood acetaldehyde concentrations in relatives of alcoholics and controls. *Science 203*:54–55.

642 Schwagmeyer, P. L. 1979. The Bruce effect: An evaluation of male/female advantages. *American Naturalist 114*:932–938.

643 Schwagmeyer, P. L. 1980. Alarm calling behavior of the thirteen-lined ground squirrel, *Spermophilus tridecemlineatus*. *Behavioral Ecology and Sociobiology 7*:195–200.

644 Searle, L. V. 1949. The organization of hereditary maze-brightness and maze-dullness. *Genetic Psychology Monographs 39*:279–335.

645 Sebeok, T. (ed.). 1977. *How Animals Communicate*. Indiana University Press, Bloomington.

646 Seeley, T. D. 1977. Measurement of nest cavity volume by the honey bee (*Apis mellifera*). *Behavioral Ecology and Sociobiology 2*:201–227.

647 Seeley, T. D., R. H. Seeley, and P. Akratanakul. 1982. Colony defense strategies of the honeybees in Thailand. *Ecological Monographs 52*:43–63.

648 Seemanova, E. 1971. A study of children of incestuous matings. *Human Heredity 21*:108–121.

649 Seibt, W. U., and W. Wickler. 1979. The biological significance of pair-bond in the shrimp *Hymenocera picta*. *Zeitschrift für Tierpsychologie 50*:166–179.

650 Selander, R. K. 1965. On mating systems and sexual selection. *American Naturalist 99*:129–141.

651 Selander, R. K. 1972. Sexual selection and dimorphism in birds. In *Sexual Selection and the Descent of Man*. B. Campbell (ed.). Aldine, Chicago.

652 Seligman, M. E. P. 1970. On the generality of the laws of learning. *Psychological Reviews 77*:406–418.

653 Seyfarth, R. M., and D. L. Cheney. 1980. The ontogeny of vervet monkey alarm calling behavior: A preliminary report. *Zeitschrift für Tierpsychologie 54*:37–56.

654 Seyfarth, R. M., D. L. Cheney, and P. Marler. 1980. Monkey responses to three different alarm calls: Evidence of predator classification and semantic communication. *Science 210*:801–803.

655 Sherman, P. W. 1977. Nepotism and the evolution of alarm calls. *Science 197*:1246–1253.

656 Sherman, P. W. 1981. Reproductive competition and infanticide in Belding's ground squirrels and other animals. In *Natural Selection and Social Behavior*. R. D. Alexander and D. W. Tinkle (eds.). Chiron Press, New York.

657 Sherman, P. W. 1981. Kinship, demography and Belding's ground squirrel nepotism. *Behavioral Ecology and Sociobiology 8*:251–259.

658 Sherman, P. W., and M. L. Morton. 1979. Four months of the ground squirrel. *Natural History 88*:50–57.

659 Shettlesworth, S. J. 1972. Constraints on learning. *Advances in the Study of Behavior 4*:1–68.

660 Shuster, S. M. 1981. Sexual selection in the Socorro isopod *Thermosphaeroma thermophilum* (Cole) (Crustacea: Peracarida). *Animal Behaviour 29*:698–707.

661 Silverin, B. 1980. Effects of long-acting testosterone treatment on free-living pied flycatchers, *Ficedula hypoleuca,* during the breeding period. *Animal Behaviour 28*:906–912.

662 Silverstein, R. M. 1981. Pheromones: Background and potential for use in insect pest control. *Science 213*:1326–1332.

663 Siniff, D. B., I. Stirling, J. L. Bengston, and R. A. Reichle. 1979. Social and reproductive behavior of crabeater seals (*Lobodon carcinophagus*) during the austral spring. *Canadian Journal of Zoology 57*:2243–2255.

664 Skinner, B. F. 1962. Two "synthetic social relations." *Journal of the Experimental Analysis of Behaviour 5*:531–533.

665 Skinner, B. F. 1966. Operant behavior. In *Operant Behavior.* W. K. Honig (ed.). Appleton-Century-Crofts, New York.

666 Slobodchikoff, C. N. 1978. Experimental studies of tenebrionid beetle predation by skunks. *Behaviour 66*:313–322.

667 Smigel, B. W., and M. L. Rosenzweig. 1974. Seed selection in *Dipodomys merriami* and *Perognatus penicillatus. Ecology 55*:328–339.

668 Smith, A. P., and J. Alcock. 1980. A comparative study of the mating systems of Australian eumenid wasps (Hymenoptera). *Zeitschrift für Tierpsychologie 53*:41–60.

669 Smith, N. G. 1968. The advantage of being parasitized. *Nature 219*:690–694.

670 Smith, R. L. 1979. Paternity assurance and altered roles in the mating behavior of a giant water bug *Abedus herberti* (Heteroptera: Belostomatidae). *Animal Behaviour 27*:716–728.

671 Smith, R. L. 1979. Repeated copulation and sperm precedence: Paternity assurance for a male brooding water bug. *Science 205*:1029–1031.

672 Smith, S. M. 1977. Coral-snake pattern recognition and stimulus generalization by naive great kiskadees (Aves: Tyrannidae). *Nature 265*:535–536.

673 Smith, S. M. 1978. The "underworld" in a territorial species: Adaptive strategy for floaters. *American Naturalist 112*:571–582.

674 Snow, B. K. 1977. Territorial behavior and courtship in the male three-wattled bellbird. *Auk 94*:623–645.

675 Snow, D. W. 1956. Courtship ritual: The dance of the manakins. *Animal Kingdom 59*:86–91.

676 Snow, D. W. 1976. *The Web of Adaptation.* Demeter Press, New York.

677 Sordahl, T. A. 1980. Antipredator behavior and parental care in the American Avocet and Black-necked Stilt (Aves: Recurvirostridae). Ph.D. thesis, Utah State University, Logan, Utah.

678 Southwick, C. H., M. A. Beg, and M. R. Siddiqi. 1965. Rhesus monkeys in North India. In *Primate Behavior.* I. DeVore (ed.). Holt, Rinehart & Winston, New York.

679 Stacey, P. B. 1979. Habitat selection and communal breeding in the acorn woodpecker. *Behavioral Ecology and Sociobiology 6*:53–66.

680 Stallcup, J. A., and G. E. Woolfenden. 1978. Family status and contributions to breeding by Florida scrub jays. *Animal Behaviour 26*:1144–1156.

681 Starr, C. K. 1979. The origin of insect sociality: A review of contemporary theory. In *Social Insects.* H. R. Hermann (ed.). Academic Press, New York.

682 Stein, R. A. 1976. Sexual dimorphism in crayfish chelae: Functional significance linked to reproductive activities. *Canadian Journal of Zoology 54*:220–227.
683 Stein, Z., M. Susser, G. Saenger, and F. Marolla. 1972. Nutrition and mental performance. *Science 178*:708–713.
684 Sternberg, D. E., D. P. van Kammen, P. Lerner, and W. E. Bunney. 1982. Schizophrenia: Dopamine β-hydroxylase activity and treatment response. *Science 216*:1423–1425.
685 Struhsaker, T. T. 1977. Infanticide and social organization in the redtail monkey (*Cercopithecus ascanius schmidti*) in the Kibale forest, Uganda. *Zeitschrift für Tierpsychologie 45*:75–84.
686 Suga, N., and W. E. O'Neill. 1979. Neural axis representing target range in the auditory cortex of the mustache bat. *Science 206*:351–353.
687 Suga, N., and T. Shimozawa. 1974. Site of neural attenuation of responses to self-vocalized sounds in echolocating bats. *Science 183*:1211–1213.
688 Summers-Smith, D. 1963. *The House Sparrow*. Collins, London.
689 Swan, L. W. 1970. Goose of the Himalayas. *Natural History 79*(Dec.):68–75.
690 Symons, D. 1979. *The Evolution of Human Sexuality*. Oxford University Press, New York.
691 Tasker, C. R., and J. A. Mills. 1981. A functional analysis of courtship feeding in the red-billed gull, *Larus novaehollandiae scopulinus*. *Behaviour 77*:222–241.
692 Tautz, J., and H. Markl. 1978. Caterpillars detect flying wasps by hairs sensitive to airborne vibration. *Behavioral Ecology and Sociobiology 4*:101–110.
693 Tenaza, R. R., and R. L. Tilson. 1977. Evolution of long-distance alarm calls in Kloss's gibbon. *Nature 268*:233–235.
694 Terrace, H. S., L. A. Petitto, R. J. Sanders, and T. G. Bever. 1979. Can an ape create a sentence? *Science 206*:891–902.
695 Thomas, E. M. 1958. *The Harmless People*. Random House, New York.
696 Thornhill, R. 1976. Sexual selection and paternal investment in insects. *American Naturalist 110*:153–163.
697 Thornhill, R. 1976. Sexual selection and nuptial feeding behavior in *Bittacus apicalis* (Insecta: Mecoptera). *American Naturalist 110*:529–548.
698 Thornhill, R. 1980. Sexual selection in the black-tipped hangingfly. *Scientific American 242* (June):162–172.
699 Thornhill, R. 1981. *Panorpa* (Mecoptera: Panorpidae) scorpionflies: Systems for understanding resource-defense polygyny and alternative male reproductive efforts. *Annual Review of Ecology and Systematics 12*:355–386.
700 Thornhill, R., and J. Alcock. 1983. *The Evolution of Insect Mating Systems*. Harvard University Press, Cambridge, Massachusetts.
701 Thornhill, R., and N. W. Thornhill. 1983. Human rape: An evolutionary perspective. *Ethology and Sociobiology 7*:137–173.
702 Thorpe, W. H. 1958. The learning of song patterns by birds, with special reference to the song of the chaffinch. *Ibis 100*:535–570.
703 Thorson, G. 1950. Reproductive and larval ecology of marine bottom invertebrates. *Biological Reviews 25*:1–45.
704 Tinbergen, N. 1951. *The Study of Instinct*. Oxford University Press, New York.
705 Tinbergen, N. 1958. *Curious Naturalists*. Doubleday, Garden City, New York.
706 Tinbergen, N. 1959. Comparative studies of the behavior of gulls (Laridae): A progress report. *Behaviour 15*:1–70.

707 Tinbergen, N. 1960. *The Herring Gull's World.* Doubleday, Garden City, New York.

708 Tinbergen, N. 1963. The shell menace. *Natural History 72*(Aug.):28–35.

709 Tinbergen, N. 1973. *The Animal in Its World: Explorations of an Ethologist,* Vols. 1 and 2. Harvard University Press, Cambridge, Massachusetts.

710 Tinbergen, N., and A. C. Perdeck. 1950. On the stimulus situations releasing the begging response in the newly hatched herring gull (*Larus argentatus* Pont.). *Behaviour 3*:1–39.

711 Tokarz, R. R., and D. Crews. 1980. Induction of sexual receptivity in the female lizard, *Anolis carolinensis*: Effects of estrogen and the antiestrogen CI-628. *Hormones and Behavior 14*:33–45.

712 Tokarz, R. R., and D. Crews. 1981. Effects of prostaglandins on sexual receptivity in the female lizard, *Anolis carolinensis. Endocrinology 109*:451–457.

713 Topoff, H. 1977. The pit and the antlion. *Natural History 86* (April):64–71.

714 Trail, P. W. 1980. Ecological correlates of social organization in a communally breeding bird, the acorn woodpecker, *Melanerpes formicivorous. Behavioral Ecology and Sociobiology 7*:83–92.

715 Trivers, R. L. 1971. The evolution of reciprocal altruism. *Quarterly Review of Biology 46*:35–57.

716 Trivers, R. L. 1972. Parental investment and sexual selection. In *Sexual Selection and the Descent of Man.* B. Campbell (ed.). Aldine, Chicago.

717 Trivers, R. L. 1974. Parent–offspring conflict. *American Zoologist 14*:249–264.

718 Trivers, R. L. 1976. Sexual selection and resource-accruing abilities in *Anolis garmani. Evolution 30*:253–269.

719 Trivers, R. L., and H. Hare. 1976. Haplodiploidy and the evolution of the social insects. *Science 191*:249–263.

720 Truman, J. W., and S. E. Reiss. 1976. Dendritic reorganization of an identified motoneuron during metamorphosis of the tobacco hornworm moth. *Science 192*:477–479.

721 Tryon, R. C. 1940. Genetic differences in maze-learning ability in rats. *Yearbook of the National Society for the Study of Education 39*:111–119.

722 Tullock, G. 1979. On the adaptive significance of territoriality: Comment. *American Naturalist 113*:772–775.

723 Turillazzi, S., and L. Pardi. 1981. Ant guards on nests of *Parischnogaster nigricans serrei* (Buysson) (Stenogastrinae). *Monitore Zoologica Italiana 15*:1–7.

724 Turnbull, C. 1962. *The Forest People.* Doubleday (Natural History Press), Garden City, New York.

725 Tutin, C. E. G. 1979. Mating patterns and reproductive strategies in a community of wild chimpanzees (*Pan troglodytes schweinfurthii*). *Behavioral Ecology and Sociobiology 6*:29–38.

726 Tuttle, M. D., and M. J. Ryan. 1981. Bat predation and the evolution of frog vocalization in the Neotropics. *Science 214*:677–678.

727 Tuttle, R. H. 1969. Knuckle-walking and the problem of human origins. *Science 166*:953–961.

728 Urquhart, F. A. 1960. *The Monarch Butterfly.* University of Toronto Press, Toronto.

729 van den Assem, J. 1967. Territory in the three-spined stickleback, *Gasteroceus aculeatus. Behaviour Supplements 16*:1–164.

730 van den Berghe, P. L. 1983. Human inbreeding avoidance: Culture in nature.

Behavioral and Brain Science 6:91–124.

731 van der Post, L. 1958. *The Lost World of the Kalahari*. Penguin Books, Baltimore.

732 van Iersel, J. J. A., and J. van den Assem. 1965. Aspects of orientation in the digger wasp *Bembix rostrata*. *Animal Behaviour Supplements 1*:145–162.

733 van Lawick, H., and J. van Lawick-Goodall. 1971. *Innocent Killers*. Houghton Mifflin, Boston.

734 van Lawick-Goodall, J. 1970. Tool-using in primates and other vertebrates. *Advances in the Study of Behavior 3*:195–249.

735 Vaughan, T. A., and S. T. Schwartz. 1980. Behavioral ecology of an insular woodrat. *Journal of Mammalogy 61*:205–218.

736 Vayda, A. P. 1976. *War in Ecological Perspective*. Plenum Press, New York.

737 Verner, J. 1977. On the adaptive significance of territoriality. *American Naturalist 111*:769–775.

738 Vetter, R. S. 1980. Defensive behavior of the black widow spider *Latrodectus hesperus* (Araneae: Theridiidae). *Behavioral Ecology and Sociobiology* 7:187–193.

739 Victoria, J. K. 1972. Clutch characteristics and egg discriminative ability of the African village weaverbird *Ploceus cucullatus*. *Ibis 114*:367–376.

740 Vité, J. P., and D. L. Williamson. 1970. *Thanasimus dubius*: Prey perception. *Journal of Insect Physiology 16*:233–239.

741 Vleck, D. 1981. Burrow structure and foraging costs in the fossorial rodent, *Thomomys bottae*. *Oecologia 49*:391–396.

742 vom Saal, F. S., and L. S. Howard. 1982. The regulation of infanticide and parental behavior: Implication for reproductive success in male mice. *Science 215*:1270–1272.

743 von Frisch, K. 1953. *The Dancing Bees*. Harcourt Brace Jovanovich, New York.

744 von Frisch, K. 1967. *The Dance Language and Orientation of Bees*. Harvard University Press, Cambridge, Massachusetts.

745 von Frisch, K. 1974. Decoding the language of the bee. *Science 185*:663–668.

746 Waage, J. K. 1973. Reproductive behavior and its relation to territoriality in *Calopteryx maculata* (Beauvois) (Odonata: Calopterygidae). *Behaviour 47*:240–256.

747 Waage, J. K. 1979. Dual function of the damselfly penis: Sperm removal and transfer. *Science 203*:916–918.

748 Waage, J. K. 1979. Adaptive significance of postcopulatory guarding of mates and nonmates by male *Calopteryx maculata* (Odonata). *Behavioral Ecology and Sociobiology* 6:147–154.

749 Waddington, C. H. 1957. *The Strategy of Genes*. Allen and Unwin, London.

750 Wade, N. 1976. Sociobiology: Troubled birth for a new discipline. *Science 191*:1151–1155.

751 Walcott, C. 1972. Bird navigation. *Natural History 81*(June):32–43.

752 Walker, A., and R. E. F. Leakey. 1978. The hominids of East Turkana. *Scientific American 239*(August):44–56.

753 Walker, T. J., Jr. 1957. Specificity in the response of female tree crickets (Orthoptera, Gryllidae, Oecanthinae) to calling songs of the males. *Annals of the Entomological Society of America 50*:626–636.

754 Wallraff, H. G., and A. Foa. 1981. Pigeon navigation: Charcoal filter removes relevant information from environmental air. *Behavioral Ecology and Sociobiology 9*:67–77.

755 Walter, H. 1979. *Eleonora's Falcon, Adaptations to Prey and Habitat in a Social Raptor.* University of Chicago Press, Chicago.

756 Ward, P., and A. Zahavi. 1973. The importance of certain assemblages of birds as "information-centres" for food finding. *Ibis 115*:517–534.

757 Warner, R. R., D. R. Robertson, and E. G. Leigh, Jr. 1975. Sex change and sexual selection. *Science 190*:633–638.

758 Washburn, S. L. 1978. Human behavior and the behavior of other animals. *American Psychologist 33*:405–418.

759 Watts, C. R., and A. W. Stokes. 1971. The social order of turkeys. *Scientific American 224*(June):112–118.

760 Weatherhead, D. J., and R. J. Robertson. 1979. Offspring quality and the polygyny threshold: "The sexy son hypothesis." *American Naturalist 113*:201–208.

761 Weber, N. A. 1972. The attines: The fungus-culturing ants. *American Scientist 60*:448–456.

762 Wecker, S. C. 1964. Habitat selection. *Scientific American 211*(Oct.):109–116.

763 Wehner, R. 1976. Polarized-light navigation by insects. *Scientific American 235*(July):106–115.

764 Wellington, W. G., and D. Cmiralova. 1979. Communication of height by foraging honey bees, *Apis mellifera ligustica* (Hymenoptera, Apidae). *Annals of the Entomological Society of America 72*:167–170.

765 Wells, K. D. 1977. The social behaviour of anuran amphibians. *Animal Behaviour 25*:666–693.

766 Wells, K. D. 1977. Territoriality and male mating success in the green frog (*Rana clamitans*). *Ecology 58*:750–762.

767 Wells, K. D. 1977. The courtship of frogs. In *The Reproductive Biology of Amphibians.* D. H. Taylor and S. I. Guttman (eds.). Plenum Press, New York.

768 Wells, K. D. 1979. Reproductive behavior and male mating success in a neotropical toad, *Bufo typhinius. Biotropica 11*:301–307.

769 Welty, J. 1975. *The Life of Birds,* 2nd edition. Saunders, Philadelphia.

770 Wenner, A. M. 1971. *The Bee Language Controversy.* Educational Programs Improvement Corporation, Boulder, Colorado.

771 Werren, J. H., M. R. Gross, and R. Shine. 1980. Paternity and the evolution of male parental care. *Journal of Theoretical Biology 82*:619–631.

772 West, M. J., A. P. King, and D. H. Eastzer. 1981. The cowbird: Reflections on development from an unlikely source. *American Scientist 69*:56–66.

773 West Eberhard, M. J. 1975. The evolution of social behavior by kin selection. *Quarterly Review of Biology 50*:1–33.

774 West Eberhard, M. J. 1979. Sexual selection, social competition, and evolution. *Proceedings of the American Philosophical Society 123*:222–234.

775 Weygoldt, P. 1980. Complex brood care and reproductive behavior in captive poison-arrow frogs, *Dendrobates pumilio.* O. Schmidt. *Behavioral Ecology and Sociobiology 7*:329–332.

776 White, T. D. 1980. Evolutionary implications of pliocene hominid footprints. *Science 208*:175–176.

777 Whitham, T. G. 1979. Habitat selection by *Pemphigus* aphids in response to resource limitation and competition. *Ecology 59*:1164–1176.

778 Whitham, T. G. 1979. Territorial defense in a gall aphid. *Nature 279*:324–325.

779 Whitham, T. G. 1980. The theory of habitat selection: Examined and extended using *Pemphigus* aphids. *American Naturalist 115*:449–466.

780 Whitten, W. K. 1966. Pheromones and mammalian reproduction. *Advances in Reproductive Physiology 1*:155–177.
781 Wickler, W. 1968. *Mimicry in Plants and Animals*. World University Library, London.
782 Wickler, W., and U. Seibt. 1981. Monogamy in Crustacea and man. *Zeitschrift für Tierpsychologie 57*:215–234.
783 Wicklund, C., and T. Jarvi. 1982. Survival of distasteful insects after being attacked by naive birds: A reappraisal of the theory of aposematic coloration evolving through individual selection. *Evolution 36*:998–1002.
784 Wiens, J. A. 1977. On competition and variable environments. *American Scientist 65*:590–597.
785 Wiley, R. H. 1973. Territoriality and non-random mating in sage grouse, *Centrocercus urophasianus. Animal Behaviour Monographs 6*:87–169.
786 Williams, G. C. 1966. *Adaptation and Natural Selection*. Princeton University Press, Princeton, New Jersey.
787 Williams, G. C. 1975. *Sex and Evolution*. Princeton University Press, Princeton, New Jersey.
788 Williams, G. C. 1980. Kin selection and the paradox of sexuality. In *Sociobiology: Beyond Nature/Nurture?* G. W. Barlow and J. Silverberg (eds.). Westview Press, Boulder, Colorado.
789 Williams, T. C., and J. M. Williams. 1978. An oceanic mass migration of land birds. *Scientific American 239*(Oct.):166–176.
790 Willows, A. O. D. 1971. Giant brain cells in mollusks. *Scientific American 224*(Feb.):68–75.
791 Willows, A. O. D., and G. Hoyle. 1969. Neuronal network triggering a fixed action pattern. *Science 166*:1549–1551.
792 Wilson, E. O. 1963. Pheromones. *Scientific American 208*(May):100–114.
793 Wilson, E. O. 1971. *The Insect Societies*. Harvard University Press, Cambridge, Massachusetts.
794 Wilson, E. O. 1975. *Sociobiology, The New Synthesis*. Harvard University Press, Cambridge, Massachusetts.
795 Wilson, E. O. 1976. Academic vigilantism and the political significance of sociobiology. *BioScience 26*(183):187–190.
796 Wilson, E. O., and B. Hölldobler. 1980. Sex differences in cooperative silk-spinning by weaver ant larvae. *Proceedings of the National Academy of Science 77*:2343–2347.
797 Wilson, R. S. 1972. Twins: Early mental development. *Science 176*:914–917.
798 Wiltschko, R., D. Nohr, and W. Wiltschko. 1981. Pigeons with a deficient sun compass use the magnetic compass. *Science 214*:343–345.
799 Wingfield, J. C. 1980. Fine temporal adjustment of reproductive functions. In *Avian Endocrinology*. A. Epple and M. H. Stetson (eds.). Academic Press, New York.
800 Wingfield, J. C., and D. S. Farner. 1980. Control of seasonal reproduction in temperate-zone birds. *Progress in Reproductive Biology 5*:62–101.
801 Wittenberger, J. F. 1981. *Animal Social Behavior*. Duxbury Press, Boston.
802 Wolf, L. L. 1975. "Prostitution" behavior in a tropical hummingbird. *Condor 77*:140–144.
803 Woolfenden, G. E. 1975. Florida scrub jay helpers at the nest. *Auk 92*:1–15.
804 Woolfenden, G. E. 1981. Selfish behavior by Florida scrub jay helpers. In *Natural Selection and Social Behavior*. R. D. Alexander and D. W. Tinkle (eds.). Chiron Press, New York.

805 Woolfenden, G. E., and J. W. Fitzpatrick. 1977. Dominance in the Florida scrub jay. *Condor 79*:1–12.

806 Woolfenden, G. E., and J. W. Fitzpatrick. 1978. The inheritance of territory in group breeding birds. *BioScience 28*:104–108.

807 Wright, S. 1980. Genic and organismic selection. *Evolution 34*:825–843.

808 Wynne-Edwards, V. C. 1962. *Animal Dispersion in Relation to Social Behaviour.* Oliver & Boyd, Edinburgh.

809 Yasukawa, K. 1981. Male quality and female choice of mate in red-winged blackbird (*Agelaius phoeniceus*). *Ecology 62*:922–929.

810 Yasukawa, K. 1981. Song repertoires in the red-winged blackbird (*Agelaius phoeniceus*): A test of the Beau Geste hypothesis. *Animal Behaviour 29*:114–125.

811 Yokoyama, K., and D. S. Farner. 1978. Induction of Zugunruhe by photostimulation of encephalic receptors in white-crowned sparrows. *Science 201*:767–779.

812 Young, V. R., and N. S. Scrimshaw. 1971. The physiology of starvation. *Scientific American 225*(Oct.):14–21.

813 Yunis, J. J., and O. Prakash. 1982. The origin of man: A chromosomal pictorial legacy. *Science 215*:1525–1530.

814 Zach, R. 1979. Shell-dropping: Decision-making and optimal foraging in northwestern crows. *Behaviour 68*:106–117.

815 Zahavi, A. 1975. Mate selection—A selection for a handicap. *Journal of Theoretical Biology 53*:205–214.

816 Zahavi, A. 1977. Reliability in communication systems and the evolution of altruism. In *Evolutionary Ecology.* B. Stonehouse and C. M. Perrins (eds). University Park Press, Baltimore.

817 Zajonc, R. B. 1976. Family configuration and intelligence. *Science 192*:227–236.

818 Zeiller, W. 1971. Naked gills and recycled stings. *Natural History 80*(Dec.):36–41.

819 Zenone, P. G., M. E. Sims, and C. J. Erickson. 1979. Male ring dove behavior and the defense of genetic paternity. *American Naturalist 114*:615–626.

820 Zippelius, H. 1972. Die Karawanenbildung bie Feld- und Hausspitzmaus. *Zeitschrift für Tierpsychologie 30*:305–320.

821 Zucker, I. 1983. Motivation, biological clocks and temporal organization of behavior. In *Handbook of Behavioral Neurobiology: Motivation.* E. Satinoff and P. Teitelbaum (eds.). Plenum Press, New York.

822 Zucker, I., P. G. Johnston, and D. Frost. 1980. Comparative physiological and biochronometric analyses of rodent season and reproductive cycles. *Progress in Reproductive Biology 5*:102–133.

Illustration Credits

Chapter 1

3 From F. J. Ayala, 1978. "The Mechanisms of Evolution," *Scientific American 239*(Sept.):57. Copyright © 1978 by Scientific American Inc. All rights reserved.

6 From H. D. Rijksen, 1981. *Behaviour 78*:138–167.

Chapter 2

3 From R. S. Wilson, 1972. *Science 175*:914–917.

5 From W. G. Quinn *et al.*, 1979. *Nature 277*:212–214.

6 From J. J. Yunis and O. Prakash, 1982. *Science 215*:1525–1530.

9 From D. W. Pyle and M. H. Gromko, 1981. *American Naturalist 117*:133–146.

12–15 From S. J. Arnold, 1980. In *Foraging Behavior,* A. Kamil and T. Sargent (eds.), Garland Press, New York.

Chapter 3

2 Adapted from S. Levine, 1966. "Sex Differences in the Brain," *Scientific American 214*(April):86–87.

6 From P. Marler and M. Tamura, 1964. *Science 146*:1483–1486.

7 From M. Konishi, 1965. *Zeitschrift für Tierpsychologie 22*:770–783.

8 From P. Marler and M. Tamura, 1964. *Science 146*:1483–1486.

11,12 From H. F. Harlow, 1962. In *Roots of Behavior,* E. L. Bliss, (ed.), Harper & Row, New York.

13,14 From W. A. Mason, 1978. In *The Development of Behavior: Comparative and Evolutionary Aspects,* G. M. Burghardt and M. Bekoff (eds.), Garland Press, New York.

16 From C. H. Waddington, 1957. *The Strategy of the Genes,* Allen & Unwin, Winchester, MA.

Chapter 4

3 **(A)** From N. Tinbergen, 1951. *The Study of Instinct,* Oxford University Press, New York.
(D) From K. Lorenz, 1952. *King Solomon's Ring,* Thomas Y. Crowell Company, New York.

7 From R. D. Alexander, 1962. *Evolution 16*:443–467.

9 From W. S. Bristowe, 1958. *The World of Spiders,* William Collins Sons & Company, London.
12 From H. Zippelius, 1972. *Zeitschrift für Tierpsychologie 30*:305–320.
15 From H. H. Kendler, 1968. *Basic Psychology,* 2nd ed., W. A. Benjamin, Menlo Park, CA.
16 From B. F. Skinner, 1962. *Journal of the Experimental Analysis of Behavior 5*:531–533.
17 From R. C. Bolles, 1969. *Journal of Comparative and Physiological Psychology 68*:355–358. Copyright © 1969 by the American Psychological Association.
20 From P. P. G. Bateson *et al.,* 1980. *Journal of Zoology,* London, *191*:61–74.
21,22 From M. J. West *et al.,* 1981. *American Scientist 69*:56–66.

Chapter 5

1 From K. D. Roeder, 1967. *Nerve Cells and Insect Behavior,* rev. ed., Harvard University Press, Cambridge, MA.
7,8 From H. Markl and J. Tautz, 1975. *Journal of Comparative Physiology 99*:79–87.
10 From G. Pollack, 1977. *Science 196*:675–677.
12 From J. A. Matsubara, 1981. *Science 211*:722–724.
13 From C. D. Hopkins, 1980. *Behavioral Ecology and Sociobiology 7*:1–13.
17 From J.-P. Ewert, 1973. *Fortschritte der Zoologie 21*:307–333.
18 From D. H. Hubel and T. N. Wiesel, 1965. *Journal of Neurophysiology 28*:229–289.
20 From J. L. Gould, 1982. *Ethology: The Mechanisms and Evolution of Behavior,* W. W. Norton & Co., New York. Copyright © 1982 by James L. Gould. Used by permission.

Chapter 6

2 From D. Crews, 1975. *Science 189*:1059–1065.
3 From R. R. Tokarz and D. Crews, 1981. *Endocrinology 109*:451–457.
4 From D. Crews, 1975. *Science 189*:1059–1065.
5 From D. Crews and N. Greenberg, 1981. *American Zoologist 21*:273–294.
6 From J.-P. Ewert, 1980. *Neuroethology,* Springer-Verlag, New York.
9 From W. J. Davis, 1973. *Science 180*:317–320.
10 From W. J. Davis *et al.,* 1974. *Journal of Comparative Physiology 90*:225–243.
13,14 From W. Loher, 1972. *Journal of Comparative Physiology 79*:173–190.
16 From R. B. Lockard, 1978. *Journal of Mammalogy 59*:563–568.
17 Adapted from E. T. Pengelley and S. J. Asmundson, 1970. *Comparative Biochemistry and Physiology 32*:155–160.
19 From D. S. Farner, 1970. *Environmental Research 3*:119–133.
21 From R. D. Ohmart, 1969. "Physiological and Ethological Adaptations of the Rufous-winged Sparrow (*Aimophila carpalis*) to a Desert Environment," Ph.D. Thesis, University of Arizona.

Chapter 7

1 From N. Tinbergen and H. Falkus, 1970. *Signals for Survival,* Oxford University Press, New York.
3 Data from H. Kruuk, 1964. *Behaviour Supplements 11*:1–129.
12 From M. D. Beecher, 1982. *American Zoologist 22*:477–490.
15 From P. W. Sherman, 1981. *Behavioral Ecology and Sociobiology 8*:251–259.

Chapter 8

1 From S. C. Wecker, 1964. "Habitat Selection," *Scientific American 211*(Oct.):109–116. Copyright © 1964 by Scientific American Inc. All rights reserved.
2 From L. Partridge, 1978. In *Behavioural Ecology,* J. R. Krebs and N. B. Davies (eds.), Sinauer Associates, Sunderland, MA.
9 From B. Hölldobler, 1980. *Science 210*:86–88.
10 From S. T. Emlen, 1975. "The Stellar-Orientation System of a Migratory Bird," *Scientific American 223*(Aug.):102–111. Copyright © 1975 by Scientific American Inc. All rights reserved.
15 From T. G. Whitham, 1979. *Nature 279*:324–325.
20 From J. R. Krebs, 1971. *Ecology 52*:2–22.
21 **(C)** Courtesy of the American Museum of Natural History.

Chapter 9

1 From R. Zach, 1979. *Behaviour 68*:106–117
2 From D. Vleck, 1981. *Oecologia 49*:391–396.
5 From N. B. Davies and A. I. Houston, 1981. *Journal of Animal Ecology 50*:157–180.
6 From T. Caraco, 1979. *Ecology 60*:621.
7 From R. C. Farentinos, 1981. *Science 213*:1273–1275.
9 From R. I. Bowman, 1961. *University of California Publications in Zoology 58*:1–302.
11 From J. C. Munger and J. H. Brown, 1981. *Science 211*:510–512.
12 From G. J. Kenagy, 1973. *Oecologia 12*:383–412.
14 Adapted from B. Heinrich, 1976. *Ecological Monographs 46*:105–128.
15 From T. W. Pietsch and D. B. Grobecker, 1978. *Science 201*:369–371.
16 From J. E. Lloyd, 1977. In *How Animals Communicate,* T. A. Seboek (ed.), Indiana University Press, Bloomington.

Chapter 10

3 From T. D. Sargent, 1976. *Legion of Night: The Underwing Moths,* University of Massachusetts Press, Amherst.
4 From B. Heinrich and S. L. Collins, in press. *Ecology,* courtesy of The Ecological Society of America.
9 From R. S. Vetter, 1980. *Behavioral Ecology and Sociobiology 7*:187–193.
10 From M. Lindauer, 1961. *Communication among Social Bees,* Harvard University Press, Cambridge, MA.
11 **(A)** Adapted from S. Turillazi and L. Pardi, 1981. *Monitore Zoologica Italiana 15*:1–7.
(C) Adapted from C. S. Henry, 1972. *Psyche 79*:1–22.

12 From S. J. Arnold, 1982. *Copeia 1982*:247–253.
14 From E. D. Brodie, 1977. *Nature 268*:627–628.
15 From N. E. Pierce and P. S. Mead, 1981. *Science 211*:1185–1187.
18 From H. E. Evans and M. J. West, 1970. *The Wasps,* University of Michigan Press, Ann Arbor.
20 From R. E. Kenward, 1978. *Journal of Animal Ecology 47*:449–460.
21 **(A)** From M. J. Ryan *et al.,* 1981. *Behavioral Ecology and Sociobiology 8*:273–278.
(B) From W. H. Calvert *et al.,* 1979. *Science 204*:847–851.

Chapter 11

4 From M. Daly and M. I. Wilson, 1978. *Sex, Evolution, and Behavior,* Duxbury Press, Boston, MA.
7 From R. Thornhill and J. Alcock, 1983. *The Evolution of Insect Mating Systems,* Harvard University Press, Cambridge, MA.
11 **(B)** From V. Geist, 1971. *Mountain Sheep,* University of Chicago Press, Chicago, IL.
12 **(A)** From T. S. McCann, 1981. *Journal of Zoology 195*:295–310.
(B) From G. Hausfater, 1975. *Contributions to Primatology 7*:1–150.
13 From R. D. Alexander *et al.,* 1979. In *Evolutionary Biology and Human Social Behavior: An Anthropological Perspective,* N. A. Chagnon and W. Irons (eds.), Wadsworth Publishing Company, Belmont, CA.
14 From N. B. Davies and T. R. Halliday, 1978. *Nature 274*:683–685.
18 From W. H. Cade, 1981. *Science 212*:563–564.
24 From S. J. Arnold, 1976. *Zeitschrift für Tierpsychologie 42*:247–300.
25 From G. B. Schaller, 1972. *The Serengeti Lion,* University of Chicago Press, Chicago, IL.

Chapter 12

7 From R. Thornhill, 1976. *American Naturalist 110*:529–548.
8 From R. D. Howard, 1978. *Evolution 32*:850–871.
9 From C. K. Catchpool, 1980. *Behavior 74*:148–166.
10 From C. R. Tasker and J. A. Mills, 1981. *Behaviour 77*:222–241.
12 From G. F. McCracken and J. W. Bradbury, 1981. *Behavioral Ecology and Sociobiology 8*:11–34.
13 From R. Thornhill and J. Alcock, 1983. *The Evolution of Insect Mating Systems,* Harvard University Press, Cambridge, MA.
16 From J. F. Downhower and K. B. Armitage, 1971. *American Naturalist 105*:355–370.
21 From D. W. Pyle and M. H. Gromko, 1978. *Experientia 34*:449–450.
23 From L. W. Oring, 1982. In *Avian Biology,* Volume VI, D. S. Farner and A. P. King (eds.), Academic Press, New York.

Chapter 13

8 From R. Thornhill and J. Alcock, 1983. *The Evolution of Insect Mating Systems,* Harvard University Press, Cambridge, MA.
9,10 From J. Alcock, 1973. *Journal of Natural History 7*:411–420.
12 From J. M. Diamond, 1982. *Science 216*:431–434.

13,15 From K. von Frisch, 1953. *The Dancing Bees,* Harcourt Brace Jovanovich, New York.
18 **(B)** From P. Marler, 1959. *Darwin's Biological Work,* P. R. Bell (ed.), Cambridge University Press, New York.
21 From J. E. Lloyd, 1966. *Miscellaneous Publications of the Museum of Zoology,* University of Michigan, *130*:1–95.
23 From E. S. Morton, 1977. *American Naturalist 109*:17–34.
24 From S. L. Gish and E. S. Morton, 1981. *Zeitschrift für Tierpsychologie 56*:74–84.

Chapter 14

2 From J. L. Hoogland, 1981. *Ecology 62*:252–272.
3 From J. L. Hoogland, 1979. *Behaviour 69*:1–35.
10 From E. O. Wilson, 1975. *Sociobiology: The New Synthesis,* Harvard University Press, Cambridge, MA.
11 From I. Rowley, 1981. *Zeitschrift für Tierpsychologie 55*:263.
14 From Y. P. Cruz, 1981. *Nature 294*:446–447.

Chapter 15

7 From W. Irons, 1979. In *Evolutionary Biology and Human Social Behavior: An Anthropological Perspective,* N. A. Chagnon and W. Irons (eds.), Wadsworth Publishing Company, Belmont, CA.

Film Index and Directory

Two of the more important distribution centers for film rentals are located at Pennsylvania State University and the University of California. It is likely that they have many of the same films. Persons wishing to rent a film may want to check with the distributor closest to them before ordering from a more distant source. Each film listed below references one or more chapters; at the conclusion of the referenced chapter(s), the reader will find brief comments about the film.

Audio-Visual Services
Pennsylvania State University
University Park, PA 16802

Aggressive Behavior in Mature Male Bison, Chapter 11
Army Ants, A Study in Social Behavor, Chapter 9
The Ax Fight, Chapter 15
The Biology of Polistes, Chapter 14
Castles of Clay, Chapter 14
Complex Behavior: Chaining, Chapter 4
Courtship Behavior of the Queen Butterfly, Chapter 11
Development of the Child: Infancy, Chapter 4
Development of the Child: Language Development, Chapter 4
DNA, Blueprint of Life, Chapter 2
The Fruit Fly: A Look at Behavior Biology, Chapter 2
Gelada, Chapter 12
The Hyena Story, Chapter 9
In a Frog's Eye, Chapter 5
Life on a Thread, Chapter 9
Miss Goodall and the Wild Chimpanzees, Chapter 15
Northern Elephant Seal, Chapter 12
Patterns for Survival, Chapter 10
Reproductive Behavior of the Black Grouse, Chapter 12
Rhesus Monkeys of Santiago Island, Chapter 11
The Uganda Kob, Territoriality and Mating Behavior, Chapter 12
Wild Dogs of Africa, Chapter 14
Year of the Wildebeest, Chapter 8

University of California Extension Media Center
Berkeley, CA 94720

Baboon Behavior, Chapter 14
Baboon Social Organization, Chapter 10
Birth of the Red Kangaroo, Chapter 4
The Desert People, Chapter 15
Evolution of Nests of the Weaverbird, Chapter 13
Imprinting, Chapter 4
Insect Parasitism, Chapter 9
Life in a Weaverbird Colony, Chapter 12
The Monarch Butterfly Story, Chapter 10
Polar Ecology, Predator and Prey, Chapter 10
To Alter Human Behavior . . . Without Mind Control, Chapter 4
Token Economy: Behaviorism Applied, Chapter 4

McGraw-Hill Films
1221 Avenue of the Americas
New York, NY 10020

Animal Landlord, Chapter 8
The Hunters, Chapter 15
The Mussel Specialist, Chapter 9
In Search of a Mate, Chapter 11
Signals for Survival, Chapters 6, 7, 13
The Social Cat, Chapters 9, 14
Survival and the Senses, Chapter 5

National Geographic Society
Washington, DC 20036

Department 1517
Life Cycle of the Honeybee, Chapters 13, 14
Spiders: Aggression and Mating Behavior, Chapter 11

Department 1523
Konrad Lorenz: Science of Animal Behavior, Chapter 4
The Tool Users, Chapter 9

Department 1636
Feeding and Food Sharing, Chapter 15
Tool-Using Species, Chapter 9

Harper & Row Media
10 East 53rd Street
New York, NY 10022

The Galapagos Finches, Chapter 9
The Pelicaniform Birds, Chapter 11
Strategy for Survival, Chapters 6, 10

Time-Life Multimedia
Time & Life Building, Room 32–48
Rockefeller Center
New York, NY 10020

Bird Brain, The Mystery of Bird Navigation, Chapter 8
The First Signs of Washoe, Chapter 4
Why Do Birds Sing?, Chapter 3

UCLA Instructional Media Library
Royce Hall No. 8
405 Hilgard Avenue
Los Angeles, CA 90024

Food Handling in Kangaroo Rats, Chapter 9
Predatory Behavior of the Grasshopper Mouse, Chapter 9
The Social Behavior of Belding's Ground Squirrel, Chapter 7

National Film Board of Canada
Learning Corporation of America
711 Fifth Avenue
New York, NY 10022

Evolution, Chapter 1

Sterling Education Films
P.O. Box 8497
Universal City, CA 91608

Adelie Penguins of the Antarctic, Chapter 8

Office of the Scientific Counselor
Australian Embassy
Massachusetts Avenue
Washington, DC 20036

The Termite Colony, Chapter 14

William Franklin
Iowa State University
Ames, Iowa 50011

The Guanaco of Patagonia, Chapter 12

Index

Action potential, 127
Acoustical signals, *see also* Song
  of budgerigar, 189
  costs of, 368–369
  of cricket, 74, 175, 376, 454
  of gull, 195
  of toads, 362
Adoption, 531–535
Adultery, 529–530
Aggressive behavior
  of aphid, 246
  of baboons, 485–486
  and body size, 360–362
  of humans, 512–522
  of males, 355–362
  and threat displays, 249
Alarm signal, *see also* Pheromones
  of deer, 319–321
  of ground squirrels, 478–481
  of gulls, 198
Alexander, R. D., 360, 531, 535
Altruism, *see also* Reciprocal altruism
  of eusocial insects, 492
  and habitat saturation, 487–489
  toward relatives, 328, 475–477, 486–487, 534
Andersson, M., 391
Annual cycle
  of anole, 160–163
  of black-headed gull, 195–198
  of golden-mantled ground squirrel, 179–180
  of white-crowned sparrow, 181–186
Anole, 157–165, 187, 375
Ants
  acacia, 308
  and butterflies, 314–315
  harvester, 248
  honey pot, 248–249
  orientation by, 236–238
  parasites of, 89
  sex ratio in colonies of, 496
  social feeding by, 289–292
  trophic eggs of, 500
Ant-lions, 284
Anti-predator behavior
  of butterflies, 302, 314–315, 325–331
  of deer, 319–321
  of fish, 288, 324, 467–469
  of frogs and toads, 324, 461
  of ground squirrels, 206, 478–479
  of lizards, 295–296, 302–303
  of moths, 131, 135, 294, 297–300, 303–305
  of monkey, 114
  of prairie dogs, 469–471
  of spider, 306–307
Aphid, 229–230
Archeopteryx, 425–426
Arnold, S., 48–50
Artificial selection, 38–42, 368
Attack deflection, 301–303
Audition
  of bat, 137–141
  of lacewing, 133–135
  of moth, 126, 129–132, 135, 137
Australopithecus, 427
Axelrod, R., 474

Baboons, 15, 357, 485–486
Bänziger, H., 428, 430
Barash, D., 506
Bat
  audition and prey capture, 137–141
  female defense polygyny in, 403

lek polygyny in, 413
little brown, 125
Mexican freetail, 139
predation on frogs, 325
sonar, 125–126
Beecher, I., 214
Beecher, M., 214
Beetle
bruchid, 278–279
douglas-fir bark, 448–449
ground, 375
rove, 89
tenebrionid, 280
Bellugi, U., 103
Belovsky, G., 271
Benzer, S., 44
Biological clock
annual, 161, 179
circadian, 174–177, 184
and navigation by birds, 233–234
and suprachiasmatic nuclei, 177
Birkhead, T. R., 212
Birth control, 535–538
Blackbird, red-winged, 407
Blowfly, 172–174
Bluegill, 324, 467–469
Bowerbird
golden, 435–437
yellow-fronted gardener, 438
Bradbury, J., 404
Brain
of bat, 137–139
of humans, 119, 508
of mantis, 167–169
of rodents, 177
of toad, 146–150
Brockway, B., 189
Brodie, E. D., Jr., 313
Brower, J., 330
Brower, L. P., 327, 329
Budgerigar, 189
Bullfrog, 396–397
Bumblebee, 280–281
Bunting
indigo, 239, 408
lark, 407–408
Butterfly
black swallowtail, 387
hairstreak, 302, 314–315
Heliconius, 374

Cade, W., 368, 369
Camouflage, *see* Cryptic coloration
Canary, 189
Caraco, T., 270, 473
Catchpool, C., 397
Chemical repellents, 307–312, 327
Chickadee, 299
Chimpanzee, 37–38, 105, 285–287, 516
Circadian rhythm, 174–177
Clutton-Brock, T., 355
Communication, *see also* Pheromones, Song
acoustical, 175, 189, 195, 312, 368–369, 376, 454–455
channels of, 453–457
by dances, 438–448
deceitful, 448, 452–453
electrical, 144–146
evolution of, 448–452
tactile, 457
visual, 163, 195–196, 455–456
Comparative method
defined, 14
and habitat selection, 229
and honeybee dances, 443–448
and human behavior, 514
and hypothesis testing, 14–17, 202–207
and social behavior, 493–494
and territoriality, 247–250
Competition
and diet diversity, 273–278
and tool use, 287
Competitive release, 277
Conflict behavior, 451
Convergent evolution
defined, 202
and feeding lures, 282–284
and homing ability, 231
and social predators, 290, 516
and threat displays, 254–255
Coolidge effect, 353, 523
Cooperation, *see also* Altruism
and communication, 452
evolution of, 473–477
in hunting, 288–292
in mate acquisition, 481–486
in wife-sharing, 531
Copulation
by anoles, 159–160
and cannibalism, 167–169, 196
by damselfly, 336
forced, 366, 371–372
by lions, 399
repeat, 372
by subordinate males, 363–364
superfluous, 375
by white rats, 55–56
Courtship
of bellbird, 389–391

of budgerigar, 189
complex, 391–393
of electric fish, 144–146
and feeding of female, 398
of gulls, 195–196
of humans, 527
of monarch butterfly, 331
of spiders, 96–97
Cowbird, 117–118
Crews, D., 157, 159, 162
Cricket, 93–95, 174–176, 376, 388
Critical period
defined, 57
in song learning, 67
Crow, 261–262
Cryptic behavior
of lizards, 295–296
of moths, 297–298
Cryptic coloration, 295, 297–298
Cuckoldry, 529
Cuckoo, 85–86, 93
Culture
adaptive change in, 510–511
diversity of, 507
evolution of, 509–511
genetic basis for, 506–508

Daly, M., 518, 530
Damselfly
mate guarding in, 370–371
reproductive behavior of, 335–336, 383–384, 396, 417
Dance, honeybee
round, 438
waggle, 439–440
Darwin, C., 5, 6, 7
Davies, N., 268–269, 362
Davis, W. J., 170
Dawkins, R., 365, 453
Deception, 120, 448, 452–453
Deer
alarm signals of, 319–321
fighting in, 355
Development
of bird song, 62–69, 96–97
of cricket song, 95–96
of maternal behavior, 60–61
of nerve cell, 71–73
of sexual behavior, 55–60, 74–77
Developmental homeostasis, 70, 73, 76
Dilger, W., 32
Dilution effect, 324–325
Dinosaurs, 425
Displacement activity, 451
Displays
courtship, 336, 432–438
mutual, 483–484
stereotypy of, 446–447, 451
threat, 249, 254–256
Divergent evolution
defined, 202
in habitat selection, 228–229
and territoriality, 248
Dominance
and body size, 366
and helping, 487
and female mate choice, 386–389
and reproductive success, 357, 481, 483
and selfish herds, 324
Dominance hierarchy
in birds, 118, 481
in mammals, 254, 357
in wasps, 499
Downes, A., 430
Drosophila, 34–36, 73, 417
Ducks, 99–100, 372
Durham, W. H., 517

Eberhard, M. J., 492
Eimas, P., 102
Electric organs, 142–146
Elephant seal, 357, 360
Emlen, S., 239, 400, 403, 491
Eusocial behavior, 492
Evans, H., 492
Evolutionary benefit, 200
Evolutionary cost, 200
Evolutionary history
of bipedal locomotion, 427
of courtship behavior, 432–437
of feeding behavior, 50, 428–430
of flight, 425–426
of honeybee communication, 442–448
method of reconstructing, 431–432
of wasp nesting behavior, 492–494
Ewert, J.-P., 146
Experimental method, 17–19
Explosive mating assemblage, 414–415
Eye spots, 304–305

Farner, D., 184
Falcon, 187
Feeding behavior
of birds, 98–99, 208–210, 228, 261–263
of insects, 172–174, 237, 248, 280–281, 290–292, 428–430

of limpets, 248
of mammals, 110, 125, 272–273, 276–278
of sea slug, 170–171
of snake, 47–50
Feeding lures, 282–284
Female choice
in humans, 524–528
and male courtship, 383, 389–393
and male genes, 385–389
and male parental care, 396–399
and nuptial presents, 393–396
and sexual selection, 350
Female incitation, 387–388
Finches
Darwin's, 274–275
woodpecker, 286–287
Fireflies, 282, 284, 355, 448–449, 455
Fish
archer, 285–286
electric, 141–146
lemon tetra, 384–385
Fisher, R. A., 390
Fixed action pattern, 89
Flower constancy, 280–281
Fly
balloon, 432–433
empidid, 432–435
fruit, 34–36, 46, 73, 417
parasitoid, 368
screwworm, 39
Food selection, 271–278
Fossil behavior, 425–427
Foster, M., 483
Frogs
communication by, 460–462
clumping by, 325
egg-eating by males of, 16
male mate choice in, 354

Garcia, J., 110
Gazelle, 301
Gecko, 303
Genes
and behavioral development, 6, 8
and courtship, 45
and human behavior, 506–508
and learning, 35
and nesting, 33–34
and mental illness, 30–32
Genetic mosiacs, 45
Ghysen, A., 73
Gilbert, L., 374
Gould, J. L., 442
Gould, S. J., 505
Grasshopper, 71
Greenstone, M., 271
Griffin, D., 125
Gromko, M., 40
Group defense
by communal nesting birds, 486, 491
by fish, 287–288, 467–469
by humans, 516–518
by insects, 315–318
by mobbing, 201–202, 205–206
Group selection
defined, 8
and dominance hierarchies, 361
and teritorriality, 254–255
Gross, M., 467
Ground squirrel
Belding's, 215–217, 478–480
California, 206
round-tailed, 479
Grunion, 178–179
Guillemot, 212
Gull
black-headed, 195–202
ground-nesting, 324
hering, 98, 211
kittiwake, 202–205, 208, 210–211, 218
laughing, 99
red-billed, 398–399
Gwinner, E., 182
Gwynne, D., 354

Habitat selection
by aphids, 229–230
by bees, 225–227
by birds, 25, 228
by mice, 223–224, 276–277
by salmon, 225
by wasps, 223–224
Hailman, J., 99
Halliday, T., 362
Hamilton, W. D., 323, 474, 476, 495, 498
Handicap principle, 391–393
Hangingfly, 393–396, 417
Hare, H., 497, 498
Harlow, H., 74
Harlow, M., 74
Harris, M., 511
Hausfater, G., 357
Hedgehog, 314
Heinrich, B., 298–299
Helpers-at-the-nest, 486–492

Heron, 256
Hölldobler, B., 237, 238
Holmes, W., 479, 480
Home range, 246
Homing
  by ants, 236–238
  by honeybees, 233–235
  by limpets, 231–232
  by pigeons, 234–235
Homosexuality, 352, 523
Honeybee
  communication, 438–448
  defensive strategies, 308–310
  genetics of, 34
  giant, 405
  habitat selection by, 225–229
  monogamy in, 400
  navigation by, 233–235
  reproductive behavior, 374
Honeyguide, 405–406
Hoogland, J., 469, 470, 471
Hopkins, C., 146
Hormones
  and feeding behavior, 170–171
  and reproductive behavior of anole, 159–160, 162–163
  and reproductive behavior of rat, 56, 58–62
  and song development, 63
Howard, L., 18, 19
Howard, R., 396
Hrdy, S., 12
Hubel, D., 151
Human
  bipedalism, 427
  chromosomes, 37–38
  mental illness, 30–32, 43
  murder, 518
  selective visual perception, 151–152
  territoriality, 516–518
  tool-using, 285–287, 514
Human development
  and developmental homeostasis, 78–80
  and identical twins, 25–31
  and IQ, 28–29, 79
Human reproduction
  and child abuse, 20–21
  and infanticide, 535–537
  and mating strategies, 522–531
  and physiology, 536
Hybrids, 33
Hypothesis testing, 14, 208–217, 320–321

Illegitimate receivers, 449–450, 454
Illegitimate signalers, 449–450
Impala, 253, 404
Imprinting
  by ducks, 99–100
  and navigation in birds, 239
  by salmon, 225
  by shrew, 100–101
Inbreeding, 385–386, 525
Incest in humans, 524–526
Inclusive fitness, 476
Indigobird, 391
Individual selection
  and cooperation, 483–485
  defined, 9
  and eusociality, 498–500
  and sexual reproduction, 339–341
  and subordinate behavior, 361
Infanticide, 12, 15–20, 372, 535–536
Innate releasing mechanism, 93
Instinct, 87
Intention movement, 451
Interneurons, 127–128
Irons, W., 527
Isopod, 214–215

Jacana, 420–421
Jackdaw, 99
Janzen, D., 308
Jay
  blue, 298, 327, 329
  pinyon, 188
  scrub, 486–487
Junco, 270, 322

Kamil, A., 298
Katz, S. H., 510
Kenward, R. E., 322
Kessel, E. L., 432, 435
Kettlewell, H. B. D., 297
Kin recognition
  in ground squirrels, 479–480
  in *Polistes* wasps, 498
Kin selection
  and alarm calls, 478–480
  defined, 476
  and eusociality, 494–498
  and mate acquisition, 481–483
Krebs, J., 253, 452, 453
Kroodsma, D., 115
Kruuk, H., 201

Labov, J., 17
Lacewing, 133
Language
  in chimpanzees, 105–106
  development of, 101–104
  neural basis of, 104–106, 509–510
  and non-verbal communication, 457
Langur, 12, 120, 542
Lateral inhibition, 152, 153
Learning
  and bar pressing, 107
  and bird song, 63–70, 115, 117–118
  and the brain, 119
  defined, 88
  and diet, 312–314, 327–328, 330
  and foraging efficiency, 299
  and gull begging behavior, 99
  and human language, 101–106
  and kin recognition, 480
  and maze running, 40–42
  and memory retention, 35
  and taste aversions, 110–111
Lederhouse, R. C., 387
Leeches, 50
Lehrman, D., 70
Lemmings, 9–11, 17
Lewontin, R., 505
Ligon, D., 491
Ligon, S., 491
Lill, A., 411, 484
Limpet, 232, 248
Lindauer, M., 226, 227, 443, 444
Lions, 15, 372, 399, 471–473, 482–483
Litte, M., 498
Lloyd, J., 282
Lockard, R., 179
Lordosis, 55
Lorenz, K., 70, 89, 93, 99, 194
Lovebird, 33

MacMillan, A., 462
Magnetic sense, 183, 234, 239
Magpies, 372
Male mate choice, 354, 528–530
Manakin
  white-bearded, 409–412
  long-tailed, 483–485
Mantis, 166–169
Marler, P., 67
Marmot, 408
Martin, P., 514
Mason, W. A., 76, 77
Mate guarding
  in birds, 371–372
  in damselfly, 370–371
  in humans, 526–527
Mating effort, 397
Mating plug, 373–374
Mating swarm, 374, 432, 435
Maynard Smith, J., 366
McCann, T. S., 357
McCracken, G., 404
Mead, P., 315
Meiosis, 27, 338–339
Membrane potential, 44
Meyerriecks, A., 256
Migration
  in birds, 181–183, 239–244
  function of, 244–245
  transoceanic, 243–244
Milkweeds, 328–330
Miller, L., 134
Mimicry
  Batesian, 312–314, 330–331
  eye spot, 305
  false head, 302–303
  of female by male, 363, 377
  by fireflies, 282
  by fish, 282–283
  Müllerian, 331
  of snake, 305
  by spider, 284
Mobbing behavior, 201–202, 205–206
Monahan, M., 407
Monkey
  rhesus, 74–77
  vervet, 114
Monogamy, 400–403, 526–527
Moose, 271–272
Mormon cricket, 353–354
Morton, E. S., 305, 459, 460
Moth
  anti-predator adaptations, 298–299
  blood-drinking by, 428–429
  evasion of bats, 125–126, 129–132
  pheromones, 454
  and wasp, 135
Motmots, 112–113
Mouse
  cactus, 389
  house, 15, 18–19, 190–191
  pocket, 276–278
  prairie deer, 223–224
Mutualism, 474, 478

Narins, P., 461
Navigation, 233, 236

Nest site selection
by gull, 196, 202
by honeybee, 225–227
by wasp, 224
Neurotransmitter, 32, 127
Nottebohm, F., 190
Nuptial present, 393–396, 432–435

Offspring recognition
in birds, 210–213
in isopod, 214–215
in ground squirrel, 215–217, 479
Oldroyd, H., 433
Olfaction, *see also* Pheromone
and feeding behavior, 48–50
and maternal behavior, 60
Operant conditioning, 107–109
Optimality theory
and constraints on foraging, 270–273
defined, 200
and foraging efficiency, 261–269
and tunneling behavior, 263
Organizing substance, 63
Orientation, 237–239
Oring, L., 400, 403, 419
Oviposition behavior, 329–330
Owings, D., 179, 206

Pair-bonding
in gulls, 195–196
in humans, 526–527
in shrimp, 402
Paramecium, 43, 44
Parasitism
of butterflies, 315
by cuckoo, 85–86, 93, 212
by rove beetle, 89–92
Parental investment
defined, 344
and mating systems, 400, 421–422
and sex ratio, 367
Parental manipulation, 499
Parker, G. A., 369
Parthenogenesis, 337–338
Partridge, L., 225, 228
Paternal care, 397, 418–419, 528–530
Penguin, 231, 244–245, 323–324
Perch selection, 295–298
Personal fitness, 7
Peters, S., 67
Pheromone
alarm, 250, 318, 458
of isopod, 215
mass-recruitment, 290–291
repellent, 374
of rove beetle, 91
sex, 355, 449, 454–455
trail, 292, 444
Pierce, N., 315
Pietrewicz, A., 298
Pigeon
navigation by, 233–236
wood, 322–323
Plant defenses, 328–329
Pleiotropy, 35, 37
Pleszcynska, W., 407
Pocket gopher, 263
Polarized light detection, 236–237
Pollinator behavior, 280–281
Polyandry
defined, 400
"prostitution," 417–418
and sex role reversal, 418–421
and sperm depletion, 416–417
Polygeny, 36–37
Polygyny
defined, 400
female defense, 403–404
in humans, 528
lek, 409–413
and female reproductive success, 407–409
resource defense, 405–409
scramble competition, 413–415
Population regulation, 254
Powell, G. V. N., 322
Prairie dogs, 469–471
Proximate cause
defined, 3, 6
of habitat selection, 224–227
of homing, 232–239
of human behavior, 507
Pseudoscorpion, 9–11
Pyle, D., 40

Quail, 188

Rape, *see also* Forced copulation
in acanthocephalan worm, 377
in ducks, 372
in humans, 523–525, 530
Rat
kangaroo, 179, 276–278, 280
white, 39, 54–61, 107–111
Razorbill, 212
Receptive field, 147

Reciprocal altruism
  and alarm calls, 478
  and communal nesting, 490–492
  defined, 474
  and human behavior, 518–522, 535
  and mate acquisition, 485–486
Recruitment
  to food, 290–292
  for group defense, 318, 458
Relatedness, 475–476
Releaser, 89
Reliability of paternity, 346, 404, 529–531
Rence, B., 453
Reproductive competition
  among females, 409
  among males, 15–20, 118, 348, 351–379, 461
Reproductive cycle
  of anole, 158–160
  of gull, 195–198
  of sparrows, 183–187
Reproductive success
  defined, 7
  and dominance, 481, 483, 499
  and female choice, 527–528
  and habitat choice, 230
  and territory ownership, 250–254, 487
  and territory quality, 407–408, 412
Robbins, R., 302
Roeder, K., 125, 129, 166

Sahlins, M., 507
Salamander, 311–314, 377
Salmon, 246
Satellite
  on feeding territory, 268–269
  in reproductive competition, 363, 368–369
Schaller, G., 472, 473
Scorpionfly, 365–366
Selection
  artificial, 38–42, 368
  frequency-dependent, 367
  group, 8, 254–255, 361
  individual, 9, 339–341, 361, 483–485, 498–500
  natural, 5–7, 10
  runaway, 398
Selfish herd, 322–324, 468
Sex determination, 493, 495
Sex ratio, 367
Sex-role reversal, 354, 418–421
Sexual arousal, 351, 523
Sexual dimorphism, 360–361
Sexual interference, 376, 468
Sexual jealousy, 528–530
Sexual receptivity
  in anole, 158–160, 164
  in humans, 526–527
  in insects, 40–42, 282, 388
Sexual reproduction, 338, *see also* Courtship, Copulation
Sherman, P., 216, 478, 479, 480
Shrew, 100–101
Shrimp, 401–402
Sign stimulus, 89, 98
Silverin, B., 185
Skinner, B. F., 107
Slug
  banana, 47–50
  sea, 93, 114, 170–171, 314
Smith, S., 112, 251
Snail, 270
Snake
  feeding behavior, 47–50
  garter, 47–50, 373–374
  hog-nosed, 304
  mating plugs of, 373–374
Snow, B., 390
Social behavior
  advantages and disadvantages of, 467–469
  of ants, 289–291
  of honeybee, 438–446
  and predator avoidance, 270, 315–325, 467–471
  and prey capture, 287–291, 471–473, 514
  of social insects, 492–500
Sociobiology, 505–506
Soldiers
  of ants, 318
  of parasitic wasp, 497
  of termites, 317–318
Sonar, 125, 129, 137–140
Song
  complexity, 397–398
  of cricket, 93–96, 453
  development, 64–69
  dialects, 64–65, 115
  distortion, 460
  and environment, 458–460
  repertoire, 452–453
  of sparrows, 64–68, 383–384
  of zebra finch, 62–64

Sordahl, T., 205
Sparrow
rufous-collared, 251–252
rufous-winged, 187
swamp, 67, 69
white-crowned, 63–68, 102, 115, 181–186
Sperm competition, 369
Spermatophore, 354, 374, 377
Spider
black widow, 306–307
lycosid, 271
Spiteful behavior, 376–377
Spotted sandpiper, 419–420
Starling, 322
Sterile castes, 494–498
Stimulus-filtering, 133, 136, 143
Stinging behavior, 308–310, 492
Stokes, A., 481
Stotting behavior, 301
Strategy
conditional, 365, 368, 524
evolutionarily stable, 365
mixed, 366
pure, 367
Stridulation, 305
Suga, N., 138
Suicidal behavior, 9, 32, 374, 492
Sunbird, 264–267
Sunfish, 324, 467–469
Supernormal stimulus, 91
Superterritory, 378
Swallow
bank, 205, 213, 371–372
rough-winged, 214
Symons, D., 522, 523
Synapse, 127

Tassel-eared squirrel, 272–273
Template, 67
Territoriality
of anole, 163
of aphid, 246–247
of birds, 250–253, 264–269, 452–453
of bluegill, 367–368
of damselfly, 335
functions of, 247
of ground squirrel, 216–217
of house mouse, 190
of humans, 516–518
of impala, 253–254
of white-crowned sparrow, 184–185
Thornhill, N., 524
Thornhill, R., 365, 366, 393, 524
Tinbergen, N., 89, 93, 98, 194, 195, 207
Tit
blue, 225, 228
coal, 225, 228
great, 253, 378
Toad
defensive behavior of, 325
feeding behavior of, 85–86, 165–166
fighting in, 361–362
vision of, 146–150
Tokarz, R., 159
Tool-using, 284–287, 514–515
Trivers, R. L., 399, 421, 474, 497, 498, 499, 520
Trophic eggs, 499
Tryon, R. C., 39–41
Turkey, 481–482
Turnbull, C., 521
Turtle, 245

Ultimate cause, 4, 6

Vetter, R., 307
Viceroy, 330
Vigilance
in gazelle, 301
and group size, 470
in flocking birds, 270, 321–322
Vision
of cat, 151
of jay, 298
of toad, 146–150
Vleck, D., 263
vom Saal, F., 18, 19
von Frisch, K., 442–445

Waddington, C. H., 81, 82
Wagtail, 267
Walcott, C., 234, 235
Warblers, 182, 183, 243
Warfare
adaptive basis of, 516–518
among ants, 248–249
among humans, 512–522
Warning coloration, 311, 327
Wasps
aggression in, 359–360
Ammophila, 1–5, 492
great golden digger, 224
nesting behavior of, 1–2, 492–494

parasitic, 359, 497
Polistes, 135, 492–494, 498
Waterbug, 372–373
Watts, R., 481
Wecker, S., 223
Wehner, R., 236
Wenner, A., 442
West, M., 117
Whales, 246
Whitham, T., 229, 230, 247
Wiesel, T., 151
Williams, G. C., 8
Williams, J., 243
Williams, T., 243
Willows, A. O. D., 93
Wilson, E. O., 505
Wilson, M., 518, 530
Wiltschko, W., 182
Wolf, L., 473
Woodhoopoe, 490–492
Woodpecker, 487–489
Woolfenden, G., 486, 497
Working hypothesis, 8–9, 508
Wren
Australian blue, 490
Carolina, 460
marsh, 115–116

Yasukawa, K., 453

Zach, R., 262

Warning Display
Fire-bellied toads perform a bizarre display when molested. By showing the brightly colored undersides of its feet, a toad warns a would-be consumer that it is poisonous. See Chapter 10 on the relationship of bright coloration, warning displays, and toxicity. Photographs by E. D. Brodie/BPS.

Nuptial Present
Mecopteran hangingflies are unusual in that males offer food presents to females in return for copulation. Here the male transfers a moth prey to his mate, who feeds while she receives his sperm. See Chapter 11 on why male animals typically provide only sperm to their partners. Photograph by the author.

Aggression and Reproductive Success
Male kangaroos fight, ultimately, in order to gain access to sexually receptive females. Males that dominate their rivals are more likely than losers to leave descendants. Fierce competition over females characterizes male behavior in many animal species (see Chapter 11). Photograph by J. N. A. Lott/BPS.